University Physics

University Physics

Rama Bhat

R. Jothilakshmi

Balasubramanian Esakki

Alpha Science International Ltd.

Oxford, U.K.

University Physics
526 pgs. | 340 figs. | 17 tbls.

Rama Bhat
Mechanical and Industrial Engineering
Concordia University
Montreal, Canada

R. Jothilakshmi
Balasubramanian Esakki
Vel Tech. Dr. R.R. and Dr. S.R. Technical University
Avadi, Chennai

Copyright © 2014

ALPHA SCIENCE INTERNATIONAL LTD.
7200 The Quorum, Oxford Business Park North
Garsington Road, Oxford OX4 2JZ, U.K.

www.alphasci.com

ISBN 978-1-84265-813-0

Printed in India

Preface

Science means knowledge and Physics is knowledge of the behaviour of material bodies and substances in this universe. Physics is the science that attempts to describe how nature works using the language of mathematics. It is often considered the most fundamental of all the natural sciences and its theories attempt to describe the behaviour of the smallest building blocks of matter, light, the Universe and everything in between.

On a less tangible level physical theories have allowed us to obtain a greater grasp of the Universe we live in. It is the theories of physics that provide us with some of our deepest notions of Space, Time, Matter and Energy. Physical theories allow us to conceptualize the workings of the building blocks of all matter. These are things we would never be able to experience in everyday life. At the other extreme the theories of cosmology tell us how the Universe began and how it could possibly end. Again this is an example of physics going beyond the limits of our experience to describe the space we live in. Although there is varied opinion on the amount of trust we can place in the theories of physics, the fact remains that these are theories produced from a rigorous and systematic method and they are constantly tested against experimental evidence. As such physical theories give us relatively concrete conceptions of notions beyond our everyday experience.

Perhaps the greatest skill a physics student develops is a sense of wonder about how things work. We are living in a technologically advanced age in which the average person relies on technology without understanding how that technology works. How many of us have looked at a DVD disc, and wondered how it can contain an entire film? Who has held an ipod and thought about how so many songs can be squeezed into a tiny space? Physics teaches us a method of systematic thinking and also the theories necessary to allow us to once again understand how the things we rely on actually work.

There are countless more examples of research in physics leading to the development of important technologies. It is hoped that today's research on nanostructures (structures a billion times smaller than a meter), quantum information or photonics (basically electronics with light) will lead to the next generation of technologies including faster and more robust computers and communication systems.

For those with an ambition to be at the forefront of developing technologies and theories that describe our reality it is necessary to study physics at school and beyond. This book on University Physics is intended primarily for an introductory level course on physics. The various topics are introduced with a brief history and the diverse stages of developments. The authors hope that the topics presented in this book are friendly approach to the reader.

Several people contributed to the final realization of this book. The idea received lot of encouragement from colleagues and friends. Many read the manuscript and gave useful comments and suggestions for improvement. We are grateful to them.

Finally, the ultimate satisfaction is when the students who will use this book, find it useful.

Rama Bhat
R. Jothilakshmi
Balasubramanian Esakki

Contents

MECHANICS

HEAT

OPTICS

MAGNETISM AND ELECTRICITY

Preliminary Ideas

1

1.1 INTRODUCTION TO PHYSICS

Physics is a branch of science that deals with the study of physical universe. The word **Physics** comes from a Greek word meaning nature. Its Sanskrit equivalent is *Bhautiki* that is used to refer to the study of the physical world. Physics, Chemistry, Biology and Mathematics can be considered as a broad classification of science in general. Physics is

the study of nature and natural phenomena. Physicists try to discover the rules that are operating in nature, on the basis of observations, experimentation and analysis. Physics deals with certain basic rules/laws governing the natural world. Physics deals with the study of the behavior of the material bodies under the action of external forces. Similarly, Chemistry deals with chemical reactions between substances and chemical compositions, Biology deals with the study of living beings, both animal and the plant world and Mathematics is an abstract science which gives a symbolic description for all the scientific phenomena.

The above is a very simplistic division of science. The study of science just for the sake of enhancing our understanding of the universe may be called as pure science. However, to benefit from the scientific phenomenon, such studies can be classified as applied science. Engineering and medicine are examples of applied sciences. It is difficult to make a clear cut demarcation between the above broad classifications. In fact there are overlapping areas between the above classifications, such as physical chemistry, biochemistry, biophysics etc.

In this book on Physics, we include the following topics:

1. Mechanics
2. Heat
3. Light
4. Optics
5. Magnetism
6. Electrostatics
7. Electricity
8. Modern Physics

This is not an exhaustive list of topics that is covered by physics. But for a first course in physics, we limit the scope to the above topics.

Mathematics is a symbolic language that can be very conveniently used to describe scientific phenomena. Hence mathematics is an integral part of any science and physics is no exception. Throughout our study of physics, mathematics will be used in describing the physical phenomena.

Historical Perspective

Principles of physics were understood by humans and even animals and hence in their own way history of physics is history of life itself in this universe. When a monkey jumps from tree to tree, it knows the effects of gravity, the need of overcoming inertia and when and how much deceleration it should apply just before landing on the other tree, how much shock of landing will be absorbed by the springiness of the branch on which it is landing etc. A hunter in the jungles of Assam or in the Vindhya forest knew the need of an aerodynamic shape for his arrows and spears. In addition, he knew with how much force he should pull the bowstring before releasing the arrow so as to effectively reach the target. However, the mathematical basis for these physical principles came much later. Early scientists understood these laws of physics when they devised scientific rules for astronomy, in the design and building of early temples or monuments etc. The mathematical principle on which sound is based and used to create music is quite amazing indeed.

In India new discoveries were not attributed to individuals. Although principles of physics were understood in developing astronomy, no individuals were given credit for this discovery exclusively. However, in Western practice discoveries were attributed to individuals although such discoveries were based on the past work of many other scientists. Hence if we put a historical perspective to physics as we know today, it was Newton who gave a mathematical background to the Laws of Gravitation, Lord Rayleigh, who contributed a lot in understanding sound. The Table 1.1 shows some of the discoveries of different principles and theories by various physicists from all over the world.

TABLE 1.1 Some physicists from different countries of the world and their major contributions

Name	Major Contribution/discovery	Country of Origin
Archimedes	Principle of buoyancy: Principle of the lever	Greece
Galileo Galilei	Law of inertia	Italy
Christiaan Huygens	Wave theory of light	Holland
Isaac Newton	Universal law of gravitation; Laws of motion; Reflecting telescope	U.K.
Michael Faraday	Laws of electromagnetic induction	U.K.
James Clerk Maxwell	Electromagnetic theory: Light —an electromagnetic wave	U.K.
Heinrich Rudolf Hertz	Generation of electromagnetic waves	Germany
J.C. Bose	Ultra short radio waves	India
W.K. Roentgen	X-rays	Germany
J.J. Thomson	Electron	U.K.
Marie Sklodowska Curie	Discovery of radium and polonium; Studies on natural radio activity	Poland
Albert Einstein	Explanation of photoelectric effect; Theory of relativity	Germany
Victor Francis Hess	Cosmic radiation	Austria
R.A. Millikan	Measurement of electronic charge	U.S.A.
Ernest Rutherford	Nuclear model of atom	New Zealand
Niels Bohr	Quantum model of hydrogen atom	Denmark
C.V. Raman	Inelastic scattering of light by molecules	India
Louis Victor de Broglie	Wave nature of matter	France
M.N. Saha	Thermal Insulation	India
S.N. Bose	Quantum statistics	India
Wolfgang Pauli	Exclusion Principle	Austria
Enrico Fermi	Controlled nuclear fission	Italy
Werner Heisenberg	Quantum Mechanics; Uncertainty Principle	Germany
Paul Dirac	Relativistic theory of electron; Quantum statistics	U.K.
Edwin Hubble	Expanding Universe	U.S.A.
Ernest Orlando Lawrence	Cyclotron	U.S.A.
James Chadwick	Neutron	U.K.

Name	Major Contribution/discovery	Country of Origin
Hideki Yukawa	Theory of nuclear forces	Japan
Homi Jehangir Bhabha	Cascade process of cosmic radiation	India
Lev Davidovich Landau	Theory of condensed matters; Liquid helium	Russia
S.Chandrasekhar	Chandrasekhar limit, structure and evolution of stars	India
John Bardeen	Transistors; Theory of super conductivity	U.S.A.
C.H.Townes	Maser; Laser	U.S.A.
Abdus salam	Unification of weak and electromagnetic interactions	Pakistan

1.2 MEASUREMENT

1.2.1 General

Physical phenomena are observable. Some may be observable with our eyes; some may be observable with our ears or through our senses. Either the physical phenomena themselves or the effect they produce on other materials are measurable also. Hence to measure the physical phenomena we need a method of measurement or a system of measurement. Measurement of physical quantities and phenomena is done through the units and dimensions. Measurement of any physical quantity involves comparison with a certain basic, arbitrarily chosen, internationally accepted reference standard called **unit**. The result of a measurement of a physical quantity is expressed by a number (or numerical measure) accompanied by a unit as shown in Fig. 1.1. Although the number of physical quantities appears to be very large, we need only a limited number of units for expressing all the physical quantities, since they are interrelated with one another.

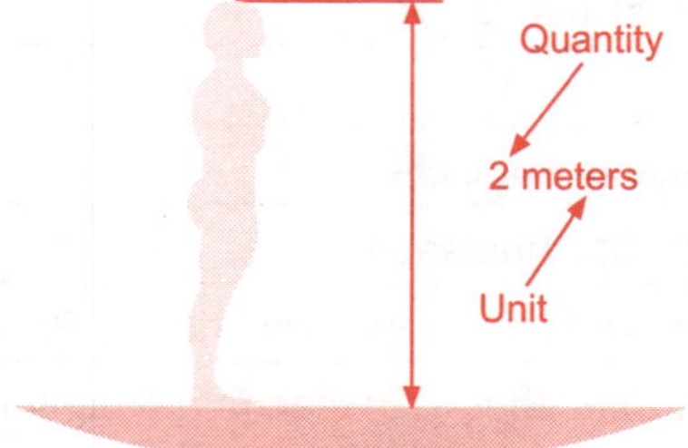

Fig. 1.1 *Measurement of Height*

Measurement of distance, time and mass was known in some way even in primitive times, because of their importance in the day-to-day life of human beings.

Whether it is laying a wooden beam across a stream for crossing or finer work of architects, carpenters and masons in olden times, measurement of length was quite essential. Different regions in India used different length units and different trades also had their own standards such as carpenters' measurement system or masons' system. Many of these systems used "Angula" as the unit which is the measurement of the fore and middle fingers placed side by side. This was not absolute unit and depended on the size of the fingers of the person making the measurement. Some standardized it by having this measurement on a wooden stick and using it for all the measurements. In making measurements of larger distances, use of the human foot or the normal step length of a person were also used. In making measurements over longer distances, units called Gavuda and Yojana were used.

Time was measured very roughly in terms of seasons of the year, days or different times of day such as morning, afternoon etc. Astrology used the unit of a Ghatika, Vighatika, Jama etc. Clocks based on the flow of water or sand through a narrow tube was used for precise measurements. Varying lengths of shadows of objects such as pillars or shadows of humans on a bright sunny day were also used.

Measurement of mass was done using balance and by comparing the substance to be weighed with another substance of known mass or weight. Goldsmiths used the mass of a small seed called gurugunji to measure gold.

With the advent of British in India, British system of units came into use, foot for length, second for time and pound for weight as fundamental units. Furlongs, miles or inches were units derived from foot, minute and hour were derived from second whereas stone, quarter, hundred weight and ton were derived from pound. Later on the Metric units came into popularity because of their convenience with the use of decimal system. With international trade and interaction in the fields of science and engineering, it was more convenient to use an International System of Units which is called "Systeme Internationale" or SI units. SI system of units uses meter for length, second for time and kilogram for mass. Units for length, mass and time were not sufficient in studying heat, electricity, magnetism etc., and SI units use additional units for current (Ampere), temperature (Kelvin), luminous intensity (Candela). In order to denote the amount of substance, a unit called mole or mol is used in chemistry.

1.2.2 Fundamental Units and Derived Units

All the physical quantities in mechanics can be expressed in terms of length, mass and time. Hence they are referred to as fundamental units. Units for quantities such as velocity, force, work etc., can be derived in terms of the fundamental units and hence they are called "Derived Units".

For example, meter, second and kilogram are fundamental units in SI system. The units of Newton for force, Pascal for pressure and Joule for work are derived units which can be expressed as combinations of fundamental units of length, mass and time as given in Table 1.2. We will be using all these units later in our discussions.

TABLE 1.2 The fundamental SI Units

Quantity	Unit name	Unit symbol
mass	kilogram	kg
time	second	s
length	meter	m
temperature	kelvin	K
Electric current	ampere	A
Amount of substance	mole	mol
Luminous intensity	candela	cd

1.2.3   System of Units

Different civilizations in the world had their own systems of units such as the Angula, Ghatika, and Tooka system prevalent in India. Even though these systems were well developed and monumental projects were accomplished using them, which stand even today, they were not standardized in a precise way. Industrialization in Europe in the sixteenth century needed more precise system of units and the foot-pound-second system (FPS System) in Britain and centimeter-gram-second system (CGS System) in France became popular. At one time, because of large British Empire, FPS system replaced locally prevalent systems in many countries including India, even though the system is quite inconvenient to use. On the other hand the CGS system was quite convenient because it was based on the decimal system.

Since many of the derived units in CGS system become inconveniently small, the meter-kilogram-second system (MKS System) came into use.

MKS system was expanded to include the units for current (Ampere), temperature (Kelvin) and luminous intensity (Candela). This new system is named the Systeme Internationale d'units or SI units in brief. This system is becoming more and more accepted in several countries for use in science and engineering, even though countries such as U.K. and U.S.A. are still clinging to the FPS units.

The Table 1.3 contains some of the SI derived units that we often encounter. In addition, some of the general symbols and their names are given in Table 1.4.

TABLE 1.3 SI derived Units

SI derived unit	Symbol	SI base unit	Alternative unit
newton	N	$kg\ m\ s^{-2}$	-
joule	J	$kg\ m^2\ s^{-2}$	N m
hertz	Hz	s^{-1}	-
watt	W	$kg\ m^2\ s^{-3}$	$J\ s^{-1}$
volt	V	$kg\ m^2\ s^{-3}\ A^{-1}$	WA^{-1}
ohm	Ω	$kg\ m^2\ s^{-3}\ A^{-2}$	VA^{-1}
pascal	Pa	$kg\ m^{-1}\ s^{-2}$	$N\ m^{-2}$

TABLE 1.4 Some units retained for general use

Name	Symbol	Value in SI Unit
Minute	Min	60s
Hour	h	60 min = 3600 s
Day	d	24h = 86400 s
Year	y	$365.25d = 3.156 \times 10^7\ s$
Degree	°	$1° = (\Pi/180)$ rad
Litre	L	$1\ dm^3 = 10^{-3}\ m^3$
Ton	t	$10^3\ kg$
Carat	c	200 mg
bar	bar	$0.1\ MPa = 10^5\ Pa$
Curie	Ci	$3.7 \times 10^{10}\ s^{-1}$
Roentgen	R	$2.58 \times 10^{-4}\ C/kg$
Quintal	q	100 kg
Barn	b	$100\ fm^2 = 10^{-28}\ m^2$
Are	A	$1\ dam^2 = 10^2\ m^2$
Hectare	ha	$1\ hm^2 = 10^4\ m^2$
Standard atmospheric pressure	atm	$101325\ Pa = 1.013 \times 10^5\ Pa$

1.3 DIMENSIONS

1.3.1 General

The fundamental units of mass, length and time are used to describe the dimensions of physical quantities. A long object such as a rod has only a length dimension, [L], the time duration to do a certain task has only a time dimension, [T], and the mass of an object has only mass dimension, [M].

The dimension of a derived quantity such as a volume, velocity or force depends on the combination of fundamental units needed to describe it. In general, any derived quantity will have the dimensions of $M^a L^b T^c$, where the values of a, b and c can be found as soon as we know how to describe the derived quantity in terms of the fundamental units. The metric prefixes for each of the quantity can be described by the order of 10^x. The list of metric prefixes is given in Table 1.5.

TABLE 1.5 Metric prefixes

Abbreviation	Prefix	Power
E	exa	10^{18}
P	peta	10^{15}
T	tera	10^{12}
G	giga	10^9
M	mega	10^6
k	kilo	10^3
da	deka	10
	(no prefix)	1
d	deci	10^{-1}
c	centi	10^{-2}
m	milli	10^{-3}
μ	micro	10^{-6}
n	nano	10^{-9}
p	pico	10^{-12}
f	femto	10^{-15}
a	atto	10^{-18}

1.3.2 Dimensional Representation of Quantities

Consider the following examples,

(i) $\qquad$ [Volume] = [Length][Breadth][Height]

$$= [L]\,[L]\,[L] = [L^3] \qquad (1.1)$$

$$= [M^0\,L^3\,T^0]$$

In Eq. (1.1) we can see right away that $a = 0$, $b = 3$ and $c = 0$.

(ii) $\qquad$ [Velocity] = [Distance]/[Time]

$$= [L]/[T] = [M^0 L^1 T^{-1}] \qquad (1.2)$$

(iii) $\quad$ [Acceleration] = [Velocity]/[Time]

$$= [M^0 L^1 T^{-1}]/[T^1] = [M^0 L^1 T^{-2}] \qquad (1.3)$$

$$(iv) \quad [\text{Force}] = [\text{Mass}] \, [\text{Acceleration}]$$
$$= [M] \, [M^0L^1T^{-2}] = [M^1L^1T^{-2}] \quad\quad (1.4)$$

$$(v) \quad [\text{Density}] = [\text{Mass}]/[\text{Volume}]$$
$$= [M]/[L^3] = [M^1L^{-3}T^0] \quad\quad (1.5)$$

$$(vi) \quad [\text{Pressure}] = [\text{Force}]/[\text{Area}]$$
$$= [M^1L^1T^{-2}]/[L^2] = [M^1L^{-1}T^{-2}] \quad\quad (1.6)$$

$$(vii) \quad [\text{Work}] = [\text{Force}] \, [\text{Distance}]$$
$$= [M^1L^1T^{-2}] \, [L^1] = [M^1L^2T^{-2}] \quad\quad (1.7)$$

$$(viii) \quad [\text{Power}] = [\text{Work}]/[\text{Time}]$$
$$= [M^1L^2T^{-2}]/[T^1] = [M^1L^2T^{-3}] \quad\quad (1.8)$$

$$(ix) \quad [\text{Momentum}] = [\text{Mass}] \, [\text{Velocity}]$$
$$= [M^1] \, [L^1T^{-1}] = [M^1L^1T^{-1}] \quad\quad (1.9)$$

$$(x) \quad [\text{Impulse}] = [\text{Force}] \, [\text{Time}]$$
$$= [M^1L^1T^{-2}] \, [T^{-1}] = [M^1L^1T^{-1}] \quad\quad (1.10)$$

1.3.3 Dimensional Equations

A dimensional equation provides the dimensions of a derived physical quantity. Dimensions of all the ten physical quantities discussed previously can be written in the form of dimensional equations just by enclosing the quantities on either side of the equations in rectangular brackets. For example, the dimensions of force are obtained using the dimensional equations as follows:

$$[\text{Force}] = [\text{Mass}] \, [\text{Acceleration}] = [M] \, [L/T^2] = [MLT^{-2}] \quad\quad (1.11)$$

If two physical quantities have the same dimensions, then we can see that these quantities are similar or in other words, these quantities are physically inter-changeable. For example, impulse and momentum both have the dimensions of $[MLT^{-1}]$. Later on we will see that they are similar and interchangeable. An impulse given to an object results in an increased momentum of the object.

1.3.4 Principle of Homogeneity of Dimensions

The dimensions of the terms on both sides of an equation must be the same. In order to compare two physical quantities, their dimensions must be identical. If the dimensions on both sides of an equation are not identical the equation is not correct.

Applications of dimensional analysis

(i) Dimensional analysis can be used to derive a physical equation.

(ii) Dimensional analysis can be used to verify, if the given equation is dimensionally correct.

(*iii*) Dimensional analysis can be used to find the dimensions of unknown parameter in the equation.

1.4 ERRORS AND APPROXIMATIONS

1.4.1 Errors

Measurement is the foundation of all experimental science and technology. The result of every measurement by any measuring instrument contains some uncertainty. This uncertainty is called error. Every calculated quantity which is based on measured values also has an error.

Error in a physical measurement can be defined as the difference between the true value and the observed value of the measured quantity. No physical measurement can provide the "exact" or "true" value. There is always some uncertainty in all physical measurements because of (*i*) type of instrument used in the measurement, (*ii*) conditions during the measurement, (*iii*) the individual or individuals making the measurement and so on. It must be emphasized here that error is not a mistake. A mistake can be avoided with due care. However, errors are not under the control of the experimenter and hence cannot be avoided.

In order to know the exact error in a measurement we should know the "true" value of the quantity to be measured. But we do not know the "true" value and that is why we want to make the measurement. Therefore it is impossible to know the exact error involved in a measurement. However, we can identify the types of errors that can occur in physical measurements and try to minimize them or deal with them effectively.

Errors may be classified into two groups, systematic errors and random errors.

(*i*) Systematic Error

Systematic errors occur because of faulty calibration or faulty operation of a measuring instrument. This type of error affects every measurement done using that instrument in the same way. Zero errors in balances or screw gauges are systematic errors. Parallax error is another example. Once the sources of these errors are identified, it is possible to apply proper corrections to remove such errors.

(*a*) **Instrumental errors:** Arise due to imperfect design or calibration of the measuring instrument, zero error in the instrument, etc. For example, the temperature graduations of a thermometer may be inadequately calibrated such that it may read 104°C at the boiling point of water at standard temperature and pressure conditions (STP) whereas it should read 100°C. In a vernier calliper, the zero mark of vernier scale may not coincide with the zero mark of the main scale, or simply an ordinary metre scale may be worn off at one end.

(b) Imperfection in experimental technique or procedure: In order to determine the temperature of a human body, a thermometer placed under the armpit will always give a temperature lower than the actual value of the body temperature. Other external conditions (such as changes in temperature, humidity, wind velocity, etc.) during the experiment may systematically affect the measurement.

(c) Personal errors: Arise due to an individual's bias, lack of proper setting of the apparatus or individual's carelessness in taking observations without observing proper precautions, etc. For example, if the experimenter, by habit, always holds the head a bit too far to the right while reading the position of a needle on the scale, that will introduce an error due to **parallax**. Systematic errors can be minimised by improving experimental techniques, selecting better instruments and removing personal bias as far as possible. For a given set-up, these errors may be estimated to a certain extent and the necessary corrections may be applied to the readings.

(ii) Random Error

Even after making due allowance for systematic errors, the results may differ from each other when the same measurement is repeated a number of times. Such errors without any known cause are termed as random errors or accidental errors. These errors are usually quite small and there is no easy means of avoiding them. Further, positive and negative errors are equally likely. These errors can be cancelled or corrected by taking the mean value of a large number of measurements.

1.4.2 Estimation of Error

Since measurements are not "exact" and some error is always present, it is quite important that a result is provided along with an indication of the accuracy of the measurement. Estimated error is normally given in terms of (i) relative error (also termed as fractional error) or (ii) percentage error.

(i) Relative error (fractional error), E_r, is given by

$$E_r = dx/x \tag{1.12}$$

where x is the value of the measured quantity and dx is the error in the measured value.

(ii) Percentage error, E_p, is given by

$$E_p = (dx/x)100 \tag{1.13}$$

If x is "exact" value, we obtain a good value for the relative and the percentage errors. Since "exact" value is not available, we must take x as the average value of the measured quantity.

EXAMPLE 1.1

The mass of an object is measured five times and the values are 523 gm, 521 gm, 525 gm, 519 gm and 522 gm. find (a) the relative error and (b) the percentage error.

SOLUTION

The average of the five readings is obtained as

$$\frac{(523+521+525+518+522)}{5} = 522 \text{ gm.}$$

The maximum error is in the third and fourth measurements and is equal to 3 gm. Therefore the result is expressed as 522 ± 3 gm.

The Relative error, $E_r = 3/522 = 0.005747$

The Percentage error, $E_p = (3/522) \times 100 = 0.5747$

The final result in an experiment may be obtained as the addition, subtraction, multiplication or division of more than one measurement. In such cases, the maximum possible error is estimated by considering the algebraic operations involved among the measured values. For example, consider an experiment where the temperature rise when a substance is heated is to be measured. Let the initial temperature be $T_i = 28 \pm 0.5°C$ and the final temperature is measured to be $T_f = 38 \pm 0.5°C$. Since the temperature rise is given by

$$dT = T_f - T_i \tag{1.14}$$

there is a subtraction operation involved between the two measurements. The two extreme possibilities in the final result are given by

$$dT_1 = (38 + 0.5) - (28 - 0.5) = 11°C \tag{1.15}$$

$$dT_2 = (38 - 0.5) - (28 + 0.5) = 9°C \tag{1.16}$$

The mean rise is given by

$$dT_{av} = 38 - 28 = 10°C \tag{1.17}$$

Hence the maximum error is given as 1°C.

If the final result is obtained by multiplying or dividing the measurements, then the total error is obtained by adding the fractional errors (or percentage errors) involved in the individual measurements. When this is extended to the case of raising a quantity to the nth power, then the total error is obtained by adding the fractional error (or percentage error) n times. This is also equivalent to multiplying the fractional error (or percentage error) by n. Consider that the final result is given as

$$f(x) = x^a \tag{1.18}$$

Using the ideas from calculus let a small variation or error in x be given by dx. The corresponding error in $f(x)$, say $df(x)$, is given by

$$df(x) = (x + dx)^a - x^a$$
$$= [x^a + ax^{a-1}dx + \text{(term with higher powers of } dx) - x^a]$$
$$\approx ax^{a-1}dx \qquad (1.19)$$

where binomial theorem is used to expand $(x + dx)^a$ and the higher powers of dx are neglected.

Hence the relative error is given by

$$df(x)/x^a = ax^{a-1}dx/x^a = adx/x \qquad (1.20)$$

In general, if the final result is obtained as

$$R = x^a y^b z^c \qquad (1.21)$$

then using the result of Eqn. (1.20) we can write the error in R as

$$dR = ax^{a-1}y^b z^c + bx^a y^{b-1}z^c + cx^a y^b z^{c-1} \qquad (1.22)$$

The above result uses chain rule of differentiation. The relative error in R is given by

$$dR/R = a(dx/x) + b(dy/y) + c(dz/z) \qquad (1.23)$$

It can be seen that the quantity having the lowest value and the highest power has the greatest influence on the final result.

When error in one quantity dominates the error in others needed for the final result, this quantity must be measured with extra care. It is unnecessary to measure other quantities with extreme care, since they do not influence the final result significantly.

The actual error alone will not give the idea of accuracy in a measurement. The error must always be compared with the measured quantity to know the severity of the error. This information is provided by the relative error or the fractional error.

1.4.3 Significant Figures

As discussed above, every measurement involves errors. Thus, the result of measurement should be reported in a way that it indicates the precision of the measurement. Normally, the reported result of measurement is a number that includes all digits in the number that are known reliably and the first digit that is uncertain. The reliable digits and the first uncertain digit are known as **significant digits** or **significant figures**.

For example, the diameter of human hair is approximately 0.00005 meters. If our measurement accuracy can be extended still further, we may be able to add one more digit to the right of 5 to show the accuracy, say 0.000054 m. In the first case, with 0.00005 m, the number of significant figures is one and in the second case, with

0.000054 m, the number of significant figures is two. The above numbers are written as 5×10^{-5} m and 5.4×10^{-5} m, respectively.

The number of significant figures gives an idea of accuracy. Larger the number of significant figures greater the accuracy. The non-zero digit to the extreme left of a number gives the most significant figure. The non-zero digit to the extreme right of a number gives the least significant figure. The total number of digits from the most significant digit to the least significant digit gives the number of significant figures.

If the numbers are not verifiable by physical measurement, then they are not reliable and are not significant. In the example of the human hair, if the accuracy of the measuring instrument is limited and we can say precisely only upto 0.00001 m accuracy only, then the last digit in 0.000054 m is not significant. The figures that are not significant can be dropped without affecting the accuracy. When dropping such non-significant digit, the rules for rounding a number are normally followed. For example, the number 5.566 can be made 5.57 with three significant figures. This is because the last digit of 6 is greater than 5 and hence while dropping this digit, the one to the left is increased to 7. If the number was 5.563, then the last digit can be dropped and it can be written as 5.56 with three significant figures.

In order to avoid any ambiguity in the number, they are normally expressed as a number between 1 and 10 and multiplied by an appropriate power of 10. The number of digits in the number gives the number of significant figures. For example, 8.532, 8.530×10^{-3}, 8.500×10^{5}, 4.566×10^{9} all have four significant figures.

The following rules may be used to arrive at the correct number of significant figures when arithmetic calculations are done on two or more numbers.

1. Addition or subtraction: The decimal location of the least significant figure among all the numbers involved, determines the number of significant figures in the final result. The addition or subtraction should not be carried beyond the decimal location of the least significant digit.

2. Multiplication or division: The result obtained after the multiplication or division operation should have the same number of significant figures as the number having the least number of significant figures involved in the above operation.

EXAMPLE 1.2

(*i*) 1.2589 + 4.5 = 5.7589, which becomes 5.8.

SOLUTION

This is because 4.5 contained only one digit to the right of the decimal point and hence the result is rounded off to one digit after the decimal point.

(*ii*) *5.649(1.1) = 6.174 which becomes 6.2 because 1.1 has only one digit after the decimal point and hence the result is rounded off to have only one digit after the decimal point.*

1.4.4 Approximations

Rounding off a number to a required number of significant figures is referred to as approximation. If we have a substance that costs ₹ 1.25 per gm, then 10.5 gm of that substance will cost $(10.5) \times 1.25 = 13.125$ in Rupees. But the Rupee denomination goes only up to $1/100^{th}$ of a Rupee and we cannot have 0.125 Rupee. Either we have to pay ₹ 13 and Paise 12 or ₹ 13 and Paise 13. In either case we are approximating it to $1/100^{th}$ place or upto two decimal digits.

Consider a simple pendulum with a cord of length 60.1 cm supporting a pendulum bob of diameter 5.24 cm. We can calculate the length of the pendulum as $60.1 + (5.24/2) = 62.72$ cm. But is this accuracy of a second decimal place really necessary? The length of the cord has been measured with an accuracy of $1/10^{th}$ of a cm. This may be because of the limitations of the measuring scale that was used which had only up to $1/10^{th}$ of a cm. If a more precise scale was used the length could have been found to be either 60.14 cm or 60.16 cm or some other number in the second decimal place. So if the accuracy in the measurement of the length of the cord is limited to only $1/10^{th}$ of a cm, then there is no increase in accuracy in adding the radius of the pendulum bob upto $1/100^{th}$ of a cm. We can hence limit the accuracy of the overall length of the simple pendulum to 62.7 cm only. Hence we are approximating the length of the pendulum to the same degree of accuracy as that used for the cord.

If the accuracy of the measurements of one of the quantities is upto the first decimal place only, there is no use in maintaining a high degree of accuracy in the calculations either using a four figure logarithmic tables or using an electronic calculator. The final result may be approximated to the same degree of accuracy as that of the least accurate value in the quantities being handled. Approximations are quite desirable because of the savings in time in doing the calculations.

Rules to identify the significant figures:

(i) All non-zero digits are significant. Powers of ten are not counted in significant figures. For example, 1.7×10^5 has 2 significant figures.

(ii) In a number with a decimal, zeroes appearing to the left of a digit are not counted in significant figures. For example, 0.002 has only one significant figure in it.

(iii) In a number with a decimal, the number of zeroes at the end is counted in significant figures. For example, 1.700 has 4 significant figures.

(iv) Shifting the position of the decimal does not change the number of significant figures. For example, 2.340 and 234.0 both have 4 significant figures.

(v) All the zeros between two non-zero digits are significant, no matter where the decimal place is, if at all. For example, 203.4 cm has 4 significant digits, 2.05 has 3 significant digits.

(*vi*) The terminal or trailing zeros in a number without a decimal point are not significant. Thus 125 m = 12500 cm = 125000 mm have three significant figures each.

(*vii*) Changing the units do not change the number of significant figures.

There are many methods that can be followed in effecting the approximations. The following examples give two methods.

EXAMPLE 1.3

Obtain the product of 64.7 × 11.003.

SOLUTION

Since the first number, 64.7, has only one digit after the decimal point, we need to get the final result also upto one decimal place. The second number can be considered as 11.003 = (11 + 0.003). The product of 64.7 with 11 will give 711.7 which have only one digit after decimal point. However, when 64.7 is multiplied with 0.003 the result is 0.1941. Rounding of this result upto one digit after decimal point, we have 0.2 only. Only this value must be added to 711.7 to get the final result for 64.7 × 11.003 = 711.9, if we want accuracy upto one decimal point.

When we multiplied 64.7 with 0.003, the result of 0.1941 was not completely utilized in rounding it off to one decimal place. The digits in the third and fourth place were not utilized at all. These two digits came because of the last two digits in the number 64.7. Hence this portion of the result could as well be obtained by taking the product of 60 × 0.003 = 0.18, which gives 0.2 when rounded off to one decimal place.

i.e. 64.7 × 11.003

$$= 64.7 \times (11 + 0.003)$$

$$= 64.7 \times 11 + 60 \times 0.003$$

$$= 711.7 + 0.18 = 711.7 + 0.2 = 711.9.$$

EXAMPLE 1.4

Obtain the result of 100.096 / 1.00224.

SOLUTION

The result can be calculated quickly and quite accurately by the application of binomial theorem as follows:

We have

$$\frac{1+x}{1+y} \approx 1 + x - y;$$

when x and y are much smaller than 1 as in the present case. Hence,

$$\frac{100.096}{1.00224} = \frac{100(1+0.00096)}{(1+0.00224)} = 100(1 + 0.00096 - 0.00224)$$

$$= 100(0.99872) = 99.872.$$

The result is accurate upto 3 decimal places, which is the same accuracy as that of the number 100.096, since this number has the least accuracy among the two numbers being multiplied.

EXAMPLE 1.5

Obtain the result $\dfrac{760(1.0036)}{1.00046}$

SOLUTION

$$\frac{750(1.0036)}{1.00046} = \frac{760(1+0.0036)}{(1+0.00046)}$$

$$= 760(1 + 0.0036 - 0.00046)$$

$$= 760(1.00314) = 760(1 + 0.00314)$$

$$= 760(1) + 760(0.003) + 760(0.0001) + 760(0.00004)$$

$$= 760 + 0.76(3) + 0.076(1) + 0.0076(4)$$

$$= 760 + 2.28 + 0.076 + 0.0304$$

$$= 762.$$

The result is to the same accuracy as that of the number having the least number of decimal places i.e. 760 having no decimal places.

EXAMPLE 1.6

Obtain $\sqrt{102}$

SOLUTION

In this case, it is possible to make use of the relation

$\sqrt{1+x} \approx \left(1+\dfrac{1}{2}x\right)$ when x is quite smaller than 1. Hence

$$\sqrt{102} = \sqrt{100(1.02)} = 10\left(\sqrt{1.02}\right) = 10(1 + 0.02/2) = 10(1.01) = 10.1.$$

When square roots are taken it is important that the accuracy of the result must have one more decimal place than the quantity whose square root is taken. In the present example, the quantity whose square root is taken is 102 and it has no digit after the decimal place, while the result of 10.1 has one decimal place. If not, one can verify that it is not possible to get back the accuracy when we square it.

1.5 SCALARS AND VECTORS

1.5.1 Scalar Quantity

Physical quantities which can be specified by a number and unit that describes only its magnitude are called scalar quantities. A scalar quantity does not have a direction. Some physical scalar quantities are mass, length, density, energy and temperature. Some of these quantities will be precisely defined later. When we refer to 12 mangoes, 100 bananas, or 5 kg sugar, 2 kg rice, or 33°C, they are all scalar quantities. Scalar quantities can be treated by the rules of algebra for addition, subtraction, multiplication and division.

1.5.2 Vectors

Physical quantities which are completely specified only when both their magnitude and direction are given are called vector quantities. Consider the change of position of a particle, which is called displacement. If a particle moves from position A to position B as shown in Fig. 1.2, we can represent this displacement by drawing a line which has a length of AB to some scale. The displacement has the magnitude of AB measured in length units and has the direction from A to B marked with an arrow in the figure. Some examples of physical vector quantities are force, velocity, acceleration, electric field strength and magnetic induction. These quantities will be defined precisely later when we discuss topics where these quantities are used.

A vector quantity is represented by a line with an arrow head. The magnitude of the vector quantity is given by the length of the line and the direction is given by the arrow head. Suppose a car moves along a straight road from position A to B. The length of the line AB, shown in Fig. 1.2, is a scalar quantity. When we specify the direction, say from A to B, with an arrowhead, then the line AB becomes a vector $\vec{a}$.

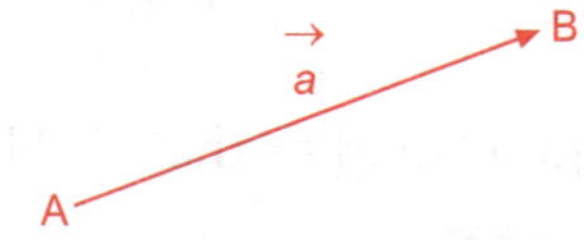

Fig. 1.2 *Representation of a vector*

The magnitude of a vector, also known as the modulus of the vector, is denoted as $|\vec{a}|$ and is a scalar quantity. The modulus is always positive.

1.6 ADDITION OF VECTORS

(*i*) Two Vectors with Same Direction

Let vectors $\vec{a}$ and $\vec{b}$ act along the same straight line in the same direction as shown in Fig. 1.3(*a*) or be parallel to each other as shown in Fig. 1.3(*b*). In either case, join the tip *B* of vector $\vec{a}$ to the root *C* of vector $\vec{b}$. The result of addition of these two vectors is

$$\vec{c} = \vec{a} + \vec{b} \tag{1.24}$$

which is given by the directed line *AD* shown in Fig. 1.3 (*c*).

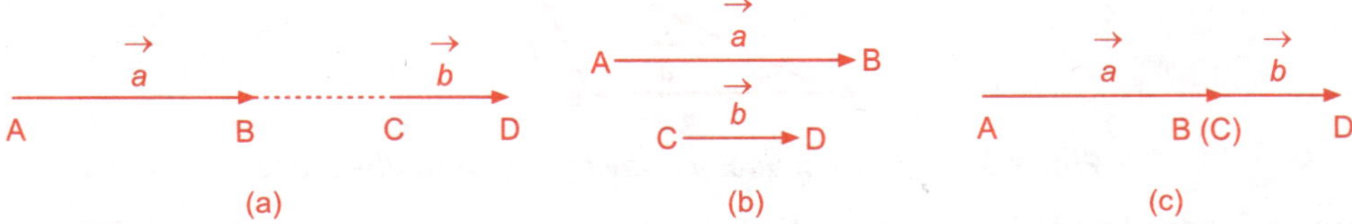

Fig. 1.3 *Addition of vectors having same direction*

If vectors *A* and *B* are along the same straight line but have opposite directions as shown in Fig. 1.4, then their sum is obtained by joining the tip *B* of vector $\vec{a}$ with the root *C* of vector $\vec{b}$. The sum of the two vectors is $\vec{c}$, represented by the directed line *AD* shown in Fig. 1.4(*b*). We assumed that the magnitude of $\vec{a}$ is larger than that of $\vec{b}$ and hence the vector $\vec{c}$ has the same direction as that of $\vec{a}$.

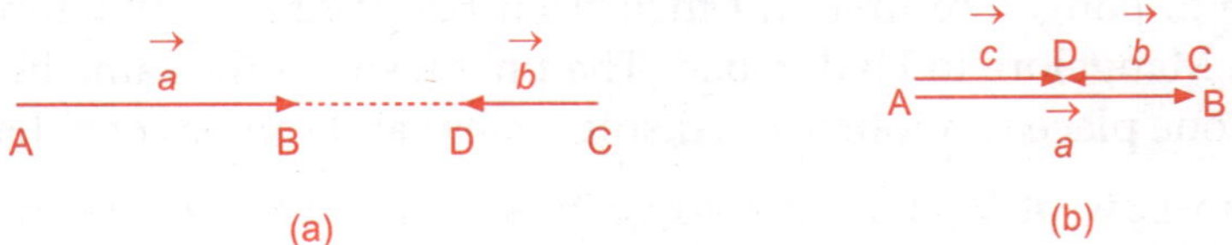

Fig. 1.4 *Addition of vectors having opposite directions*

We have seen that the displacement is a vector quantity. Let a car go from *A* to *B* .which is represented by vector $\vec{a}$. If another displacement of the car, say *CD*, is to be added to the first displacement, then we have to put the two displacements together to obtain the final displacement *AD* where the points *B* and *C* coincide.

(*ii*) Two Vectors Inclined To Each Other, Law of Triangles

Consider a displacement of the particle from *A* to *B*. Let the particle undergo a subsequent displacement from *B* to *C* as shown in Fig. 1.5. The net effect of the two displacements will be the same as a displacement from *A* to *C*. Then *AC* is the sum or resultant of the displacements *AB* and *BC*. Notice that if we add the magnitudes of the lengths *AB* and *BC* we will not get the magnitude of *AC*.

If we denote the vector quantities with an overhead arrow, say the vector of displacement from *A* to *B* is represented as $\vec{a}$, and so on as shown in Fig. 1.5, then we have the vector sum of $\vec{a}$ and $\vec{b}$ given by

$$\vec{a} + \vec{b} = \vec{c}.$$ (1.25)

The actual vector addition can be performed geometrically as follows. On a diagram drawn to scale, draw a line representing the vector $\vec{a}$ in both magnitude and direction as shown in Fig. 1.5. Then draw $\vec{b}$ with its tail at the head of $\vec{a}$ and then draw a line from the tail of $\vec{a}$ to the head of $\vec{b}$ to construct the vector sum $\vec{c}$. This procedure can be generalized to obtain the sum of any number of successive displacements.

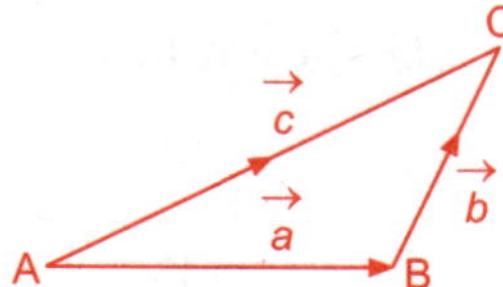

Fig. 1.5 *Addition of vectors having opposite directions*

Law of Triangles: When two vectors $\vec{a}$ and $\vec{b}$ are arranged in such a way that the root of $\vec{b}$ is placed at the tip of $\vec{a}$ then the sum of the two vectors is obtained from the third side of the triangle which is the line joining the root of $\vec{a}$ to the tip of $\vec{b}$ in both magnitude and direction.

This is illustrated in Fig. 1.5. The above procedure can be thought of as a travel plan. Let the point A represent Mangalore. Let the points B and C represent Bangalore and Hyderabad, respectively. In order to reach Hyderabad from Mangalore, one can go from Mangalore to Bangalore first and then from Bangalore to Hyderabad, or one can go directly from Mangalore to Hyderabad. The final result is the same in both the cases. Travelling from one place to another is a displacement and can be considered as a vector.

Parallelogram Law of Vectors: If two vectors to be added are represented both in magnitude and direction by the two adjacent sides of a parallelogram with their common tails meeting at a point, then their resultant or their vector sum is represented both in magnitude and direction by the diagonal of the parallelogram passing through the common tail.

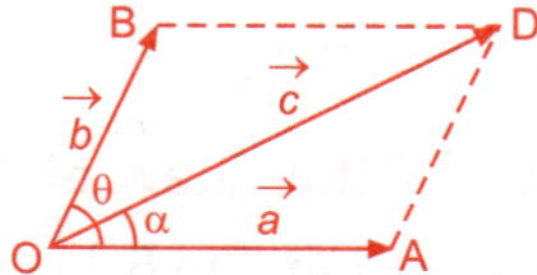

Fig. 1.6 *Parallelogram law of vectors*

Let OA and OB in Fig. 1.6, represent the vectors $\vec{a}$ and $\vec{b}$ in both magnitude and direction. Then we have

$$c = \sqrt{a^2 + b^2 + 2ab\cos\theta}$$ (1.26)

where a, b and c are the magnitudes of the vectors $\vec{a}$, $\vec{b}$ and $\vec{c}$ respectively. (Note: This result can easily be obtained if a perpendicular is dropped from D on line OA produced, to form a right angle triangle). The direction of resultant $\vec{c}$ is given by

$$\tan \alpha = \frac{b\sin\theta}{a + b\cos\theta} \tag{1.27}$$

The resultant of the two vectors under the two special cases of (*i*) the two vectors in the same direction, and (*ii*) the two vectors being in opposite direction can be obtained using the Eqs. (1.26) and (1.27) very easily.

(*i*) Vectors in the Same Direction: In this case, we have $\theta = 0$, and consequently $\cos\theta = 1$ and $\sin\theta = 0$. Hence, from Eq. (1.26) we have

$$c = \sqrt{a^2 + b^2 + 2ab} = (a+b)^2 \tag{1.28}$$

and the resultant is inclined to the vector $\vec{a}$ at an angle $\alpha = 0$.

(*ii*) Vectors in the Opposite Direction: In this case, we have $\theta = 180°$ and hence $\cos\theta = -1$ and $\sin\theta = 0$. Substituting this in Eqs. (1.26) and (1.27) we get

$$c = \sqrt{a^2 + b^2 - 2ab} = (a-b)^2 \tag{1.29}$$

and $\alpha = 0$.

Law of Polygon of Vectors: If a number of vectors are represented in magnitude and direction by the sides of a polygon taken in order, then the resultant is represented in both magnitude and direction by the closing side of the polygon taken in the opposite order.

Consider the vectors $\vec{a}$, $\vec{b}$, $\vec{c}$, $\vec{d}$ and $\vec{e}$ as shown in Fig. 1.7. Starting from a reference point, say the origin O, we place the root of the first vector $\vec{a}$ on O. Then we connect the root of the next vector, $\vec{b}$ on the tip of vector $\vec{a}$. Continuing, we place the root of the vector to be added on the tip of the previously added vector. When all the vectors to be added are positioned, if the polygon closes, that means that the resultant is zero. If the polygon does not close, then the resultant is given by the line joining the point O with the tip of the last vector, say, $\vec{e}$ in the present case. The line OE, denoted by the vector $\vec{r}$ gives the resultant of the system of vectors.

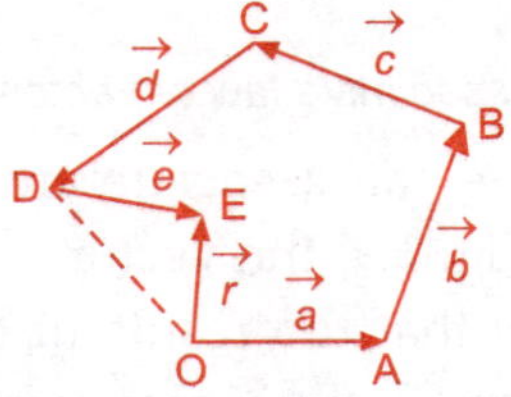

Fig. 1.7 *Law of polygon of vectors*

1.6.1   Laws of Vector Addition

Like algebraic addition of scalar quantities, vectors also obey the commutative and associative laws of addition.

1.6.1.1 Commutative Law of Vector Addition

In Fig. 1.8, two vectors $\vec{a}$ are $\vec{b}$ added in two different ways from which we can see that vector addition follows commutative law

$$\vec{a} + \vec{b} = \vec{b} + \vec{a} \tag{1.30}$$

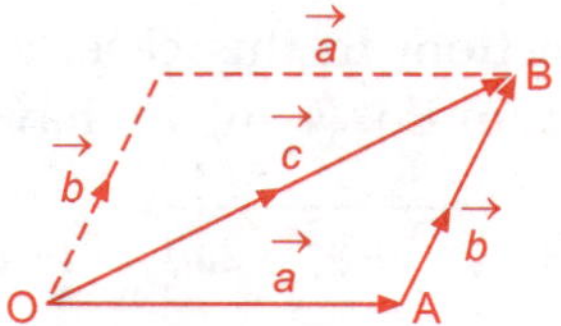

Fig. 1.8 Commutative law of vector addition

Commutative law implies that we can add vectors in any order or we can commute (interchange) vectors while adding. Note that subtraction of vectors does not follow commutative law, which is quite obvious, because

$$\vec{a} - \vec{b} \neq \vec{b} - \vec{a} \tag{1.30}$$

1.6.1.2 Associative Law of Vector Addition

We can verify easily that vector addition follows the associative law, shown in Fig. 1.8, given by

$$\vec{a} + \left(\vec{b} + \vec{c}\right) = \left(\vec{a} + \vec{b}\right) + \vec{c} \cdot \tag{1.32}$$

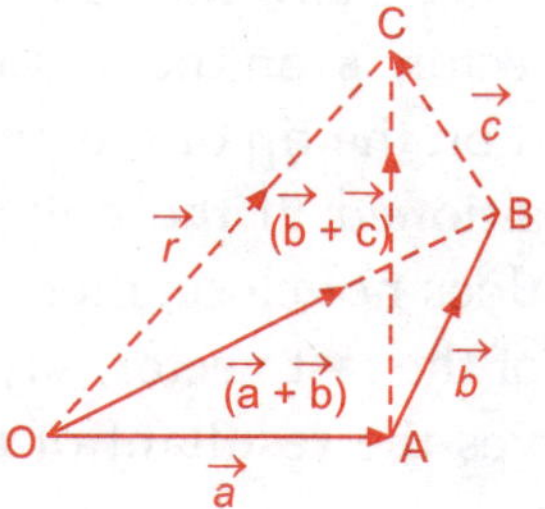

Fig. 1.9 *Associative law of vector addition*

Associative law implies that we can associate any vectors initially and add them subsequently to other vectors. In Fig. 1.9, the vector $(\vec{b} + \vec{c})$ is represented by the line AC and from triangle OAC we see that the resultant of the vectors $\vec{a}$ and $(\vec{b} + \vec{c})$ is given by $\vec{r}$. Similarly the vector sum $(\vec{a} + \vec{b})$ is represented by the line OB and from the triangle OBC we see that the resultant of the vectors $(\vec{a} + \vec{b})$ and $\vec{c}$ is also given by $\vec{r}$.

1.6.2 Subtraction of Vectors

Subtraction of vectors can be performed by defining the negative of a vector to be another vector of equal magnitude but opposite direction. Then

$$\vec{a} - \vec{b} = \vec{a} + (-\vec{b})$$
(1.33)

This is illustrated in the following Fig. 1.10.

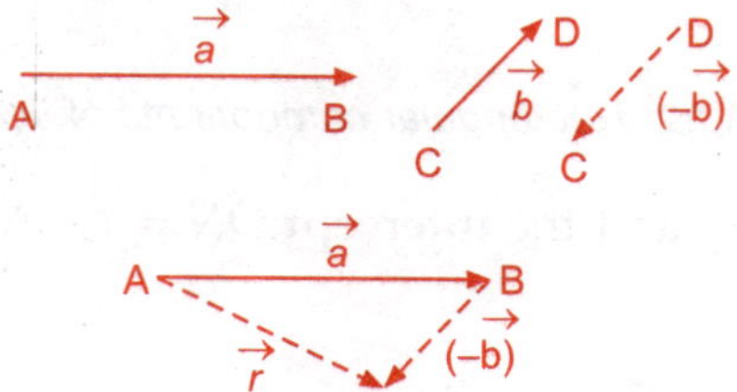

Fig. 1.10 *Subtraction of vectors*

1.6.3 Resolution of Vectors

Any two or more vectors can be added using the law of vector addition. The reverse process of splitting up of a given vector into a number of vectors whose resultant will be the original vector itself, is called resolution of a vector and each resolved vector is known as a component vector.

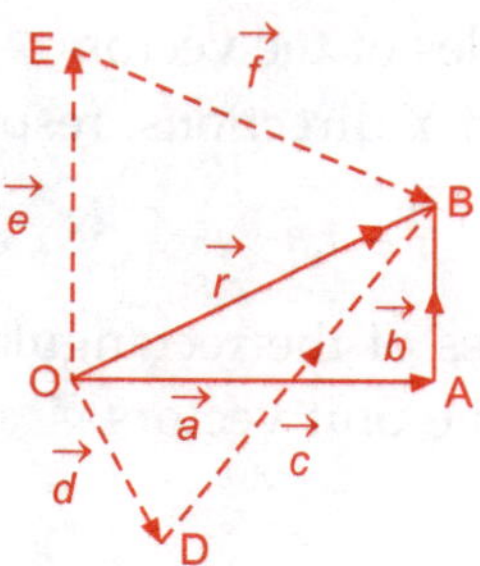

Fig. 1.11 *Several ways of splitting a given vector $\vec{r}$*

A single vector $\vec{r}$ can be equivalent to any two component vectors, $\vec{a}$ and $\vec{b}$ or $\vec{c}$ and $\vec{d}$ or $\vec{e}$ and $\vec{f}$ as shown in Fig. 1.11.

1.6.3.1 Rectangular Components

It is quite convenient to resolve a given vector into its components along two mutually perpendicular directions. Consider a vector $\vec{r}$ represented by the line OP in magnitude and direction in Fig. 1.12. Draw Cartesian coordinate axes OX and OY with origin at the root of the vector $\vec{r}$ at O. Let θ be the angle between the vector $\vec{r}$ and the X-axis. From P draw PM and PN perpendicular to the X and Y axes, respectively, as shown in Fig. 1.12.

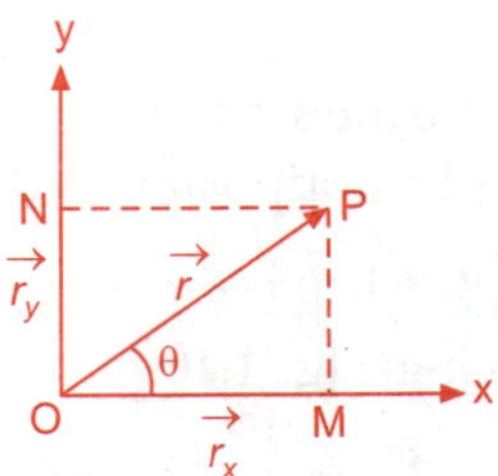

Fig. 1.12 *Rectangular components of vectors*

Let the intercept $OM = \vec{r}_x$ and the intercept $ON = \vec{r}_y$. According to the triangle law of vectors we have

$$\vec{r} = \vec{r}_x + \vec{r}_y \tag{1.34}$$

Hence $\vec{r}_x$ and $\vec{r}_y$ are the resolved components of vector along the X and Y axes. These are known as the rectangular components of vector $\vec{r}$.

Analytically we can represent the rectangular components as follows. From the right angled triangle OPM, we have the ratio, $OM/OP = \cos\theta$ and $ON/OP = \sin\theta$. Hence, we have

$$r_x = r\cos\theta, \; r_y = r\sin\theta \tag{1.35}$$

where r, r_x and r_y are the magnitudes of the vectors $\vec{r}$, $\vec{r}_x$ and $\vec{r}_y$, respectively. If $\vec{i}$ and $\vec{j}$ are unit vectors along the X and Y directions, respectively, we can write

$$\vec{r} = r_x\vec{i} + r_y\vec{j} \tag{1.36}$$

where r_x and r_y are the magnitudes of the rectangular components along the X and Y directions, respectively, whereas the unit vectors $\vec{i}$ and $\vec{j}$ denote the directions of the two components.

1.6.3.2 Addition of Vectors using Resolved Components

If θ is the angle made by the vector $\vec{a}$ with the x axis as shown in Fig. 1.13, then we have

$$a_x = a\cos\theta, \; a_y = a\sin\theta$$

$$a = \sqrt{a_x^2 + a_y^2} \tag{1.37}$$

The quantity a is the magnitude of $\vec{a}$ i.e., $|\vec{a}|$.

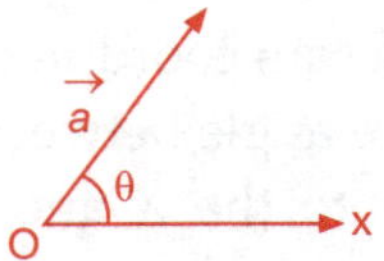

Fig. 1.13 *Resolving component of a vector*

A vector can be very conveniently represented as the product of its magnitude and a unit vector which indicates the direction. A unit vector has a magnitude of unity and a specific direction. For example, if unit vectors $\vec{i}$, $\vec{j}$ and $\vec{k}$ are assumed to be along x, y and z axes, respectively, then a vector in three-dimensional space can be represented as

$$\vec{a} = a_x\vec{i} + a_y\vec{j} + a_z\vec{k} \tag{1.38}$$

where a_x, a_y and a_z are the resolved components of along the x, y and z axes, respectively.

1.6.4 Vector Products

Vectors are used to denote physical quantities which require specification of both magnitude and direction. By properly defining the vector products it is possible to describe certain physical phenomena involving vectors. Depending on the context, three types of vector products can be used: (*i*) Multiplication of a vector by a scalar, (*ii*) Scalar product or dot product of two vectors and (*iii*) Vector product or cross product of two vectors.

1.6.4.1 Multiplication of a Vector by a Scalar

When a vector is multiplied by a scalar, k, the result is another vector with the same direction, but whose magnitude is equal to the product of the magnitude of the original vector and the scalar quantity. The vector is scaled down or up depending on whether k is less than or greater than 1, respectively. Let $\vec{a}$ be a vector which is multiplied by a scalar quantity k. Then the product $k\vec{a}$ is obtained as

$$\vec{b} = k\vec{a} = k|\vec{a}|\left(\frac{\vec{a}}{|\vec{a}|}\right) \tag{1.39}$$

$$k = \frac{OB}{OA}$$

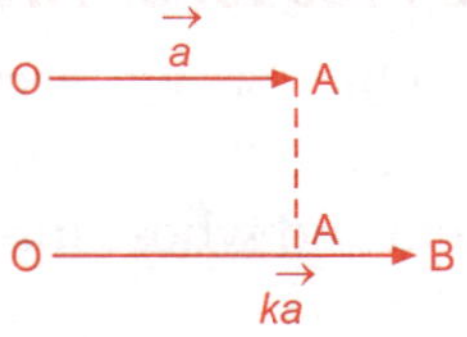

Fig. 1.14 *Multiplication of vector by scalar*

where the quantity within the parenthesis is a unit vector along the vector $\vec{a}$. This is shown in Fig. 1.14.

1.6.4.2 Scalar Product of Two Vectors

The scalar product of two vectors is a scalar quantity which is obtained by multiplying the product of their magnitudes by the cosine of the included angle between them.

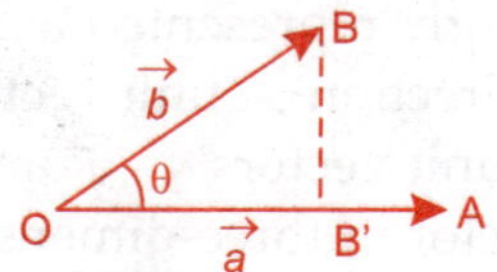

Fig. 1.15 *Scalar product of two vectors*

Referring to Fig. 1.15, if θ is the angle between the two vectors, then their scalar or dot product is obtained as

$$\vec{a}.\vec{b} = ab \cos \theta \tag{1.40}$$

where the product is characterized with a dot in between the two vectors and a and b are the magnitudes of the two vectors. Writing the dot product in the form

$$\vec{a}.\vec{b} = a\,(b \cos \theta) = b\,(a \cos \theta) \tag{1.41}$$

the result can be assumed as the product of the magnitude of the resolved component of vector $\vec{b}$ along the direction of $\vec{a}$, with the magnitude of $\vec{a}$, or as the product of the magnitude of the resolved component of $\vec{a}$ along the direction of $\vec{b}$ with the magnitude of $\vec{b}$. Some special cases are considered below.

(*i*) When $\theta = 0$, $\cos \theta = 1$ and hence $\vec{a}.\vec{b} = ab$. Here vectors $\vec{a}$ are $\vec{b}$ collinear.

(*ii*) When $\theta = 90°$, $\cos \theta = 0$ and hence $\vec{a}.\vec{b} = 0$. The vectors are at right angles to each other.

(*iii*) When $\theta = 180°$, $\cos \theta = -1$ and hence $\vec{a}.\vec{b} = -ab$. The two vectors are along the same line but opposite directions.

Work done by a force $\vec{F}$, acting on a body and displacing $\vec{s}$ it by a displacement is obtained by taking the dot product of these two vectors as

$$W = \vec{F}.\vec{s}. \tag{1.42}$$

1.6.4.3 Vector Product or Cross Product of Two Vectors

The cross product or the vector product of two vectors $\vec{a}$ and $\vec{b}$ is a single vector $\vec{c}$ whose magnitude is equal to the product of the magnitudes of the two vectors multiplied by the sine of the angle between them and whose direction is perpendicular to the plane containing the two vectors.

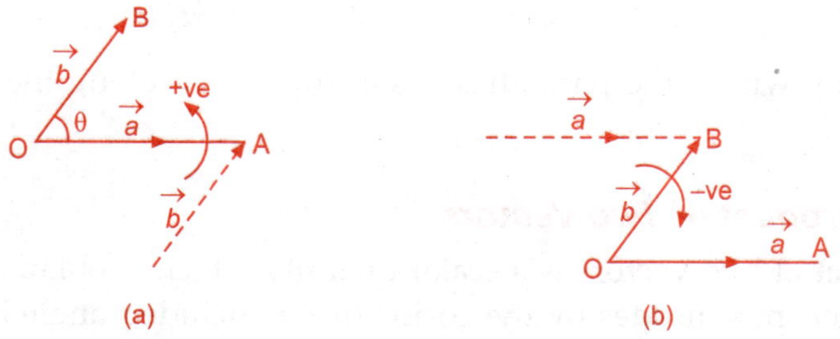

(a)   (b)

Fig. 1.16 *Vector product of two vectors*

Referring to Fig. 1.16(a) we have

$$\vec{a} \times \vec{b} = ab \sin \theta \; \hat{n} \tag{1.43}$$

where $\hat{n}$ is a unit vector perpendicular to the plane containing the vectors $\vec{a}$ and $\vec{b}$.

Physically the cross product operation, $(\vec{a} \times \vec{b})$, can be thought of as the rotation of lever $\vec{a}$ about a hinge at O, when it is acted on by the force $\vec{b}$, shown in dotted lines, in Fig. 1.16(a). The result of this operation is the rotation of $\vec{a}$ about its hinge at O. The product is positive in this case since, the rotation is anticlockwise. In Fig. 1.16(b) the vector product operation $(\vec{b} \times \vec{a})$ is shown. This is the rotation of the lever $\vec{b}$ about a hinge at O, when acted on by a force $\vec{a}$. The resulting rotation of lever $\vec{b}$ is clockwise and hence the vector product is negative.

Examples of the vector product are:

(*i*) The moment of a force $\vec{F}$ when it is rotating the lever represented by the vector $\vec{r}$ in both magnitude and direction, as shown in Fig. 1.17(a),

$$\vec{M} = \vec{r} \times \vec{F} \tag{1.44}$$

The moment is positive if the moment acts in the anticlockwise direction.

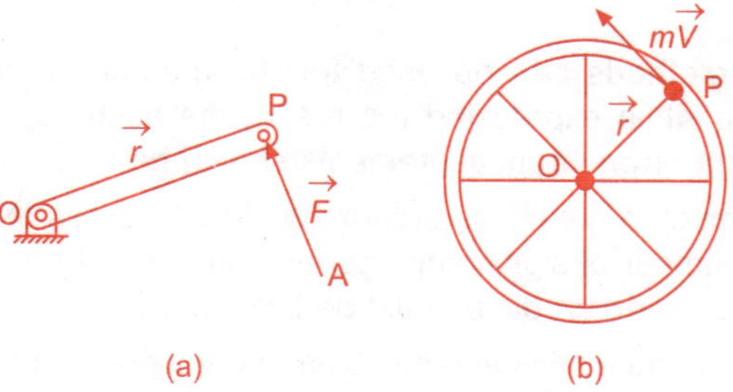

Fig. 1.17 *Examples of vector products*

(*ii*) The angular momentum which is the cross product of the linear momentum of a particle when it is rotating about a point O with the position vector of the particle, as shown in Fig. 1.17(b),

$$\vec{H} = \vec{r} \times (m\vec{v}). \tag{1.45}$$

SUMMARY

1. Physics is a quantitative science, based on measurement of physical quantities. Certain physical quantities have been chosen as fundamental or base quantities (such as length, mass, time, electric current, thermodynamic temperature, amount of substance and luminous intensity).

2. Each base quantity is defined in terms of a certain basic, arbitrarily chosen but properly standardized reference standard called unit (such as metre, kilogram, second, ampere, kelvin, mole and candela). The units for the fundamental or base quantities are called fundamental or base units.

3. Other physical quantities, derived from the base quantities can be expressed as a combination of the base units and are called derived units. A complete set of units, both fundamental and derived is called a system of units.

4. The International System of Units (SI) based on seven base units is at present internationally accepted unit system and is widely used throughout the world.

5. The SI units are used in all physical measurements, for both the base quantities and the derived quantities obtained from them. Certain derived units are expressed by means of SI units with special names (such as joule, newton, watt, etc.).

6. The SI units have well defined and internationally accepted unit symbols (such as m for metre, kg for kilogram, s for second, A for ampere, N for newton etc.).

7. Physical measurements are usually expressed for small and large quantities in scientific notation, with powers of 10. Scientific notation and the prefixes are used to simplify measurement notation and numerical computation, giving indication to the precision of the numbers.

8. Certain general rules and guidelines must be followed for using notations for physical quantities and standard symbols for SI units, some other units and SI prefixes for expressing properly the physical quantities and measurements.

9. In computing any physical quantity, the units for derived quantities involved in the relationship(s) are treated as though they were algebraic quantities till the desired units are obtained.

10. Direct and indirect methods can be used for the measurement of physical quantities. In measured quantities, while expressing the result, the accuracy and precision of measuring instruments along with errors in measurements should be taken into account.

11. In measured and computed quantities proper significant figures only should be retained. Rules for determining the number of significant figures, carrying out arithmetic operations with them, and 'rounding off' the uncertain digits must be followed.

12. The dimensions of base quantities and combination of these dimensions describe the nature of physical quantities. Dimensional analysis can be used to check the dimensional consistency of equations, deducing relations among the physical quantities, etc. A dimensionally consistent equation need not be actually an exact (correct) equation, but a dimensionally wrong or inconsistent equation must be wrong.

Dynamics–Kinematics $\quad$ 2

2.1 MOTION OF A PARTICLE

2.1.1 Introduction

Motion is the continual change in the position of objects and is experienced by everything in the universe. We walk, run and ride a bicycle. Even when we are sleeping, air moves into and out of our lungs and blood flows in arteries and veins. We see leaves falling from trees and water flowing down a dam. Automobiles and planes carry people from one place to the other. The earth rotates about its own axis once every twenty-four hours and revolves around the sun once in a year. The sun itself is in motion in the Milky way, which is again moving within its local group of galaxies.

When a person carries out tasks without interest, we say that he/she is doing the task mechanically or like a machine. Mechanical systems are lifeless systems unlike biological systems such as animals or human beings. Mechanical systems can move from one place to another only by the help of external forces applied on them, whereas, biological systems can move by themselves.

Mechanics is the study of motion of material bodies or mechanical systems caused by external forces acting on them. If the external forces acting on the body or system of bodies are arranged in such a way that the bodies are held at rest with zero motion, the study deals with Statics. If the external forces make the bodies to move, the corresponding study deals with Dynamics. Hence the vast area of Mechanics is divided into Statics and Dynamics for convenience.

Inside the Chapter

- Motion of a Particle
- Motion along a Straight Line
- Analysis of Motion
- Motion in Two Dimensions
- Simple Harmonic Motion
- Summary

In Dynamics, if the study is restricted to the motion only and does not include the forces causing the motion, we call it Kinematics. If the motion is studied including the forces causing the motion also, we call it Kinetics. For example, when an arrow is released from a bow, if the study deals with only the displacement, velocity and acceleration of the arrow, that is referred to as the kinematics part of the motion. If the study includes the force with which the bow string was pulled before releasing the arrow and what is the impact with which the arrow strikes the target, that is referred to as the study of kinetics.

2.1.2 Historical Perspectives

Ideas on Motion in Ancient Indian Science

Ancient Indian thinkers had arrived at an elaborate system of ideas on motion. Force, the cause of motion, was thought to be of different kinds: force due to continuous pressure (nodan), as the force of wind on a sailing vessel; impact (abhighat), as when a potter's rod strikes the wheel; persistent tendency (sanskara) to move in a straight line (vega) or restoration of shape in an elastic body; transmitted force by a string, rod, etc. The notion of (vega) in the Vaisesika theory of motion perhaps comes closest to the concept of inertia. Vega, the tendency to move in a straight line, was thought to be opposed by contact with objects including atmosphere, a parallel to the ideas of friction and air resistance. It was correctly summarised that the different kinds of motion (translational, rotational and vibrational) of an extended body arise from only the translational motion of its constituent particles. A falling leaf in the wind may have downward motion as a whole (patan) and also rotational and vibrational motion (bhraman, spandan), but each particle of the leaf at an instant only has a definite (small) displacement. There was considerable focus in Indian thought on measurement of motion and units of length and time. It was known that the position of a particle in space can be indicated by distance measured along three axes. Bhaskara (1150 A.D.) had introduced the concept of 'instantaneous motion' (*tatkaliki gati*), which anticipated the modern notion of instantaneous velocity using Differential calculus. The difference between a wave and a current (of water) was clearly understood; a current is a motion of particles of water under gravity and fluidity while a wave results from the transmission of vibrations of water particles.

Astronomy was a well developed science in India. Astronomical predictions of planetary positions, eclipses etc., required precise mathematical calculations. It is amazing to imagine that without the aid of present day computers, ancient Indian scientists computed planetary positions and celestial events quite accurately. The principles of mechanics were understood very clearly using which monumental buildings and temples were built. Rigorous mathematical background must have existed to do these tasks.

In the recent past the European and Greek scientists who contributed to the study of mechanics as it is today are :

1. Newton
2. Archimedes
3. Copernicus
4. Galileo
5. Kepler
6. Tycho Brahe
7. Bernoulli
8. Einstein

2.1.3 Concept of Particles, Rigid Bodies and Flexible Bodies

Particle

In our day-to-day lives we constantly deal with material bodies and substances. The stones or bricks used in the buildings, the tools we use like hammer, pick axe, or pen and pencils, the connecting rod or the flywheel in an automobile, the airplane wing are all material bodies. All these bodies have physical dimensions of length in one, two or three dimensional space. Because of their finite size, they can have also the rotational motion, along with the translational motion (motion without rotation). When we discuss translational motion of bodies in a straight or a curved path in space, it is convenient to think of a very tiny body which has negligible size. Such a body having all its mass concentrated at a point and has no shape or size is defined as a particle. Consider a sphere of mass m and radius R as shown in Fig. 2.1. Imagine that this sphere is subjected to uniform pressure all around it and is compressed tremendously such that $R \rightarrow 0$. Then we can consider such a compressed entity as a particle of mass m.

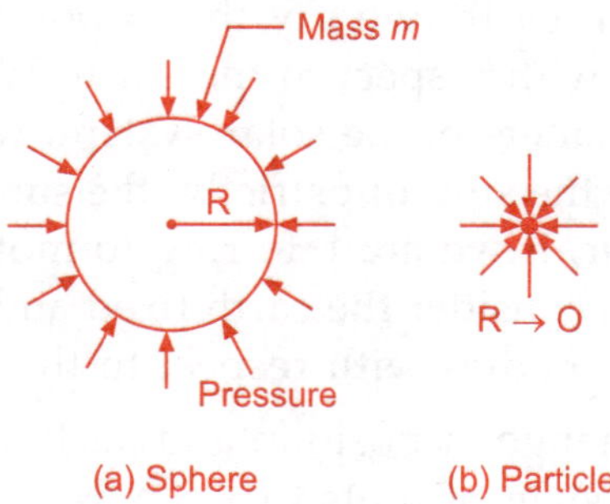

Fig. 2.1 *Concept of a particle*

Rigid Body

When several particles are held together, the resulting body will have a size and shape. The individual particles are not capable of any relative motion among themselves, such a body is referred to as a rigid body. Stones, bricks, steel balls etc., can be considered as rigid bodies, for all practical purposes.

Flexible Body

When the particles forming the body are not rigidly held together, but can move relative to each other under the action of external forces, it is defined as a flexible body. Consider a steel beam fixed at one end to a rigid wall and free at the other end. If a force is acting at the free end, the beam will bend as shown in Fig. 2.2. Since the end A does not move at all while the end B moves down, the beam is not rigid, it is flexible.

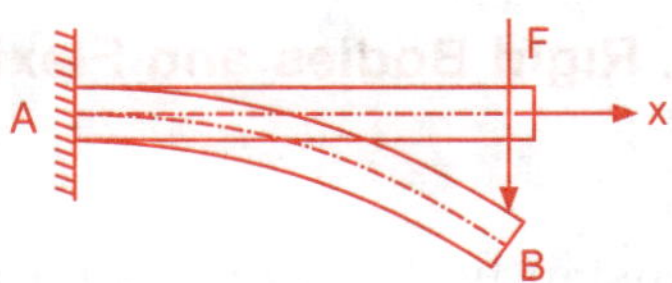

Fig. 2.2 *Example of a flexible body*

2.2 MOTION ALONG A STRAIGHT LINE

2.2.1 Motion

A body is said to be in motion when it changes its position with respect to time. If the body is not moving then the body is said to be at rest. The condition of motion or rest must be defined with respect to a reference point, whose condition is known absolutely. Since everything in this universe keeps on moving, it is impossible to get such a reference point whose condition is completely known. Hence we say that an object is at rest with respect to another object, or that the object is moving with respect to another object.

For motions of bodies on earth, we say that a point on earth is at rest and we define the motion of bodies with respect to the earth. When we want to refer to the motion of earth and other planets in the solar system, we say that the sun is at rest and planets move relative to the sun. But strictly, the sun is also moving with respect to the distant stars. However, if we are referring to motions of bodies on earth, for all practical purposes we can consider the earth fixed and to be at rest, so that we can describe the motion of other bodies with respect to the earth.

Since motion involves change of position and such a change of position requires time, we need to define a system of units that can express motion quantities.

We have already discussed units in Chapter 1. The basic units are to measure the length, mass and time. In SI units, the length is measured in meters, the mass in kilograms and the time in seconds. As we can see later on, these units can be used to define several more derived units.

Position

The position of a particle is defined with respect to a reference point, say the origin of a Cartesian coordinate system. If a train is travelling from New Delhi to Chennai, the origin can be fixed at New Delhi and the distance along the track to the point where the train is currently, defines the position of the train as shown in Fig. 2.3.

Fig. 2.3 *Position of a particle*

For illustration purposes, the track from New Delhi to Chennai is represented by a straight line. As the train moves towards Chennai, the distance from New Delhi to the train changes continuously, which means that as the train moves the position of the train changes continuously.

Displacement

When the position of the particle changes, it is said to have undergone a displacement. Displacement of a particle, when it is travelling along a straight line, is the change of its position in a particular direction. In Fig. 2.4, the particle P is initially at A and its final position is at B and hence its displacement is given by the distance AB.

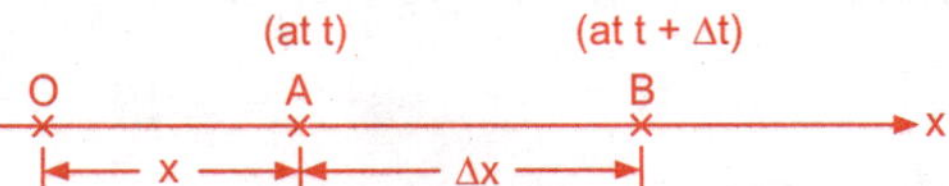

Fig. 2.4 *Displacement of a particle*

When the displacement takes place in the positive x direction, the displacement is positive and if it takes place in the negative x direction, the displacement is negative. Displacement is a vector quantity, having both magnitude and direction.

Velocity

The rate of displacement or the rate of change of position of a particle is the velocity of the particle. Velocity has both magnitude and direction and hence it is a vector quantity. If a small displacement of Δx takes place in a time interval between t and $t + \Delta t$, then the average velocity is given by

$$\text{Average velocity,} \qquad v_{av} = \frac{\Delta x}{\Delta t} \qquad\qquad (2.1)$$

If Δt is quite small, then this average velocity may be said to be between time t and $t + \Delta t$. Here it is assumed that the velocity is constant between time t and $t + \Delta t$. The instantaneous velocity at time t is obtained by making the observation time Δt as small as possible, and is given by

$$v(t) = \lim_{\Delta t \to 0} \frac{\Delta x}{\Delta t} = \frac{dx}{dt} \qquad\qquad (2.2)$$

A constant velocity is referred to as uniform velocity. Since velocity is a vector having both magnitude and direction, the velocity changes when there is a change either in its magnitude or in direction or both.

EXAMPLE 2.1

A train travels the distance of 45 km from Mangalore to Kasaragod in 1 hr. If the track between these two stations is quite straight then what is the average velocity of the train?

SOLUTION

From Eq. (2.1) the average velocity is given by

$$\frac{45(1000)}{1(60 \times 60)} = 1.25 \text{ m/sec.}$$

EXAMPLE 2.2

The displacement of a particle is given as a function of time as $x(t) = 5t^2 + 5t + 2$ in meters. Find the instantaneous velocity at times (i) $t_1 = 10$ sec, (ii) $t_2 = 20$ sec, and at (iii) $t_3 = 60$ sec.

SOLUTION

From Eq. (2.2), $v(t) = dx/dt = 10t + 5$ m/sec. Hence

$\quad$ (i) $v(10) = 105$ m

$\quad$ (ii) $v(20) = 205$ m

and $\quad$ (iii) $v(60) = 605$ m.

Speed

The rate of change of position of a particle, without considering the direction is called the speed. If in Example 2.1, the track from Mangalore to Kasaragod was not a straight line, but was changing directions quite often along the way, the velocity would be changing even if the train covered equal distances during equal intervals of time. However, the speed would be constant.

EXAMPLE 2.3

An athlete is running along an oval track which has a distance of 250 m for one complete round. If she completes 40 rounds in 30 minutes, what is her average speed? Provide the answer to an accuracy of tenth of a meter per second.

SOLUTION

The speed is obtained from Eq. (2.1) without any consideration to the direction of the motion. Hence the speed is given by

$$\text{Average speed} = \frac{250(40)}{30(60)} = 33.3 \text{ m/sec}$$

Acceleration

The rate of change of velocity is acceleration. Since velocity is a vector quantity, velocity changes when there is a change in magnitude of the velocity or in direction or both. Acceleration is also a vector quantity. If the velocity at time t is v and at time $t + \Delta t$ the velocity changes to $v + \Delta v$, then the average acceleration of the particle during the time between t and $t + \Delta t$ is given by

$$\text{Average acceleration, } a_{av} = \frac{\Delta v}{\Delta t} \tag{2.3}$$

The instantaneous acceleration at time t is given by making the observation time Δt as small as possible, and is given by

$$a(t) = \lim_{\Delta t \to 0} \frac{\Delta v}{\Delta t} = \frac{dv}{dt} \tag{2.4}$$

A constant acceleration is termed as uniform acceleration. The acceleration due to gravity, which is the acceleration with which bodies fall down close to the surface of the earth, is uniform acceleration and has a value of 9.81 m/s^2.

EXAMPLE 2.4

An automobile starts from rest and accelerates uniformly. If its velocity after 10 seconds is 15 m/s, what is the average acceleration of the automobile?

SOLUTION

The total change in velocity in 10 seconds is given by $(15 - 0)$ m/s. From Eq. (2.3), the average acceleration is given by

$$a_{av} = \frac{15 - 0}{10} \text{ m/s}^2.$$

EXAMPLE 2.5

The displacement of a particle was given in Example 2.2 as $x(t) = 5t^2 + 5t + 2m$. What is the acceleration of the particle at (i) 10 seconds, (ii) 20 seconds and (iii) 60 seconds?

SOLUTION

We have $v(t) = dx/dt = 10t + 5$ m/s and

$$a(t) = dv/dt = d^2x/dt^2 = 10 \text{ m/s}^2. \text{ Hence}$$

(*i*) $\qquad a(10) = 10 \text{ m/s}^2$

(*ii*) $\qquad a(20) = 10 \text{ m/s}^2, \text{ and}$

(*iii*) $\qquad a(60) = 10 \text{ m/s}^2.$

The acceleration is constant, and therefore, the motion is uniformly accelerated motion.

2.3 ANALYSIS OF MOTION

A motion is completely known when the acceleration, velocity and displacements are known at any time. Later on, we will see that motion is caused by force. In fact, Newton's second law of motion states that acceleration is proportional to the force applied. Hence, in many situations, by knowing the nature of the force, we know the nature of acceleration also. Analysis of motion consists of finding the corresponding velocities and displacements of the moving body.

When there is no force acting on the body, and if it is already moving, it will continue to move with the same velocity. Motion with constant velocity is known as uniform motion. When a constant force is acting on the body, the resulting acceleration is constant, and the motion is known as uniformly accelerated motion. The applied force can be proportional to the displacement or velocity of the motion or it may be a function of time. In such cases, the analysis of motion consists of integrating the corresponding acceleration expressions once to obtain the velocity and twice to obtain the displacement of the motion. However, the following expressions are derived without using integration since the students may not have studied calculus at this level. Appendix I derives the same expressions using calculus. In the following we will analyse uniform motion and uniformly accelerated motion.

The following notations are used in all the relations given below:

u is the initial velocity

v is the final velocity

a is the uniform acceleration

t is the time interval.

2.3.1 Uniform Motion

When the velocity of motion is constant, the motion is called uniform motion. Hence we have

$$\text{Velocity} = \frac{\text{Displacement}}{\text{Time}} = \text{Constant} \qquad (2.5)$$

i.e.
$$v = \dot{x} = \frac{x - x_0}{t}$$

Hence $$x = x_0 + vt \tag{2.6}$$

where x_0 is the position at time $t = 0$, and x is the position at time t.

2.3.2 Uniformly Accelerated Motion

1. Velocity at any time t:

From Eq. (2.2), acceleration a is given as

$$a = \frac{\text{Change in Velocity}}{\text{Time}} = \frac{v - u}{t} \tag{2.7}$$

Hence $$v = u + a.t \tag{2.8}$$

2. Distance x travelled in time t:

By definition

$$\text{Average Velocity, } v_{av} = \frac{\text{Distance Travelled}}{\text{Time}} = \frac{x}{t} \tag{2.9}$$

Since the acceleration is constant, the velocity is changing linearly with time according to Eq. (2.8). Hence

$$\text{Average velocity, } v_{av} = \frac{(u+v)}{2} \tag{2.10}$$

Substituting Eq. (2.10) into Eq. (2.9), and using Eq. (2.8) we get

$$\frac{x}{t} = \frac{(u+v)}{2} = \frac{(u+u+at)}{2} \tag{2.11}$$

Rearranging Eq. (2.11), we obtain x as

$$x = ut + \frac{1}{2}at^2 \tag{2.12}$$

3. Velocity as a Function of Distance:

From Eq. (2.11) we have

$$\frac{x}{t} = \frac{u+v}{2} \tag{2.13}$$

From Eq. (2.8), we can obtain an expression for the time taken for the change from velocity u to velocity v, as

$$t = \frac{v-u}{a} \tag{2.14}$$

Substituting for t in Eq. (2.13) from Eq. (2.14), we get

$$v^2 = u^2 + 2ax. \tag{2.15}$$

EXAMPLE 2.6

A particle is moving along a straight line with a constant acceleration a. If the initial velocity is u, and x_1 and x_2 are the distances travelled in two consecutive equal intervals of time t each, find a relation between u, x_1, x_2 and t.

SOLUTION

From Eq. (2.12) we have

$$x_1 = ut + \frac{1}{2}at^2 \tag{2.16}$$

$$x_1 + x_2 = u(2t) + \frac{1}{2}a(2t)^2 = 2ut + 2at^2 \tag{2.17}$$

Multiplying Eq. (2.16) by 2 and subtracting it from Eq. (2.17), we get

$$x_1 + x_2 - 2x_1 = at^2. \tag{2.18}$$

Simplifying Eq. (2.18), we get the relation

$$a = \frac{(x_2 - x_1)}{t^2}. \tag{2.19}$$

Substituting in Eq. (2.16), we get $3x_1 - x_2 = 2ut$.

EXAMPLE 2.7

A particle is moving with a constant acceleration a. Its initial velocity is given as u. What is the distance travelled by the particle during the nth second of the motion?

SOLUTION

Distance travelled during the first n seconds is obtained from Eq. (2.12) as

$$x_n = un + \frac{1}{2}an^2 \tag{2.20}$$

Distance travelled during the first $(n - 1)$ seconds is given by

$$x_{n-1} = u(n-1) + \frac{1}{2}a(n-1)^2 \tag{2.21}$$

Hence the distance travelled during the nth second is given by

$$x_n - x_{n-1} = u + an - \frac{1}{2}a = u + a\left(n - \frac{1}{2}\right). \tag{2.22}$$

2.3.3 Velocity-Time Graph

A velocity-time graph is the plot of velocity versus time as shown in Fig. 2.5. Time is represented along the x-axis and the corresponding velocity at different times is plotted along the y-axis.

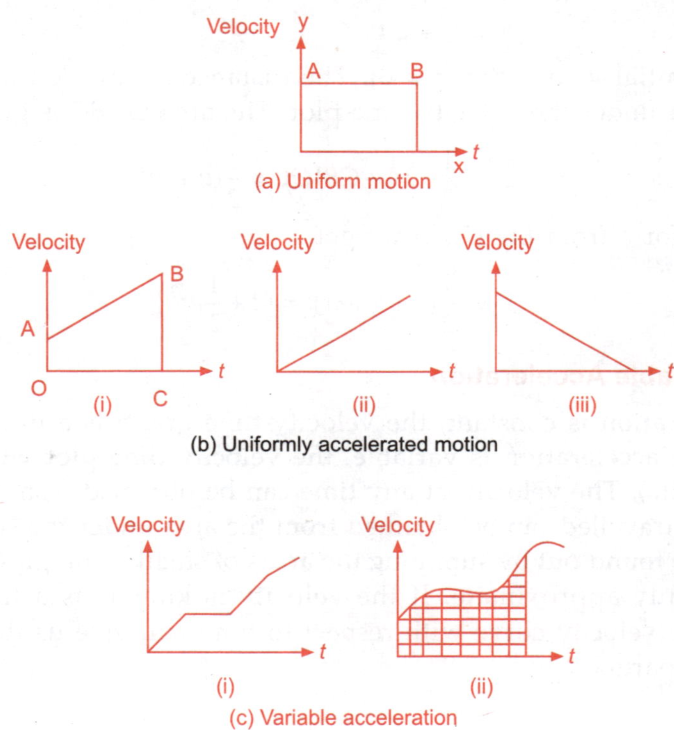

Fig. 2.5 *Velocity-time graphs*

Uniform Motion

The velocity-time graph of a particle executing a uniform motion, i.e., moving with constant velocity, is a line parallel to the x-axis as shown in Fig. 2.5(a). Since velocity is constant, we have from Eq. (2.2)

$$v = \frac{\Delta x}{\Delta t} = \frac{x}{t} \tag{2.23}$$

Hence the distance travelled in time is given by

$$x = v.t \tag{2.24}$$

But $v.t$ is the area under the velocity-time plot. If x_0 is the initial displacement, then the present position of the particle is given by

$$x = x_0 + v \cdot t \tag{2.25}$$

Uniformly Accelerated Motion

The velocity-time plot for a particle executing a uniformly accelerated motion is shown in Fig. 2.5(b). In this case the velocity is increasing constantly, which appears as a line inclined to the x-axis with a slope of a. The velocity at any time is obtained by the equation for the line and is given by

$$v = u + a.t \tag{2.26}$$

where u is the initial velocity (at $t = 0$). The distance x travelled by the particle is given by the area under the velocity-time plot. The area $OABC$ is given by

$$x = \frac{1}{2}(OA + CB)OC = \frac{1}{2}(u + v)t \tag{2.27}$$

Substituting for v from Eq. (2.26) we get

$$x = \frac{1}{2}(u + u + at)t = ut + \frac{1}{2}at^2. \tag{2.28}$$

Motion with Variable Acceleration

When the acceleration is constant, the velocity-time graph is a line inclined to the x-axis. When the acceleration is variable, the velocity-time plot will be a curve as shown in Fig. 2.5(c). The velocity at any time can be obtained from the graph itself, and the distance travelled can be obtained from the area under the curve at any time t. The area can be found out by summing the areas of small rectangles as shown in the figure. This is only approximate. If the velocity is known as a function of time, integration of the velocity curve with respect to time will give us the exact distance travelled by the particle.

EXAMPLE 2.8

A car travels from Mangalore to Udupi at a speed of 50 km/hr and returns from Udupi to Mangalore at a speed of 60 km/hr. What is the average velocity of the whole journey?

SOLUTION

The average speed is given by

$$\text{Average Speed} = \frac{\text{Total distance travelled}}{\text{Total time}}$$

If x km is the distance from Mangalore to Udupi, the total distance travelled is $2x$ km.

$$\text{The time taken for the whole journey} = \frac{x}{50} + \frac{x}{60} = \frac{11x}{300} \text{ hr.}$$

$$\text{Hence the average speed for the whole journey} = \frac{2x}{(11x/300)} = \frac{600}{11} = 54.55 \text{ km/hr.}$$

This answer is given to an accuracy of two digits after decimal place. The exact answer would be 54.5454545454.... km/hr.

Reaction time: When a situation demands our immediate action, it takes some time before we actually respond. Reaction time is the time a person takes to observe, think and act. For example, if a person is driving and suddenly a boy appears on the road, then the time elapsed before she slams the brakes of the car is the reaction time. Reaction time depends on the complexity of the situation and on an individual.

2.4 MOTION IN TWO DIMENSIONS

When the particle is moving along a straight line, the direction of the applied force and the direction of motion are along the same line. However, if the direction of the applied force is different from that of the motion, then the particle will change its direction and follow a curved path. Motion along a curved path is called curvilinear motion.

We see several examples of such curvilinear motion in our daily lives. When we throw a stone away from us with a certain angle with respect to the horizontal, we can see the stone following a parabolic path. The force of gravity acts always vertically downward which forces the initially given motion downward continuously so as to follow a curved path. In general, particles in the atmosphere under the influence of gravitational forces follow a curved path and their motion is referred to as projectile motion. A satellite orbiting around the earth, or the moon orbiting around the earth, or the earth orbiting around the sun are all examples of curvilinear motion.

2.4.1 Projectile Motion

A body travelling freely under the action of gravity is referred to as a projectile. A bullet fired from a gun, a stone tossed into the air, are examples of projectiles. The direction of throw may have any angle from 0° to 90° with respect to the horizontal. If a ball is given a horizontal velocity from a certain point above the ground and another ball is just dropped vertically from the same point, both balls will reach the ground at the same time, but at different points (provided the air resistance is very small). Hence, the vertical motion is not influenced by the horizontal motion. In analyzing the motion of a projectile, the horizontal and vertical motions are solved separately. The final result may be combined again using vector addition.

Consider a particle projected with a velocity u, at an angle θ with respect to the horizontal as shown in Fig. 2.6. In order to solve the horizontal and vertical motions separately, we need to get the horizontal and vertical components of the initial velocity, which are, respectively, given by $u \cos \theta$ and $u \sin \theta$.

Horizontal Motion: The component $u \cos \theta$ remains constant since there is no other force acting on the body in the horizontal direction and hence, the distance x travelled along OX is given by

$$x = (u \cos \theta)t \tag{2.29}$$

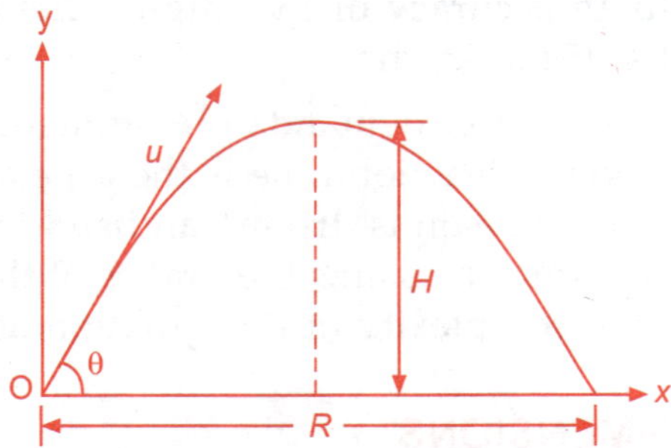

Fig. 2.6 *Projectile Motion*

Vertical Motion: In the vertical direction, the distance travelled by the particle is given by

$$y = (u\sin\theta)t - \frac{1}{2}gt^2 \tag{2.30}$$

The values of x and y refer to the x and y coordinates of the particles at any time t. Substituting the value of t from Eq. (2.29) into Eq. (2.30), we get

$$y = u\sin\theta\left(\frac{x}{u\cos\theta}\right) - \frac{1}{2}g\frac{x^2}{u^2\cos^2\theta} \tag{2.31}$$

This equation can be put in the form

$$y = x\tan\theta - \frac{1}{2}\left(\frac{g}{u^2\cos^2\theta}\right)x^2 \tag{2.32}$$

Eq. (2.32) is in the form of a parabola,

$$y = ax + bx^2 \tag{2.33}$$

where $a = \tan\theta$, and $b = -\dfrac{1}{2}\left(\dfrac{g}{u^2\cos^2\theta}\right)$

The following quantities may be calculated conveniently in connection with a projectile motion.

1. Maximum Height Reached by the Projectile

This can be obtained by analyzing the vertical part of the motion. At the maximum height, the vertical velocity of the projectile is zero. From Eq. (2.15), which relates the initial and final velocities with the uniform acceleration and the distance travelled in the direction of motion, we have

$$0 = (u\sin\theta)^2 - 2gh_m \tag{2.34}$$

Hence the maximum height h_m is obtained as

$$h_m = \frac{u^2 \sin^2 \theta}{2g}.$$ (2.35)

2. Time of Flight

The time of flight is the time when the projectile is launched until it touches the ground. When the projectile touches the ground we have $y = 0$. Substituting this value in Eq. (2.30) we have

$$0 = (u\sin\theta)t - \frac{1}{2}gt^2 = t\left(u\sin\theta - \frac{g}{2}t\right)$$ (2.36)

This equation is satisfied if $t = 0$, or $t = \dfrac{2u\sin\theta}{g}$. Since $t = 0$ is a trivial value, (the

projectile is touching the ground just when the projectile is launched also, which is not the answer that we are interested!), the time of flight is given by

$$t = \frac{2u\sin\theta}{g}$$ (2.37)

3. Range of Projectile

The maximum distance travelled by the projectile in the horizontal direction is called the range, R, of the projectile. Substituting the time of flight from Eq. (2.37) in Eq. (2.29) we get the range, R, as

$$R = u\cos\theta\left(\frac{2u\sin\theta}{g}\right) = \frac{u^2 \sin 2\theta}{g}$$ (2.38)

4. Maximum Range of the Projectile

From Eq. (2.38), we see that the range is maximum when $\sin 2\theta$ is maximum. The maximum value of $\sin 2\theta$ is 1. Hence $2\theta = 90°$ and hence $\theta = 45°$. Hence, for maximum range, the projectile must be launched at an angle of 45° to the horizontal.

2.4.2 Uniform Circular Motion

Circular motion is a particular kind of curvilinear motion. Uniform circular motion is the motion of bodies along a circular path with constant speed. Even though the speed along the circular path is constant, the direction changes continuously and hence the velocity of the body in circular motion changes continuously. Any change of velocity, whether in magnitude, or in direction or both, introduces acceleration of the body.

Velocity is always tangential to the path. When the body has moved by a small displacement along the circle, the new velocity changes the direction slightly as shown in Fig. 2.7.

Drawing these velocity vectors from the same origin, as shown in Fig. 2.8, we see that the change in velocity vector, Δv, is directed along a normal to the velocity at A. In general, the change in velocity is directed towards the centre of the circular path. This acceleration, which is always directed towards the centre of the circular path, is called as centripetal acceleration or radial acceleration.

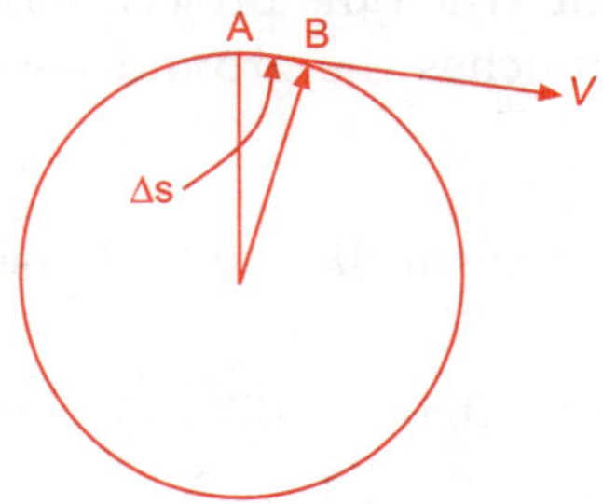

Fig. 2.7 *Velocity in Circular Motion*

Centripetal Acceleration: Consider a particle moving along a circular path of radius R and with a uniform speed v as shown in Fig. 2.8. The velocity of the particle is tangential to the path at point A. Let the particle cover a distance Δs along the circle in time Δt. The arc Δs subtends an angle of $\Delta \theta$ at the centre of the circle. Hence we have

(2.39)

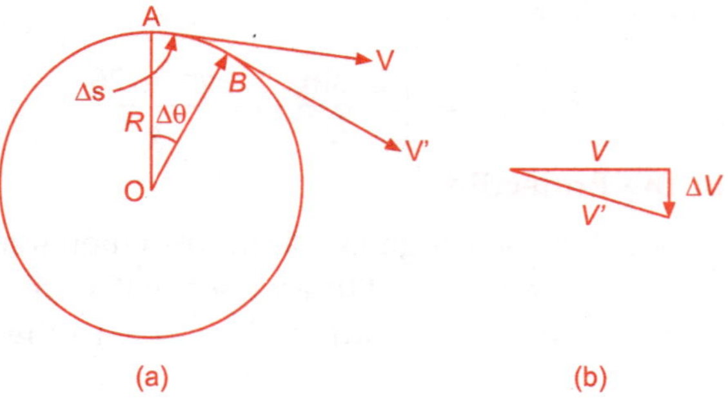

(a)   (b)

Fig. 2.8 *Centripetal Acceleration*

The velocity vectors v and v' at points A and A' are drawn from the same origin in Fig. 2.8. The angle between the two velocity vectors, v and v' is $\Delta \theta$. Hence the change in the velocity vector, Δv is given by

$$\Delta v = v\Delta \theta = v\Delta s/R. \tag{2.40}$$

The rate of change of velocity, which is the acceleration, is given by

$$\frac{\Delta v}{\Delta t} = \frac{v}{R} \cdot \frac{\Delta s}{\Delta t} = \frac{v^2}{R} \tag{2.41}$$

since we know that the quantity, $\Delta s/\Delta t = v$, being the rate of change of displacement along the circle. Hence the radial acceleration is obtained as

$$a = v^2/R \tag{2.42}$$

and it is directed towards the centre of the circular path.

Certain quantities which are characteristic of circular motion will be discussed below.

(*i*) **Period:** Period, T is the time taken for one complete revolution around the circular path. Since the length of travel around the circle is $2\pi R$ and the speed of travel is v, the period is given by

$$T = 2\pi R/v. \tag{2.43}$$

(*ii*) **Angular speed or angular velocity:** Angular speed, ω is the angle, either in radians or in degrees, covered in unit time. Since the body covers an angle of 2π radians when it goes around the circle once, the angular speed is given by

$$\omega = \frac{2\pi}{(2\pi R/v)} = \frac{v}{R} \tag{2.44}$$

Normally, ω is expressed in radians per second. It is related to the number of revolutions per second, f, by

$$f = \omega/2\pi \tag{2.45}$$

Number of degrees per second is given by

$$\dot\theta = 360 f = \frac{180\omega}{\pi}. \tag{2.46}$$

Centripetal Force: Newton's second law states that $F = m.a$. The acceleration a is caused by the force F. That force which causes the centripetal acceleration of $a = v^2/R$, is called the centripetal force. Centripetal force is radial force acting on a body which is moving in a circular path with a constant speed and is directed towards the centre of the circle, just as the centripetal acceleration is directed towards the centre of the circular path.

The force F causing a linear motion can be visualized very easily. The centripetal force is a little bit difficult to visualize. However, we come across this force quite often, and it is not really difficult to understand that. Tie a string to a ball and give a revolving motion with the hand holding the end of the string. Then the ball will start moving along a circular path as shown in the Fig. 2.9. The hand is exerting a force on the string to keep the ball moving along the circular path. If at any instant the string breaks, the ball would move off along a tangent to the path at the instant the string breaks. Hence to keep the ball along the circular path, or in other words, to make the velocity of the ball change its direction continuously, the hand pulls the string towards itself. The hand is at the centre of the circular path. This force which pulls the ball towards the centre is the centripetal force. When the string breaks or if the hand releases the string, the centripetal force disappears and the ball will move along a tangent to the circle.

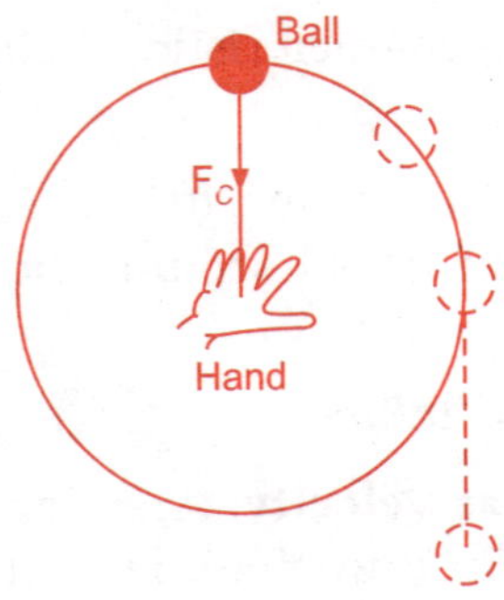

Fig. 2.9 *Centripetal Force Directed Toward Centre of Circle*

In a grinding machine, the particles which are ground by the grinding wheel will fly off from the wheel in a tangential direction. Since the force which was holding these particles to the material being ground has been broken by the grinding process, there is no more centripetal force that can keep these particles to adhere to the circular path.

The same principle holds good when an athlete executes a hammer throw. As soon as the athlete whirls the hammer around and lets it go, it will lose the centripetal force to keep it in the circular path, and will go along a tangent. When earth is going around the sun, the centripetal force is provided by the gravitational pull exerted by the sun on earth.

Centrifugal Reaction: In the example of the ball tied to a string and revolved around by the hand, while the hand pulls the string towards the centre, the hand also experiences an equal and opposite force exerted on it by the string. This is the centrifugal reaction. Newton's third law states that every action has an equal and opposite reaction. The agent which exerts the centripetal force is subjected to a reaction force, which is called the centrifugal reaction. The centrifugal reaction is directed away from the centre of the circular path.

Centrifugal Force: We have already discussed that the rectilinear motion of a body can be discussed exactly following the statement of Newton's second law, or from the point of view of dynamic equilibrium. The effective force, which is the product of mass and acceleration, is the centripetal force. If we look at the motion of a body along a circular path, the corresponding application of dynamic equilibrium concept needs an inertia force. The corresponding inertia force is termed as centrifugal force. Centrifugal force is equal and opposite to centripetal force.

Examples of Centripetal Force

1. **Bicycle Negotiating a Curve:** When a cyclist goes around a curve, he or she leans towards the centre of curvature, so that the horizontal component of the ground reaction provides the necessary centripetal force in order to change the velocity (or to provide the necessary radial acceleration). This is shown in Fig. 2.10.

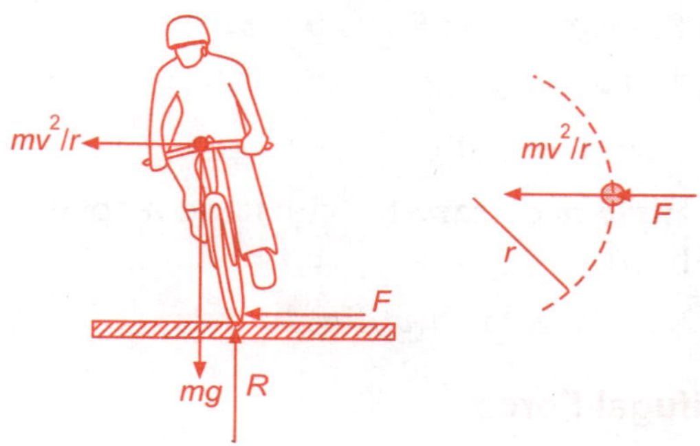

Fig. 2.10 *Bicycle Turning Along a Curve*

1. **Banking of Roads and Railway Tracks:** Since four-wheeled vehicles cannot incline themselves when negotiating curves as the two wheelers can do, it is common practice to provide "banking" for road curves. If the road is level around a curve, the necessary centripetal force is supplied only through friction between the wheels and the road surface. Since the force of friction is limited to μN, the speed of the vehicle must also be limited such that the centripetal force necessary for negotiating the curve is less than or equal to the friction force. If the velocity exceeds this limit, the vehicle will slip away. To overcome this problem and to enable the vehicle to negotiate the curves with higher speeds the road surfaces are inclined slightly in a radial direction, raising the outer edge and lowering the inner edge of the curve. This prevents skidding and damage to tires. This is illustrated in Fig. 2.11, which shows a car moving round a curve along a road banked at an angle θ. The banking angle needed to sustain a maximum speed v of the vehicle is calculated as follows. Figure 2.11(*b*) shows an equivalent particle model for the car. In this model the reaction

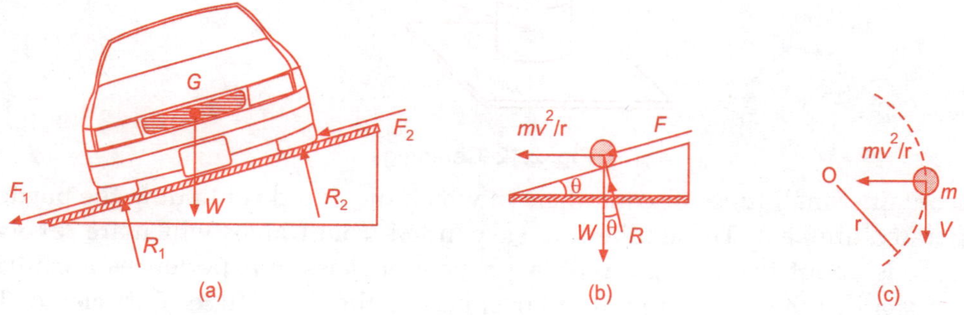

Fig. 2.11 *Banking of Roads*

force, R, is the resultant of the reaction forces acting on all the four tires of the car. Considering all the forces acting on the particle model of the car, from the horizontal components we have

$$R \sin \theta = mv^2/R \tag{2.47}$$

Considering the vertical components, we get

$$R \cos \theta - mg = 0 \text{ or } R \cos \theta = mg \tag{2.48}$$

From Eqs. (2.47) and (2.48) we get

$$\tan \theta = v^2/Rg \tag{2.49}$$

Hence the banking angle necessary to withstand a speed of v when the radius of curvature is R is given by

$$\theta = \tan^{-1}(v^2/Rg). \tag{2.50}$$

Applications of Centrifugal Force

As mentioned earlier, if a particle is moving around a circular path and it is constrained, (Ex: the string constraining the ball), then centripetal force acts through the string to balance the effective force which is the product of the mass and centripetal acceleration. If the particle is not constrained, then the centripetal force is absent. Hence the particle physically accelerates radially outwards, with an acceleration equal and opposite to the centripetal acceleration. This phenomenon is used in several industrial devices described below.

1. **Centrifuge:** Centrifuge is a machine which rotates at very high speeds and is used to separate small particles of different densities. This is schematically shown in Fig. 2.12.

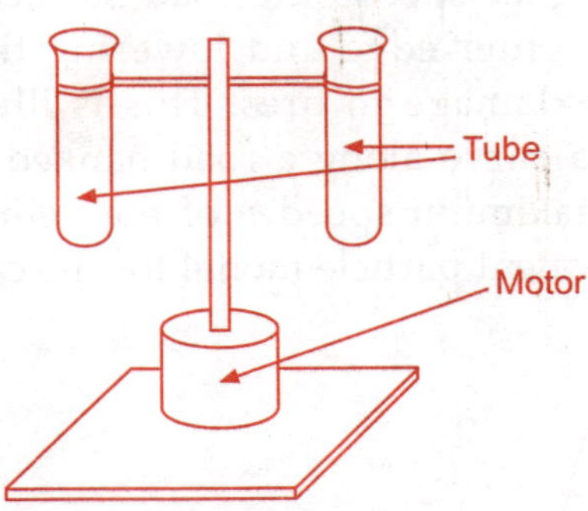

Fig. 2.12 Centrifuge

It contains small tubes or containers in which the liquid containing the particles to be separated are kept. These tubes are suspended from frames which are revolved at high speeds about the vertical axis. A particle of mass m experiences a centrifugal force of $m\omega^2R$, where ω is the angular speed of the centrifuge. Particles of larger densities experience larger centrifugal forces and move away from the axis, whereas particles of smaller densities remain closer to the axis of rotation. Centrifuge is used in clinical laboratories to separate proteins, hormones, viruses etc., in different liquid samples. The acceleration in a centrifuge, $\omega^2 R$, is expressed in terms of g, the acceleration due to gravity. Acceleration in some centrifuges can reach values in the order of $10^6 g$.

2. **Centrifugal Drying Machine:** This is used to dry clothes. It consists of a cylindrical vessel with perforated walls. When damp clothes are kept inside and the vessel is rotated about its axis, the water particles are forced out through the perforations.

3. **Centrifugal Compressors and Pumps:** Air under lower pressure enters the vanes at the centre, and the vanes rotate about an axis as shown in Fig. 2.13. The air acquires large velocities and when it leaves the vanes this kinetic energy is converted into potential energy in the form of higher pressure.

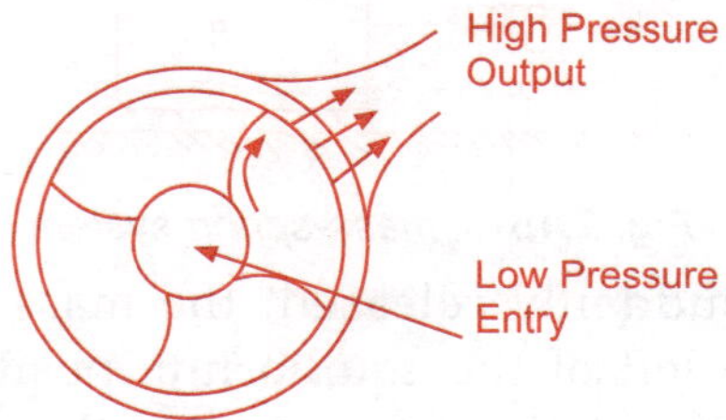

Fig. 2.13 *Centrifugal Compressor*

Centrifugal pumps pump liquids from a lower level to a higher level. Liquid enters the vanes at the centre and flow through the vanes. As it moves it acquires higher velocities and gains kinetic energy. This kinetic energy gets converted into potential energy as the liquid leaves the vanes.

4. **Centrifugal Governors:** A centrifugal governor is shown in Fig. 2.14. It consists of two balls fixed to a linkage which is restrained by collar that can move freely up or down a central shaft. As the shaft rotates, the balls rise due to centrifugal force. When the speed of rotation increases, the height of rise of the balls also increases. If

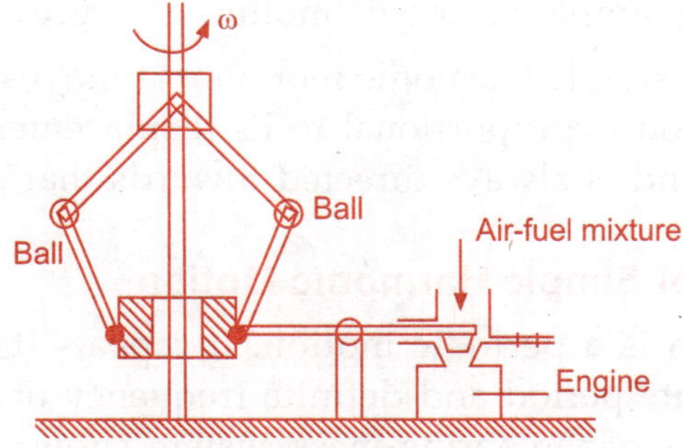

Fig. 2.14 *Centrifugal Governor*

the speed rises above a particular value or goes below a specified value, a controller mechanism connected to the governor will control the speed of the engine or the motor that drives the shaft.

2.5 SIMPLE HARMONIC MOTION

2.5.1 Introduction

Consider a helical spring with one end fixed to a support and the other end connected to a mass m as shown in Fig. 2.15. If a force F is applied to the mass and it is pulled in the x direction as shown, the spring exerts an equal force on the mass pulling it back to its equilibrium position.

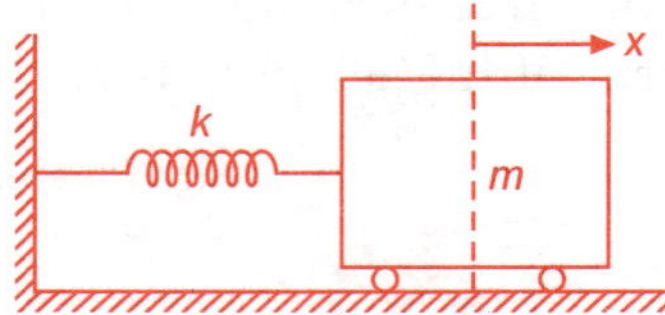

Fig. 2.15 *A mass-spring system*

If now the force F is suddenly released, the mass will be pulled back to its original position, go to the left of the equilibrium position and then return to the equilibrium position. It will continue to perform these to and fro motions until compelled by an external force to stop the motion. The to and fro motion of the mass is periodic in nature and it is called a simple harmonic motion.When a body, which is subjected to elastic constraints of any kind, is displaced from its equilibrium position against any such constraint, it develops a restoring force tending to return it towards the equilibrium position. If the elastic constraint, such as the helical spring in the above example, follows the Hooke's law, then the restoring force acting on the body is proportional to the displacement from the equilibrium position. Newton's second law of motion states that the externally applied force on a body accelerates the body with an acceleration given by $F = ma$. This means that the acceleration of the body is proportional to its displacement from the equilibrium position and is directed always towards it. The resulting simple harmonic motion is of great importance in Physics.

A particle executes a simple harmonic motion if it moves in a straight line and at any instant its acceleration is proportional to its displacement at that instant from a fixed point in the line, and is always directed towards that point.

2.5.2 Characteristics of Simple Harmonic Motion

Simple harmonic motion is a periodic motion. It repeats itself at regular intervals, and hence it has a definite period and definite frequency of repetition. Consider the simple harmonic motions of the spring-mass system shown in Fig. 2.15. Taking the origin at the position of rest of the system, let us denote the displacement of the mass by x, which changes with time. The mass can reach a maximum displacement of $x_{max} = A$, to the right, where it reverses its direction and travels to the left. It can go to a maximum displacement of A to the left of the equilibrium position. Starting at $t = 0$, when the mass has reached the equilibrium position during its travel to the right, the displacement of the mass as a function of time is shown in Fig. 2.16. The characteristics of the simple harmonic motion are shown in the Fig. 2.16, and described below.

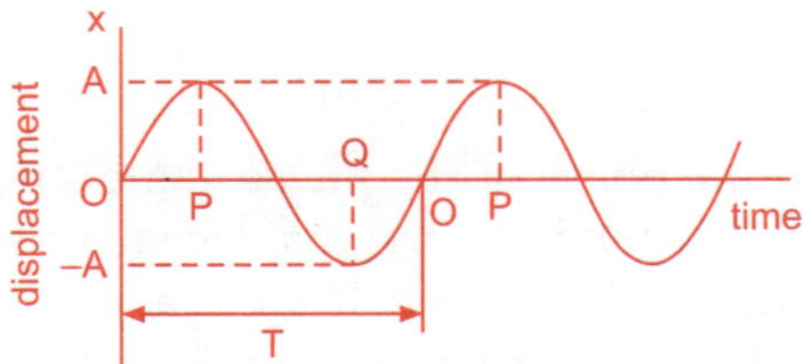

Fig. 2.16 *Displacement-time curve during simple-harmonic motion*

Period of oscillation T, is defined as the time interval between two successive passages of the same point in the same direction. In Fig. 2.16, the mass starts from the equilibrium position at O, travels to the right extreme at P, travels to the left and passes the point O to go upto the extreme left point Q and then returns to the point O. This completes one complete cycle and the time taken for this complete cycle is called the period of oscillation. The time for the mass to travel from P to Q and back to P is also equal to one period.

Amplitude of oscillation A, is the maximum displacement covered by the mass. This is given by the distance OA in Fig. 2.16.

Consider Fig. 2.17 which shows the equilibrium position O, the right extreme A and the left extreme A during the simple harmonic motion of the mass. The position of the mass at any time is indicated by x.

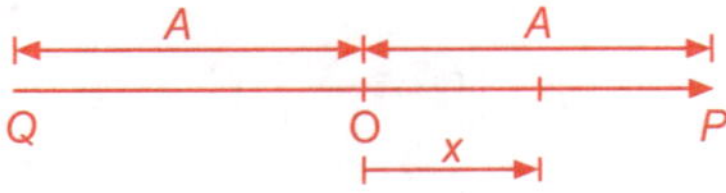

Fig. 2.17 *The domain of simple-harmonic motion*

Acceleration at a given displacement: From the definition of the simple harmonic motion, the acceleration is proportional to the displacement from the mean position and is directed towards it, and hence

$$a = -\omega^2 x \qquad (2.51)$$

where the minus sign indicates that the acceleration is always towards the centre, while x is measured outwards away from centre. The quantity ω^2 is the constant of proportionality, written as a square to emphasize that its value is positive always.

Velocity at any Displacement: The velocity v, at any displacement x, is given by

$$v = \pm\omega\sqrt{A^2 - x^2} \tag{2.52}$$

where the double sign is used because each value of x is passed through twice during one oscillation - once on the outward and once on the inward journey. The velocity is zero when $x = x_{max} = A$, at the extreme limit of the path, and the velocity is a maximum $v = v_{max} = \omega A$, when $x = 0$ at the centre.

Displacement at a Given Time: The displacement x at any time t is given by the relation

$$x = A \sin(\omega t + \theta) \tag{2.53}$$

where $(\omega t + \theta)$ is called the phase or the phase angle, and θ is known as the phase difference or the epoch, and depends on the displacement when $t = 0$. If we start the motion at $t = 0$ with $x = 0$, then we have $\theta = 0$. If $x = x_0$ at $t = 0$, then we have $x_0 = A \sin \theta$, and hence $\theta = \sin^{-1}(x_0/A)$. If $x = A$ at $t = 0$, then $A = A \sin \theta$, and hence $\sin \theta = 1$. Consequently, $\theta = \pi/2$.

Period and Frequency: In one complete oscillation, the phase $(\omega t + \theta)$ increases by 2π. Since θ is a constant, as $(\omega t + \theta)$ increases by 2π, ωt must increase by 2π, and hence t increases by $(2\pi/\omega)$. Thus the period, T, which is the time interval between successive passings of the same point on the path in the same direction, is given by

$$T = 2\pi/\omega. \tag{2.54}$$

The number of oscillations per second is called the frequency, denoted by n, and is given by

$$n = \frac{1}{T} = \frac{\omega}{2\pi} \tag{2.55}$$

or $\omega = 2\pi n$. Substitution of ω in the displacement-time relation gives some other forms of useful relations such as:

$$x = A \sin(2\pi n t + \theta), \text{ or} \tag{2.56}$$

$$x = A\sin\left(\frac{2\pi}{T}t + \theta\right).$$

2.5.3 Projection of Uniform Circular Motion on a Diameter

Projection of a uniform circular motion on a diameter becomes a simple harmonic motion. Uniform circular motion is a circular motion with constant angular velocity. Consider a particle which moves along a circle of radius A, with uniform angular velocity ω. Let us start the time when it just passes point P, as shown in Fig. 2.18.

Then at any time t the angle covered is $\theta = \omega t$. PQ forms a diameter and OP is along x axis. In the Fig. 2.18 $OX = x$ is the displacement of the particle in the x-direction at any time. Then we have

$$x = A \cos \theta = A \cos \omega t \tag{2.57}$$

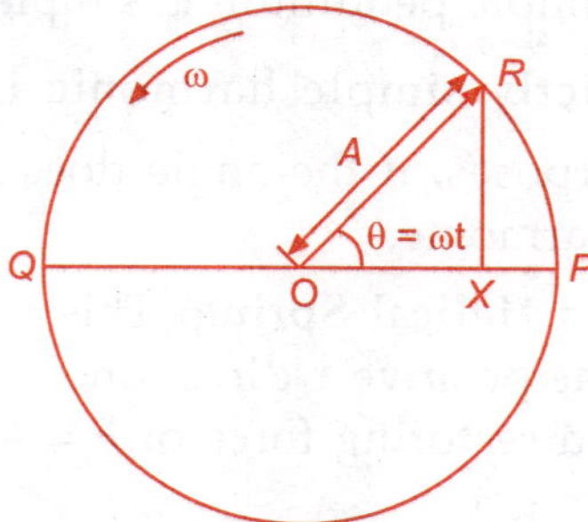

Fig. 2.18 *Uniform circular motion as simple harmonic motion as a diameter*

Thus the particle moves along the line *POQ* with simple harmonic motion.

2.5.4 Examples of Simple Harmonic Motion

(*i*) **Small Oscillations of a Simple Pendulum:** A simple pendulum consists of a small heavy bob of mass *m* at the end of a small inextensible string of length *L*, firmly secured at the end *O* as shown in Fig. 2.19. When pulled aside and released it oscillates to and fro.

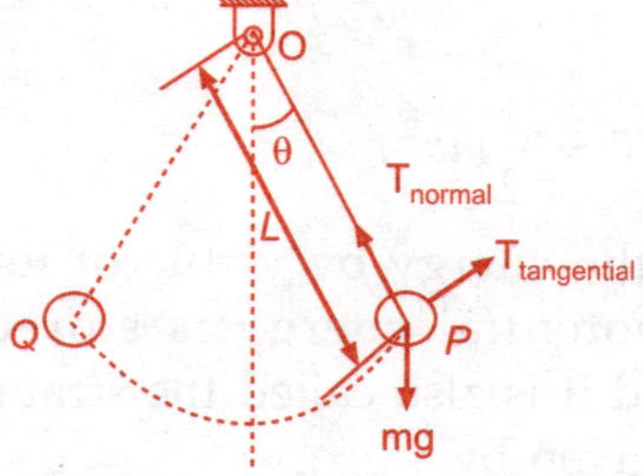

Fig. 2.19 *A simple pendulum*

Consider a small angular displacement of θ as shown in the Fig. 2.19. The external forces acting on the pendulum bob are the string tension *T*, and the weight *mg* of the bob acting vertically downwards. Resolving the weight *mg* along the string and perpendicular to it (normal and tangential directions, respectively), we see that the component (*mg* cos θ) acts radially and balances the tension *T*. The component (- mg sin θ) acting along the tangential direction is unbalanced. When the angle θ is very small, the tangential direction on an average is same as the horizontal x-direction. In this case the restoring force (−*mg* sin θ) acts towards the vertical line through the point *O* and for small angles, sin θ ≈ θ. Hence the restoring force is equal to −*mg*θ. If length of travel corresponding to the angular displacement θ is *x*, and we have *x* = *L*θ. The acceleration is obtained by dividing the restoring force by the mass and hence acceleration, *a*, is equal to

$$a = -g\theta = -gx/L \tag{2.58}$$

Hence the motion of the simple pendulum is simple harmonic with period of $T = 2\pi\sqrt{L/g}$. The motion is strictly simple harmonic if the angle θ is very small. However, for all practical purposes, if the angle does not exceed a few degrees the motion is essentially simple harmonic.

(*ii*) **Mass Constrained by a Helical Spring:** This is shown in Fig. 2.15. Consider a small displacement of x in the positive x-direction. If the spring constant is given as k N/m, then the spring exerts a restoring force of $F = - kx$ on the mass. Since $F = ma$ where a is the acceleration, we get

$$a = -\frac{k}{m}x \tag{2.59}$$

Thus the motion of the mass is simple harmonic and the period is given by $T = 2\pi\sqrt{m/k}$.

2.5.5 Energy of Simple Harmonic Motion

The energy in the simple harmonic motion consists of kinetic energy and potential energy. Kinetic energy is that stored in a system by virtue of its motion and is equal to

$$T = \frac{1}{2}mv^2. \tag{2.60}$$

The potential energy is the energy by virtue of its position. In the case of the mass-spring system, the potential energy is stored in the spring due to the deformation in the spring and it is also called the strain energy in the spring. In this case the potential energy is given by

$$v = \frac{1}{2}kx^2. \tag{2.61}$$

In the case of the simple pendulum, the potential energy is by virtue of its elevation. As the pendulum reaches the position directly below O, the bob is at the lowest position and the corresponding potential energy can be considered as zero (or minimum). When it reaches the extreme position of P or Q, the bob is at its highest position and the corresponding potential energy is the highest and is equal to

$$V = mgh \tag{2.62}$$

where h is the height of the position P or Q from the horizontal drawn at the lowest point below O.

Since the velocity at any displacement x is given by Eq. (2.52), substitution of this relation in the expression for kinetic energy in Eq. (2.60) will result in

$$T = \frac{1}{2}m\omega^2\left(A^2 - x^2\right). \tag{2.63}$$

We see that when $x = 0$, the kinetic energy is a maximum and when $x = A$, the kinetic energy is zero. This is because when $x = A$, the mass is at the extreme position, where it is reversing its direction and the velocity at that instant is zero.

SUMMARY

1. An object is said to be in *motion*, if its position changes with time. The position of the object can be specified with reference to a conveniently chosen origin. For motion in a straight line, position to the right of the origin is taken as positive and to the left as negative.

2. *Path length* is defined as the total length of the path traversed by an object.

3. *Displacement* is the change in position: $\Delta x = x_2 - x_1$. Path length is greater or equal to the magnitude of the displacement between the same points.

4. An object is said to be in *uniform motion* in a straight line, if its displacement is equal in equal intervals of time. Otherwise, the motion is said to be *non-uniform*.

5. *Average Speed* is the ratio of total path length traversed and the corresponding time interval. The average speed of an object is greater or equal to the magnitude of the average velocity over a given time interval.

6. The path length traversed by an object between two points is in general, not the same as the magnitude of displacement. The displacement depends only on the end points; the path length (as the name implies) depends on the actual path. In one dimension, the two quantities are equal only if the object does not change its direction during the course of motion. In all other cases, the path length is greater than the magnitude of displacement.

7. In view of point 1 above, the average speed of an object is greater than or equal to the magnitude of the average velocity over a given time interval. The two are equal only if the path length is equal to the magnitude of displacement.

8. The origin and the positive direction of an axis are a matter of choice. You should first specify this choice before you assign signs to quantities like displacement, velocity and acceleration.

9. If a particle is speeding up, acceleration is in the direction of velocity; if its speed is decreasing, acceleration is in the direction opposite to that of the velocity. This statement is independent of the choice of the origin and the axis.

10. The sign of acceleration does not tell us whether the particle's speed is increasing or decreasing. The sign of acceleration (as mentioned in point 3) depends on the choice of the positive direction of the axis. For example, if the vertically upward direction is chosen to be the positive direction of the axis, the acceleration due to gravity is negative. If a particle is falling under gravity, this acceleration, though negative, results in increase in speed. For a particle thrown upward, the same negative acceleration (of gravity) results in decrease in speed.

11. The zero velocity of a particle at any instant does not necessarily imply zero acceleration at that instant. A particle may be momentarily at rest and yet have non-zero acceleration. For example, a particle thrown up has zero velocity at its uppermost point but the acceleration at that instant continues to be the acceleration due to gravity.

12. The kinematic equations for uniform acceleration do not apply to the case of uniform circular motion since in this case the magnitude of acceleration is constant but its direction is changing.

13. An object subjected to two velocities v_1 and v_2 has a resultant velocity $v = v_1 + v_2$. Care must be taken to distinguish it from velocity of object 1 relative to velocity of object 2: $v_{12} = v_1 - v_2$. Here v_1 and v_2 are velocities with reference to some common reference frame.

14. The resultant acceleration of an object in circular motion is towards the centre only if the speed is constant.

15. The shape of the trajectory of the motion of an object is not determined by the acceleration alone but also depends on the initial conditions of motion (initial position and initial velocity). For example, the trajectory of an object moving under the same acceleration due to gravity can be a straight line or a parabola depending on the initial conditions.

TABLE 2.1 Various physical quantities and their symbols and dimensions

Physical Quantity	Symbol	Dimensions	Unit	Remark
1. Position vector	r	[L]	m	Vector. It may be denoted by any other symbol as well.
2. Displacement	Δr	[L]	m	-do-
3. Velocity				
(a) Average	$\bar{v}$		ms^{-1}	$\dfrac{\Delta r}{\Delta t}$
		$[LT^{-1}]$		
(b) Instantaneous	v			$\dfrac{dr}{dt}$
4. Acceleration				
(a) Average	$\bar{a}$			$\dfrac{\Delta V}{\Delta t}$
		$[LT^{-2}]$	ms^{-2}	
(b) Instantaneous	a			$\dfrac{dv}{dt}$
5. Projectile motion				
(a) Time of max. height	t_m	[T]	s	$\dfrac{V_0 \sin\theta_0}{g}$
(b) Max. height	h_m	[L]	m	$\dfrac{(V_0 \sin\theta_0)^2}{2g}$
(c) Horizontal range	R	[L]	m	$\dfrac{(V_0^2 \sin 2\theta_0)}{g}$
6. Circular motion				
(a) Angular speed	ω	$[T^{-1}]$	rad/s	$\dfrac{\Delta\theta}{\Delta t} = \dfrac{v}{r}$
(b) Centripetal acceleration	a_c	$[LT^{-2}]$	ms^{-2}	$\dfrac{v^2}{r}$

Dynamics–Kinetics   3

3.1 FORCES

3.1.1 Force as the Cause of Motion

We discussed the motion of bodies in the previous Chapter. Motion takes place when a body changes its position or undergoes a displacement. The displacement of a body is caused by force. When force is applied on a body, it responds with a displacement in the direction of the force. Since force has both a magnitude and a direction, it is a vector quantity.

Aristotle's Fallacy

The Greek thinker, Aristotle (384 B.C.–322 B.C.), held the view that if a body is moving, something external is required to keep it moving. According to this view, for example, an arrow shot from a bow keeps flying since the air behind the arrow keeps pushing it. The view was part of an elaborate framework of ideas developed by Aristotle on the motion of bodies in the universe. Most of the Aristotelian ideas on motion are now known to be wrong and need not concern us. For our purpose here, the Aristotelian law of motion may be phrased thus: An external force is required to keep a body in motion.

Aristotelian law of motion is flawed, as we shall see. However, it is a natural view that anyone would hold from common experience. Even a small child playing with a simple (non-electric) toy-car on a floor knows intuitively that it needs to constantly drag the string attached to the toy car with some force to keep it going. If the child releases the string, the toy car comes to rest. This experience is common to most terrestrial motions. External forces seem to be needed to keep bodies in motion. Left to themselves, all bodies eventually come to rest. What is the flaw in Aristotle's argument? The answer is: a moving toy car comes to rest because the external force of friction on the car by the floor opposes its motion. To counter this force, the child has to apply an external force on the car in the direction of motion. When the car is in uniform motion, there is no net external force acting on it: the force by the child cancels the force (friction) by the floor. The corollary is: if there were no friction, the child would not be required to apply any force to keep the toy car in uniform motion. The opposing forces such as friction (solids) and viscous forces (for fluids) are always present in the real world. This explains why forces by external agencies are necessary to overcome the frictional forces to keep bodies in uniform motion. Now, we understand where Aristotle went wrong. He coded this practical experience in the form of a basic argument. To get at the true law of nature for forces and motion, one has to imagine a world in which uniform motion is possible with no frictional forces opposing. This is what Galileo did.

3.1.2 Review of Different Types of Forces

(*i*) Mechanical Forces:

Mechanical forces are those which are applied through mechanical contact with the body. A helical spring, secured at one end to a fixed support at O and with a weight W pulling it down at the other end A, is shown in Fig. 3.1. Since the weight W is causing a displacement of the end A, the weight W is a force. It is a mechanical force because it is in physical contact with the end A of the spring.

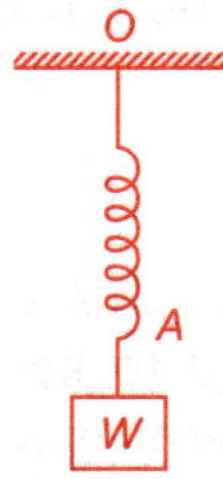

Fig. 3.1 *Helical spring with a load*

A body B is being pushed up the incline in Fig. 3.2.

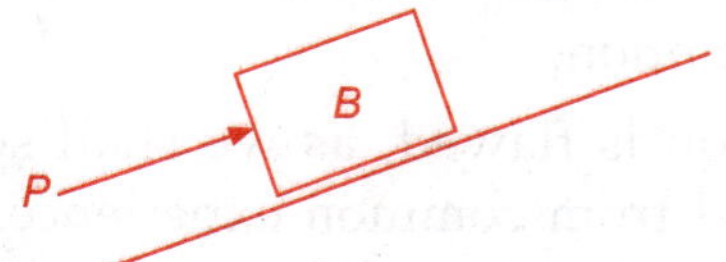

Fig. 3.2 *A body pushed up the inclined plane*

In order to push the body up, the muscles exert a force P on the body. Since the hand pushing the body B is in contact with the body, the muscle force is similar to a mechanical force.

(ii) Force due to Friction:

When two bodies are in contact and one moves relative to the other, the motion is opposed by a force F, called force of friction. Referring to Fig. 3.3, a mechanical force P is applied to body A, which is placed over body B, at rest. The force P tries to displace body A relative to B.

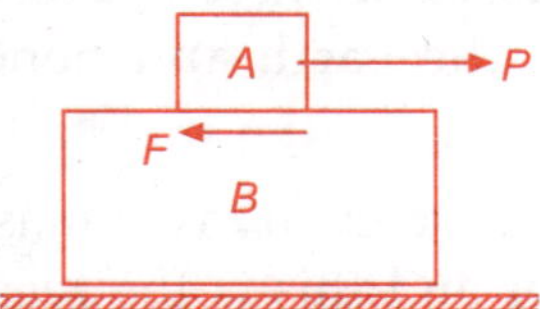

Fig. 3.3 *Relative motion of two solid blocks involving friction*

The force F increases as we increase P. The force F will go on increasing to a maximum value of F_{max}, which depends on the materials of the bodies A and B. If the applied mechanical force P is less than the force of friction F_{max}, it is not possible to move A. When P is just equal to F_{max}, the body A just starts sliding relative to B. As soon as there is a relative motion, the force of friction comes down from F_{max} to a lower value and stays constant irrespective of the relative velocity between the two bodies.

(iii) Cohesive force:

Matter consists of atoms, and atoms held cohesively together form material bodies. The force of attraction between the different atoms making up the body is the cohesive force.

(iv) Adhesive Force:

Two bodies can be bonded together by using a thin intermediate layer of glue holding the two bodies. In such a case the force holding the two bodies together is referred as adhesive force.

(v) Elastic Force:

In the example of the spring and the mass shown in Fig. 3.1, as the weight W pulls the spring down, there is an equal and opposite force applied by the spring on the weight, pulling it up. As a result, the system assumes a steady state position of equilibrium. The force exerted by the spring is referred to as elastic force.

3.1.3 Forces at a Distance

Unlike mechanical forces, there are forces which can be applied from a distance. Examples are the gravitational force, electrostatic force, magnetic force, nuclear force etc.

(*i*) Gravitational Force:

Two masses are attracted to each other with a force which is proportional to the product of their masses and inversely proportional to the square of the distance between them. When the two masses are in the vicinity of a very large mass, for example two masses on earth, the force of attraction between the smaller masses may not be appreciable. However, when one is small but the other is large, as for example a mass on the surface of the earth, we can observe this force very clearly. A mass held at a height above the surface of the earth and released, will fall down because of the force exerted by earth on the mass. The force of attraction between the mass and the earth is not sufficient to move the earth and hence the smaller mass will come towards the earth's surface.

The gravitational force acts between any two masses. In the solar system, the sun and the planets attract each other and this is the reason why planets revolve around the sun.

(*ii*) Electrostatic Force:

Like charges (either two positive charges or two negative charges) repel each other and unlike charges attract each other. The force of repulsion or attraction among electrical charges is referred to as electrostatic force. The electrostatic force is proportional to the product of the two charges and is inversely proportional to the square of the distance between them.

(*iii*) Magnetic Force:

When a magnet is placed near a ferromagnetic material such as a piece of iron, the magnet attracts the piece of iron with a force which is called the magnetic force. If two bar type magnets are held together along a line with a small gap between the ends facing each other, they may attract or repel each other. When the two ends repel we call them like poles and when they attract we call them as unlike poles. The poles are referred to as north and south poles, because earth itself forms a large magnet and the north pole and the south pole form the two ends of this magnet.

(*iv*) Nuclear Force:

An atom of a material consists of a central nucleus and several electrons which orbit around the nucleus. In order for the electrons to revolve around the nucleus, there must be a force of attraction between them and this force is referred to as nuclear force.

3.1.4 Vector Property of a Force

A vector is defined as a quantity having both magnitude and direction. A force is a vector because it has both magnitude and direction.

A force can be represented by a directed line as shown in Fig. 3.4. The length of the line denotes the magnitude of the force to a suitable scale.

Fig. 3.4 *Vector representation of a force*

Later on we will discuss methods of adding and resolving forces. Vector analysis methods can be applied to forces also and this simplifies the treatment of forces in several applications.

3.2 LAWS OF MOTION

3.2.1 Introduction

Newton enunciated the much celebrated Newton's Laws of Motion in the 17th century. He expounded his laws in his book Principia in 1687. Newton's laws gave a new impetus to the study of motion and explained several motion phenomena on earth and planetary motions also. Newton recognized the importance of Momentum, which is the product of the mass and velocity of a body in motion. He also recognized that to initiate the motion of a body from a state of rest, or to change the motion of a body which is already moving, it is necessary to apply Force. Before we discuss the Newton's Laws of Motion, let us examine these two quantities, which are so important to the motion.

(*i*) Force:

A body can be set into motion in a particular direction, or a body which is already in motion can be made to move faster or slower, only by the application of an external agent, which Newton called, Force. Since force is related to the motion parameters, the units of force will be given later, after we discuss Newton's laws of motion.

(*ii*) Momentum:

Momentum is defined as the product of the mass of the body in motion and its velocity.

$$P = mv \tag{3.1}$$

The unit of momentum is given from Eq. (3.1) as the product of the units of mass and velocity as kg.meter/second, or kg.m/s. Since mass of a body can be considered as constant, momentum changes when the velocity is changed.

In the following, the three laws of Newton will be given and will be discussed one by one.

Newton's First Law: Everybody continues to be in its state of rest or of uniform motion along a straight line unless it is compelled to change that state by an external force.

Newton's Second Law: The rate of change of momentum of a body is directly proportional to the impressed force and takes place in the direction of the force.

Newton's Third Law: For every action, there is an equal and opposite reaction.

The first law defines the force, which is the external agent causing motion. The second law provides a relation between the force and the change in momentum, which can be used to measure force. The third law recognizes the fact that the force has to be supplied from another body, and when it applies a force on the body which starts moving, the moving body exerts an equal and opposite force on the body which applied the force in the first place. Let us discuss each law in detail now.

Newton's First Law:

Everybody continues to be in its state of rest or of uniform motion, unless it is compelled to change that state by an external force.

A state of rest indicates that there is no motion. Uniform motion means that the body moves with constant velocity. Hence a body which is already moving with constant velocity does not need any external agent to continue its state of uniform motion. Only when we want to change its velocity, either to increase the velocity, or to reduce it, we need an external agent to do it.

A car which is moving with constant velocity can be stopped only by applying the brakes. The brakes supply the force needed to change the velocity (or the momentum) of the car. When a car is moving with constant velocity, if we want to increase the speed, we must press on the accelerator pedal. This results in increased power from the engine, which results in an increased force to speed up the car.

The property of the body by virtue of which it cannot change its state of rest or of uniform motion along a straight line is called Inertia. Newton's first law is referred to as law of inertia.

A passenger in an automobile can fall forward when the brakes are applied without his/her knowledge. This is due to the property of inertia of the passenger's body mass. When the car is travelling with constant velocity, the passenger's body also acquires the same velocity. When the car slows down suddenly, the legs and hip of the passenger slow down along with the automobile, however, the head and the upper portion of the body, which are not constrained to the car, continue with the original velocity. This is the reason for the passenger's upper part of the body falling forward. Use of passenger seat belts constrain the passenger's upper part of the body to the car, and prevent such falling when the automobile is stopped suddenly. Some countries have made the use of seat belts in passenger cars mandatory, to prevent accidents.

Newton's Second Law:

The rate of change of momentum of a body is directly proportional to the impressed force and takes place in the direction of the force.

The rate of change of the momentum is the change in momentum divided by the time taken for the change. Since the mass of a body remains constant, change in momentum results in change in velocity of the body. When a body is moving along a straight line from left to right, if the force is applied on the body in the direction of the motion, i.e., to the right, then the momentum of the body will increase, or the velocity of the body will increase. If the force is in a direction opposite to that of the motion, then the velocity of the body will decrease.

According to the second law of motion

$$\text{force} \propto \text{rate of change of (momentum)}$$

$$\propto \text{rate of change of (mass} \times \text{velocity)}$$

$$\propto \text{mass} \times \text{(rate of change of velocity); since mass is constant}$$

$$\propto \text{mass} \times \text{acceleration.}$$

We have seen in Chapter 2 that the rate of change of velocity is acceleration. If a force F acting on a body of mass m produces an acceleration a, then we have

$$F \propto m.a \tag{3.2}$$

Using a constant of proportionality, K, we can write

$$F = K.m.a. \tag{3.3}$$

Unit of force is defined in such a way that the constant of proportionality, K, becomes unity. Hence, defining the unit force as that which sets a unit mass into unit acceleration, the constant of proportionality becomes 1. Consequently,

$$F = m.a \tag{3.4}$$

Unit of Force:

The unit of force is derived using the relation in Eq. (3.3) or (3.4). In SI units, the unit of mass is kg and the unit of acceleration is meter/(second)^2 or m/s^2. Hence the unit force is that which gives an acceleration of 1 m/s^2 to a 1 kg mass and is expressed in kg.m/s^2. This unit of force is an absolute unit and does not change from place to place. It is same whether used on earth or on moon or any other planet. In honour of Newton, the unit of force is named as Newton, or it is given the symbol N.

$$1 \text{ N} = 1 \text{ kg.m/s}^2$$

We need 10 N force to give an acceleration of 1 m/s^2 to a mass of 10 kg.

Newton is a large force and sometimes a smaller unit called dyne is used in physical measurements. Dyne is the force needed to give an acceleration of 1 cm/s^2 to a mass of 1 gm (one thousandth of a kg). Hence $1N = 10^5$ dynes.

Impulse of a Force:

Impulse of a force is the product of force and time of application of the force. Impulse of a force is equal to the change of momentum produced in a body due to the application of the force during a certain interval of time. When a force F acts on a body of mass m, for a time t, changing its velocity from u to v, then the impulse of the force is given by

$$I = Ft = mat = m\left(\frac{v-u}{t}\right)t = m(v-u) \qquad (3.5)$$

Since the impulse of the force $I = m\,(v - u)$, the impulse is equal to the change in momentum.

A large force acting on a body for a short interval of time is called an impulsive force. Effect of such an impulsive force is measured by the impulse of the force. When a bell is rung, the rod hitting the bell applies an impulsive force. A sharp knock, collision of two bodies, force of an explosion, the impact of a cricket bat hitting a ball etc., are examples of impulsive forces.

Impulsive forces generate very large accelerations in the bodies on which they are applied, resulting in a sudden change in velocities in a very short interval of time. Hence the accelerated part of the motion takes place in a very short interval of time, and the noticeable effect is normally a change in velocity, or the change in momentum. The impulsive force imparted on a cricket ball can be measured by knowing the mass of the ball and by measuring the speed with which the ball issues forth when it is hit.

Newton's third Law:

For every action, there is an equal and opposite reaction.

The third law underscores the fact that the forces occur in pairs, whether the bodies acted upon by the forces are at rest or in motion. When a body is at rest, its weight pulls it down, however, an equal and opposite force is exerted upwards by the support on which it is resting, say, the table or the ground, keeping it in equilibrium. This is an example of a problem in Statics. For static equilibrium, all the forces acting on a body must add upto zero. Only one force cannot be acting on a body if the body is in static equilibrium. If the body is in motion, and a force is applied on it resulting in a change in its momentum, then the body pushes the medium exerting this force with an "inertia" force, which is equal to ($m.a.$). If a hand is exerting this force on the body, then the hand will experience an equal and opposite force exerted by the accelerating body back on it. We will list examples which illustrate this phenomenon.

    (*i*) When a rocket is launched into space, the pressurized gases produced by the burning propellant in the engine escape through the nozzle at the back end of the rocket. These gases are pushed back from the rocket with enormous force. The escaping gases in turn exert large force on the rocket pushing it forward into space which is shown in Fig. 3.5.

Fig. 3.5 *Rocket launch vehicle*

(*ii*) In jet planes, as shown in Fig. 3.6, the pressurized gases produced by the burning fuel leave the engines and go backward with tremendous force. These gases in turn exert equal force of reaction on the plane pushing it forward.

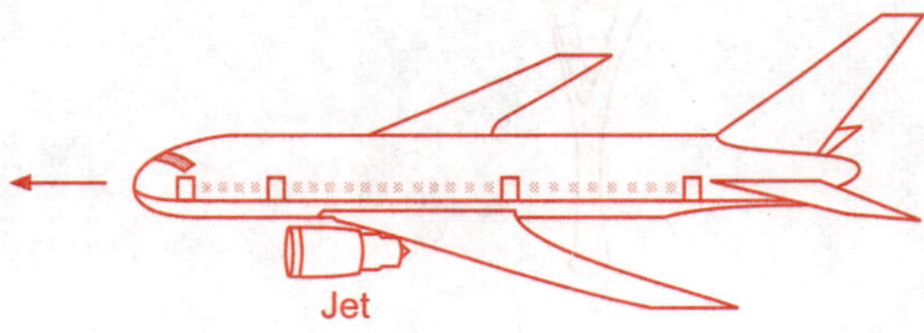

Fig. 3.6 *Jet Plane*

(*iii*) The engine in an automobile rotates the wheels. When the wheels try to turn, the ground opposes its rotation because of friction. As a result, the automobile gets a push in the forward direction and it moves forward. This is shown in Fig. 3.7.

Fig. 3.7 *Surface friction pushing the automobile*

(*iv*) When a gun is fired, the pressurized gases push the bullet with tremendous force. The bullet goes forward with very high velocity because of its small mass. The reaction force on the gun makes it to recoil. An experienced shooter will be prepared for this recoil and adjusts his/her balance in the process. This is shown in Fig. 3.8.

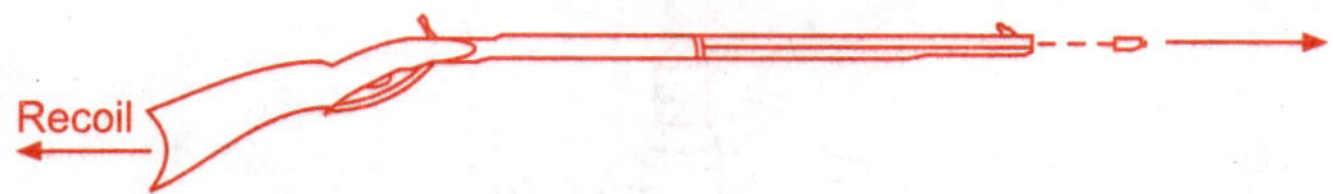

Fig. 3.8 *Recoil of a Gun*

(*v*) When a player hits the ball either in a tennis game or in cricket, the bat imparts an impulsive force on the ball as a result of which the ball moves away with high velocity. The reaction force on the bat is transmitted on to the bat, which in turn is transmitted to the hand. This is shown in Fig. 3.9. If the player is not careful, the reaction may cause injury to the hand.

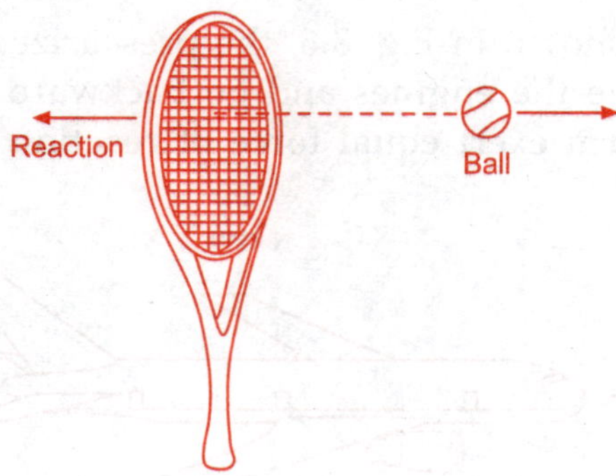

Fig. 3.9 *Reaction from ball on the Racket*

(*vi*) When a person tries to step down from a boat on the shore, the boat seems to be pushed away from the person. The person exerts a push on the boat in order to move towards the shore, which is responsible for this motion. This is shown in Fig. 3.10.

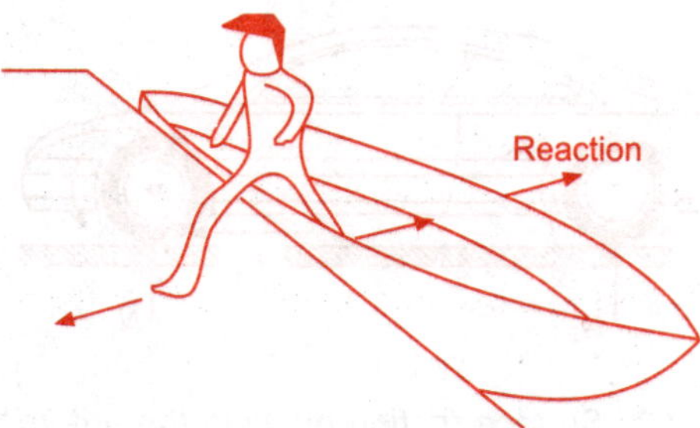

Fig. 3.10 *Reaction of the boat while stepping Ashore*

(*vii*) A swimmer pushes the water backwards. An equal and opposite reaction force is exerted on the swimmer by the water which makes the swimmer to move forward. This is shown in Fig. 3.11.

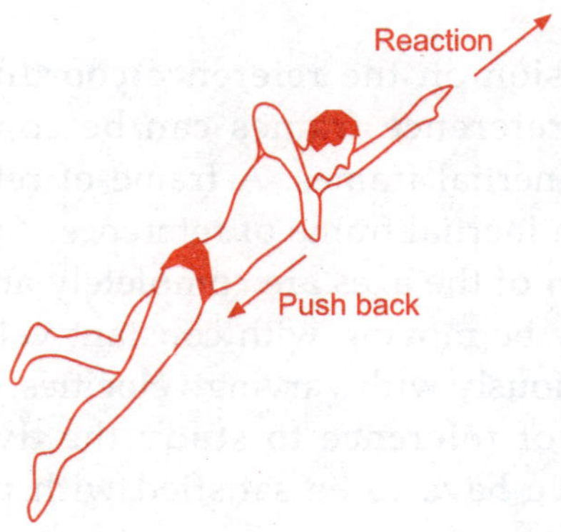

Fig. 3.11 *Swimmer forcing the water backwards*

(*viii*) Flying birds push the air in a downward and backward direction. The air pushes the bird with an upward force to keep it afloat in air and a forward force to move it forward. This is shown in Fig. 3.12.

Fig. 3.12 *Birds flapping wings creating lift*

Conservation of Momentum

Newton's second law states that the rate of change of momentum is equal to the impressed force. If a body is in motion and there is no force acting on it, then, according to the second law, the body does not experience any change in momentum. Hence the momentum of the body remains constant. When there is no force acting on a moving body, we say that the momentum of the body is conserved. This is the law of conservation of momentum. The product of mass and velocity of a body in motion along a straight line is the linear momentum. Later on we will also study about the angular momentum which is the moment of momentum about a point. Since we are discussing the motion of bodies along a straight line now, we will discuss linear momentum in more detail in the following.

Frames of Reference

Elaborating on the discussion on the reference coordinate system mentioned in Chapter 2, two types of reference frames can be considered, namely, inertial reference frames and non-inertial frames. A frame of reference in which Newton's Laws hold good is called an inertial frame of reference. An inertial reference frame is one in which the orientation of the axes are completely and absolutely fixed, and the origin may be fixed or may be moving with constant velocity. Since every object in this universe moves continuously with varying velocities, it will be impossible to find a completely fixed frame of reference to study the dynamics of bodies. Hence, depending on the study, we have to be satisfied with reference frames which are almost like an inertial frame.

A frame of reference fixed on the surface of the earth, may be considered as an inertial reference for most of the dynamic phenomena that take place on earth. A frame fixed at the centre of the earth will be slightly more accurate than the one fixed on the surface of earth. However, since earth rotates about its own axis and orbits around the sun, fixing the frame of reference at the centre of Sun is still closer to an inertial frame. For many astronomical phenomena, even this frame may not be inertial enough and it may be necessary to fix the frame with the origin on the distant stars.

Reference frames which move with acceleration or whose axes rotate about the origin are non-inertial.

Assume that a non-inertial frame S' moves with a constant acceleration a' along the x axis, with respect to an inertial frame S. For a mass m moving with an acceleration a along the x axis, with respect to the inertial frame S, Newton's second law holds good and we have

$$F = ma \tag{3.6}$$

Acceleration a_r, of the mass relative to the non-inertial frame S' is given by

$$a_r = a - a' \tag{3.7}$$

Hence the force F_r measured with respect the frame S' is given by

$$F_r = ma_r = m(a - a') = F + F' \tag{3.8}$$

The force $F = ma$ is the actual force on the body and $F' = - m.a'$ is a force that must be added to the actual force to obtain F_r. The quantity F_r is a frame dependent force and is also called as a fictitious force or a pseudo force. Even though such a fictitious force is not real, its effect can be felt.

For example, a person in a lift feels as though his/her weight has come down when the lift starts to move down. This feeling lasts only until the lift acquires a constant velocity. As soon as the lift attains a constant speed, this feeling is gone.

3.2.2 Conservation of Linear Momentum

The product of mass and velocity of a body in motion along a straight line is the linear momentum. Newton's second law states that the rate of change of momentum is equal to the impressed force. If a body is in motion and there is no force acting on it, then, according to the second law, the body does not experience any change in momentum. Hence the momentum of the body remains constant. When there is no force acting on a moving body, we say that the momentum of the body is conserved. This is the law of conservation of momentum.

From Eq. (3.8), we see that if $F = 0$, then

$$m\,(v - u) = 0 \tag{3.9}$$

Hence $mv = mu$, and hence, the initial momentum of the body stays unchanged. This is called as conservation of momentum.

Later on we will also study about the angular momentum which is the moment of momentum about a point. Since we are discussing the motion of bodies along a straight line now, we will discuss linear momentum in more detail in the following.

Equilibrium of a Particle

Equilibrium of a particle in mechanics refers to the situation when the net external force on the particle is zero. According to the first law, this means that, the particle is either at rest or in uniform motion. If two forces F_1 and F_2, act on a particle, equilibrium requires

$$F_1 = -\,F_2 \tag{3.10}$$

i.e., the two forces on the particle must be equal and opposite. Equilibrium under three concurrent forces F_1, F_2 and F_3 requires that the vector sum of the three forces is zero.

$$F_1 + F_2 + F_3 = 0 \tag{3.11}$$

In other words, the resultant of any two forces, say F_1 and F_2, obtained by the parallelogram law of forces must be equal and opposite to the third force, F_3. As seen in Fig. 3.13, the three forces in equilibrium can be represented by the sides of a triangle with the vector arrows taken in the same sense. The result can be generalized to any number of forces. A particle is in equilibrium under the action of forces F_1, F_2,... F_n, if they can be represented by the sides of a closed n-sided polygon with arrows directed in the same sense. Equation (3.11) implies that

$$\begin{aligned}
F_{1x} + F_{2x} + F_{3x} &= 0 \\
F_{1y} + F_{2y} + F_{3y} &= 0 \\
F_{1z} + F_{2z} + F_{3z} &= 0
\end{aligned} \tag{3.12}$$

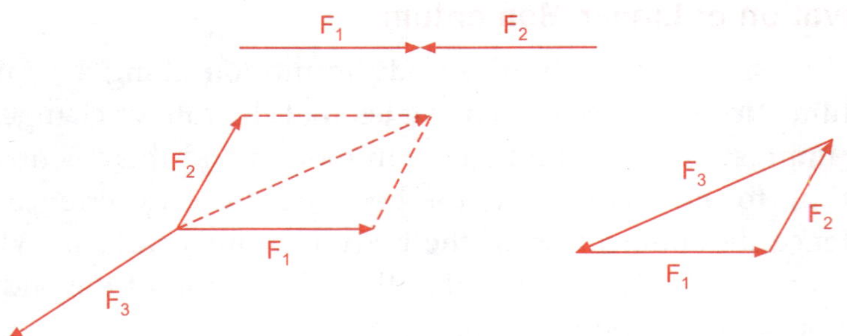

Fig. 3.13 *Equilibrium of concurrent forces*

where F_{1x}, F_{1y} and F_{1z} are the components of F_1 along x, y and z directions, respectively.

3.2.3 Impact or Collisions of Bodies

The collision or impact between two bodies is said to be direct or head-on collision if the direction of motion of each body is along the common normal at the point of impact. The collision is said to be oblique, if the direction of motion of either of the bodies is not along the common normal at the point of impact.

Consider two bodies A and B of masses m_1 and m_2, respectively, moving with velocities u_1 and u_2, respectively, along the positive x direction in a straight line. If u_1 is greater than u_2, then mass m_1 will collide with mass m_2 after a while, as shown in Fig. 3.14. Let the velocities of masses m_1 and m_2 after collision be v_1 and v_2, respectively. Considering the two bodies, A and B as a system, we see that there are no forces external to this system acting on them. A will exert a force on B and B will exert a force on A. However these forces are within the system. Hence the sum of the momenta of the two body system is conserved. Consequently, we have

$$m_1 u_1 + m_2 u_2 = m_1 v_1 + m_2 v_2 \tag{3.13}$$

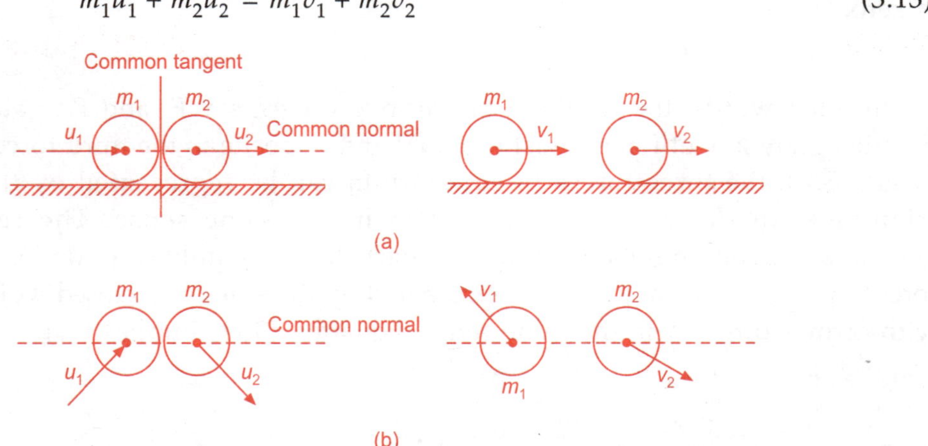

Fig. 3.14 *(a) Direct Central Impact (b) Oblique Central Impact*

Alternatively, we can also consider the bodies A and B separately, and apply Newton's third law to illustrate the conservation of momentum. After collision the change in momentum of body A is given by

$$\Delta P_A = m_1(v_1 - u_1) \tag{3.14}$$

and the change in momentum of body B is given by

$$\Delta P_B = m_2(v_2 - u_2) \tag{3.15}$$

From Newton's third law, the force exerted by body A on B is equal and opposite to the force exerted by body B on A. Let this force be F. The duration of application of both forces is same and is, say, Δt. Therefore $F.\Delta t$ acts on m_1 whereas $-F.\Delta t$ acts on m_2. Then from Eqs. (3.14) and (3.15), we know that the change in momentum is given by the impulse of the force acting on the body, and hence

$$F.\Delta t = m_1(v_1 - u_1) \text{ and } -F.\Delta t = m_2(v_2 - u_2) \tag{3.16}$$

and this can be written after rearranging in the form

$$m_1 u_1 + m_2 u_2 = m_1 v_1 + m_2 v_2 \tag{3.17}$$

3.2.4 Coefficient of Restitution

It is experimentally observed that in the case of direct collision between two bodies the relative velocity after collision bears a constant ratio to the relative velocity before collision and is in the opposite direction. This constant ratio is known as the coefficient of restitution, or coefficient of resilience, and is denoted by the letter e. Let u_1 and u_2 be the initial velocities of the two colliding bodies before impact, and v_1 and v_2 be the velocities after impact. Then we have

$$\frac{v_1 - v_2}{u_1 - u_2} = -e \tag{3.18}$$

The value of e is between 0 and 1. When $e = 1$ we have a perfectly elastic impact and when $e = 0$, we have a perfectly plastic impact. For all other types of impacts we have

$0 < e < 1$. For example,

$e = 0.94$ for glass-glass impact,

$e = 0.81$ for ivory-ivory impact,

$e = 0.20$ for lead-lead impact,

$e = 0.91$ for cast iron-glass impact,

$e = 0.78$ for brass-ivory impact, and

$e = 0.25$ for lead-glass impact.

Case (*i*) Perfectly Elastic Collision: In the case of perfectly elastic collision, the value of $e = 1$. Hence, from Eq. (3.18) we have

$$v_1 - v_2 = -(u_1 - u_2) \tag{3.19}$$

This can be rearranged and written as

$$v_1 + u_1 = v_2 + u_2 \tag{3.20}$$

Substituting this relation in Eq. (3.17), which expresses the conservation of linear momentum of the two body system, we obtain

$$m_1\left(v_1^2 - u_1^2\right) = m_2\left(v_2^2 - u_2^2\right) \tag{3.21}$$

Rearranging this equation we get

$$m_1 u_1^2 + m_2 u_2^2 = m_1 v_1^2 + m_2 v_2^2 \tag{3.22}$$

Multiplying both sides of this equation by (1/2), we get

$$\frac{1}{2}\left(m_1 u_1^2 + m_2 u_2^2\right) = \frac{1}{2}\left(m_1 v_1^2 + m_2 v_2^2\right). \tag{3.23}$$

Equation (3.23) states that the sum of the kinetic energies of the two bodies before impact is same as that after the impact. Hence the kinetic energy is conserved in an elastic impact.

Case (ii) Perfectly Plastic Impact: In the case of a perfectly plastic collision, the value of $e = 0$. Hence from Eq. (3.18), we have

$$v_1 - v_2 = 0 \tag{3.24}$$

Hence, after impact, $v_1 = v_2$, and the two bodies move together. Let this common velocity be v. Then from Eq. (3.17), we have

$$(m_1 u_1 + m_2 u_2) = (m_1 + m_2)v \tag{3.25}$$

Loss in kinetic energy during a perfectly plastic impact is given by

$$\Delta E = \frac{1}{2}\left(m_1 u_1^2 + m_2 u_2^2\right) - \frac{1}{2}(m_1 + m_2)v^2 \tag{3.26}$$

Substituting for v from Eq. (3.25) in Eq. (3.26), we get an expression for the loss in kinetic energy during a perfectly plastic collision as

$$\Delta E = \frac{1}{2}\frac{m_1 m_2}{(m_1 + m_2)}(u_1 - u_2)^2 \tag{3.27}$$

Case (iii) General Direct Impact: When the value of, $e \neq 0$ or $e \neq 1$ or then the loss of kinetic energy during impact is given by

$$\Delta E = \frac{1}{2}\left(m_1 u_1^2 + m_2 u_2^2\right) - \frac{1}{2}\left(m_1 v_1^2 + m_2 v_2^2\right) \tag{3.28}$$

Substituting from Eq. (3.17) and from Eq. (3.18), it is possible to put the loss of kinetic energy in the form

$$\Delta E = \frac{1}{2}\frac{m_1 m_2}{(m_1 + m_2)}(u_1 - u_2)^2 (1 - e^2) \tag{3.29}$$

From this equation also it can be seen that when the coefficient of restitution $e = 1$, the loss in kinetic energy is zero. Hence energy is conserved during a perfectly elastic impact.

3.2.5 Determination of Coefficient of Restitution

Coefficient of restitution depends on both the colliding bodies. It can be determined by dropping a ball made out of one of the colliding materials from a height h_1, upon a plane surface of the other material. The height of rebound of the ball, h_2, is measured. The velocity of the ball when it strikes the surface is given by

$$u = \sqrt{2gh_1} \tag{3.30}$$

The velocity of rebound of the ball is related to the height of rebound by the relation

$$-v = \sqrt{2gh_2} \tag{3.31}$$

The velocities of the plane surface before and after the impact are zero. Hence the coefficient of restitution of the impacting bodies is obtained by using the Eq. (3.18) as

$$\frac{v}{u} = -e \tag{3.32}$$

Using Eqs. (3.31) and (3.32), we get the value of e as

$$e = \sqrt{h_2/h_1}\,. \tag{3.33}$$

3.3 FRICTION

Consider a solid block resting on a solid surface, acted upon by an external force in order to move the block relative to the solid surface, as shown in Fig. 3.15. The resulting rubbing action produces a force called the force of friction which acts in the plane of contact between the two surfaces in such a direction as to oppose any relative motion. As the external force is increased, the force of friction also will increase proportionately upto a certain value. Beyond that value, the force of friction cannot increase and hence the external force will succeed in causing the motion. Depending on the relative strengths of the external force, P, and the maximum force of friction, F_{max}, three possibilities exist.

1. The external force tending to cause motion is less than the maximum force of friction that can be developed between the rubbing surfaces. In this case, force of friction is equal and opposite to the components of applied forces in the direction of anticipated motion.

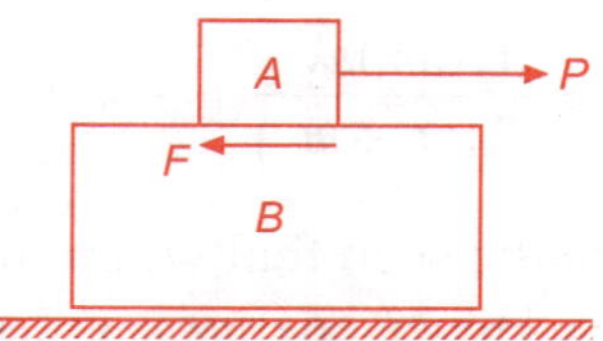

Fig. 3.15 *Friction between two bodies*

2. The surfaces rubbing together are on the verge of executing relative motion. This happens when the applied force is just equal to the maximum force of friction, or the limiting friction, that can be developed between the rubbing surfaces.

3. Relative motion between the rubbing surfaces takes place. As soon as the motion starts, the force exerted by friction against motion becomes less than the value of the limiting friction, and is called the force of sliding friction. The resultant force in the direction of motion is the difference between the applied force in that direction and the force of sliding friction.

Laws of Friction: The following laws of friction are applicable to dry solid surfaces,

1. When the rubbing surfaces are at rest, friction can take any value upto a certain maximum value called the limiting friction, F_L.

2. For a given pair of surfaces, the limiting friction F_L is proportional to the normal force R between the surfaces, at right angles to the plane of contact. The ratio F_L/R is called the coefficient of static friction, or static coefficient of friction, μ_L, between the two surfaces,

$$F_L = \mu_L R \tag{3.34}$$

3. In the case of ordinary surfaces, the limiting friction is independent of the area of the surface in contact.

4. When motion occurs, the sliding frictional force, F, is proportional to the normal force, R, between the surfaces. The ratio F/R is called the coefficient of sliding friction, or the dynamical coefficient of friction, μ_d, between the two surfaces. Hence,

$$F_L = \mu_d R \tag{3.35}$$

5. In general, the sliding friction is independent of the area of contact between the sliding surfaces, and is also independent of velocity of relative motion. This is known as Coulomb's law.

3.3.1 Measurement of Coefficient of Friction

The principle used is to measure the minimum force required to overcome the frictional force in order to just start relative motion between the two surfaces in contact. Depending on the method of application of the force required to overcome the frictional force we have the following two methods.

Method 1: As shown in Fig. 3.16, movable block of one material A is kept on a fixed horizontal surface or bed of the other material B. One end of block A is secured to a cord which passes over a frictionless pulley and containing a scale-pan that can hold weights. The weight of block A is weighed and its weight can be increased by adding additional weights on its top. The normal force, R, is equal to the total weight of the block A and any weights placed on it.

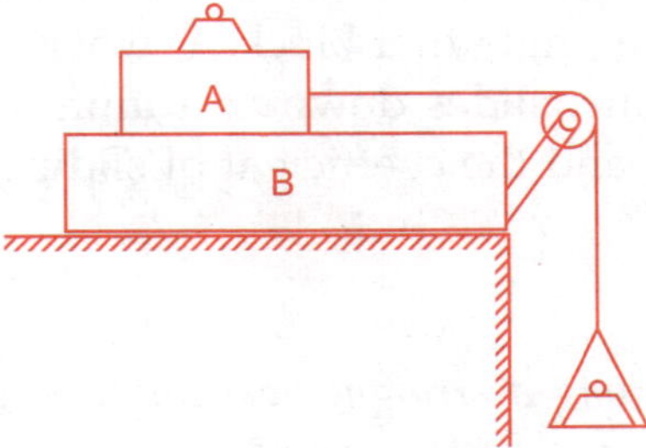

Fig. 3.16 *Measurement of coefficient of friction*

(a) **Coefficient of Static Friction:** Weights are added on the scale-pan very slowly until the block A just starts to move. The weight in the scale-pan now is equal to F_L and the coefficient of static friction is given by $\mu_L = F_L/R$. The experiment is repeated several times and the average value of μ_L is taken.

(b) **Coefficient of Sliding Friction:** Weights are added slowly to the scale-pan **until when displaced slightly,** the block A slides over B at a uniform velocity. Then the load in the scale-pan is just overcoming the force of sliding friction, and hence is just equal to F. The coefficient of sliding friction is obtained as $\mu_d = F/R$. Several readings are taken with different weights on A and the average value of μ_d is taken.

Method 2: The block A is placed on the bed B, which can be tilted to any position, to form an inclined surface as shown in Fig. 3.17.

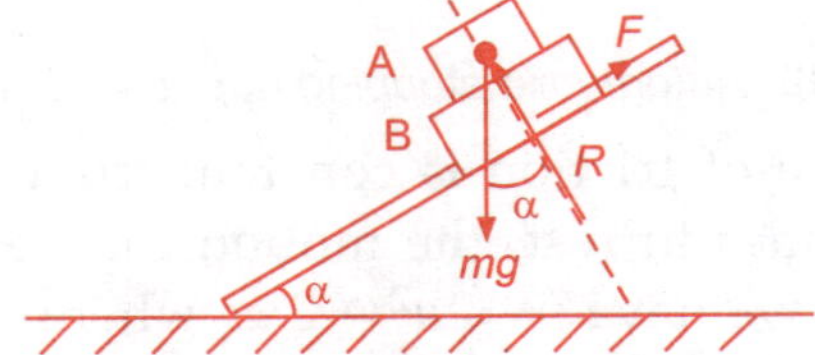

Fig. 3.17 *Relative motion of coefficient of friction by inclined plan method*

(a) **Coefficient of Static Friction:** The block A is weighed and the weight of block A and any additional weights placed on it will be equal to the total weight W_A. The bed B is slowly tilted until the block A just starts sliding downwards. The angle α just when the motion of A ensues is noted. The experiment is repeated and the average value of α is obtained. The normal reaction between the block A and the bed B is given by

$$R = W_A \cos \alpha \qquad (3.36)$$

and the component of the W_A along the inclined plane which pulls the block downwards is given by

$$F_L = W_A \sin \alpha \qquad (3.37)$$

Hence the coefficient of static friction is given by $\mu_L = F_L/R$.

(b) **Coefficient of Sliding Friction:** The bed B is tilted until it is inclined to the horizontal at such an angle that block A, when started from rest by a gentle push down the incline, slides down at a uniform velocity. We have $F = W_A \sin \alpha$, $R = W_A \cos \alpha$, and the coefficient of sliding friction given by $\mu_d = F/R$.

EXAMPLE 3.1

An automobile is travelling along a straight horizontal road at a velocity of 54 km/hour (15 m/s). If the coefficient of sliding friction between the tires and the road is 0.2, what is the shortest distance in which the automobile can be stopped by applying the brakes hard and letting it skid forward?

SOLUTION

The forces acting on the automobile, considered as a particle, are shown in Fig. 3.18. The automobile is assumed to be travelling in the x-direction. If the brakes are applied very hard, the automobile will skid forward.

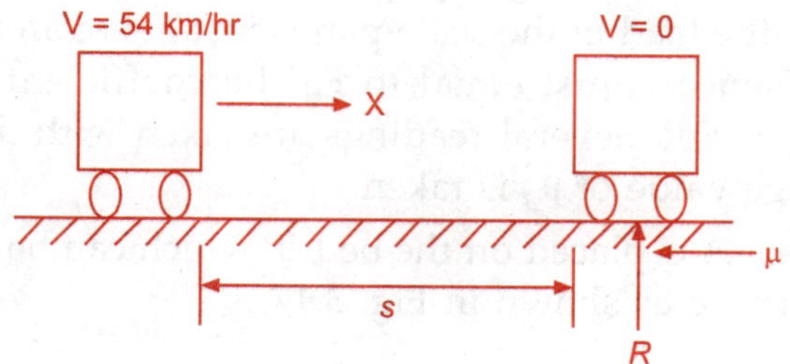

Fig. 3.18 *Automobile stopping due to friction force*

Assuming that the force of friction is constant from the time of application of brakes until the vehicle comes to rest, the motion can be assumed to be uniformly decelerated. Consequently, we have $v^2 = u^2 + 2as$ where the final velocity $v = 0$, the initial velocity $u = 15$ m/s and the distance $s = -15^2/2a$. Applying Newton's second law of motion we have

$$F = ma$$

The normal reaction between the automobile and the road is

$$R = mg$$

and the frictional force F is

$$F = -\mu_d R = -\mu_d mg = ma$$

Hence we have

$$a = -\mu_d g = -0.2 \times 9.81 = -1.962 \text{ m/s}^2$$

Using this value in the equation for s we get

$$s = -15^2/2a = -15^2/(-2 \times 1.962) = 57.34 \text{ m.}$$

3.4 GRAVITATION

3.4.1 Introduction

Motion of planets in the solar system fascinated our ancestors. Astronomy, which was very popular in ancient times in India, could predict the occurrence of solar and lunar eclipses with accuracy. Even though specific persons cannot be associated with the development of the science of planetary motions in India, it was studied to a high degree of accuracy. Ancient Greeks had also made a systematic study of the motion of planets, and specific scientists there can be associated to certain discoveries. Ptolemy, a Greek astronomer in the 2nd century A.D. proposed the geocentric theory, which suggested that the universe is centered around the earth, and all the planets moved around it. Copernicus challenged this theory and formulated the heliocentric theory, which considered that the sun is the centre of the solar system, and all the planets orbited around the sun. A Danish astronomer, Tycho Brahe, studied the position of the planets in the solar system in the 16th century. Kepler formulated three laws in the early 17th century, based on Tycho Brahe's observations. Newton reduced these three laws into a single law which is true not only for planets in the solar system, but also for bodies on earth. Since this holds good for all heavenly bodies, such as stars, the law is called the universal law of gravitation.

3.4.2 Newton's Law of Gravitation

Every body of matter in the universe attracts every other body with a force which is directly proportional to the product of their masses and inversely proportional to the square of the distance between them.

Let m_1 and m_2 be the masses of the two bodies and let d be the distance between them. The force of attraction, F, acts along the line joining the two bodies and can be represented mathematically as

$$F \propto \frac{m_1 m_2}{d^2} \tag{3.38}$$

Using a constant of proportionality, say G, we can represent the force as

$$F = \frac{Gm_1m_2}{d^2} \qquad (3.39)$$

where G is called as the gravitational constant. An inspection of the dimensions of the quantities in Eq. (3.39) will show that G is not a dimensionless constant. The dimensions of G can be obtained quite easily by writing the dimensions of quantities in Eq. (3.39) as follows,

$$[MLT^{-2}] = \frac{[G][M][M]}{[L]^2} \qquad (3.40)$$

Hence the dimensions of G are obtained as

$$[G] = [M^{-1}L^3T^{-2}]. \qquad (3.41)$$

From Eq. (3.41), we can see that the units of the gravitational constant G are $\left(\dfrac{m^3}{kg.s^2}\right)$ in SI units. From Eq. (3.41), we can see that if $m_1 = 1$, $m_2 = 1$ and $d = 1$, then $F = G$. Hence the gravitational constant can be interpreted as the force of attraction between two bodies of unit masses at a unit distance apart. Since the constant is independent of the location of the two bodies in the universe, it is known as UNIVERSAL GRAVITATION CONSTANT. The value of G is 6.673×10^{-11} m^3/kg.s^2. The law of gravitation is valid for small or large masses, and is independent of the medium surrounding the two bodies and in between them. It is independent of other environmental factors such as temperature, pressure etc. Hence the law is known as UNIVERSAL LAW OF GRAVITATION. While the gravitational forces exist between any pair of bodies, their effect is appreciable only when one of the bodies has a very large mass. The effect of gravitational forces is apparent in the case of motion of a planet around the sun, of satellites orbiting around the earth, or of bodies falling on the surface of the earth.

3.4.3 Acceleration due to Gravity

All bodies on the surface of the earth fall down, if they are free to move, because of the force of attraction between the body and the earth. Since the mass of earth is considerably larger than the mass of the bodies, these bodies move towards the centre of earth, while the earth remains relatively unmoved. The gravitational force of attraction between the earth and these bodies is called gravity. The force of attraction from earth acting on a body of mass m, produces an acceleration in the body according to Newton's second law of motion. This is the acceleration with which the body will fall down towards earth when it is let free, and it is called the acceleration due to gravity, represented by g. Hence we have

$$f = mg \qquad (3.42)$$

When bodies are close to the surface of the earth, the distance between the body and the earth's centre is almost same as the radius of the earth, R. Hence the force of attraction between the earth and the mass can be written from Eq. (3.39) as

$$f = \frac{GMm}{R^2} \tag{3.43}$$

where M is the mass of the earth, m is the mass of the body and R is the distance between them. A comparison between Eqs. (3.42) and (3.43) shows that

$$g = \frac{GM}{R^2}. \tag{3.44}$$

This equation suggests that the acceleration due to gravity, g, does not depend on the mass of the body on the surface of the earth, and is a constant at a given place. All falling bodies move with the same acceleration towards the centre of the earth. This phenomenon was first demonstrated by Galileo from his celebrated experiment from the leaning tower of Pisa in Italy.

Since the earth is not truly spherical, the distance from the centre of the earth to the body on the surface of the earth, R, varies from place to place. Value of g at sea level varies from 9.781 m/s^2 at the equator to 9.833 m/s^2 at the poles. On an average, the radius of the earth can be taken as $R = 6370$ km and an average value of $g = 9.81$ m/s^2.

It is said that the universal law of gravitation was discovered by Newton when he observed an apple falling from a tree, and he reflected that the earth must attract an apple and the moon in the same way. This basic concept of continuity of gravitational attraction is more easily understood now.

After studying the planetary motions Kepler developed an equation relating the size of the planet and the time taken by it for one revolution around the sun. He proposed that the ratio (T^2/a^3) is a constant for any planet, where

a = Semi-major axis in units of 10^{10} m.

T = Time period of revolution of the planet in years(y).

Q = The quotient (T^2/a^3) in units of 10^{-34} T^2 m^{-3}.

The data in Table 3.1 proves Kepler's law of ratios.

TABLE 3.1 Measurement of planetary motions

Planet	a	T	Q
Mercury	5.79	0.24	2.95
Venus	10.8	0.615	3.00
Earth	15.0	1	2.96
Mars	22.8	1.88	2.98

Planet	a	T	Q
Jupiter	77.8	11.9	3.01
Saturn	143	29.5	2.98
Uranus	287	84	2.98
Neptune	450	165	2.99
Pluto	590	248	2.99

EXAMPLE 3.2

Find the mean density of the earth if g is 9.81 m/s², R is 6370 km and the universal gravitational constant, G, is given as 6.673 × 10⁻¹¹ m³/kg.s².

SOLUTION

From Eq. (3.44) the mass of the earth can be written as

$$M = gR^2/G \tag{3.45}$$

Mass of the earth is also equal to

$$M = Vd \tag{3.46}$$

where V is the volume of the earth and d is the density of the earth. Assuming that the earth is a perfect sphere of radius R, the volume is given by

$$V = (4/3)\pi R^3 \tag{3.47}$$

From Eqs. (3.45), (3.46) and (3.47), we get

$$M = gR^2/G = (4/3)\pi R^3 d \tag{3.48}$$

and hence we get d as

$$d = \frac{3}{4}\frac{g}{\pi RG} \tag{3.49}$$

Substituting the values we get d as

$$d = 5510 \text{ kg/m}^3.$$

As can be seen from Eq. (3.44), the acceleration due to gravity, g, depends on the radius of the earth at the point of interest. If the body is on the surface of the earth, then the assumption that the distance between the two bodies is equal to R is correct. However, if the body is at a height h from the surface of the earth, or at a depth h below the surface of the earth, then the acceleration due to gravity for that body will be slightly different. Further, earth is rotating about its own axis, and hence anybody on the surface of the earth is also moving along a circular path. This motion introduces a centripetal acceleration directed towards the axis of rotation. This additional acceleration will modify the acceleration due to gravity. We will discuss these effects in more detail in the following.

(*i*) Variation of *g* with Height:

Let g be the value of the acceleration due to gravity on the surface of the earth, and g' be the value at a point P, which is at a height h from the surface of the earth, as shown in Fig. 3.19. Consequently, we have $g = GM/R^2$, and $g' = GM/(R + h)^2$.

Dividing g' by g, we get

$$\frac{g'}{g} = \frac{R^2}{(R+h)^2} = \frac{1}{\left(1+\dfrac{h}{R}\right)^2} \tag{3.50}$$

Hence the value of g' is obtained as

$$g' = g\left(1+\frac{h}{R}\right)^{-2} \tag{3.51}$$

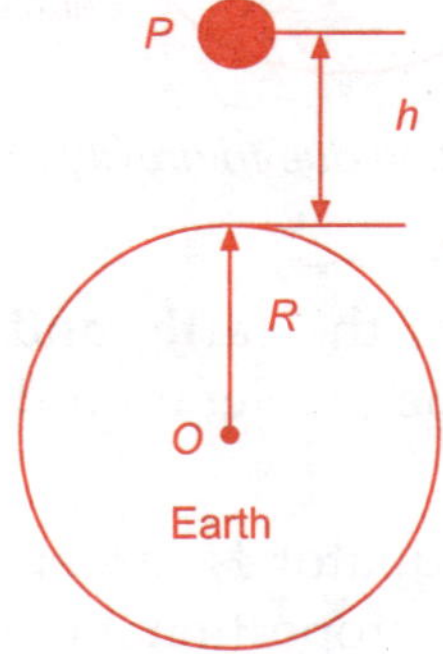

Fig. 3.19 *Acceleration due to gravity above the surface of the earth*

If the body is close to the surface of the earth, then h is quite small. Consequently, the quantity $(h/R) \ll 1$, and we can write g' from Eq. (3.51) as

$$g' = g\left(1-\frac{2h}{R}\right) \tag{3.52}$$

From Eq. (3.52) it can be seen that the value of g' decreases as the height h increases.

(*ii*) Variation of *g* with Depth:

Let the body be at P, which is at a depth h below the surface of the earth as shown in Fig. 3.20.

The body is being attracted by the shaded portion of the earth having a mass $M' = (4/3)\pi(R - h)^3\, d = g'(R - h)^2/G$. On the surface of the earth, the value of g is given by $g = (4/3)\pi RGd$. Hence the ratio of g' to g is given by

$$\frac{g'}{g} = \left(\frac{R-h}{R}\right) \tag{3.53}$$

Consequently, we have

$$g' = g\left(1 - \frac{h}{R}\right) \tag{3.54}$$

The value of g' decreases as the depth increases.

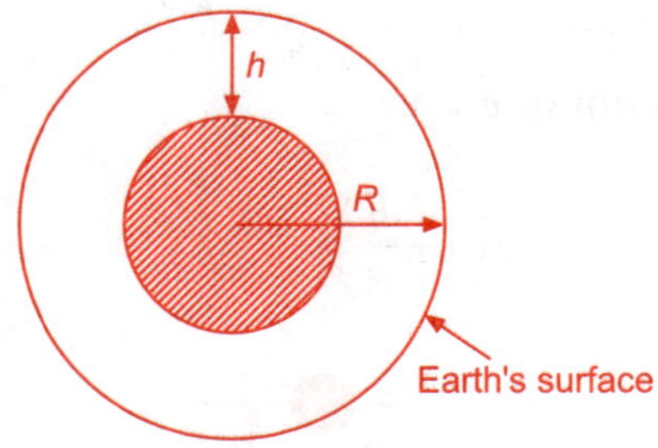

Fig. 3.20 *Acceleration due to gravity below earth's surface*

(*iii*) Variation of *g* with Latitude:

Due to the ellipsoidal shape of the earth, and also due to rotation of the earth about its own axis, the value of the acceleration due to gravity varies from point to point.

(*a*) Earth's diameter at the equator is larger than its polar diameter by about 40 km. Since g is inversely proportional to the distance, it will be larger at the poles than at the equator. It varies from a value of 9.781 m/s^2 at the equator to 9.833 m/s^2 at the poles. The value of the acceleration due to gravity increases with the latitude of the given place.

(b) Effect of Earth's Rotation on g: As the earth rotates about its own axis, everybody on its surface moves along a circular path. The centripetal force necessary to keep the body in that circular path is provided by a portion of the gravitational attraction. Consequently, the observed value of g at a place will be smaller. Since the centripetal acceleration is maximum at the equator, the effective value of the acceleration due to gravity is minimum at the equator, and increases with increase in latitude. Referring to Fig. 3.21(*a*), P is a point on earth with latitude *l*. *OA* and *OB* are the equatorial and polar radii of earth, respectively. Earth's gravitational attraction force mg on a body of mass m at point P acts along PO. The body of mass m at point P describes a circle of radius $PC = r = R \cos \lambda$, with an angular velocity ω. A centripetal force of magnitude $(m\omega^2 r)$ acts from point P towards C. The forces are represented in the form of a parallelogram in Fig. 3.21(*b*). The force mg acts along PO and is given by the length PS. The force $m\omega^2 r$ acts along PC and is given by PQ.

We have the apparent force mg' given by $QS = PT$. From the figure, we have

$$PT = PS - SU = PS - ST \cos \lambda \qquad (3.55)$$

Since $ST = QP$, we can write from Eq. (3.55)

$$mg' = mg - m\omega^2 r \cos \lambda \qquad (3.56)$$

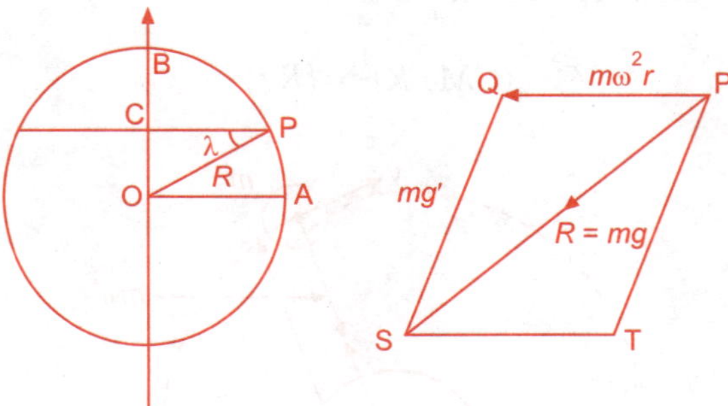

Fig. 3.21 *Effect of earth's rotation on g*

Simplifying Eq. (3.56) and using the relation $r = R \cos \lambda$, we get

$$g' = g - \omega^2 R \cos^2\lambda \qquad (3.57)$$

3.4.4 Satellites

Satellites are celestial bodies that revolve around a planet. Moon is a satellite of the earth and Mars and Jupiter have two and twelve satellites, respectively.

When a satellite such as the moon is orbiting around the earth, the necessary centripetal force is provided by the gravitational attraction force between the satellite and the earth. Referring to Fig. 3.22, we have

$$\frac{mv^2}{r} = \frac{GMm}{r^2} \qquad (3.58)$$

where m is the mass of the satellite, M is the mass of the earth and r is the radius of the satellite's orbit measured from the centre of the satellite to the centre of the earth, and v is the velocity of the satellite around the earth and is called the orbital velocity of the satellite.

Artificial satellites are put in orbits around the earth for communication, remote sensing etc. The principle used in such artificial satellites is the same as that followed by satellites. If a body is sent up high in the sky, and then given a horizontal velocity, then it will revolve around the earth. The necessary horizontal velocity or the orbital velocity v is found from Eq. (3.58) as

$$V = (Gm/r)^{\frac{1}{2}} = [GM/(R+h)]^{\frac{1}{2}} \qquad (3.59)$$

where h is the height of the satellite above the earth's surface. This velocity is independent of the mass of the satellite.

The period of revolution, T, of the satellite is given by

$$T = 2\pi r/v = 2\pi\left(r^3/GM\right)^{\frac{1}{2}} = 2\pi\left[(R+h)^3/GM\right]^{\frac{1}{2}} \tag{3.60}$$

For a satellite close to the surface of the earth, h is quite small compared to R and hence Eqs. (3.59) and (3.60) reduce to

$$V = (GM/R)^{\frac{1}{2}} = (Rg)^{\frac{1}{2}} \tag{3.61}$$

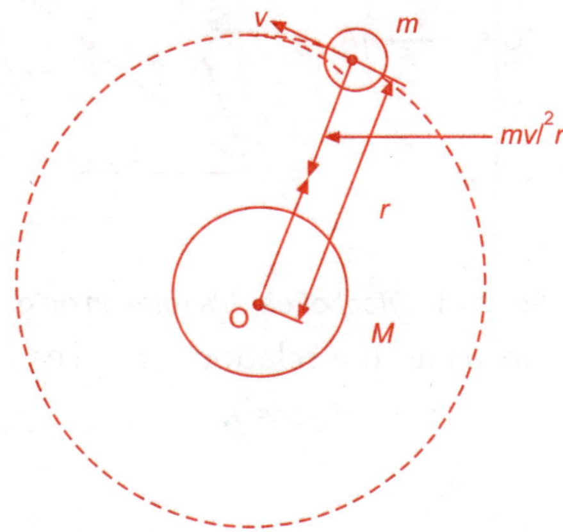

Fig. 3.22 *Satellite orbiting around a bigger plant*

and the period is given by

$$T = 2\pi\left(R^3/GM\right)^{\frac{1}{2}} = 2\pi(R/g)^{\frac{1}{2}} \tag{3.62}$$

where g is substituted for GM/R^2.

The first Indian artificial satellite, Aryabhata, was put into orbit on April 19, 1975. The very first artificial earth satellite was Sputnik I, which was put into orbit by the USSR in October 1957. A satellite must move at a high speed to avoid being drawn to earth. Satellites are normally launched by huge rockets, with three to four stages. Each stage increases the velocity of the stages until the desired velocity is attained. Finally, the satellite is separated from the rocket, and is given an appropriate velocity so as to put it into the required orbit.

Artificial satellites are equipped with a number of instruments, sensors, infrared cameras etc., in order to obtain environmental information for weather forecasting purposes and for use in mining, agriculture, forestry etc. Satellites also help in determining exact distances and locations, and precise geographical features of the land.

In India, the Indian Space Research Organization oversees the launching and maintenance of satellites in orbits.

3.4.5 Escape Velocity

Escape velocity is the minimum velocity with which a body must be projected vertically upwards from the surface of the earth so as to overcome the gravitational pull of the earth and escape into space. As the body moves against the gravitational pull, work is done on the body. The body must possess enough kinetic energy to compensate for this work of gravitational force if it must escape the gravitational force field into space. An expression for the escape velocity is given as

$$V_e = \sqrt{2GM/R} = \sqrt{2gR} \qquad (3.63)$$

Escape velocity is independent of the mass of the body being projected and the angle of projection.

EXAMPLE 3.3

Find the escape velocity for a body to escape the gravitational force field of the earth.

SOLUTION

The escape velocity is given by Eq. (3.63). The required quantities for earth are

$$g = 9.81 \text{ m/s}^2, R = 6370 \text{ km}$$

Hence $V_e = (2 \times 9.81 \times 6370 \times 1000)^{1/2} = 11200$ m/s or 11.2 km/s.

Gas molecules move constantly with velocities which increase with temperature and pressure. When gas is heated it gains energy which is converted into kinetic energy giving the gas molecules very high velocities. If the velocities of the atmospheric gas molecules exceed the escape velocity, they can free themselves from the gravitational force field and move out into space.

For example, the root mean square velocity (a measure of the average velocity) of hydrogen molecules at 0°C is about 1/6 of the escape velocity on earth. When it is heated, some molecules may acquire the escape velocity. An explanation for the rarity of hydrogen and helium gases in the earth's atmosphere is that when the earth was very hot in the past, these gases which are quite light, acquired the escape velocity and escaped from earth's atmosphere.

Same explanation holds good for the absence of gaseous atmosphere on moon and satellites of other planets. Since satellites have very small masses, their g value is much smaller and hence they have very small escape velocities.

Some idea of the trajectories followed by projectiles or satellites is given below, by comparing their velocities to the escape velocity.

In Fig. 3.23, three possible trajectories are shown for a projectile.

    (*a*) Hyperbolic trajectory: If the velocity $v > v_e$, then the trajectory is a hyperbola and the body escapes into space.

(b) Parabolic trajectory: If the velocity $v = v_e$, the trajectory is parabolic. This is the limiting case of the body just having enough velocity to escape into space.

(c) Elliptical Orbit: If the velocity v is in the range $(v_e/\sqrt{2}) < v < v_e$, then the path is an ellipse with earth at one focus.

(d) Circular Orbit: If the velocity is $v = (v_e/\sqrt{2})$, then the orbit is a circle.

(e) Return to Earth: If the velocity $v < (v_e/\sqrt{2})$, then the body will return to earth.

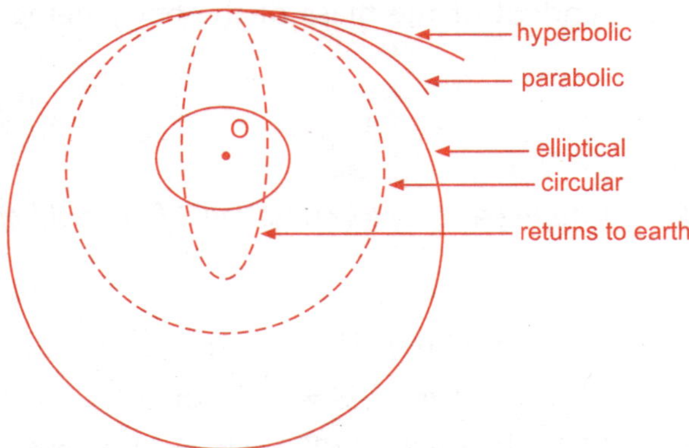

Fig. 3.23 *Satellite orbiting different velocities*

3.4.6 Weightlessness

While going downwards in elevators or riding down steeply in roller coasters, we feel that our weight has reduced considerably. The reaction that we experience from the floor is smaller than our weight that we experience while standing on a rigid floor. This condition of negligible reaction from the moving floor is referred to as weightlessness. Astronauts in space vehicles orbiting the earth experience weightlessness. In order to get used to this type of environment, they train themselves on earth using huge revolving rides and inclined platforms. It is possible to create weightlessness when a jet plane dives and swoops through an arc with its centre away from earth (convex side towards earth). Weightlessness in space is highly inconvenient and uncomfortable to the astronaut. They cannot move around as they do on earth and they float to move from one place to another in space. Pouring liquids and drinking of liquids is not possible in space. There is neither up nor down in space in this type of environment. Prolonged weightlessness may induce nausea or other adverse health effects.

In a satellite revolving around the earth, by virtue of its velocity, every part and parcel of the satellite has acceleration towards the centre of the earth. This acceleration is exactly the value of earth's acceleration due to gravity at that position.

Thus in the satellite everything inside it is in a state of free fall. This is just as if we were falling towards the earth from a height. Thus, in a manned satellite, people inside experience no gravity. Gravity for us defines the vertical direction and thus for them there are no horizontal or vertical directions, all directions are the same. Weightlessness may be advantageously used in some industrial manufacturing operations, such as in making of perfectly spherical ball bearings.

Space Technology in India

India launched the low orbit satellite Aryabhatta in 1975. In the early years of its space programme, Indian made satellites were launched using Soviet launch vehicles. Indigenous launch vehicles were employed in the early 1980's to send the Rohini series of satellites into space. The programme to send polar satellites into space began in late 1980's. A series of Indian Remote Sensing Satellites (IRS) have been launched for use in surveying, weather prediction and for carrying out experiments in space. India launched an experimental communications satellite (GSAT-1) into space in 2001 which was a geostationary satellite. In 1984 Rakesh Sharma became the first Indian astronaut to orbit in space. The Indian Space Research Organisation (ISRO) is the umbrella organisation that has a number of centres. Its main launch centre at Sriharikota (SHAR) is 100 km north of Chennai. The Indian Satellite Centre (ISAC) designs and builds satellites. The National Remote Sensing Agency (NRSA) is near Hyderabad. Its national centre for research in space and allied sciences is the Physical Research Laboratory (PRL) at Ahmadabad.

3.5 ELASTICITY

3.5.1 Concept of Elasticity

Consider a helical spring as shown in Fig. 3.24. One end of the spring is secured to the support point at O and the other end is free. The undeformed length of the spring is given by L. If the free end of the spring is loaded with weights, the spring will undergo deflection which depends on the load applied. If the load is increased, the deflection also increases. A force versus deflection graph is shown in the same figure,

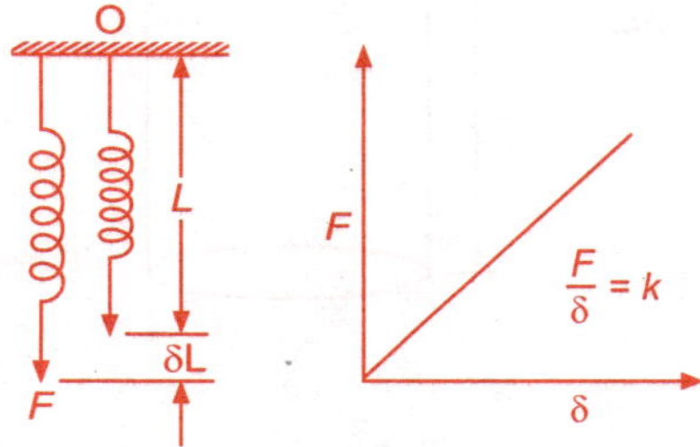

Fig. 3.24 *Force displacement relation from an elastic spring*

which is almost a straight line. If the load is reduced, the deflection in the spring also comes down. Under ideal conditions, when the load is completely removed, the spring loses the deflection completely and regains its original undeformed length, L.

When the applied loads are small, most of the metals deform proportional to the applied loads, and when the load is removed they regain their original length. If the load applied is large, then the deflection behaviour may move into the nonlinear region and when the load is removed completely, the original shape may not be completely recovered. The property of recovering completely from the deflected state when the load causing the deflection is removed is called elasticity and the deformation caused is known as **elastic** deformation. However, if force is applied to a lump of putty or mud, they have no gross tendency to regain their previous shape, and they get permanently deformed. Such substances are called **plastic** and this property is called **plasticity.** Putty and mud are close to ideal plastics.

3.5.2 Stress and Strain

Deformation of materials subjected to loads depends on the force per unit area rather than on the force alone. For example, a given load will elongate a thin metallic wire more than a thick one, because the area of cross section of the thin wire is less and therefore the force per unit area larger.

Stress is defined as force per unit area of the material. If the area considered is normal or perpendicular to the force, it is called normal stress whereas if the area is parallel to the force, the material experiences a shear stress.

There are three ways in which a solid may change its dimensions when an external force acts on it. These are shown in Fig. 3.25. In Fig. 3.25 (a), a cylinder is stretched by two equal forces applied normal to its cross-sectional area. The restoring force per unit area in this case is called **tensile stress**. If the cylinder is

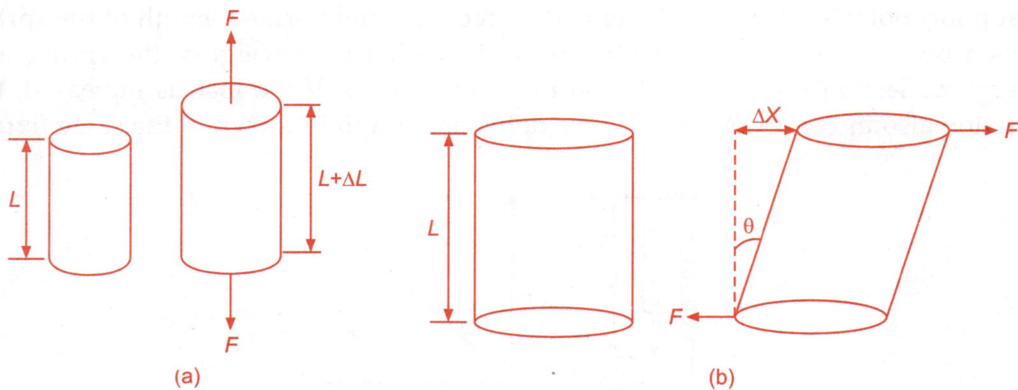

Fig. 3.25 *(a) Cylinder subjected to tensile stress stretches it by an amount ΔL*
(b) A cylinder subjected to shearing (tangential) stress deforms by an angle θ

compressed under the action of applied forces, the restoring force per unit area is known as **compressive stress**. Tensile or compressive stress can also be termed as longitudinal stress.

Strain is defined as the change in dimension per unit dimension, or the deformation of a specimen of unit dimensions.

When the material is subjected to small loads, the deflection in the material is proportional to the load and the load versus deflection curve is a line. When the load reaches a certain value, the linear relation between load and deflection changes into a nonlinear relation. The stress corresponding to the limiting load where this change occurs is known as the elastic limit.

For small applied stresses and strains within the elastic limit, the strain is proportional to the applied stress producing it. Under these conditions, the ratio of stress to strain is called the modulus of elasticity of the material, denoted by the symbol E. Hence

$$E = \sigma/\varepsilon \tag{3.64}$$

where σ is the stress and ε is the strain.

Three Types of Stress and Strain

Three types of stress and strain are considered as follows:

1. **Longitudinal:** Consider a slender cylindrical specimen of the material, as shown in Fig. 3.26, subjected to forces F on either side. The area of cross section of the specimen is A and the longitudinal stress is given by $\sigma = F/A$ N/m^2. This stress is the same at all points along the specimen. If an elongation δL is produced in a specimen of length L, the longitudinal strain is $\varepsilon = \delta L/L$. For small elastic displacements, the Young's modulus of elasticity is given by $E = \sigma/\varepsilon$.

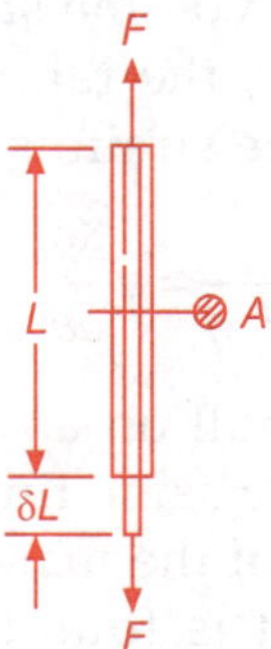

Fig. 3.26 *Longitudinal stress and strain*

2. **Shear:** Shear or tangential forces applied on the planes of the area concerned, as shown in Fig. 3.27, constitute a shear stress acting through the material. If F is the

magnitude of one of the forces, and the area of the surface on which the force acts is A, the shear stress is given by

$$\tag{3.65}$$

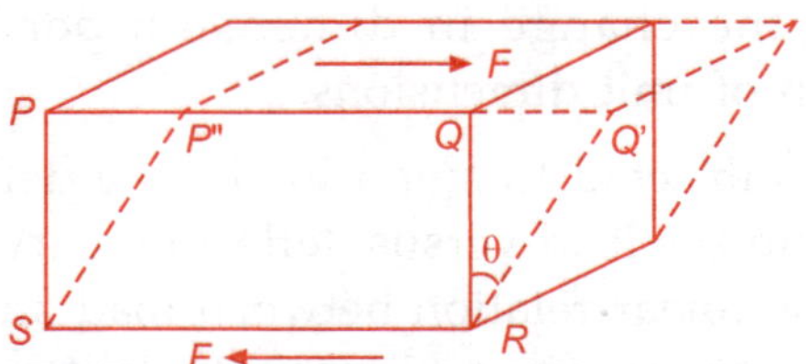

Fig. 3.27 *Shear stress and strain*

The deformation alters the shape of the rectangular figure seen in elevation as $PQRS$ to $P'Q'RS$, displacing P to P' and Q to Q', and rotating SP and RQ through θ radians to SP' and RQ'. The shear strain is the ratio of the lateral deformation in the direction of the stress to the original dimension, i.e., $\delta x/y$ in Fig. 3.27. As any elastic displacement is necessarily small, $\delta x/y$ is for all practical purposes the shear strain, or the angle of shear, θ. Within the region of perfect elasticity, the ratio of the shear stress τ, to shear strain, ε, is called the modulus of rigidity, G.

Hence,

$$G = \tau/\varepsilon. \tag{3.66}$$

3. **Bulk or Volume:** When a solid is immersed in a fluid, the fluid exerts a uniform pressure on the solid from all directions. This uniform pressure, which is force per unit area, constitutes a bulk or volume stress. Similarly, if a small elemental volume of a fluid is considered, the rest of the fluid exerts a uniform pressure on this elemental volume from all directions.

Let p be the pressure or the volume stress acting on a deformable solid. Let the volume stress change by a small amount, δp. The solid undergoes deformation resulting in a change in its volume, δv. The ratio of the change in volume to the original volume, v, is called a bulk or volume strain, $\delta v/v$. For small volume changes, within the region of perfect elasticity, the ratio of the incremental volume stress to the corresponding incremental volume strain is called the bulk modulus, k, given by

$$k = \frac{\delta p}{(\delta v/v)} = v\frac{\delta p}{\delta v}. \tag{3.67}$$

Poisson's Ratio: A longitudinal pull on an elastic body causes a deformation in the longitudinal direction, while at the same time it experiences a lateral contraction. Consider a flat rectangular specimen of the material with a length of a and a width of b. For small elastic displacements it is found that the ratio of the lateral strain, $(\delta b/b)$ to the longitudinal strain, $(\delta a/a)$, is called the Poisson's ratio, denoted by the symbol, μ,

$$\mu = \frac{\delta b/b}{\delta a/a} \tag{3.68}$$

3.5.3 Hooke's Law for an Ideal Elastic Material

For small elastic deformations, the deformation is directly proportional to the force producing it. In the case of a slender metallic wire, the ratio, (load/deflection), is a constant. This direct proportional relation is known as Hooke's law. In all such cases, the ratio, (stress/strain) is a constant. The load versus deflection graph for a metallic wire is shown in Fig. 3.28.

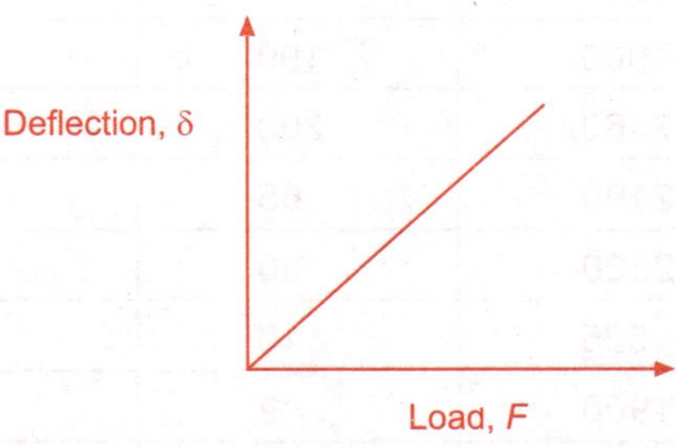

Fig. 3.28 *Load-deflection graph*

3.5.4 Strain Energy

When the load is applied on an elastic body, the load does work on the body and the body undergoes strain. The external work done on the body is stored in the body in the form of strain energy.

Consider a metallic wire of length L, area of cross section A and loaded with a tensile load F. The corresponding elongation in the wire is x. The load-extension graph is shown in Fig. 3.29. The strain energy stored is given by the area under the load-extension graph, given by the triangle OPQ.

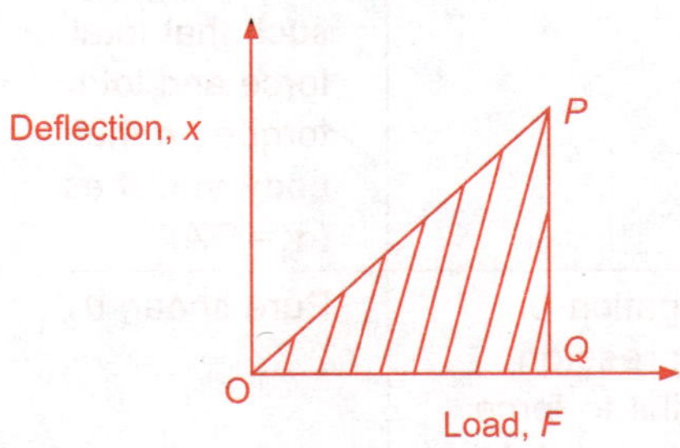

Fig. 3.29 *Strain energy stored in elongated metallic mire*

Hence, the strain energy is given by

$$U = \frac{1}{2}Fx.$$ (3.69)

We have $F = \sigma A$, where $\sigma = F/A$, is the stress and $x = \varepsilon L$, where $\varepsilon = \delta L/L$, is the strain. Substituting in Eq. (3.69) we get the expression for the strain energy stored as

$$U = \frac{1}{2}\sigma\varepsilon AL = \frac{1}{2}\sigma\varepsilon V$$ (3.70)

TABLE 3.2   Young's moduli and yield strengths of some materials

Substance	Density ρ (Kg m^{-3})	Young's Modulus Y (10^9 N m^{-2})	Ultimate Strength, S_u (10^6 N m^{-2})	Yield Strength S_Y (10^6 N m^{-2})
Aluminium	2710	70	110	95
Copper	8890	110	400	200
Iron (Wrought)	7800-7900	190	330	170
Steel	7860	200	400	250
Glass	2190	65	50	-
Concrete	2320	30	40	-
Wood	525	13	50	-
Bone	1900	9	170	-
Polystyrene	1050	3	48	-

TABLE 3.3   Stress, strain and various elastic moduli

Type of Stress	Tensile or Compressive	Shearing	Hydraulic
Stress	Two equal and opposite forces perpendicular to opposite faces ($\sigma = F/A$)	Two equal and opposite forces parallel to opposite surfaces (forces in each case such that total force and total torque on the body vanishes ($\sigma_s = F/A$)	Forces perpendicular everywhere to the surface, force per unit area same everywhere
Strain	Elongation or compression parallel to force direction ($\Delta L/L$) (longitudinal strain)	Pure shear, θ	Volume change (compression or elongation) ($\Delta V/V$)
Change in Shape	Yes	Yes	No
Change in Volume	No	No	Yes
Elastic Modulus	$Y = (F{*}L/A\Delta L)$	$G = (F{*}\theta)/A$	$B = -p/(\Delta V/V)$
Name of Modulus	Young's modulus	Shear modulus	Bulk modulus
State of Matter	Solid	Solid	Solid, liquid and gas

The values of the Young's modulus, density, ultimate strength and the yield strength for different materials are given in Table 3.2. Also, the various definitions such as tensile or compressive stress and shear stress are given in Table 3.3.

3.6 PRINCIPLE OF WORK AND ENERGY

3.6.1 Work

Work is done when a force, acting on a body, moves through a distance. If the force is constant, and the displacement is in the same direction as that of the force, the work of the force is the product of the force and the distance through which it acts. Work is done by a force irrespective of the direction of the force, and consequently, work has only magnitude. Hence work is a scalar quantity.

When the force and displacement are in the same direction, the work is given by

$$W = FS \qquad (3.71)$$

If the force is applied at an angle of θ with respect to the horizontal, as shown in Fig. 3.30, then the expression for the work is given by

$$W = FS \cos \theta \qquad (3.72)$$

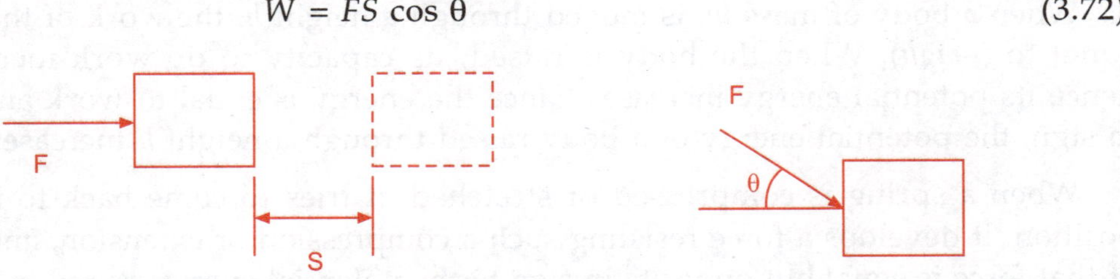

Fig. 3.30 *Work of a force*

Hence work can be considered as the product of the force and the component of the displacement in the direction of the force. Or it can also be considered as the product of the displacement and the component of the force in the direction of the displacement.

Consider a mass m being moved up through a height h. The mass m is being pulled down towards the centre of the earth with a force equal to its weight, mg. This force moves through a distance of h, but in a direction opposite to that of the force. Hence the work done by the weight is given by

$$W = mg\,(-h) = -mgh. \qquad (3.73)$$

Unit of Work:

From Eq. (3.71) we can see that the unit of work is given by the product of the units of force and distance. In SI units, the unit of work is

$$[W] = [Nm] \qquad (3.74)$$

The unit of work is Joule which is equal to Newton. Meter, (N.m).

Since we have discussed about a smaller unit of force, i.e., dyne, the corresponding unit of work is one dyne.cm which is termed as one erg. 1 Joule = 10^7 ergs.

3.6.2 Energy

Energy is the capacity to do work. It is equal to but opposite in sign to work and has the same units as that of work. Energy can be in several forms such as, mechanical energy, electrical energy, chemical energy, thermal energy, sound energy etc. Mechanical energy is the energy possessed by a body by virtue of its position, configuration or motion. Mechanical energy has two forms, (*i*) potential energy, which is due to its position or configuration, and (*ii*) kinetic energy, which is due to its motion.

Potential Energy

Potential energy of a body is the energy possessed by the body by virtue of its position or configuration. It is equal to the amount of work that is necessary to take the body through a displacement or to take the body through a change in configuration.

When a body of mass m, is moved through a height h, the work of the weight is equal to $(-mgh)$. When the body is raised, its capacity to do work increases, and hence its potential energy increases. Since the energy is equal to work and opposite in sign, the potential energy of a body raised through a height h increases by mgh.

When a spring is compressed or stretched, it tries to come back to its original position. It develops a force resisting such a compression or extension, and the work of that force is equal but opposite in sign to the potential energy stored in the spring. Hence, when the hairspring of a watch is wound it stores energy that can run the watch for a certain length of time.

Water stored at a higher altitude possesses potential energy, and it has the capacity to do work. When it is made to flow through turbines, it can rotate them at high speeds and appear as kinetic energy of the turbines. When the turbine shafts turn generator shafts it produces electrical energy.

Kinetic Energy

Kinetic energy of a body is the energy possessed by the body by virtue of its motion. Kinetic energy of a body in motion is equal to the work done by forces acting on the body which change the state of rest of the body to a state of motion. Kinetic energy of a body can be found by knowing the amount of work done on the body to accelerate it to the present state of motion.

If a constant force F, acting on a body at rest produces a constant acceleration a, resulting in a final velocity v for the body, then we have the relation $F = ma$ and also $v^2 = u^2 + 2as$. Since $u = 0$, we have $v^2 = 2as$. The work done by the force F is given by

$$W = FS = ma(v^2/2a) = (1/2)mv^2 \qquad (3.75)$$

This work by the effective force $m.a$ is termed as kinetic energy of the body. This is kinetic energy due to the translational motion.

TABLE 3.4 Typical kinetic energies (K)

Object	Mass (Kg)	Speed (ms^{-1})	K (J)
Car	2000	25	6.3×10^5
Running athlete	70	10	3.5×10^3
Bullet	5×10^{-2}	200	10^3
Stone dropped from 10m	1	14	10^2
Rain drop at terminal speed	3.5×10^{-5}	9	1.4×10^{-3}
Air molecule	10^{-26}	500	10^{-21}

3.6.3 Law of Conservation of Energy

There are several forms of energy such as mechanical energy, thermal energy, chemical energy, electrical energy. One form of energy can be converted to another form, like mechanical energy into electrical energy as in a generator, or thermal energy into mechanical energy as in internal combustion engine.

Law of conservation of energy states that in the absence of external energy input to the system, the total energy of the system is conserved. One form of energy may be converted to another form, however, the total energy in whatever form, remains constant.

With the advent of relativity theory, we know that energy and matter are interconvertible and energy can be created using matter. However, this requires special conditions and this is not possible under normal circumstances. Hence, the law of conservation of energy holds good under normal circumstances.

During motion, potential energy in the body may be converted into kinetic energy, which results in an increase in the velocity of the body. Sometimes, the kinetic energy may be converted into potential energy, in which case the velocity of the body comes down. In some forms of motion, there is a continuous exchange of potential and kinetic energies from one form to another. The following examples will illustrate the three types of motion.

(*i*) A Body in Free Fall in Atmosphere

A body of mass m at a height h from ground level possesses a potential energy of mgh with respect to the ground. If the body is allowed to fall freely under the influence of gravity, it acquires velocity continuously. We assume that the resistance by air to the motion is negligible. It is losing potential energy because it is losing height; however, because it is increasing in velocity, it acquires more and more kinetic energy. The increase in kinetic energy is at the expense of its potential energy. This is shown in Fig. 3.31.

When the body has fallen through a distance y from its original position, its potential energy becomes $mg(h - y)$.

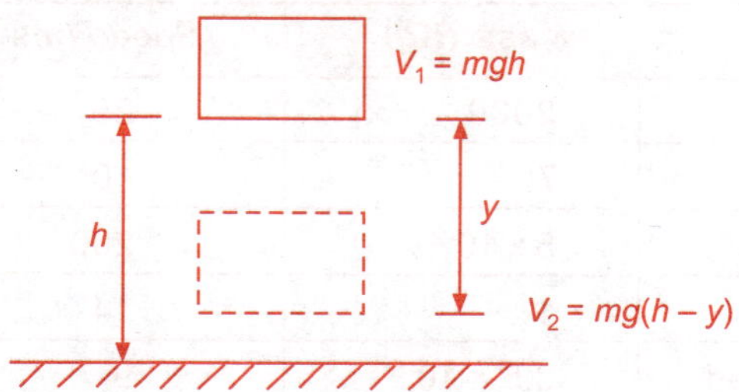

Fig. 3.31 Variation of potential energy with height

The loss in potential energy, mgy, is converted into kinetic energy. If we denote the potential and kinetic energies in the original position as V_1 and T_1, respectively, and the potential and kinetic energies when it has come down by as V_2 and T_2, respectively, then from the law of conservation of energy, we have

$$T + V = \text{constant at any instant} \tag{3.76}$$

Hence

$$T_1 + V_1 = T_2 + V_2 \tag{3.77}$$

In the present case, we have

$$T_1 = 0; \; V_1 = mgh; \; T_2 = (1/2)mv^2; \; V_2 = mg(h - y) \tag{3.78}$$

Hence, we have

$$mgh = (1/2)mv^2 + mg(h - y) \tag{3.79}$$

The velocity of the body when it comes down by y is given from Eq. (3.79) by

$$v = (2gy)^{1/2} \tag{3.80}$$

(ii) A Body moving up an Inclined Plane:

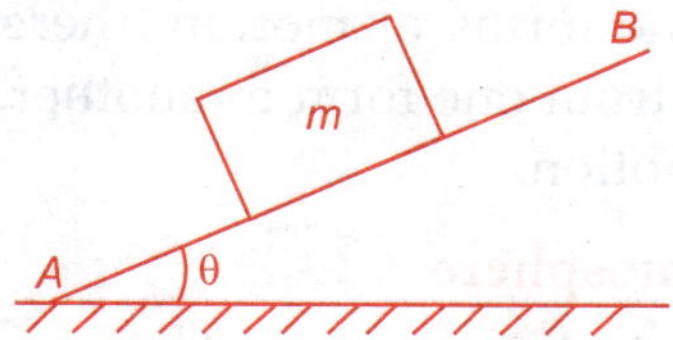

Fig. 3.32 *A body moves up in an inclined plane*

Consider a body of mass m moving up an inclined plane with an inclination of angle θ with respect to the horizontal, as shown in Fig. 3.32. Initially the body is at point A, where it has a velocity of u. If we consider this as the ground level, the potential energy of the body $V_1 = 0$. Since the velocity at A is u, the kinetic energy of the body, $T_1 = (1/2)mu^2$.

When the body has moved through a distance of s along the inclined plane to point B, the vertical component of its motion is sin $\theta = y$. If the potential and kinetic energies at point B are, respectively, $V_2 = mgy$, and $T_2 = (1/2)mv^2$, then from the law of conservation of energy, we have

$$(1/2)mu^2 + 0 = (1/2)\ mv^2 + mgy \tag{3.81}$$

Hence, the final velocity, v, is given by

$$v^2 = u^2 + 2gy \tag{3.82}$$

(*iii*) Oscillations of a Simple Pendulum:

During the oscillations of a simple pendulum, there is a continuous exchange of potential and kinetic energies. A simple pendulum is shown in Fig. 3.33. When the pendulum bob is at the extreme positions B, either to the right as shown or to the left, the bob is instantaneously at rest and hence the kinetic energy is zero. But in those positions the bob is at the maximum height from the lowest position at A. Hence the potential energy is maximum at B, either at the right or at the left.

At the lowest position A, the potential energy is minimum, and may be considered as zero if that position is taken as the reference. The bob has the largest velocity at A and hence the kinetic energy is the maximum.

Considering that the motion starts at B, we have $T_1 = 0$, and $V_1 = mgh$. At A, we have $T_2 = (1/2)mv^2$, and $V_2 = 0$. Hence using the conservation of energy principle, we have

$$mgh = (1/2)\ mv^2 \tag{3.83}$$

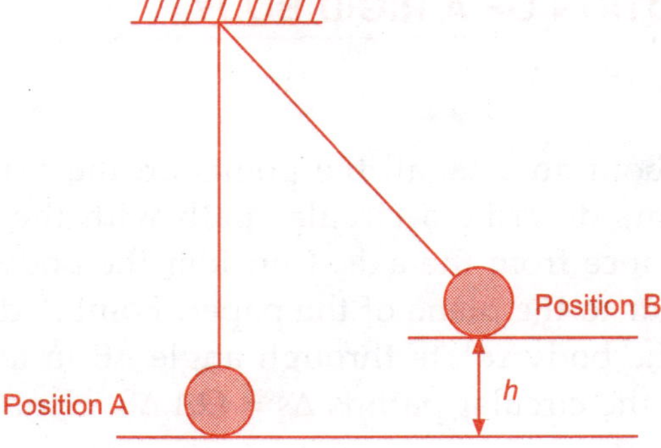

Fig. 3.33 *Simple pendulum in oscillation*

Hence the maximum velocity at A is obtained as $v = (2gh)^{1/2}$. Aproximate energy associated with various phenomena are given in Table 3.5.

TABLE 3.5 Approximate energy associated with various phenomena

Description	Energy (J)
Big Bang	10^{68}
Radio energy transmitted by the galaxy through its lifetime	10^{55}
Rotational energy of the milky way	10^{52}
Energy released in a supernova explosion	10^{44}
Ocean's hydrogen in fusion	10^{34}
Rotational energy of the earth	10^{29}
Annual solar energy incident on the earth	10^{29}
Annual global energy usage by human	3×10^{29}
Annual electrical output of large generating plant	10^{16}
Thunderstorm	10^{15}
Energy released in burning 1000 Kg of coal	3×10^{10}
Kinetic energy of large jet aircraft	10^{9}
Daily food intake of a human adult	10^{7}
Work done by a human heart per beat	0.5
Turning this page	10^{-3}
Flea hop	10^{-7}
Discharge of a single neuron	10^{-10}
Typical energy of a proton in a nucleus	10^{-13}
Typical energy of an electron in an atom	10^{-18}
Energy to break one bond in DNA	10^{-20}

3.7 ROTATIONAL MOTION OF A RIGID BODY

3.7.1 General

When a body rotates about an axis, all the points on the axis of rotation have zero velocity. All other points describe a circular path with the axis as the centre and radius equal to the distance from the axis. Consider the body rotating about an axis through O, perpendicular to the plane of the paper. Point A describes a circular path with O as centre. Let the body rotate through angle $\Delta\theta$ in a small time interval Δt. The distance AA' along the circular path is $\Delta s = OA.\Delta\theta$. Hence the speed is given by

$$v = \Delta s/\Delta t = OA\,(\Delta\theta/\Delta t) = OA\omega \qquad (3.84)$$

From Eq. (3.84), it can be seen that the velocity of point B is smaller than that of A. As the point is farther from the axis of rotation, its speed is larger. In Fig. 3.34, $\Delta\theta$ is referred to as the angular displacement of the body. It is the angle through which a line in the body perpendicular to the body rotates.

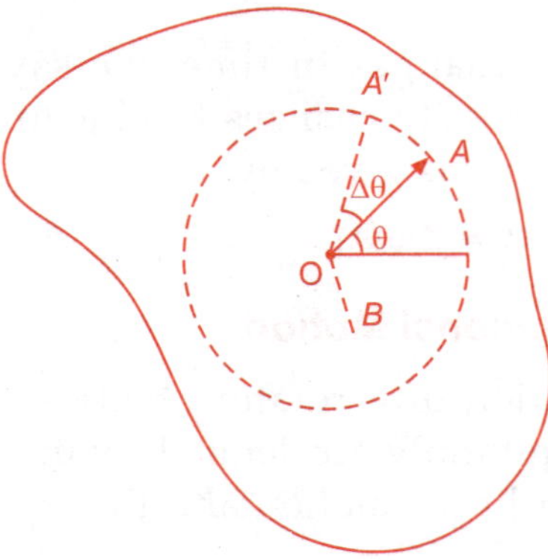

Fig. 3.34 *A body rotating about an axis through O*

The angle covered in unit time, or the ratio of the angular displacement to the time taken for this displacement, is called the angular velocity. Angular velocity of a body is a vector and its direction can be obtained by the motion of a right handed corkscrew. When the corkscrew is rotated in the direction of rotation of the body, then the direction of travel of the corkscrew indicates the direction of the angular velocity vector. Angular velocity is positive if it is anticlockwise and negative if it is clockwise, according to the above convention. The direction of the angular velocity vector can be found by the direction of the thumb of a right hand which grasps the axis of rotation with fingers encircling the axis in the direction of rotation which is shown in Fig. 3.35.

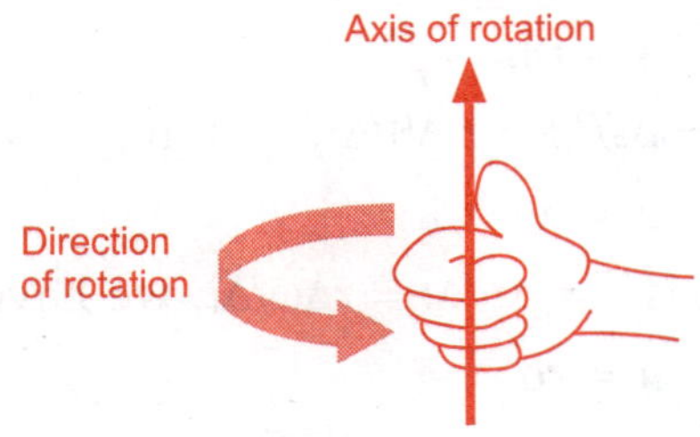

Fig. 3.35 *Direction of angular velocity*

Uniform Rotation

The rotational motion is said to be uniform if the angular velocity of the body is constant. This is analogous to the uniform motion in the case of translational motion. This means that the body describes equal angles in equal intervals of time. If the angular velocity is given by ω, then the angular displacement is given by

$$\theta = \omega t \tag{3.85}$$

It can be recalled that this is similar to the displacement in the case of uniform linear motion, $x = v.t$.

Angular Acceleration

When the angular velocity changes in time, we say that the body has angular acceleration. If the angular velocity changes from ω to $\omega + \Delta\omega$, during the time t to $t + \Delta t$, then the angular velocity is given by

$$\alpha = \Delta\omega/\Delta t. \tag{3.86}$$

Uniformly Accelerated Rotational Motion

When the angular acceleration of a rotating body stays constant throughout the motion, then it is called a uniformly accelerated rotational motion. This is analogous to the uniformly accelerated linear motion and the equations describing the motion are given below.

$$\omega = \omega_0 + \alpha t \tag{3.87}$$

$$\theta = \omega_0 t + \frac{1}{2}\alpha t^2 \tag{3.88}$$

$$\omega^2 = \omega_0^2 + 2\alpha\theta \tag{3.89}$$

where ω is the angular velocity at any time t, ω_0 is the angular velocity at time $t = 0$, α is the angular acceleration, and θ is the angular displacement in time t.

Relation Between Angular Motion and Linear Motion

Referring to Fig. 3.34, it can be seen that the displacement AA' is given by $\Delta s = OA.\Delta\theta = r.\Delta\theta$. In general, we have

$$s = r\theta. \tag{3.90}$$

The linear velocity is $v = \Delta s/\Delta t = r.\Delta\theta/\Delta t$, and in general

$$v = r\omega. \tag{3.91}$$

Since linear acceleration is $a = \Delta v/\Delta t = r\Delta\omega/\Delta t$, we have

$$a = r\alpha. \tag{3.92}$$

3.7.2 Moment of Inertia

When a body is rotating about an axis, as shown in Fig. 3.34, the linear acceleration of a point A is given by $a = r.\alpha$, where r is the distance of A from the axis of rotation. The force acting on a small mass Δm is given by

$$\Delta f = \Delta ma = \Delta mr\alpha. \tag{3.93}$$

The moment of this force about the axis of rotation is

$$\Delta M = \Delta mr^2.\alpha. \tag{3.94}$$

If such moments for the entire body are summed up, we get

$$M = (\Sigma\Delta mr^2)\alpha. \tag{3.95}$$

This is analogous to the relation $F = ma$ in translational motion. Therefore the quantity $(\Sigma \Delta m.r^2)$ in rotational motion is analogous to mass m in translational motion. Mass is a measure of the inertia in the case of translational motion. Similarly, $(\Sigma \Delta m.r^2)$ is a measure of the rotational inertia of the body and is called the moment of inertia of the body about the axis of rotation. Hence, moment of inertia of a body about an axis is defined as $(\Sigma \Delta m.r^2)$, where Δm is the mass of a small element of the body at a distance r from the axis of rotation. Alternatively, the sum of the product of mass and the square of the distance from the axis about which we want the moment of inertia, for each element of the body is called the moment of inertia. We have

$$I = (\Sigma \Delta m.r^2) \qquad (3.96)$$

Moment of inertia of a body depends on the mass distribution of the body and the location of the axis about which moment of inertia is taken. The unit of moments of inertia is kg.m^2 or mass units times the square of the unit for distance.

3.7.3 Radius of Gyration

Sometimes it is convenient to specify the radius of gyration of a body which provides an idea about the mass distribution of the body. Radius of gyration of a body is related to its moment of inertia by

$$I = mK^2 \qquad (3.97)$$

where K is the radius of gyration. Radius of gyration may be visualized as the radius of a uniform ring of mass m and the axis about which moment of inertia is taken is passing through the centre of the ring perpendicular to the plane of the ring.

3.7.4 Kinetic Energy of Rotation

When a body is rotating with an angular velocity ω, the linear speed of a point A, is given by $v = r\omega$. If a small element of mass Δm is considered at A, then the kinetic energy of this elemental mass is given by

$$\Delta T = \frac{1}{2}\Delta m(r\omega)^2 \qquad (3.98)$$

The kinetic energy of the whole body is given by summing up the kinetic energies of all such elemental masses which make up the body, and is given by

$$T = \frac{1}{2}(\Sigma \Delta m r^2)\omega^2 \qquad (3.99)$$

Since the angular speed ω is same for the whole body, it does not appear inside the summation. Since $(\Sigma \Delta m.r^2) = I$ is the moment of inertia of the body about the axis of rotation, the kinetic energy of the body is given by

$$T = \frac{1}{2}I\omega^2 \qquad (3.100)$$

3.7.5 Angular Momentum

In analogy with translational motion, where the linear momentum is the product of mass and velocity, we define the angular momentum as the product of the moment of inertia and the angular velocity, as

$$H = I\omega \tag{3.101}$$

Angular momentum is a vector quantity and has the direction of the angular velocity. Newton's second law can be recast for the rotational motion as "Rate of change of angular momentum is directly proportional to the applied moment and takes place in the direction of the moment"

The moment or the torque is given by

$$M = \frac{d}{dt}(I\omega) \tag{3.102}$$

Similar to the conservation of linear momentum in translational motion, we have conservation of angular momentum in rotational motion. A diver, for example, takes advantage of the principle of the conservation of angular momentum to perform the act. He/she jumps with the arms and legs extended and with a small angular velocity. When the arms and legs are pulled in during the act, the moment of inertia reduces and the angular velocity increases, keeping the angular momentum constant.

TABLE 3.6 Comparison of Linear and Rotational Motions

Linear	Rotational
Distance, s	Angle, θ
Velocity, $v = s/t$	Angular Velocity, $\omega = \theta/t$
Acceleration, $a = 2s/t^2$	Angular Acceleration, $\alpha = 2\theta/t^2$
Mass, m	Moment of Inertia, I
Linear Momentum, $P = m.v$	Angular Momentum, $H = I.\omega$
Force, $F = m.a$	Moment, $M = I.\alpha$
Work, $W = F.s$	Work, $W = M.\theta$
Kinetic energy, $T = \frac{1}{2}mv^2$	Kinetic Energy, $T = \frac{1}{2}I\omega^2$

SUMMARY

1. Mechanical Forces are classified into, Force due to Friction, Cohesive force, Adhesive Force, Elastic Force.

2. The forces which can be applied from a distance are Gravitational Force, Electrostatic Force, Magnetic Force and Nuclear Force.

3. Newton's first law of motion: Everybody continues to be in its state of rest or of uniform motion in a straight line, unless compelled by some external force to act otherwise.

4. Newton's Second Law: The rate of change of momentum of a body is directly proportional to the impressed force and takes place in the direction of the force.

5. Newton's Third Law: For every action, there is an equal and opposite reaction.

6. Equilibrium of a particle in mechanics refers to the situation when the net external force on the particle is zero.

7. Momentum (p) of a body is the product of its mass (m) and velocity (v) : $p = m\,v$.

8. Impulse is the product of force and time which equals change in momentum. The notion of impulse is useful when a large force acts for a short time to produce a measurable change in momentum.

9. *Law of Conservation of Momentum:* The total momentum of an isolated system of particles is conserved.

10. Angular momentum conservation leads to Kepler's second law. However, it is not special to the inverse square law of gravitation. It holds for any central force.

11. The collision or impact between two bodies is said to be direct or head-on collision, if the direction of motion of each body is along the common normal at the point of impact.

12. The collision is said to be oblique, if the direction of motion of either of the bodies is not along the common normal at the point of impact.

13. The force of friction is equal and opposite to the components of applied forces in the direction of anticipated motion.

14. An astronaut experiences weightlessness in a space station. This is not because the gravitational force is small at that location in space. It is because both the astronaut and the station are in "free fall" towards the Earth.

15. Escape velocity is the minimum velocity with which a body must be projected vertically upwards from the surface of the earth so as to overcome the gravitational pull of the earth and escape into space.

16. It is experimentally observed that in the case of direct collision between two bodies the relative velocity after collision bears a constant ratio to the relative velocity before collision and is in the opposite direction. This constant ratio is known as the coefficient of restitution.

17. The total mechanical energy of an object is the sum of its kinetic energy (which is always positive) and the potential energy. Relative to infinity (i.e. if we presume that the potential energy of the object at infinity is zero), the gravitational potential energy of an object is negative. The total energy of a satellite is negative.

18. The commonly encountered expression *mgh* for the potential energy is actually an approximation to the difference in the gravitational potential energy.

19. The external work done on the body is stored in the body in the form of energy is called as strain energy.

20. Energy is the capacity to do work. Energy can be in several forms such as, mechanical energy, electrical energy, chemical energy, thermal energy, sound energy etc.

21. Although the gravitational force between two particles is central, the force between two finite rigid bodies is not necessarily along the line joining their centres of mass.

22. The gravitational force on a particle inside a spherical shell is zero. However, (unlike a metallic shell which shields electrical forces) the shell does not shield other bodies outside it from exerting gravitational forces on a particle inside. *Gravitational shielding is not possible*.

23. Stress is the restoring force per unit area and strain is the fractional change in dimension. In general there are three types of stresses (*a*) tensile stress — longitudinal stress (associated with stretching) or compressive stress (associated with compression), (*b*) shearing stress and (*c*) hydraulic stress.

24. For small deformations, stress is directly proportional to the strain for many materials. This is known as Hooke's law. The constant of proportionality is called modulus of elasticity. Three elastic moduli *viz.,* Young's modulus, shear modulus and bulk modulus are used to describe the elastic behaviour of objects as they respond to deforming forces that act on them.

25. A class of solids called elastomers does not obey Hooke's law.

26. When an object is under tension or compression, the Hooke's law takes the form $F/A = Y\Delta L/L$ where $\Delta L/L$ is the tensile or compressive strain of the object, F is the magnitude of the applied force causing the strain, A is the cross-sectional area over which F is applied (perpendicular to A) and Y is the Young's modulus for the object.

27. A pair of forces when applied parallel to the upper and lower faces, the solid deforms so that the upper face moves sideways with respect to the lower. This type of deformation is called shear deformation and the corresponding stress is called as shearing stress.

28. In general, a deforming force in one direction can produce strains in other directions also. The proportionality between stress and strain in such situations cannot be described by just one elastic constant. For example, for a wire under longitudinal strain, the lateral dimensions (radius of cross section) will undergo a small change, which is described by another elastic constant of the material (called *Poisson's ratio*).

29. When an object undergoes hydraulic compression due to a stress exerted by a surrounding fluid, the Hooke's law takes the form $p = B\,(\Delta V/V)$, where p is the pressure (hydraulic stress) on the object due to the fluid, $\Delta V/V$ (the volume strain) is the absolute fractional change in the object's volume due to that pressure and B is the bulk modulus of the object.

30. The *work-energy theorem* states that the change in kinetic energy of a body is the work done by the net force on the body. $T_f - T_i = W_{net}$.

31. A force is *conservative* if (*i*) work done by it on an object is path independent and depends only on the end points $\{x_i, x_j\}$, or (*ii*) the work done by the force is zero for an arbitrary closed path taken by the object such that it returns to its initial position.

32. The principle of conservation of mechanical energy states that the total mechanical energy of a body remains constant if the only forces that act on the body are conservative.

Statics 4

4.1 CONCURRENT FORCES

4.1.1 Introduction

Forces which concur at a point, or meet at a point, are called concurrent forces. When a body is pushed up an inclined plane, the force used to push it passes through the centroid of the body. The weight of the body also passes through the centroid of the body. Hence the force pushing the body and the weight of the body constitute a pair of concurrent forces. This is illustrated in Fig. 4.1.

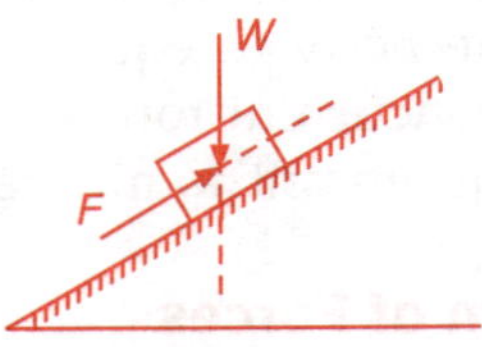

Fig. 4.1 *Concurrent forces*

When a disk is rolling along a horizontal straight path, the friction force acts at the point of contact between the disk and the plane. The weight of the disk also acts vertically down through this point of contact. Hence these two forces are concurrent forces.

In studying the motion of a body, it is important to know the resultant of all the forces acting on the body so as to determine the direction of the ensuing motion and the resulting acceleration. It is also necessary to resolve a given force along a particular direction in order to study the motion along that direction. In this chapter, we will study addition of concurrent forces and resolution of a given force into components along different directions.

Graphical Representation of a Force

A general description of a force requires its (*i*) magnitude, (*ii*) direction of action and (*iii*) the point of application. A force is a vector quantity since it has both magnitude and direction. Two forces F_1 and F_2 are represented in Fig. 4.2. The magnitudes of

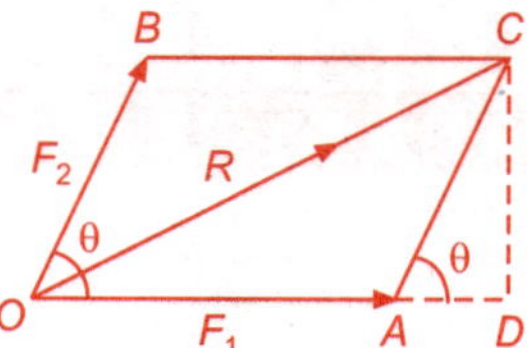

Fig. 4.2 *Resultant using parallelogram of forces*

F_1 and F_2 are represented by the lengths of the lines *OA* and *OB*, respectively. Similarly, the directions of the forces are shown by the arrows at the tips of these lines and the point of application for both the forces is at *O*. Incidentally, forces F_1 and F_2 are concurrent forces because they meet at *O*.

Resultant, Equilibrant and Components

The effect of several concurrent forces acting on a body can also be produced by a single force. The resultant of a system of concurrent forces is that single force that has the same effect on the body as the total effect of the individual forces. The individual forces are referred to as the components of the single resultant force.

A force which is equal and opposite to the resultant of a system of concurrent forces is referred to as the equilibrant. Hence equilibrant can nullify the effect of all the other forces and can keep the body in equilibrium. If a body is in equilibrium under the action of a system of concurrent forces, then any one of the forces among them can be considered as the equilibrant to all the remaining forces.

4.1.2 The Law of Parallelogram of Forces

Resultant of two concurrent forces can be obtained using the law of parallelogram of forces. Figure 4.2 shows two concurrent forces, F_1 and F_2, represented by lines *OA* and *OB* respectively, with an angle θ between them. With *OA* and *OB* as two sides, a parallelogram is constructed. The diagonal from the point *O* where the two forces meet, to the point *C* where the two parallel sides meet, is denoted as *OC*. The law of parallelogram states that:

If two concurrent forces are represented both in magnitude and direction by the two adjacent sides of a parallelogram, their resultant is represented both in magnitude and direction by the diagonal of the parallelogram, drawn from the intersecting point of the two adjacent sides.

Referring to Fig. 4.2, the magnitude and direction of the resultant R, represented by the diagonal of the parallelogram can be obtained as follows:

Magnitude of Resultant: Extend OA upto D such that CD is perpendicular to OD. Considering the right angle triangle ODC, we have

$$OC^2 = OD^2 + CD^2 = (OA + AD)^2 + CD^2$$
$$= OA^2 + 2.OA.AD + AD^2 + CD^2$$
$$= OA^2 + 2.OA.AD + AC^2$$
$$= OA^2 + 2.OA.(AC \cos \theta) + AC^2. \tag{4.1}$$

In the above we have used the relations, $AC^2 = AD^2 + CD^2$, and also $AD = AC \cos \theta$. It can be seen that $\angle CAD = \theta$. Since $F_1 = OA$, $F_2 = OB$ and $R = OC$, we can write Eq. (4.1) as

$$R^2 = F_1^2 + 2F_1F_2 \cos\theta + F_2^2 \tag{4.2}$$

This result is identical to that obtained by using the cosine law for the triangle OAC. In triangle OAC, $OA = F_1$ and $AC = F_2$ and the included angle $\angle OAC = 180° - \theta$. Hence using the cosine law we have

$$R^2 = F_1^2 + F_2^2 - 2F_1F_2 \cos(180-\theta) \tag{4.3}$$

$$= F_1^2 + F_2^2 + 2F_1F_2 \cos\theta$$

Direction of Resultant: The resultant makes an angle α with the force F_1. If we can find this angle, then the direction of the resultant can be defined. From the right angle triangle ODC, we have

$$\tan \alpha = \frac{CD}{OD} = \frac{CD}{(OA + AD)} \tag{4.4}$$

From the right angle triangle ADC, we have

$$CD = AC \sin \theta \tag{4.5}$$

and

$$AD = AC \cos \theta \tag{4.6}$$

Hence, we get

$$\tan \alpha = \frac{F_2 \sin\theta}{F_1 + F_2 \cos\theta} \tag{4.7}$$

Finally, we can write

$$\alpha = \tan^{-1}\left(\frac{F_2 \sin\theta}{F_1 + F_2 \cos\theta}\right) \tag{4.8}$$

or we can write

$$\alpha = \tan^{-1}\left[\frac{F_2 \sin\theta}{F_1 + F_2 \cos\theta}\right]. \tag{4.9}$$

EXAMPLE 4.1

Find the resultant of two concurrent forces P and Q, if the angle between them is (i) 0°, (ii) 180° and (iii) 90°.

SOLUTION

(*i*) $\theta = 0°$

If the angle between two concurrent forces is zero, they act along the same line and in the same direction. Hence their resultant is the arithmetic sum of the two forces and the direction is exactly same as before. This result can also be obtained using the expressions that were developed in connection with two arbitrarily oriented concurrent forces, where the resultant is given in magnitude and direction by Eqs. (4.3) and (4.9).

When $\theta = 0$, we have $\cos\theta = 1$, and $\sin\theta = 0$, and we get from Eq. (4.3)

$$R = (P^2 + Q^2 + 2PQ)^{1/2} = P + Q$$

The direction is obtained from Eq. (4.9) as

$$\alpha = \tan^{-1}\left(\frac{0}{P+Q}\right) = 0$$

Hence the angle made by the resultant with the force P is 0°.

(*ii*) $\theta = 180°$

Let force P be directed to the right and Q be at an angle of 180° with P. Since P and Q are in opposite directions, we have the resultant $R = P - Q$, and $\alpha = 0°$ if

$P > Q$, and $\alpha = 180°$ if $P < Q$. This result can also be obtained from Eqs. (4.3) and (4.9) as follows.

When $\theta = 180°$, we have $\cos 180° = -1$, and $\sin 180° = 0$. From Eq. (4.3)

$$R = \sqrt{P^2 + Q^2 - 2PQ} = \sqrt{(P-Q)^2} = \pm(P-Q).$$

We take the positive root because the magnitude is always positive. Hence we have $R = (P - Q)$. From Eq. (4.9) we have

$$\alpha = \tan^{-1}\left(\frac{0}{P-Q}\right) \text{ or } 180°$$

The direction is 0° if $P > Q$, and 180° if $P < Q$.

(*iii*) $\theta = 90°$

When $\theta = 90°$, we have sin 90° = 1, and cos 90° = 0. Hence, from Eq. (4.3), we have

$$R = \sqrt{P^2 + Q^2}$$

and from Eq. (4.9) we have

$$\alpha = \tan^{-1}\frac{Q}{P}.$$

4.1.3 Law of Triangle of Forces

The resultant of two concurrent forces may also be found by drawing a triangle instead of a parallelogram. Let P and Q be two forces acting at O, represented by lines OA and OB in magnitude and direction. Shift OB so as to bring the base of the force Q to the tip of force P as shown in Fig. 4.3. Then join the base of force P to the tip of force Q to form the triangle OAC. The line OC will give now the resultant of forces P and Q in both magnitude and direction.

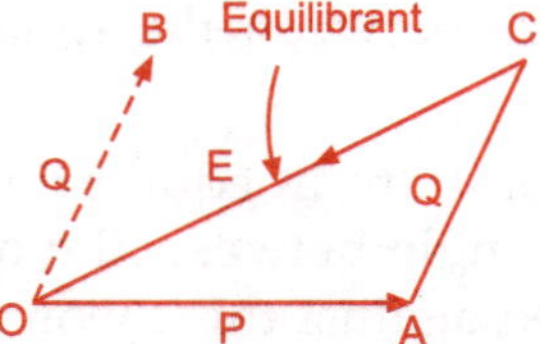

Fig. 4.3 *Law of triangle of forces*

Consider the forces P and Q where the base of force Q is placed at the tip of force P as shown in Fig. 4.3. Now introduce a force E represented by CO, which is equal to the resultant R but opposite in direction. The set of forces P, Q, and E will be in equilibrium now. This phenomenon is stated in the form of a law of triangle of forces as:

If three forces acting at a point are represented in both magnitude and direction by the sides of a triangle, taken in order, then the forces are in equilibrium.

The converse of the above law is stated as:

If three forces acting at a point are in equilibrium, then they can be represented both in magnitude and direction by the three sides of a triangle taken in order, drawn with its sides parallel to the lines of action of the forces.

Consider a weight W being supported by two struts exerting forces P and Q on the weight, as shown in Fig. 4.4. They can be represented by the three sides of a triangle as shown. We have

$$\frac{P}{EF} = \frac{Q}{FG} = \frac{W}{GE} \tag{4.10}$$

Lami's Theorem

If three forces acting at a point are in equilibrium, then we can apply the sine rule to the system of forces as

$$\frac{P}{\sin\alpha} = \frac{Q}{\sin\beta} = \frac{W}{\sin\gamma} \tag{4.11}$$

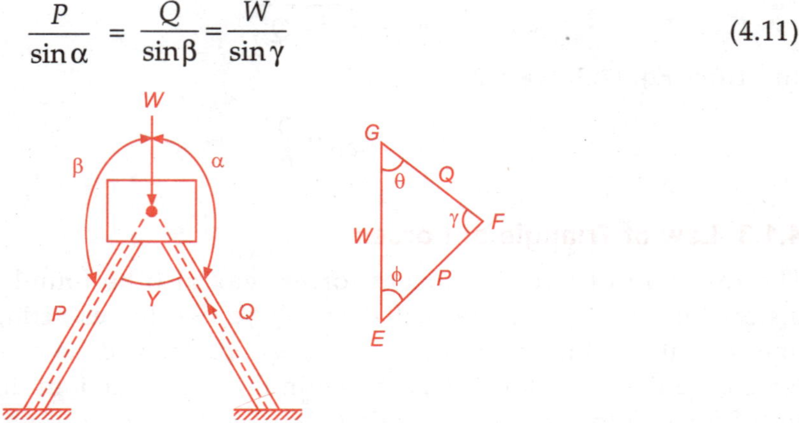

Fig. 4.4 *A weight supported by two bars*

where α, β and γ are the angles between the forces as shown in Fig. 4.4. Lami's theorem states that:

If three forces acting at a point are in equilibrium, then each of them is directly proportional to the sine of the angle between the other two forces. The expression given in Eq. (4.11) can be verified against this theorem by noting that $\theta = 180 - \alpha$, $\phi = 180 - \beta$, and $\psi = 180 - \gamma$.

An extension of the law of triangle of forces leads to the law of polygon of forces. If a number of forces acting at a point can be represented both in magnitude and direction by the sides of a polygon taken in order, then the forces are in equilibrium.

Experimental Verification of Law of Parallelogram of Forces

Consider three masses M_1, M_2 and M_3 which are supported by three strings and pulleys as shown in Fig. 4.5. The masses are in equilibrium due to the action of three forces, tensions T_1, T_2 and T_3 in the three strings. Angles θ_1 and θ_2 made by strings T_1 and T_2, respectively, with the vertical are noted. The forces $T_1 = M_1 g$, $T_2 = M_2 g$ and $T_3 = M_3 g$ are also noted.

The forces T_1, T_2 and T_3 are marked on a sheet of paper fixed behind the strings with T_1 and T_2 inclined to the vertical with angles θ_1 and θ_2, respectively, and T_3 being vertical as shown in Fig. 4.5, choosing an appropriate scale. The parallelogram is completed. The length of the diagonal OD and the angle $<COD$ are measured. Since the three forces are in equilibrium, force T_3 is the equilibrant of forces T_1 and T_2. According to the law of parallelogram of forces $OD = R$ is the resultant of forces T_1 and T_2 and OD must be vertical, i.e.; $<COD = 180°$. This can be verified from the measured results. The experiment may be repeated with different values of M_1, M_2 and M_3.

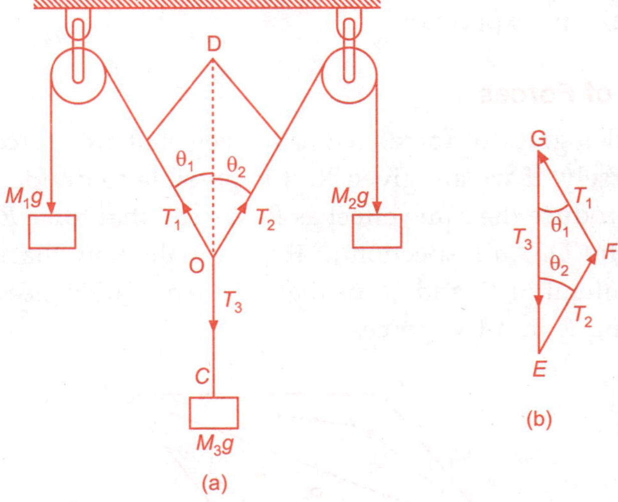

Fig. 4.5 *Verification of law of parallelogram of forces*

Verification of the Law of Triangle of Forces

From the previous experiment, the forces T_1, T_2 and T_3 can be represented as shown in Fig. 4.5(*b*). Sides *EF*, *FG* and *GE* of triangle *EFG* are measured. According to the law of triangle of forces, we should obtain

$$\frac{T_1}{FG} = \frac{T_2}{EF} = \frac{T_3}{GE}.$$

The experiment may be repeated for different masses M_1, M_2 and M_3 and noting the corresponding forces T_1, T_2 and T_3.

Verification of Lami's Theorem :

In the previous experiment the forces T_1, T_2 and T_3 are represented as concurrent forces meeting at a point O as shown in Fig. 4.6. The angles α, β and γ are measured. According to Lami's theorem we must obtain

$$\frac{T_1}{\sin\alpha} = \frac{T_2}{\sin\beta} = \frac{T_3}{\sin\gamma}.$$

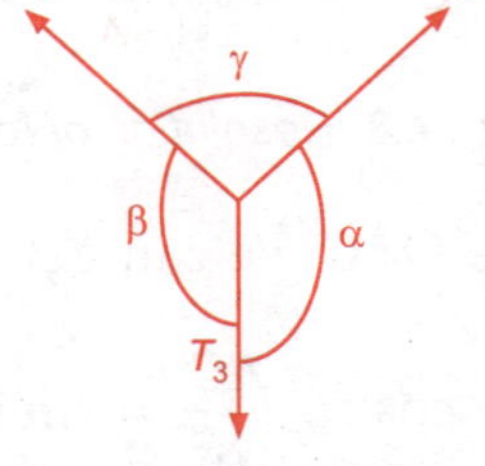

Fig. 4.6 *Representation of forces in equilibrium*

The experiment may be repeated for different values of forces T_1, T_2 and T_3 as obtained in the previous experiment.

4.1.4 Resolution of Forces

In the law of parallelogram of forces we have seen that two forces P and Q have a resultant R. Conversely, if we are given R, it is possible to divide this force into two forces P and Q to produce the same effect as R. We say that force R has been resolved into two forces P and Q. An inspection of Fig. 4.7 will show that the same R can be obtained as the resultant of P_1 and Q_1 or that of P_2 and Q_2. Hence there is more than one way of resolving R into two forces.

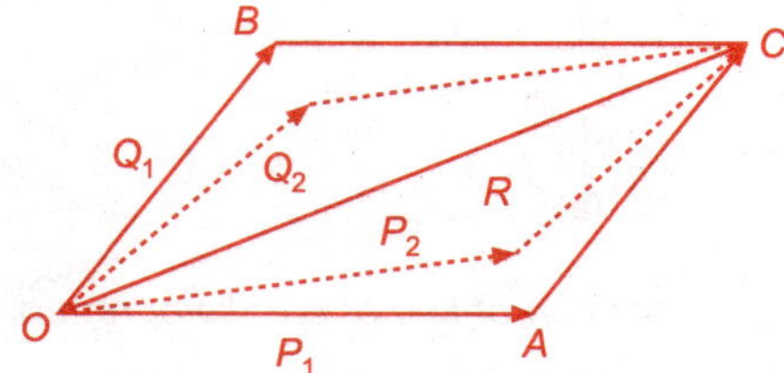

Fig. 4.7 *Resolution of forces*

In fact we can resolve R in many ways and we can get the effect of R in a given direction. The process of finding the effect of a force in a specified direction is called the resolution of force. The effect of a force in a given direction is called the component of the force in that direction. In particular, if the components are considered along any two mutually perpendicular directions, they are called rectangular components.

Referring to Fig. 4.8, the sides OA and OB represent the rectangular components of R.

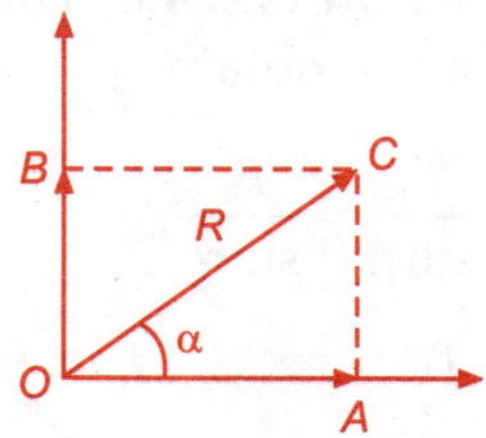

Fig. 4.8 *Resolution of forces*

From the right angle triangle OAC we can get

$$\frac{OA}{OC} = \cos\alpha, \quad \frac{AC}{OC} = \sin\alpha$$

Denoting $OA = P$, $OB = AC = Q$ and $OC = R$, we have $P = R \cos \alpha$ and $Q = R \sin \alpha$. It can be seen that OA is the graphical projection of OC on the line OA. Hence the component of any force in any given direction can be regarded as the graphical projection of that force in that direction.

This idea of graphical projection to obtain the components of a force in a given direction is very convenient. We can verify the law of parallelogram of forces very easily using this concept. From the law of parallelogram of forces we get the resultant of forces P and Q as R. Draw AG as perpendicular to OC at G. Let us resolve the forces P and Q along OC and perpendicular to OC. We have OG as the component of force P on the line OC as shown in Fig. 4.9. Similarly, GC is the component of $AC = OB = Q$ on OC. Component of P perpendicular to OC is given by GA, and component of $AC = CB = Q$ perpendicular to OC is given by AG. Since AG and GA are equal and opposite in direction, they cancel each other. Hence the components of P and Q along the direction OC are given by $OG + GC = OC$ and perpendicular to the direction OC is $AG + GA = AG - AG = 0$. Hence the total effect of forces P and Q is only along the direction OC and is equal to $OC = R$.

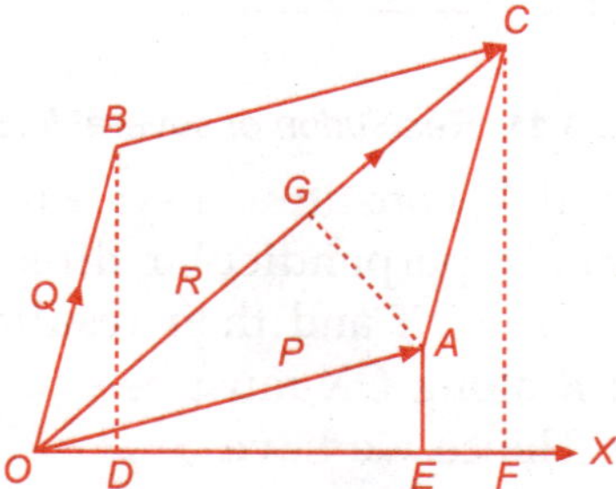

Fig. 4.9 *Resolution of resultant*

Similarly, considering the resolved parts of forces P and Q along the horizontal OX, we have

$$OE + OD = OE + EF = OF.$$

We can see that OF is also the resolution of $R = OC$ on the horizontal line OX. Hence the algebraic sum of the resolved parts of two forces in a particular direction is equal to the resolved part of their resultant in that direction.

4.1.5 Resultant of Concurrent Coplanar Forces

If a number of concurrent forces acting at a point can be represented both in magnitude and direction by the sides of a polygon taken in order, then the forces are in equilibrium. If the lines do not close into a polygon, then the system of forces is not in equilibrium and the line which closes the polygon is the equilibrant. The equilibrant starts at the tip of the last force and joins the root of the first force as shown in Fig. 4.10 by DE. The force which is equal and opposite to the equilibrant, represented by ED, is the resultant R of the system of forces.

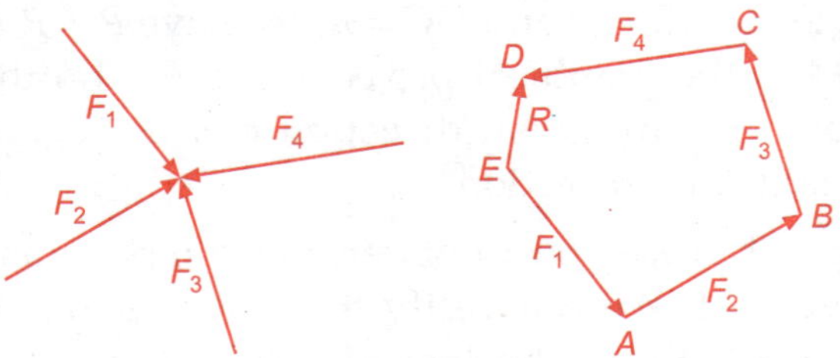

Fig. 4.10 *Polygon of forces*

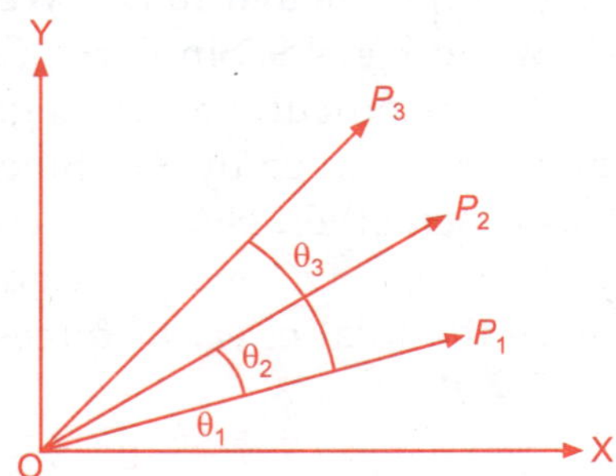

Fig. 4.11 *Resolution of several forces*

In Fig. 4.11, forces P_1, P_2 and P_3 constitute a system of concurrent forces acting at O. OX and OY are two mutually perpendicular directions. If θ_1, θ_2, θ_3,.... are the angles made by P_1, P_2, P_3, ... with OX and their resultant R makes an angle θ with OX, then the resolved part of R along OX must be equal to the resolved parts of the given forces in that direction. Hence we have,

$$R \cos \theta = P_1 \cos \theta_1 + P_2 \cos \theta_2 + P_3 \cos \theta_3 + \ldots$$

$$= \sum P_i \cos \theta_i = X \tag{4.12}$$

where i = 1, 2, 3,... Similarly along OY,

$$R \sin \theta = P_1 \sin \theta_1 + P_2 \sin \theta_2 + P_3 \sin \theta_3 + \ldots$$

$$= \sum P_i \sin \theta_i = Y \tag{4.13}$$

Squaring and adding Eqs. (4.12) and (4.13) we get

$$R^2 \cos^2 \theta + R^2 \sin^2 \theta = X^2 + Y^2 \tag{4.14}$$

But

$$R^2 \cos^2 \theta + R^2 \sin^2 \theta = R^2 (\cos^2 \theta + \sin^2 \theta) = R^2 \tag{4.15}$$

Hence, from Eqs. (4.14) and (4.15) we have

$$R^2 = X^2 + Y^2 \text{ or } R = \sqrt{X^2 + Y^2} \tag{4.16}$$

Dividing Eq. (4.13) by Eq. (4.12) we get the angle θ as follows

$$\tan\theta = \frac{Y}{X}, \text{ or } \theta = \tan^{-1}\left(\frac{Y}{X}\right). \qquad (4.17)$$

We have the magnitude and the direction of the resultant in Eqs. (4.16) and (4.17), respectively.

If a set of concurrent coplanar forces is in equilibrium, then the resultant $R = 0$, or $X^2 + Y^2 = 0$. X^2 and Y^2 are always positive and hence, if $X^2 + Y^2$ must be zero then both X and Y must be zero. Thus the condition for the equilibrium of a set of concurrent coplanar forces can be stated as follows:

If a set of concurrent coplanar forces are to be in equilibrium, then the algebraic sum of their resolved parts in any two mutually perpendicular directions should vanish separately.

The converse of the above statement given below is also true:

If the algebraic sums of the resolved parts of a set of concurrent coplanar forces in any two perpendicular directions vanish separately, then they are in equilibrium.

4.2 PARALLEL FORCES, MOMENTS AND COUPLES

4.2.1 Parallel Forces

Forces acting along parallel lines are called parallel forces.

If two forces are parallel and they have the same direction, as shown in Fig. 4.12 (a), then they are called like parallel forces. Forces acting along parallel lines in opposite directions, as shown in Fig. 4.12 (b) are called unlike parallel forces.

Fig. 4.12 *Parallel forces*

We cannot use the law of parallelogram of forces to find the resultant of two like parallel forces or two unlike parallel forces because they are not concurrent and hence it is not possible to form a parallelogram. In order to find the resultant of parallel forces we have to use the following two principles.

1. Principle of Transmissibility of Forces: The point of application of a force may be considered anywhere on its line of action provided this point is rigidly connected with the body.

Referring to Fig. 4.13, the force P is shown in magnitude and direction by the line OA. The point of application of the force is at O. Line OA is extended upto some arbitrary point B and the line OB is the line of action of the force. Instead of the force P starting at O and ending at A, it can as well be starting at O' and ending at A' provided $O'A' = OA$. For the latter case the point of application of the force P is at O'.

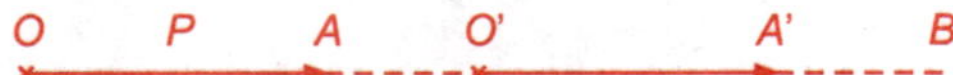

Fig. 4.13 *Transmissibility of force along line of action*

2. Principle of Superposition of Forces: A second system of forces can always be superposed upon or removed from a system of forces without affecting the system, provided the second system itself is in equilibrium.

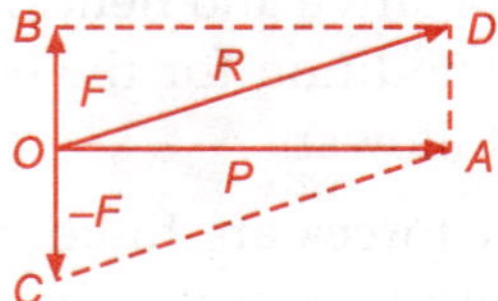

Fig. 4.14 *Superposition of forces*

As shown in Fig. 4.14, originally we have the force P. If we add two forces, F and $-F$, represented by lines OB and OC, where $OB = OC$ in magnitude, the total system formed by P, F, and $-F$ will be equal to the single force P, according to the above principle. In order to verify this, let us add forces P and F to obtain R using the law of parallelogram of forces. The corresponding parallelogram is $OADB$ and the resultant is given by $R = OD$. Now let us add R and $-F$ again using the law of parallelogram of forces. The corresponding parallelogram is $OCAD$. The resultant is given by the diagonal of the parallelogram $OCAD$ which is $OA = P$. Hence the addition of the two equal and opposite forces F and $-F$ did not make any change in the original force P.

Resultant of Two Parallel Forces

Let the lines AB and CD in Fig. 4.15, represent two parallel forces P and Q. A force S acting at A and directed to the left as shown by line AE in the Fig. 4.15, is introduced along with an equal but opposite force acting at C and directed to the right, shown by the line CF. In view of the principle of superposition of forces, introduction of these two forces which are equal and opposite along the same line of action will not change the physical conditions of the body.

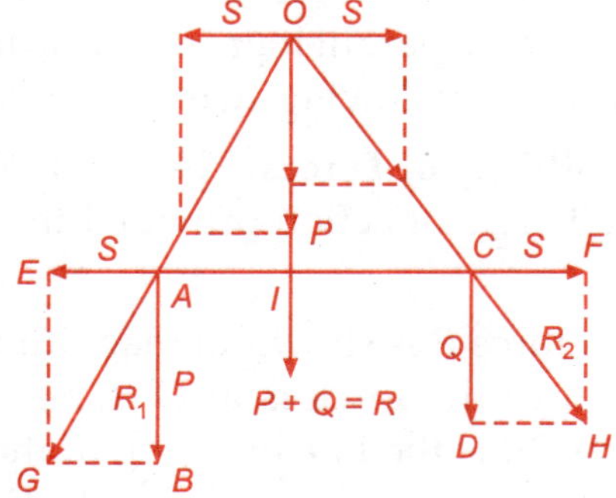

Fig. 4.15 *Resultant of two parallel forces*

The parallelogram $ABGE$ and $CDHF$ are completed. Let R_1 be the resultant of forces P and S. Let R_2 be the resultant of Q and S. Using the principle of transmissibility of forces we can transmit the forces R_1 and R_2 along their respective lines of action to meet at point O. At O we can remove the two equal and opposite forces S from R_1 and R_2, and we are left with only P and Q. Because P and Q act along the same direction, we can add them to obtain

$$P + Q = R \qquad (4.18)$$

which is shown Fig. 4.15.

The point of application can be O itself for both forces. However, if the rigid body does not extend upto O, then we will have difficulty in physically applying the resultant R at O. Using the principle of transmissibility of forces, the resultant may be transmitted to anywhere on its line of application. One possibility is point I which is the point of intersection of lines AC and the resultant R. From similar triangles ABG and OIA, we have

$$\frac{AB}{OI} = \frac{BG}{IA} \qquad (4.19)$$

Since we have $AB = P$ and $BG = AE = S$, we get the relation

$$P.IA = S.OI \qquad (4.20)$$

Similarly, from similar triangles CDH and OIC we get

$$\frac{CD}{OI} = \frac{DH}{IC} \qquad (4.21)$$

Again we have $CD = Q$ and $DH = CF = S$, and hence

$$Q.IC = S.OI \qquad (4.22)$$

Dividing Eq. (4.22) by Eq. (4.20), we obtain the relation

$$P.IA = Q.IC \qquad (4.23)$$

Eq. (4.23) can be expressed as

$$\frac{P}{Q} = \frac{IC}{IA} \qquad (4.24)$$

Hence the point I divides the line AC internally in the inverse ratio of the magnitude of the forces. Thus the following conditions can be formulated as the laws of resultant of two like parallel forces,

   (*i*) The magnitude of the resultant is equal to the sum of the magnitude of the two forces,

   (*ii*) The direction of the resultant is the same as that of the given forces, and

   (*iii*) The point of action of the resultant divides the line joining the points of action of the individual forces internally in the inverse ratio of the forces.

Resultant of Two unlike Parallel Forces

A similar procedure as used in the case of like parallel forces is used here also. Referring to Fig. 4.16, the two unlike parallel forces P and Q are represented by the lines AB and CD. Two forces of magnitude S are introduced at A and C in opposite directions as shown in the given figure.

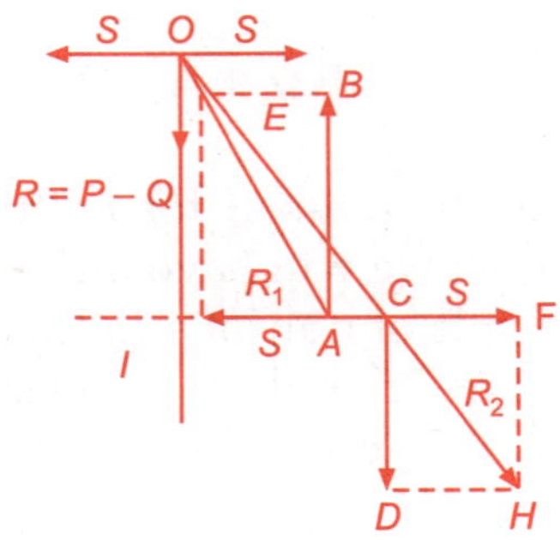

Fig. 4.16 *Resultant of two unlike parallel forces*

Resultant of P and S is given by R_1 and that of Q and S is given by R_2. Both R_1 and R_2 are extended along their respective lines of action to meet at point O. Cancelling the equal and opposite forces S at O, we are left with P and Q at O acting parallel to their original directions. If we assume that the magnitude of P is greater than that of Q then the resultant.

$R = P - Q$ will act in the same direction as that of P.

Again, we can consider the point of action of the resultant R at point I which is the intersection point of the resultant R with the line AC extended. From similar triangles ABE and OIA we have

$$\frac{AB}{OI} = \frac{BE}{IA} \tag{4.25}$$

Since $AB = P$ and $BE = AI = S$, we have from Eq. (4.25)

$$P.IA = S.OI \tag{4.26}$$

Similarly, from similar triangles CDH and OIC we have

$$\frac{CD}{OI} = \frac{DH}{IC} \tag{4.27}$$

Since $CD = Q$ and $DH = CF = S$, we have from Eq. (4.27)

$$Q.IC = S.OI \tag{4.28}$$

Dividing Eq. (4.26) by Eq. (4.28) gives

$$P.IA = Q.IC \tag{4.29}$$

Hence, we get

$$\frac{P}{Q} = \frac{IC}{IA} \tag{4.30}$$

This implies that the point I divides the line AC externally in the inverse ratio of the forces.

The following conditions can be stated as the laws of resultant of unlike parallel forces.

For two unlike parallel forces,

(*i*) The magnitude of the resultant is equal to the difference in magnitudes of the individual forces,

(*ii*) The direction of the resultant is the same as that of the larger of the two forces, i.e., the line of action of the resultant is parallel to that of the two forces, and

(*iii*) The point of action of the resultant divides the line joining the points of action of the two forces externally in the inverse ratio of the forces.

4.2.2 Moments

Just as force acting on a body causes translational motion, moments cause rotational motion of bodies. Consider a pulley with a cord wrapped around it. When the cord is pulled on one side of the pulley, the pulley rotates. The cord is pulled with a force applied on it. If the cord was passing through the centre of the pulley, then a force applied on one end of the cord will not produce any rotational motion of the pulley. Hence, for the force to produce any rotational motion, it must be applied at a distance away from the centre of rotation. The rotational motion indicates the presence of a moment. Hence, we can say that whenever a force is acting away from the centre of rotation there is a moment produced, which causes the rotational motion.

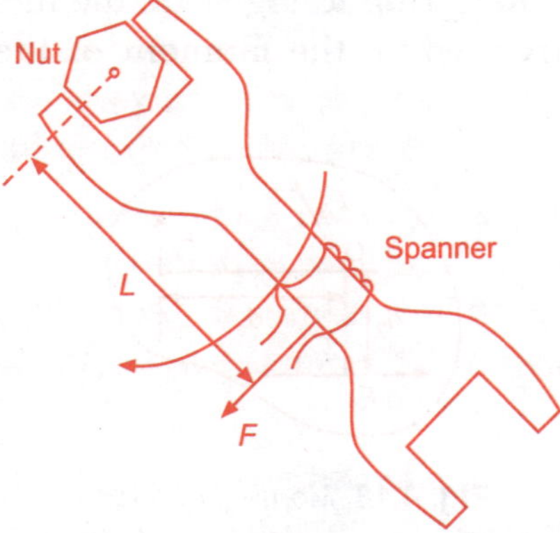

Fig. 4.17 *Tightening a nut using a hand applied moment*

When a nut is being tightened we apply a force on the spanner at a distance away from the centre of the bolt and nut, as shown in Fig. 4.17. The nut rotates about its centre and the force is applied away from the centre. Hence, there is a moment applied on the spanner in order to cause a rotational motion of the nut. It is easier to loosen a tight nut with a long spanner than with a short spanner. That is because when the same force is applied farther away from the centre, a larger moment is applied on the spanner which can cause a rotational motion more easily.

Moment of a Force

The rotational effect or the turning effect of force is measured by its moment or its torque. The rotational effect of a force depends on

    (*i*) The magnitude of the force, and

    (*ii*) The perpendicular distance of the line of action of the force from the axis of rotation.

This is shown in Fig. 4.18, where a force F is applied at a distance d from the centre of rotation. The distance d is also called the moment arm. Hence the moment of a force about a point or axis may be defined as the capacity of the force to turn the body on which it is acting about the point or axis, and it is measured by the product of the force and the perpendicular distance of the line of action of the force from the point or axis. The moment of a force is independent of the point of application provided the point lies on the line of action of the force. If the line of the force passes through the point or axis, then $d = 0$, and hence the moment of the force about that point is zero. Consequently, the force has no capacity to rotate the body about that point or axis.

The rotation produced by a moment may have two senses. The rotation may be in a clockwise sense or in an anticlockwise sense. It is common practice to consider the anticlockwise (or counterclockwise) direction as positive, and hence the moment causing an anticlockwise rotation is said to be positive and a moment is negative if it causes a clockwise rotation. Referring to Fig. 4.18, the moment of force F_1 about O is anticlockwise and is positive, while the moment of the force F_2 is negative, and moment due to F_3 is zero.

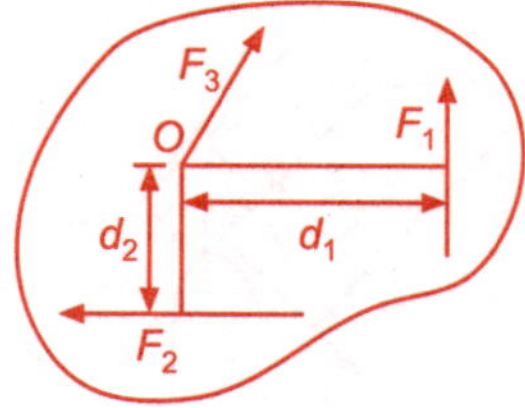

Fig. 4.18 *Moment of a force*

The moment of a force about a point is equal to twice the area of the triangle formed by joining the two ends of the line representing the force to the point of rotation as apex. Referring to Fig. 4.19, the line AB represents the force F and O is the point of rotation. Joining the points A and B to O we get the triangle OAB. The perpendicular distance from the force F to the point of rotation O is d which is the moment arm. Hence the moment of the force about O is given by

$$M_0 = Fd = AB.OD = 2\left(\frac{1}{2}AB.OD\right) \tag{4.31}$$

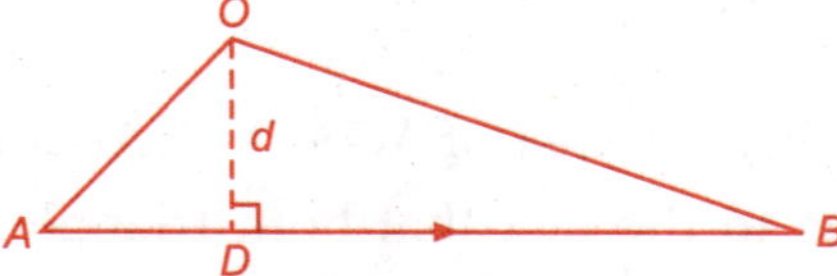

Fig. 4.19 *Moment of a force equal to twice the area of triangle formed by force and centre of rotation*

Clearly, the expression in Eq. (4.31) is equal to twice the area of the triangle OAB.

Theorem of Moments

When several forces act simultaneously on a body, the total turning effect produced is equal to the turning effect due to the resultant of the forces acting on the body. This result is stated as the theorem of moments as follows.

The algebraic sum of the moments of a number of forces about a point is equal to the moment of their resultant about the same point.

This theorem holds good for concurrent forces as well as parallel forces. The theorem is illustrated for the two types of forces below.

(*i*) Moment of Two Concurrent Forces

Two forces P and Q, represented by lines OA and OB, respectively, are acting at a point O as shown in Fig. 4.20. The diagonal of the parallelogram, OC, represents the resultant R.

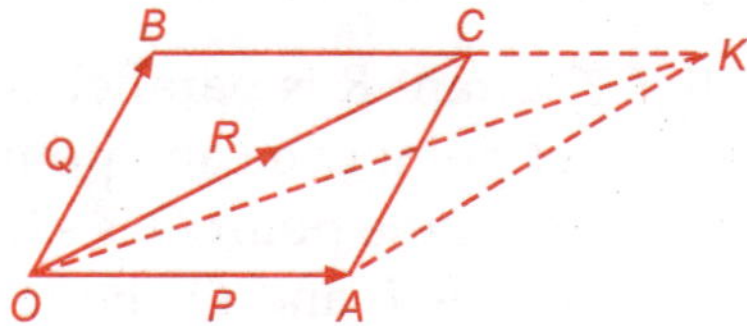

Fig. 4.20 *Moment of concurrent forces equal to that of their resultant*

The side BC of the parallelogram is extended upto the point K. Let M_P and M_Q represent the moments of the forces P and Q about K, respectively. Then using Eq. (4.31) we have

$$M_P = 2\ (\text{Area of } \Delta OAK) \tag{4.32}$$

Similarly,

$$M_Q = 2 \text{ (Area of } \Delta OBK)$$ (4.33)

M_Q is negative because the moment of force Q about K is clockwise. Adding the two moments algebraically we get

$$M_P + M_Q = 2(\text{Area of } \Delta OAK - \text{Area of } \Delta OBK)$$ (4.34)

We have

$$(\text{Area of } \Delta OBK) = (\text{Area of } \Delta OBC + \text{Area of } \Delta OCK)$$

Further we also have

$$(\text{Area of } \Delta OBC) = (\text{Area of } \Delta OAC) = (\text{Area of } \Delta OAK).$$

The above result is due to the fact that both triangles OAC and OAK have the same base and the same altitude. Hence, we have

$$(\text{Area of } \Delta OBK) = (\text{Area of } \Delta OAK + \text{Area of } \Delta OCK)$$

Using the above result in Eq. (4.34) we get

$$M_P + M_Q = 2 \, (- \text{ Area of } \Delta OCK)$$ (4.35)

Clearly the result in Eq. (4.35), is the moment of the resultant R about K. Note that moment of P about K is positive and the moments of Q and R about K are negative.

(*ii*) Moments of Two Parallel Forces

Two like parallel forces P and Q are shown in Fig. 4.21, by the lines AB and CD, such that the line AC is perpendicular to the lines of action of both the forces. Their resultant R is also shown acting at I. From Eq. (4.23), we have

$$P.IA = Q.IC$$ (4.36)

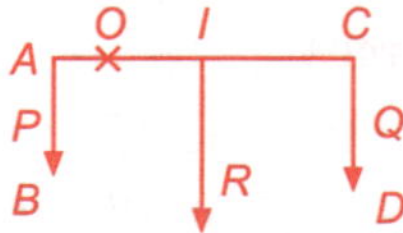

Fig. 4.21 *Moment of parallel forces*

Since the line of action of the resultant R is parallel to those of the forces P and Q, the line AC is perpendicular to R. Let point O be any point on the line AC. We want to show that the moment of P and Q about the point O is same as the moment of R about O. Denoting M_0 as the moment of forces P and Q about the point O, we have

$$M_o = P.AO - Q.CO$$ (4.37)

From Fig. 4.21, we have $AO = AI - OI$ and $CO = CI + OI$. Substituting this into Eq. (4.37) and using Eq. (4.36) we get

$$M_o = P(AI - OI) - Q \, (CI + OI)$$

$$= -(P + Q).OI = R.OI$$ (4.38)

since $P.AI = Q.CI$. The result of Eq. (4.38) is the moment of R about O.

4.2.3 Couples

A system of two equal and unlike parallel forces constitute a couple. Since the resultant of the system of two equal and unlike parallel forces is zero, it does not cause any translational motion. The effect of the couple on a body is to produce a rotational motion only.

We use such couples to do several routine tasks. When we open the screw cap of a bottle, we apply two equal and unlike parallel forces using our thumb and forefinger or thumb on one side and all the other fingers together on the other side. We do the same thing when we turn a door knob, or when we turn on a water tap. The perpendicular distance between the two unlike parallel forces is called the moment arm of the couple.

The turning effect of the couple:

The turning effect of the couple is also known as the moment of the couple or the torque.

Let two forces of magnitude F each, form an equal and unlike parallel force system as shown in Fig. 4.22. Let the force system be represented in such a way that the line AB is perpendicular to both the forces.

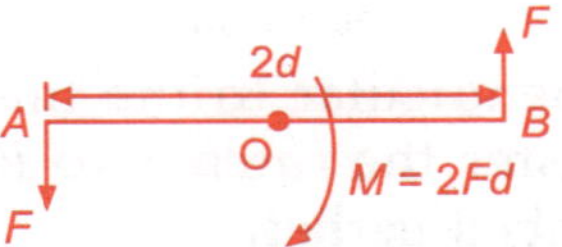

Fig. 4.22 *Couple formed by equal and opposite forces*

Point O is any point on the line AB. The moment of the two forces about the point O is given by

$$F.OA + F.OB = F(OA + OB) = F.AB \qquad (4.39)$$

The result is the moment of the couple. The moment of the couple is a constant quantity, and does not change when the point of action of the forces are changed. Moment of a single force or a system of forces will change depending on the point about which we are taking the moment, however, moment of the couple is the same at any point on the plane containing the two forces.

The units of the couple are the same as that of a moment, i.e., Newton. meters or N.m. A couple can be balanced by another couple of equal moment in the opposite sense, in the same plane. For example, a couple of 10 N.m in the anticlockwise sense can be balanced by another couple in the same plane of magnitude 10 N.m but having clockwise sense.

The effect of a number of couples in a plane is equal to the algebraic sum of the moments of all the individual couples. Hence all such couples may be replaced with a single couple having the magnitude as the algebraic sum of all the couples.

4.2.4 Effect of a General System of Forces

When a number of forces act on a rigid body, the system can be reduced to a single force acting at any point together with a couple. This is irrespective of whether the forces are parallel or nonparallel. This result follows from the fact that a force acting at any point can be replaced at another point by the same force together with a couple of moment equal to the moment of the force in the original location about that point.

The above result can be illustrated as follows. Let the general system of forces have a resultant R acting at O as shown in Fig. 4.23. In this case there is no couple or we can say that the moment of the associated couple is zero. If we want to have the force at another point A, which is not on the line of action of force R, we can transfer the force to point A along with a couple according to the above principle.

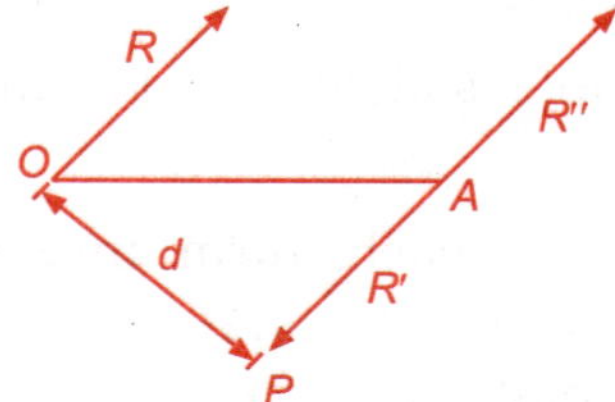

Fig. 4.23 *Couple formed by equal and opposite forces*

Let us draw line OP, perpendicular to the line of action of force R. Let the distance $OP = d$. We can transfer the force R to P according to the Principle of Transmissibility of forces described earlier.

Now, let us introduce two equal and opposite forces R' and R'', at P as shown in Fig. 4.23, with the magnitudes of R' and R'' being equal to that of R. This does not change the equilibrium of the body because they cancel each other anyway. Consider the force R at O and the force R' at P forming a couple of magnitude $M = R.d$. Finally, we are left with the force R'' and the moment of couple M. The force R'' has the same magnitude as that of R and the same direction. Hence it is equivalent to transferring the force R to point P and in the process the couple of moment M must be included.

In general, a system of forces is reduced to a system of concurrent forces at a given point, together with a system of couples whose moments are the same as the moments of the forces about that point. The concurrent forces reduce to a single resultant force R and the system of couples reduces to a single couple. Thus a system of forces can be reduced to a single force at any given point together with a single couple.

4.2.5 Equilibrium of a Rigid Body

For a body to be in equilibrium, it should not have (*i*) translational motion, and (*ii*) rotational motion. We know now that translational motions are caused by forces and rotational motions are caused by moments. Hence, if the forces acting on a body are zero and the moments about any point on the body are zero, then the body will be in equilibrium.

Limiting ourselves to the case of coplanar forces, we can state the conditions of equilibrium as follows. For a body under the action of a system of coplanar forces to be in equilibrium,

1. The algebraic sum of the resolved components of the forces along any two mutually perpendicular directions in the plane of the forces must separately vanish, and

2. The algebraic sum of the moments of the forces about any point in their plane must be zero.

In the case of coplanar parallel forces the conditions of equilibrium can be simplified to the following form.

1. The algebraic sum of the forces must be zero.

2. The algebraic sum of the moments of the forces about any point in their plane must be zero.

SUMMARY

1. A force which is equal and opposite to the resultant of a system of concurrent forces is referred to as the equilibrant.

2. The law of parallelogram states that:

   If two concurrent forces are represented both in magnitude and direction by the two adjacent sides of a parallelogram, their resultant is represented both in magnitude and direction by the diagonal of the parallelogram, drawn from the intersecting point of the two adjacent sides.

3. Law of triangle of forces states that:

   If three forces acting at a point are represented in both magnitude and direction by the sides of a triangle, taken in order, then the forces are in equilibrium.

4. The converse of the above law is stated as:

   If three forces acting at a point are in equilibrium, then they can be represented both in magnitude and direction by the three sides of a triangle taken in order, drawn with its sides parallel to the lines of action of the forces.

5. If a number of concurrent forces acting at a point can be represented both in magnitude and direction by the sides of a polygon taken in order, then the forces are in equilibrium.

6. If a set of concurrent coplanar forces are to be in equilibrium, then the algebraic sum of their resolved parts in any two mutually perpendicular directions should vanish separately.

7. Principle of Transmissibility of Forces: The point of application of a force may be considered anywhere on its line of action provided this point is rigidly connected with the body.

8. Principle of Superposition of Forces:

   A second system of forces can always be superposed upon or removed from a system of forces without affecting the system, provided the second system itself is in equilibrium.

9. Theorem of moments as follows:

   The algebraic sum of the moments of a number of forces about a point is equal to the moment of their resultant about the same point.

10. For a body to be in equilibrium, it should not have (*i*) translational motion, and (*ii*) rotational motion.

Hydrostatics and Surface Tension 5

5.1 HYDROSTATICS

5.1.1 Introduction

In the previous Chapters, we discussed the behaviour of rigid solid bodies under the action of external forces. In this Chapter we will consider the behaviour of fluids under the action of external forces.

Hydrostatics deals with the study of liquids at rest. The study is limited to perfect liquids. A perfect liquid is one which is incompressible and which offers no force of friction. We should realize that all real liquids are compressible to some extent and friction forces influence their behaviour. The effect of friction will be studied in a later Chapter.

5.1.2 Pressure

When a liquid is poured on a flat surface it tries to spread itself in all directions. It cannot do that when it is contained in a vessel. However, the liquid exerts a force on the walls and the bottom of the vessel. The force exerted by a liquid per unit area of the surface is called the liquid pressure. Such a pressure is exerted on the container wall by all fluids, i.e.; liquids and gases, in general. If the liquid exerts a force of ΔF normal to a small area ΔA surrounding a point, then the pressure at that point is given by

$$p = \frac{\Delta F}{\Delta A} \tag{5.1}$$

Pressure is expressed in Pascals in honour of Blaise Pascal, the eminent French mathematician, physicist, and philosopher who is noted for his experiments with barometers that can measure air pressure. The dimension of Pascal is Newton/square meter or N/m^2.

The characteristic features of the pressure exerted by liquids at rest are the following.

1. The pressure exerted by liquids at rest is normal, i.e.; at right angles, to the surface with which the liquid is in contact. Likewise, when a body is immersed in a liquid, its outside surface experiences pressure in a perpendicular direction, i.e.; normal to the surface.

2. The pressure exerted by a liquid at rest at a point within the liquid acts in all directions with equal magnitude.

3. Pressure at a point within a liquid at rest depends on the depth of the point below the free surface of the liquid. Pressure p at a point within a liquid of density ρ, at a depth h below the free surface of the liquid is given by

$$p = \rho g h \qquad (5.2)$$

   This is illustrated in Fig. 5.1. All points on the same level within a liquid at rest experience equal pressures.

4. The free surface of a liquid at rest is horizontal.

5.1.3 Centre of Pressure

When a flat body or a lamina is immersed in a liquid at rest, with the plane of its surface inclined to the free surface of the liquid at an angle, different points on the surface of the body will be at different depths below the liquid surface. From Eq. (5.2), it can be seen that different points of this body will be subjected to different pressures. From the definition of the pressure given in Eq. (5.1), pressure over an area results in a force. The total normal force exerted by a liquid over a surface is called thrust.

Considering the flat body as shown in Fig. 5.1, the plane surface area A of the lamina can be considered to be made up of a large number of small areas. If ΔA is the area of one such small element at a depth of h below the liquid surface, the thrust ΔF on that area is given by

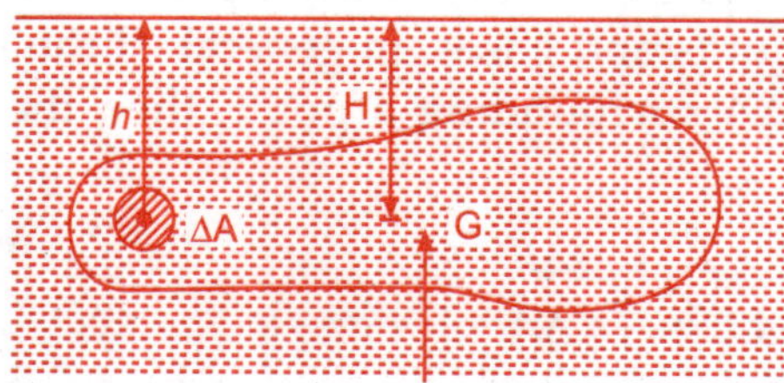

Fig. 5.1 *Centre of pressure*

$$\Delta F = \rho gh.\Delta A \tag{5.3}$$

The total thrust on the surface of the lamina is given by

$$T = \sum \rho gh.\Delta A \tag{5.4}$$

The resultant of the thrust acts through a particular point on the surface such that

$$T = \rho gH.A \tag{5.5}$$

where H is the depth of that point below the liquid surface. This point is called the centre of pressure. Hence we can define the centre of pressure as that point on the surface through which the resultant thrust of a surface immersed in a liquid at rest acts. The resultant thrust on a plane surface immersed in a liquid at rest is given by the product of pressure at the centre of pressure and the area of the surface.

It can be easily visualized that for a plane surface, centre of pressure is the same as the centre of gravity of the surface.

5.1.4 Archimedes' Principle

The Archimedes' principle states that when a body is immersed in a liquid, it appears to lose a part of its weight, and the apparent loss of weight is equal to the weight of the liquid displaced by the body.

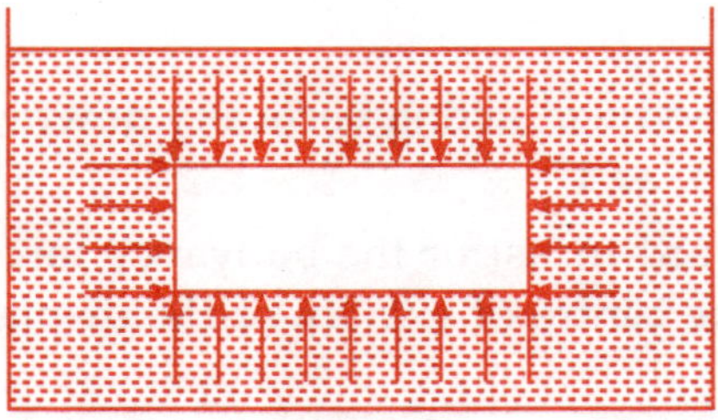

Fig. 5.2 *Principle of Archemedes*

This can be proved very easily using the illustration shown in Fig. 5.2. When the body is immersed in the liquid, the upward thrust on the lower surface of the body is greater than the downward thrust on the upper surface of the body. This is because of the larger depths of the points on the lower surface compared to those of the upper surface points. Consequently, there is a resultant upward thrust on the body. Before the body was immersed, that particular space was occupied by the liquid. That portion of the liquid was in equilibrium because the upward thrust on its lower surface was exactly equal to the weight of the liquid mass. Now, when the body has displaced that portion of the liquid, the upward thrust of the liquid on the lower surface of the body is same as that when the liquid was in that space. Hence, the upward thrust is equal to the weight of the liquid displaced by the body. Therefore, when a body is immersed in a liquid (partially or completely), it experiences an upward thrust whose magnitude is equal to the weight of the liquid displaced by the immersed portion of the body. The thrust depends on the surface area and not upon the material of the body. This leads to the principle of Archimedes.

5.1.5 Buoyancy

As already mentioned above, a body which is immersed partially or completely in a liquid experiences a resultant upward thrust whose magnitude is equal to the weight of the liquid displaced by the immersed portion of the body. The thrust depends on the surface area and not upon the material of the body. This resultant upward thrust acting on a body immersed partially or completely in a liquid is commonly called buoyancy. This buoyancy force acts through the centre of gravity of the displaced liquid and hence the centre of gravity of the displaced liquid is called the centre of buoyancy.

When helium is filled in balloons, the balloons float up in the air. Density of helium is lower than that of air and hence the buoyancy of air makes the balloons float in air.

5.1.6 Flotation

When a body is wholly or partly immersed in a liquid, the body experiences a vertical upward thrust equal to the weight of the liquid displaced by the submerged portion of the body. Hence when a body is placed in a liquid, two forces act on it, namely,

(*i*) The weight of the body, w, acting vertically downwards through the centre of gravity of the body, and

(*ii*) The resultant upward thrust or the buoyancy force acting through the centre of buoyancy, or the centre of gravity of the displaced liquid, B.

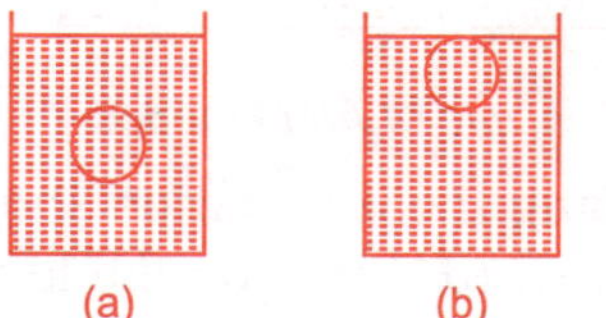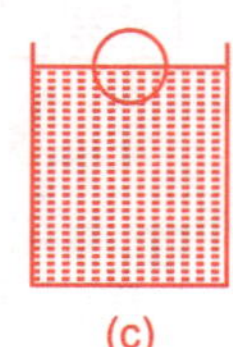

(a)   (b)   (c)

Fig. 5.3 *Flotation of bodies in liquid*

Depending on whether the weight of the body is greater than that of the displaced liquid or not, the body may sink down in the liquid or float on it. This is illustrated in Fig. 5.3. We will consider these cases in more detail in the following.

(*i*) $w > B$: This means that the weight of the body is greater than that of the displaced liquid. In this case the body's weight dominates and makes the body sink down in the liquid.

(*ii*) $w = B$: This means that the weight of the body and that of the displaced liquid are the same. The body will be in the limiting case of sinking and floating.

(*iii*) $w < B$: This means that the weight of the body is less than that of the displaced liquid. The body will float on the surface of the liquid, with a portion of the body above the liquid surface. In this case the weight of the body is equal to the weight of the liquid displaced by the submerged portion of the body. Only a portion of the body has to be under the liquid, because if the body were completely submerged then more liquid would be displaced and the corresponding upward thrust would be more than the weight of the body. Since the body floats freely, and is in equilibrium in that state, the weight of the body and the buoyancy must be equal in magnitude and act along the same vertical line.

Conditions for Flotation of a Body

A body will float freely in a liquid under the following conditions:

(*i*) The weight of the body is equal to the weight of the liquid displaced by the submerged portion of the body.

(*ii*) The centre of gravity of the body and the centre of buoyancy must lie on the same vertical line.

A substance which has a much higher density than that of the liquid may be made to float on the liquid by giving it a vessel like shape. That is how steel vessels or steel ships can float on water even though the density of steel is much larger than that of water. In contrast, a block of ice or a piece of cork will float on water because their densities are lower than that of water.

Stability of Floating Body

A body may float in a liquid under (*i*) stable, (*ii*) unstable or (*iii*) neutral equilibrium.

(*i*) *Stable Equilibrium:* When a slight tilting of the floating body changes the position of the centre of buoyancy, the amount of displaced liquid remaining the same, the lines of action of the weight of the body and the buoyancy will not be identical. They will become equal and unlike parallel forces producing a couple. If this couple tends to restore the body to its undisturbed position, then the equilibrium is said to be stable. This situation is shown in Fig. 5.4(*a*).

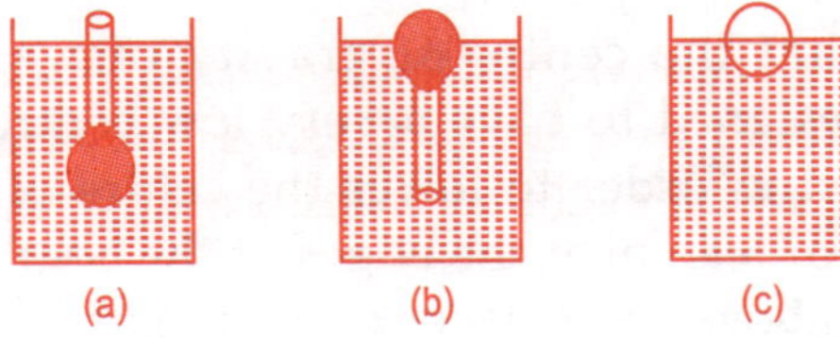

Fig. 5.4 *Stable, unstable and neutral equilibriums*

(*ii*) *Unstable Equilibrium:* When the body is disturbed from its initial position, if the couple formed by the weight of the body and the buoyancy tends to displace the body still further away from the original position, the equilibrium is said to be unstable. This situation is shown in Fig. 5.4(*b*).

(*iii*) Neutral Equilibrium: When a slight tilting of the body does not produce any change of position of the centre of buoyancy, as in the case of a sphere, the body is said to be in neutral equilibrium. This is shown in Fig. 5.4(*c*).

5.1.7 Metacentre

The vertical line passing through the centre of gravity *G* of a floating body and the centre of buoyancy *B* is called the median line or centre line of the body. When the body is slightly disturbed if the centre of buoyancy changes, the new vertical line passing through the centre of buoyancy B in the displaced position meets the central line at a point *M* as shown in Fig. 5.5. This point is called the metacentre of the body. Hence the metacentre of a floating body may be defined as the point where the new vertical line passing through the centre of buoyancy in the displaced position meets the centre line of the floating body. The distance of the metacentre from the centre of gravity of the body is called the metacentric height. The metacentric height is shown as the line *GM* in Fig. 5.5.

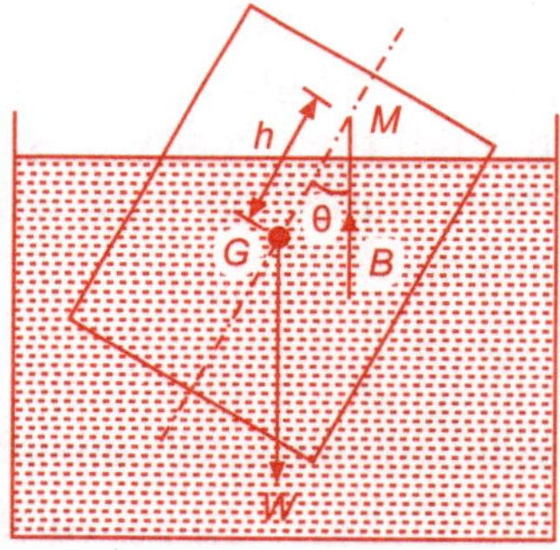

Fig. 5.5 *Metacentric Height*

The equilibrium of the body may be described in terms of the position of the metacentre with respect to the centre of gravity. The equilibrium of the body is stable if *G* is below *M*. The body is unstable if *G* is above *M*. The equilibrium is neutral if *G* and *M* coincide.

In order to ensure that the centre of gravity of a ship is always below its metacentre, the ship is designed to have a very low centre of gravity. Moreover, a ship often carries a ballast in order to lower the centre of gravity. Passengers in a boat should not stand up or lean over the edges of the boat, because if they do so the centre of gravity may go above its metacentre making the boat unstable.

Determination of the Metacentric Height in a Ship

The weight *w* of the ship may be found by the displacement method. When the ship is floated in water, the weight of the displaced water will give the weight of the ship.

A known weight w of several tons is moved across the deck of the ship, from D_1 to D_2, as shown in Fig. 5.6. Let D_1D_2 be equal to x. This causes the ship to tilt through a small angle θ, which may be measured by a plumb line. The metacentric height, h, can be obtained from the relation

$$h = \frac{wx}{w\tan\theta} \approx \frac{wx}{w\theta}$$

(5.6)

The above simplified relation is true for small θ, since for small angles we have $\tan\theta \approx \theta$ in radians.

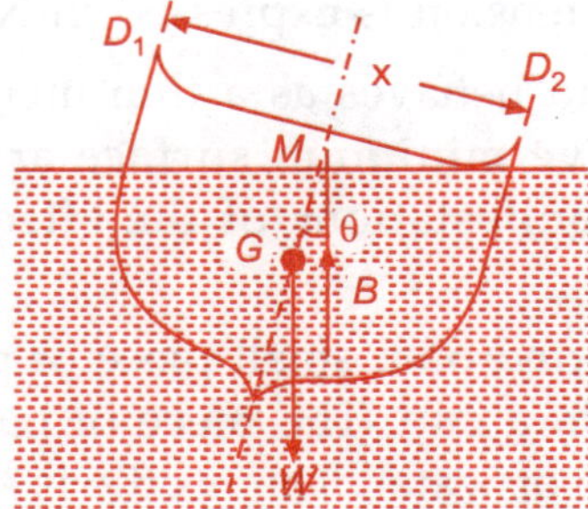

Fig. 5.6 *Measuring metacentric height in a ship*

The shifting of the weight w is equivalent to an upward force w at D_1 and an equal downward force at D_2. These forces constitute a couple of moment equal to $C_1 = wx \cos\theta$. The weight w of the ship at G and the upward thrust w at the new centre of buoyancy B', constitute the restoring couple of moment, $C_2 = wh \sin\theta$. For equilibrium, the two couples must balance each other, i.e.,

$$wh \sin\theta = wx \cos\theta$$

(5.7)

Hence, the metacentric height is given as

$$h = \frac{wx}{wh\tan\theta} \approx \frac{wx}{w\theta} \text{ for small } \theta.$$

(5.8)

5.2 SURFACE TENSION

5.2.1 Introduction

The free top surface of a liquid is generally under a state of tension. The force of tension or stretching force that exists over the surface of a liquid or a thin liquid film is commonly known as surface tension. Surface tension is measured by the tangential force per unit length acting normally on either side of a line drawn on the surface of the liquid, tending to pull it apart. If the surface is flat, the force across the line in one direction is just equal in magnitude to that in the opposite direction as shown in Fig. 5.7.

The dimensions of surface tension can be obtained by using the dimensional formula

$$[T] = [\text{force}]/[\text{distance}] = [ML^0T^{-2}]$$

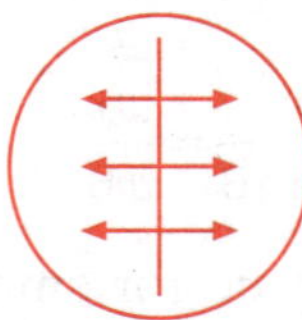

Fig. 5.7 *Force across a line on the liquid surface*

Hence, the unit of surface tension is expressed in Newton/metre or in N/m.

The free surface of a liquid behaves as a thin membrane. Liquid surface has a tendency to contract and have minimum surface area, as in a stretched rubber membrane. The following manifestations of the effect of surface tension may be easily observed.

(*i*) A small needle supported on a small piece of paper is carefully placed over the surface of water in a cup. The piece of paper sinks leaving the needle floating on the surface of water. The needle is supported on the surface with a slight depression as shown in Fig. 5.8. The needle floats in spite of the fact that the density of the material of the needle is larger than that of water.

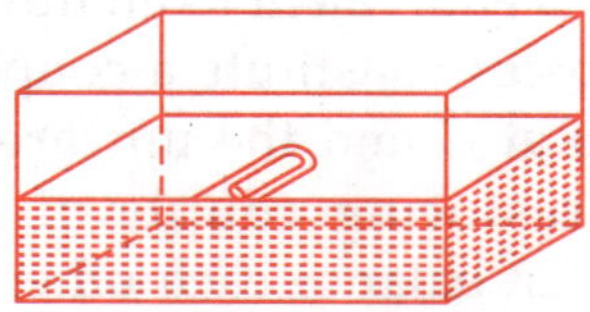

Fig. 5.8 *Force across a line on the liquid surface*

(*ii*) A thread with a small loop is tied across a wire frame as shown in Fig. 5.9 (*a*). The frame is dipped in a soap solution and taken out. The frame is completely covered with the soap film and the loop may have any arbitrary shape. However, if the portion of the soap film within the loop is punctured with a pin, the loop immediately becomes circular as shown in Fig. 5.9 (*b*). For a given perimeter, the circle has the minimum area and hence the area of the film is minimum.

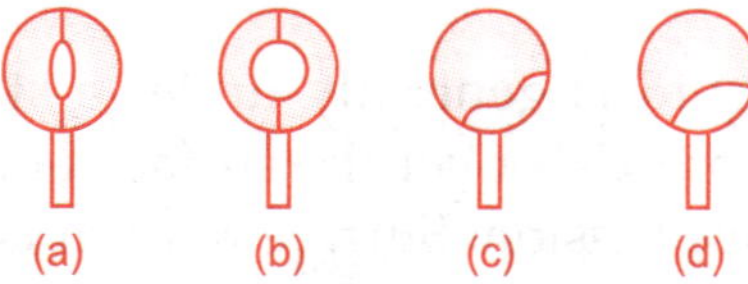

Fig. 5.9 *A thread loop acquiring a circular shape due to soap film tension*

(*iii*) As in (*ii*) above, a thread is tied slightly loosely to the frame and a thin soap film is formed by dipping the frame in the soap solution. When the frame is taken out, the thread may stay in any arbitrary position as in Fig. 5.9 (*c*), however, on piercing the soap film on one side of the thread, the thread will be drawn into a curved shape as shown in Fig. 5.9 (*d*). The film on the other side is thus contracted to occupy minimum surface area.

(*iv*) If a paint brush with camel hair is dipped into water, the hairs of the brush are seen to be spread out, but when the brush is taken out the hairs cling together because the water surface between the hairs will tend to shrink as shown in Fig. 5.10.

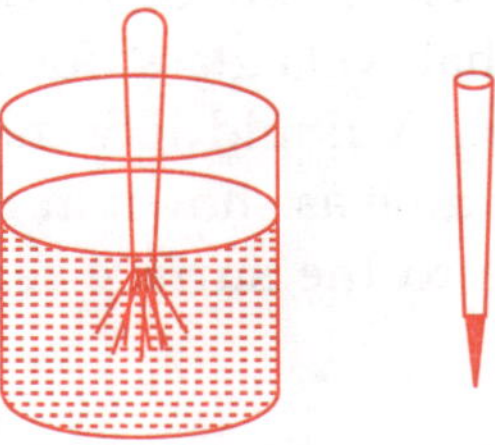

Fig. 5.10 *Clinging hairs of paint brush due to surface tension*

The surface tension is also influenced by the forces of cohesion and adhesion. Attraction between the molecules of the same kind is called cohesion, whereas that between molecules of different kinds is called adhesion. These forces have very short ranges.

When mercury is kept on a glass surface, it forms small spherical drops or larger flattened drops. This is because the cohesion between mercury molecules is greater than the adhesion between mercury and glass molecules. Sometimes the forces of adhesion between two different kinds of molecules may be greater than the cohesion among the same kind of molecules. This is the case between water and glass molecules. The adhesion of water to glass is stronger than the cohesion between water molecules. Hence if water is put on a glass surface, it wets the glass and spreads into a thin film.

5.2.2 Angle of Contact between Liquid and Solid

Surface of a liquid in contact with a solid will be curved at the point of contact. The angle made by the tangent to the liquid surface at the point of contact with the solid surface inside the liquid is called the angle of contact between the liquid and the solid. This is shown in Fig. 5.11.

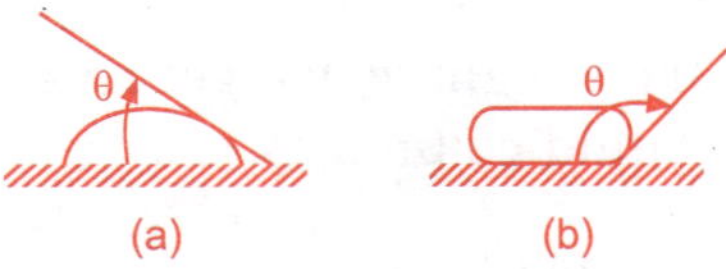

Fig. 5.11 *Angle of contact*

A meniscus is the convex or concave upper surface of a column of liquid caused by surface tension. A concave meniscus makes an acute angle of contact whereas a convex meniscus makes an obtuse angle of contact. Mercury makes an obtuse angle of contact as shown in Fig. 5.11 (*b*). In general, liquids which wet the surface have angles of contact less than 90°. The angle of contact of water with glass is almost zero, whereas that between mercury and glass is about 135°.

5.2.3 Capillary Effect

A capillary tube is a tube of very small bore or inner diameter. When a capillary tube of glass is dipped in any liquid that wets glass, for example water, the liquid rises in the tube as shown in Fig. 5.12(*a*). A liquid like mercury which does not wet glass produces a depression inside the tube as shown in Fig. 5.12(*b*). Such a rise or fall of a liquid in a capillary tube is due to the surface tension property and is known as capillarity or capillary effect.

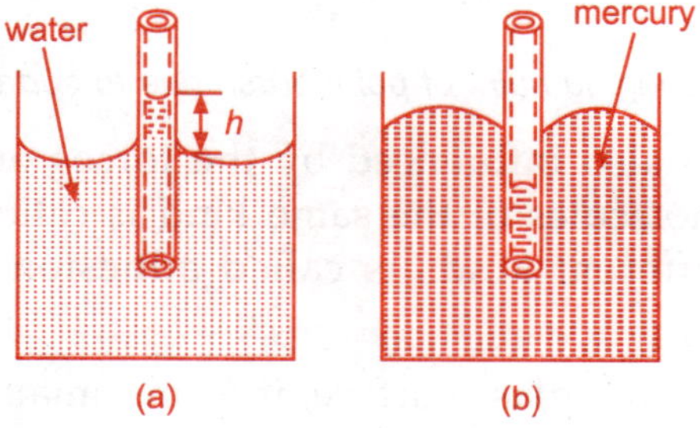

Fig. 5.12 *Capillary effect*

When a capillary glass tube is dipped vertically in water, water rises to a certain height as shown in Fig. 5.12 (*a*). This is because of the surface tension which produces an upward force in the tube. If the water column rises to a height h above the water level outside, then the weight of the water column in the tube is equal to the upward force due to surface tension.

If surface tension is T, ρ is the density of the liquid, and r is the radius of the capillary tube, then the upward force in the tube due to surface tension is given by

$$F = 2\pi r T \cos \theta \tag{5.9}$$

In the case of water we have $\rho = 1$ and $\theta = 0$. Further, the weight of the liquid column of length h is

$$w = \pi r^2 h \rho g \tag{5.10}$$

Equating the force F and the weight w, we get an expression relating the surface tension and the height of the liquid column as

$$T = \frac{\rho g r h}{2} \tag{5.11}$$

Some Attributes and Manifestations of Surface Tension Effect:

1. A liquid in small quantities tries to take a spherical shape. This is because any system to be in equilibrium must have minimum potential energy, and the potential energy of a spherical liquid drop due to surface tension will be minimum because of the smallest surface area for a spherical shape. Rain drops and mercury globules try to take spherical shapes unless flattened due to gravity.

2. In a wick lamp, oil rises in the narrow interstices between the fibers of cotton with which the wick is made.

3. In plants, sap rises through the fibers, which may be considered to be capillary tubes, through capillary action.

4. Normally the ground water table (level of water underground) is below the ground level. However, the vegetation on the ground gets the required water through capillary action in the wet subsoil. Water is drawn from the wet subsoil onto the top layers through the fine pores of soil. Water cannot rise in sandy soil through capillary action because of the large gaps between sand particles.

5. Lead shots are made by allowing molten lead to fall into a pool of water when the molten lead separate and assume the spherical shape.

6. Sharp edges of glass rods and tubes are rounded by heating until they soften, when surface tension pulls the edges into a spherical shape.

7. The spreading quality of paint or solder is due to their surface tension.

An umbrella keeps out the rain water from passing through it due to the surface tension of water that exists in contact with the umbrella fabric. In tents sometimes water may accumulate and may sag down, but still due to the surface tension effect water is prevented from leaking. However, touching the inside of the sagging portion breaks the surface film of water and water then begins to drip through.

5.2.4 Determination of Surface Tension of Water

Capillary rise method is used to measure the surface tension of water. A clean, uniform capillary tube of glass is held vertically in a stand. A bent pin or paper clip is fixed on the tube as shown in Fig. 5.13.

A beaker with water is kept on a table of adjustable height and is placed under the capillary tube such that the capillary tube dips in water. The table height is adjusted such that the bent pin just touches the water surface in the beaker. A travelling microscope is focused on the water meniscus in the capillary such that the horizontal cross wire is tangential to the water meniscus.

The reading of the microscope is noted. Now, the beaker of water is removed without disturbing the capillary tube. The microscope is focused on the bent pin such that the horizontal cross wire of the microscope just touches the tip of the pin. The reading is again noted. The difference between the two readings gives the capillary rise, h. The experiment is repeated several times and the mean value of the capillary rise, h, is determined.

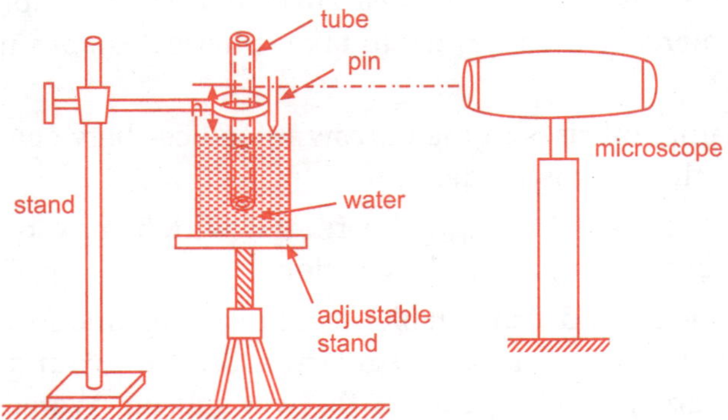

Fig. 5.13 *Measurement of surface tension*

Mean internal radius r of the capillary tube is obtained using the travelling microscope and measuring the radius at both ends of the tube. The surface tension T may now be calculated using the expression given in Eq. (5.11), i.e.; $T = \rho grh/2$.

5.2.5 Factors Affecting Surface Tension

Impurities present in a liquid appreciably affect surface tension. A highly soluble substance such as salt increases the surface tension whereas sparingly soluble substances such as soap decrease the surface tension. The surface tension decreases with rise in temperature. The temperature at which the surface tension of a liquid becomes zero is called critical temperature of the liquid.

5.2.6 Applications of Surface Tension

(*i*) During stormy weather, oil is poured into the sea around the ship. As the surface tension of oil is less than that of water, it spreads on the water surface.

(*ii*) Due to the decrease in surface tension, the velocity of the waves decreases. This reduces the violent lashing of the waves on the ship.

(*iii*) Lubricating oils spread easily to all parts because of their low surface tension.

(*iv*) Dirty clothes cannot be washed with water unless some detergent is added to water. When detergent is added to water, one end of the hairpin shaped molecules of the detergent get attracted to water and the other end, to molecules of the dirt. Thus the dirt is suspended surrounded by detergent molecules and this can be easily removed. This detergent action is due to the reduction of surface tension of water when soap or detergent is added to water.

(*v*) Cotton dresses are preferred in summer because cotton dresses have fine pores which act as capillaries for the sweat.

SUMMARY

1. The basic property of a fluid is that it can flow. The fluid does not have any resistance to change of its shape. Thus, the shape of a fluid is governed by the shape of its container.

2. A liquid is incompressible and has a free surface of its own. A gas is compressible and it expands to occupy all the space available to it.

3. If F is the normal force exerted by a fluid on an area A then the average pressure

$$p = \frac{\Delta F}{\Delta A}$$

4. The pressure exerted by a liquid at rest at a point within the liquid acts in all directions with equal magnitude.

5. The Archimedes' principle states that when a body is immersed in a liquid, it appears to lose a part of its weight, and the apparent loss of weight is equal to the weight of the liquid displaced by the body.

6. Conditions for Flotation of a Body:
    (*i*) The weight of the body is equal to the weight of the liquid displaced by the submerged portion of the body.
    (*ii*) The centre of gravity of the body and the centre of buoyancy must lie on the same vertical line.

7. Surface tension is a force per unit length (or surface energy per unit area) acting in the plane of interface between the liquid and the bounding surface. It is the extra energy that the molecules at the interface have as compared to the interior.

8. Cohesive force is the force of attraction between the molecules of the same substance. This cohesive force is very strong in solids, weak in liquids and extremely weak in gases. Adhesive force is the force of attraction between the moelcules of two different substances.

9. Sphere of influence is a sphere drawn around a particular molecule as centre and molecular range as radius. The central molecule exerts a force of attraction on all the molecules lying within the sphere of influence.

Hydrodynamics and Viscosity 

6.1 HYDRODYNAMICS

6.1.1 Introduction

Earlier we dealt with liquids at rest in Hydrostatics. Hydrodynamics deals with liquids in motion. The study of hydrodynamics uses Bernoulli's theorem which will be discussed in the next section. The motion of a fluid may broadly be classified into two categories; (*i*) steady flow or streamline flow, and (*ii*) turbulent flow.

Streamline Flow

When a fluid is flowing through a tube, if the velocity at a point has constant magnitude and direction, the flow is said to be a streamline flow. Consider the fluid flowing through a pipe as shown in Fig. 6.1.

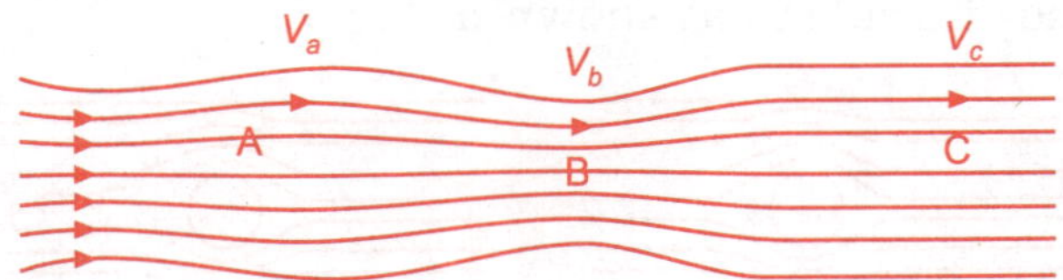

Fig. 6.1 *Streamline flow*

Arbitrary points *A*, *B*, and *C* are indicated inside the pipe. Let the velocities at these points be given by V_a, V_b, and V_c, respectively. If these velocities remain constant at any time, in both magnitude and direction, then the flow is said to be steady or streamline flow. The direction of the flow at any given point is along the streamline at that point. A tubular region consisting of streamlines is called a tube of flow or stream tube as shown in Fig. 6.2.

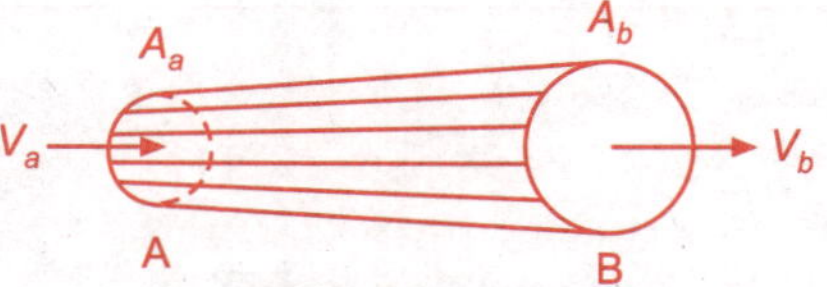

Fig. 6.2 *Stream tube*

Let the cross sectional areas be A_a and A_b, respectively, at points A and B in the stream tube, and the corresponding velocities be V_a and V_b, respectively. The quantity of flow at any point is given by the product of the velocity and the cross sectional area at that point. The velocity is the same at all points on a cross section. Since the same volume of fluid crosses each section of the tube in unit time, we have

$$A_a V_a = A_b V_b \tag{6.1}$$

It can be easily seen that the velocity of flow reduces in a pipe when the area of cross section of the pipe increases.

When the velocity of flow is quite small in a pipe, the flow is generally streamlined.

Turbulent Flow

When the velocity of flow in a pipe carrying a fluid is above a certain limit, then the flow loses its orderly streamline nature, and becomes turbulent. In turbulent flow, the paths of the particles change continuously and are irregular. Portions of the liquid acquire a rotary motion forming what are called eddies or vortices.

Turbulent flow can be formed in a pipe having a streamline flow by introducing a solid object obstructing the path. Because of the solid object, the areas available for the fluid to flow are reduced and as per Eq. (6.1) the velocity increases. If the increased velocity is still below a certain value, the flow may remain streamline. However, if the velocity increases above a certain threshold value or a critical value, then the flow becomes turbulent, as shown in Fig. 6.3.

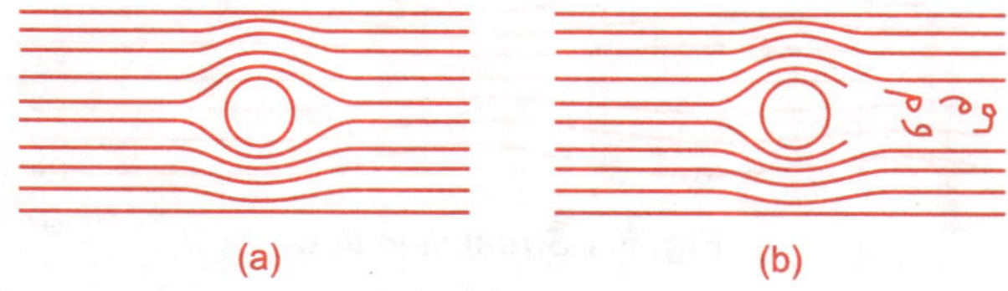

(a)   (b)

Fig. 6.3 *Streamline and turbulent flow*

Vortices are continuously shed from the downstream side of the object one after another and travel down the flow. Normally, vortices are shed from both upside and down side of the downstream portion of the solid object, and form what is called as Karman Vortex Street.

6.1.2 Energy of a Liquid in Motion

A liquid in motion has three kinds of energy, (*i*) potential energy by virtue of its position, (*ii*) pressure energy by virtue of the pressure to which it is subjected, and (*iii*) kinetic energy by virtue of its motion. We have seen in the study of motion of solid bodies in the earlier Chapters that a solid body possesses potential energy and kinetic energy. The fluid is also subjected to pressure which gives the fluid more energy. Consider the water filled in a container as in Fig. 6.4.

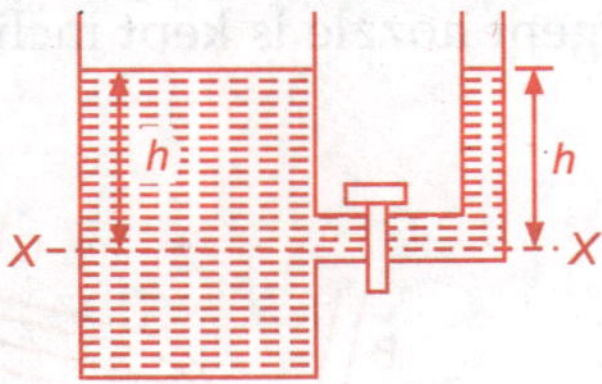

Fig. 6.4 *Pressure energy and potential energy*

A unit volume of water at the top surface has a potential energy of ρgh with respect to the level XX indicated in the container. Consider a unit volume of water at the XX level. Its potential energy is zero, however, it is subjected to the weight of water column above it and is subjected to a pressure of $p = \rho gh$. Let us open the valve on the tube at the side of the container with one end of the tube connected to the container at the level XX and open at the other end. The unit volume of water from the container at the level XX will rush through the tube and rise through a height h. When it has reached the top of the tube, it will possess a potential energy of gh. Hence the pressure energy of p at level XX has been converted into potential energy at the top. This shows that pressure is a form of energy and all three forms of energy are mutually convertible.

Bernoulli's Theorem

In an ideal fluid in steady flow, the total energy given by the sum of pressure energy, potential energy and kinetic energy per unit mass remains constant at any point along a streamline.

This theorem is due to the Swiss physicist, Daniel Bernoulli, which he formulated in 1738. When expressed per unit volume this theorem gives

$$p + \rho gh + \frac{\rho v^2}{2} = C_1 \tag{6.2}$$

where C_1 is a constant. In order to express this relation for unit mass we have to divide the Eq. (6.2) by ρ which gives

$$\frac{p}{\rho} + gh + \frac{1}{2}v^2 = C_2 \tag{6.3}$$

where $C_2 = C_1/\rho$, is another constant. When the liquid is flowing horizontally, h is constant and hence in Eq. (6.3) we can take the term gh to the right hand side of the equation and put as $C_3 = (C_2 + gh)$, which is another constant. Multiplying the resulting equation by ρ we can write for a horizontal flow

$$p + \frac{1}{2}v^2 = C_4 \tag{6.4}$$

where $C_4 = C_3/\rho$, is a constant. Thus in the case of horizontal flow of a liquid, the pressure and the velocity can increase only at the expense of one another. Points of maximum pressure correspond to those of minimum velocity and vice versa.

Experimental Verification of Bernoulli's Theorem

A pipe with a convergent divergent nozzle is kept inclined to the horizontal as shown in Fig. 6.5.

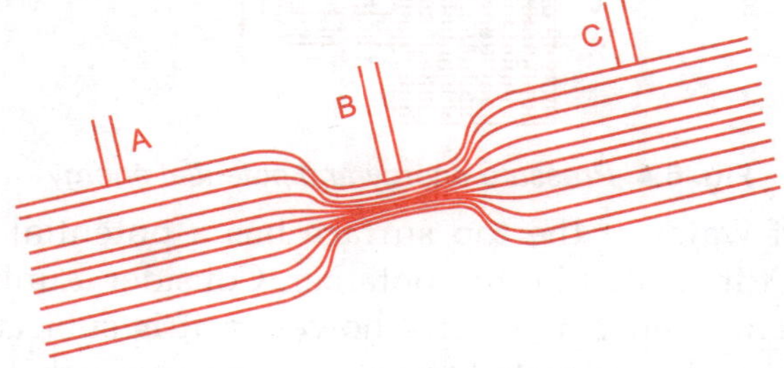

Fig. 6.5 *Verification of Bernoulli's theorem*

Water is allowed to flow through the pipe and the pressures at the points A, B and C can be measured using pitot tubes, as shown. The velocities at those points also can be measured. The height of points B and C with respect to point A can be measured. It can be verified experimentally that the total energy at points A, B or C is exactly equal, as given in Eq. (6.2).

6.1.3 Some Applications based on Bernoulli's Theorem

1. **Water Flow Meter or Venturimeter:** Quantity of flow or the rate of flow of a liquid in a pipe are measured using venturimeters.

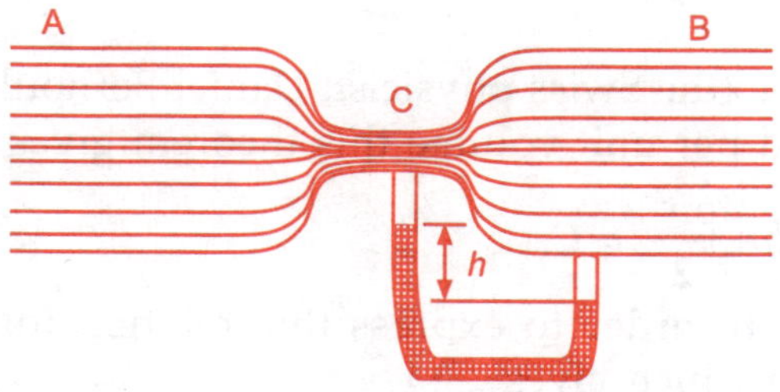

Fig. 6.6 *Venturimeter or flow meter*

A venturimeter is shown in Fig. 6.6, which has a horizontal tube with a constriction or throat. When the liquid flows through the tube which has a varying cross section, the velocity and pressure vary along the tube. A U tube manometer with vertical limbs with one limb at the constriction in the pipe and the other at the wider part of the pipe helps to measure the pressure difference between these points. If h is the difference in the levels of liquid in the two limbs of the U tube, then the quantity of liquid flowing per second, Q, is given by

$$Q = A_1 A_2 \left\{ \frac{2\rho g h}{A_1^2 - A_2^2} \right\}^{1/2} \tag{6.5}$$

where A_1 and A_2 are the cross sectional areas of the pipe and the throat, respectively, and ρ is the density of the liquid.

2. **Carburettor in Internal Combustion Engines:** A carburettor is used in automobile internal combustion engines to provide a proper mixture of petrol and air into the combustion chamber in the engine. A schematic sketch of a carburettor is shown in Fig. 6.7.

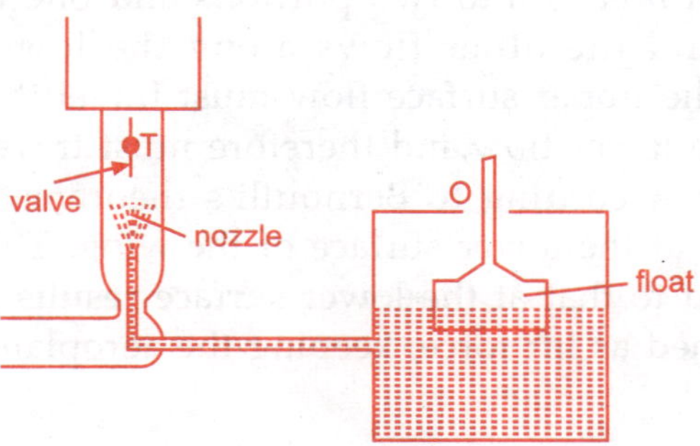

Fig. 6.7 *Carburettor*

It consists of a convergent divergent nozzle through which air is sucked into the combustion chamber during the suction stroke of the engine. In the convergent divergent nozzle a fine tube can introduce petrol from a chamber where petrol from the tank is drawn and stored at a constant level.

Petrol from the tank enters through the aperture O into the chamber. When the petrol level rises in the chamber, the float in the chamber rises and closes the aperture. When the level falls, the float comes down and opens the aperture allowing the petrol to be drawn from the tank. A connecting tube from the float chamber terminates in a jet opening at the throat of the convergent divergent tube of the carburettor. When the atmospheric air is sucked into the engine during the working of the engine, air flows past the jet with a large velocity. This causes a lowering of the pressure at the throat, and consequently, petrol is sucked into the carburettor in the form of fine spray. The petrol vapor mixes with the air and enters into the combustion chamber of the engine. The amount of the petrol-air mixture entering the engine is controlled by the throttle T. The property of pressure being low when the velocity is high when air is flowing in the convergent divergent pipe is used in the carburettor to suck petrol into the air stream.

3. **Lift Force in the Wings of Aeroplanes:** The wings of an aeroplane are given an aerofoil shape as shown in Fig. 6.8.

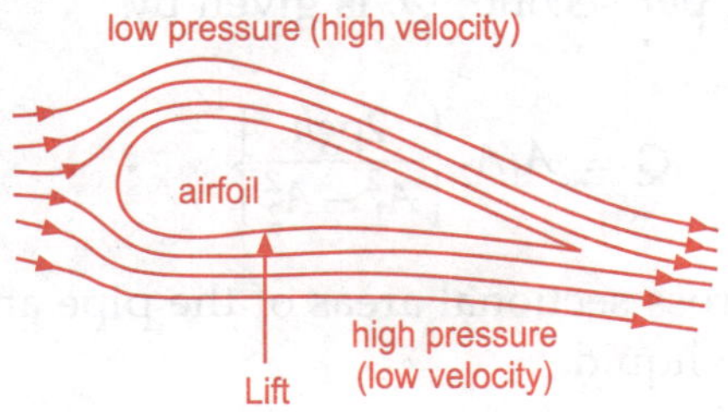

Fig. 6.8 *Flow past an airfoil*

The upper surface is made convex and the lower surface concave. When the aeroplane is moving forward towards the left, atmospheric air flows past the wing. At the nose, the stream breaks into two portions and one travels along the upper surface of the wing and the other flows along the lower surface. In order to recombine at the tail, the upper surface flow must travel the longer path along the convex surface than the lower flow,and therefore must travel with a larger velocity than the lower stream. According to Bernoulli's theorem, the pressure at the top surface falls below that at the lower surface of the wing. This lower pressure at the upper surface compared to that at the lower surface results in an upward thrust on the wing, which is termed as lift force, keeping the aeroplane floating in the air.

6.2   VISCOSITY

6.2.1 Introduction

When a liquid flows on a horizontal surface, the particles of liquid in contact with the bottom surface will have zero velocity, whereas the particles above have finite velocities. In a streamline flow over a horizontal surface, the layer of liquid in contact with the bottom surface will remain stationary and as we go up towards the top surface, the velocity of the particles will increase progressively. This results in a relative motion between successive layers of the liquid.

The streamline having the larger velocity tries to pull the layer below having the lower velocity. Similarly, the lower streamline, moving slowly, will try to slow down the upper layer which is moving faster. Hence internal forces are developed which oppose any relative motion between layers. The property of liquid by virtue of which it opposes relative motion between its different layers is called viscosity.

Thick liquids such as honey, castor oil or tar have higher viscosity and hence the lower streamlines offer higher resistance to the flow of top layers. As a result, if they are made to flow on horizontal surfaces, they flow very slowly. Thin liquids such as water, alcohol, kerosene oil, have much lower viscosity and hence their upper layers can move faster while the lower layers are impeded by the bottom surface.

Higher viscosity of a liquid helps in maintaining a streamline flow, whereas lower viscosity causes turbulent flow. Viscosity of liquids decreases rapidly with increase in temperature of the liquid. Viscous forces come into effect only when the liquid starts moving and does not act when the liquid is at rest. Thus viscous force is similar to friction between two surfaces in relative motion. Hence viscosity is sometimes referred to as internal friction. When water in a container is stirred to move, it comes to rest eventually due to internal friction between different layers.

6.2.2 Coefficient of Viscosity

When a liquid flows in a streamline flow inside a pipe, the velocity of the layers increase as we move away from the pipe inner surface towards the centre of the pipe as shown in Fig. 6.9.

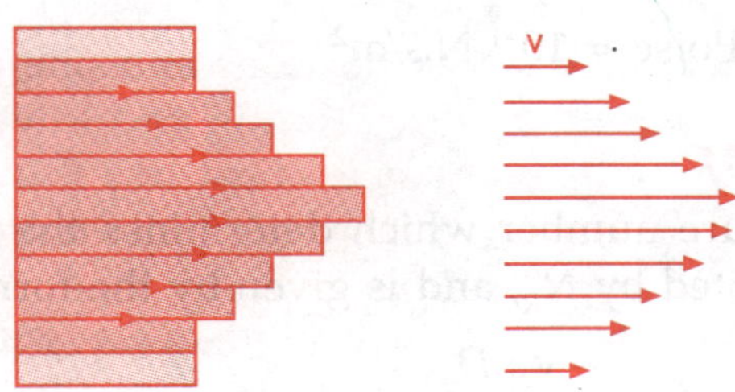

Fig. 6.9 *Velocity distribution for viscous flow in pipe*

The rate of change of velocity with distance, measured normal to direction of flow is called the velocity gradient. The velocity gradient between two layers, which are at a distance of dx apart and moving with velocities v and $v + dv$, as shown in Fig. 6.10 is given by dv/dx.

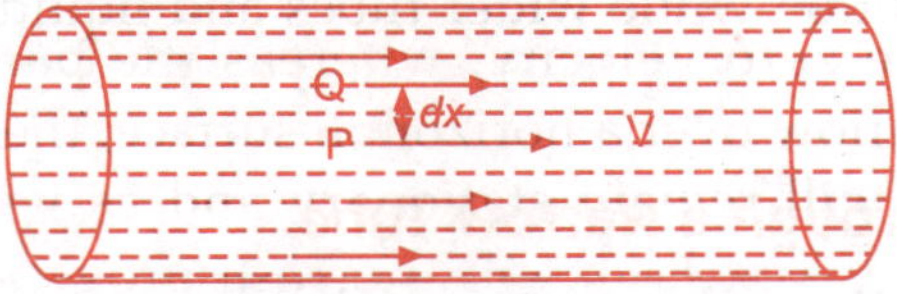

Fig. 6.10 *Velocity gradient between two layers*

The viscous force acts tangentially on a liquid layer and is proportional to

  (*i*) The area of the layers, A, and

  (*ii*) The velocity gradient, dv/dx.

Using a coefficient of proportionality, the shear force due to viscous friction is obtained as

$$F = \eta \, A \, (dv/dx) \tag{6.6}$$

The proportionality constant, η, is called the coefficient of viscosity. If $A = 1$, and $dv/dx = 1$, then $F = \eta$.

Hence, the coefficient of viscosity of a liquid may be defined as the viscous drag force acting tangentially on unit area of the surface of the liquid in motion having a unit velocity gradient.

The dimensions of the coefficient of viscosity can be found by balancing the dimensions of both the sides of Eq. (6.6) as

$$[\eta] = \frac{[\text{force}]}{[\text{area}][\text{velocity gradient}]} = \frac{[MLT^{-2}]}{[L^2][LT^{-1}/L]} = [ML^{-1}T^{-1}] \tag{6.7}$$

The dimensions of the coefficient of viscosity are, hence, Kg/(meter. second). In order to express in force units, it can be expressed as Newton. second/meter^2 or N.s/m^2. A unit of Poise is used to express the units of the coefficient of viscosity which has the dimensions of

$$1 \text{ Poise} = 10^{-1} \text{ N.s/m}^2$$

6.2.3 Reynold's number

Reynolds number is a pure number which determines the type of flow of a liquid through a pipe. It is denoted by N_R, and is given by the formula

$$N_R = \frac{v_c \rho D}{\eta} \tag{6.8}$$

where v_c is the critical velocity, ρ is the density, η is the coefficient of viscosity of the liquid and D is the diameter of the pipe. If N_R lies between 0 and 2000, the flow of a liquid is said to be streamline. If the value of N_R is above 3000, the flow is turbulent. If N_R lies between 2000 and 3000, the flow is neither streamline nor turbulent, it may switch over from one type to another. Narrow tubes and highly viscous liquids tend to promote streamline motion while wider tubes and liquids of low viscosity lead to turbulence.

6.2.4 Flow of Liquid through a Narrow Tube

A streamline liquid flow through a narrow tube due to pressure difference between its two ends will have cylindrical layers of liquid one inside the other as shown in Fig. 6.11.

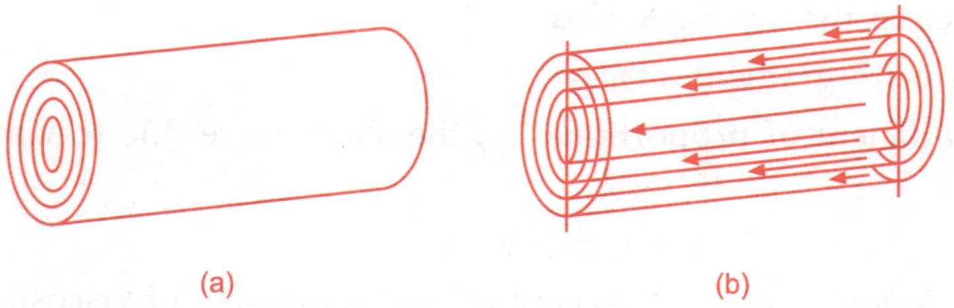

Fig. 6.11 *Flow of liquid through narrow tube*

The imaginary laminae inside the tube are shown in Fig. 6.11(*a*) and the cross section of the tube with the different laminae moving at different speeds is shown in Fig. 6.11(*b*). Those closest to the wall of the tube are moving slowly while those near the center are moving faster.

The layer of liquid in contact with the wall of the tube is at rest and the liquid moves the fastest along the axis of the tube. When the pressure difference is constant and the velocity of flow is small, the flow will be steady and streamline. In 1849, Jean Louis Marie Poiseuille, a French physician and physiologist, showed that for a liquid flowing through a narrow tube, in streamline motion, the volume *V* of the liquid flowing per second varies (*i*) as the fourth power of the radius of the tube, a, (*ii*) as the pressure difference between its ends, *p*, and (*iii*) inversely as the length of the tube, *l*. He obtained an expression relating these quantities as

$$V = \frac{\pi \rho a^4}{8 \eta l} \tag{6.9}$$

where η is the coefficient of viscosity of the liquid. This is called Poiseuille's formula. For streamline flow, the tube should be narrower when the liquid has lower viscosity as can be seen from Eq. (6.9).

Poiseuille's Method for Determination of Coefficient of Viscosity

This method is suitable for low viscosity liquids. The apparatus is shown in Fig. 6.12.

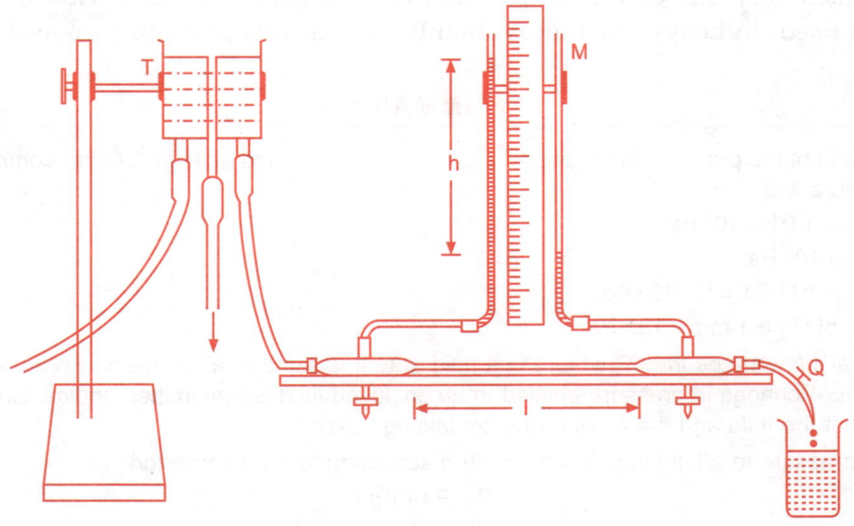

Fig. 6.12 *Poiseuille's apparatus for measuring coefficient of viscosity*

It consists of a horizontal capillary tube connected through a rubber cork bush to a bottle *A*. A piece of thread is tied to the end of the tube to make the liquid trickle down the thread and avoid backflow. The bottle is filled with the liquid whose viscosity is to be measured. Initial level of the liquid is marked. The liquid is allowed to flow through the tube which is collected in a beaker. The time of flow, t, is measured using a stopwatch. The flow is stopped and the final level is marked on the bottle. The difference in the initial level and the final level gives the height h. The volume of the liquid collected in the beaker, Q, can be measured directly or by determining the mass, m, of the liquid collected and then obtaining $Q = m/\rho$. The experiment is repeated to measure the quantities h, t and Q each time.

The mean radius, a, of the capillary tube is measured using a travelling microscope and measuring the radii at both the ends of the tube. The length, l, of the capillary tube is measured. The coefficient of viscosity is obtained as

$$\eta = \frac{\pi \rho g a^4}{8t} \frac{ht}{Q} \qquad (6.10)$$

6.2.5 Viscosity-Practical Applications

The importance of viscosity can be understood from the following examples.

(*i*) The knowledge of coefficient of viscosity of organic liquids is used to determine their molecular weights.

(*ii*) The knowledge of coefficient of viscosity and its variation with temperature helps us to choose a suitable lubricant for specific machines. In light machinery thin oils (Example, lubricant oil used in clocks) with low viscosity is used. In heavy machinery, highly viscous oils (example, grease) are used.

SUMMARY

1. The unit of the pressure is the pascal (Pa). It is also specified as $N \cdot m^{-2}$. Other common units of pressure are

   1 atm = 1.01×10^5 Pa

   1 bar = 10^5 Pa

   1 torr = 133 Pa = 0.133 kPa

   1 mm of Hg = 1 torr = 133 Pa

2. *Pascal's law* states that: Pressure in a fluid at rest is same at all points which are at the same height. A change in pressure applied to an enclosed fluid is transmitted undiminished to every point of the fluid and the walls of the containing vessel.

3. The pressure in a fluid varies with depth h according to the expression.

   $$P = Pa + \rho gh$$

   where ρ is the density of the fluid, assumed uniform.

4. The volume of an incompressible fluid passing any point every second in a pipe of non uniform cross section is the same in the steady flow.

   vA = constant (v is the velocity and A is the area of cross section)

   The equation is due to mass conservation in incompressible fluid flow.

5. *Bernoulli's principle* states that as we move along a streamline, the sum of the pressure (P), the kinetic energy per unit volume ($\rho v^2/2$) and the potential energy per unit volume ($\rho g y$) remains a constant.

$$P + \rho v^2/2 + \rho g y = \text{constant.}$$

The equation is basically the conservation of energy applied to non viscous fluid motion in steady state. There is no fluid which has zero viscosity, so the above statement is true only approximately. The viscosity is like friction and converts the kinetic energy to heat energy.

6. Though shear strain in a fluid does not require shear stress, when a shear stress is applied to a fluid, the motion is generated which causes a shear strain growing with time. The ratio of the shear stress to the time rate of shearing strain is known as coefficient of viscosity, η.

7. The onset of turbulence in a fluid is determined by a dimensionless parameter called the *Reynolds number* given by $R_e = \rho v d/\eta$ where d is a typical geometrical length associated with the fluid flow and the other symbols have their usual meaning.

Heat

7

7.1 THERMOMETRY

7.1.1 Heat is a Form of Energy

When we feel cold, we rub our hands together to keep them warm. The cold hands become warm because they have been supplied with heat, which was generated by expending mechanical energy during the rubbing of hands. Heat is produced by the expenditure of mechanical energy. In the brake drums of a motor car, heat is produced when the brake shoes rub against the brake drum. The bearings of a rotating machine get heated up when the rotating shaft rubs against the bearing surface on which it is supported.

Heat is also produced when an electric current flows in an electrical conductor. Chemical energy stored in fuels such as coal, petrol is converted to heat when they are burnt.

Heat can be converted into other forms of energy. In the internal combustion engine of an automobile, the chemical energy in the fuel is converted into heat which in turn is converted into mechanical energy.

Since other forms of energy can be converted into heat and heat in turn can be converted into mechanical energy or other forms of energy, we can conclude that heat is also a form of energy. James Joule (1818-1889) in England showed by experiment that a given quantity of mechanical energy is always converted to the same quantity of heat. Thus, the equivalence of heat and mechanical work as two forms of energy was definitely established.

7.1.2 Heat and Temperature

Let us take two buckets, one of capacity 1 liter and the other of 10 liters capacity. Let us fill them both with water from the same tank (so that waters in both buckets give the same hot or cold sensation to touch). Let us immerse identical heater coils in both and heat the waters in both the buckets. After half an hour or so, if we touch the waters in the two buckets we feel that the water in the smaller bucket feels much hotter than that in the larger bucket. Since identical heater coils have been used for identical durations in both the buckets, the electrical energy that is converted to heat is same in both the buckets. However, the same quantity of heat was able to give a hotter sensation to touch in the smaller bucket. "Hotness" and "Coldness" are qualities ascribed to the things which produce these sensations on contact with our bodies. The scientific name for hotness is temperature. We say that the temperature of the water in the smaller bucket is higher than that of the water in the larger bucket.

The sense of touch is the simplest way to distinguish hot bodies from cold bodies. By touch we can arrange bodies in the order of their hotness, deciding that A is hotter than B, B is hotter than C etc.

Let an object A which feels cold to the hand and an identical object B which feels hot be placed in contact with each other. After a sufficient length of time, A and B give rise to the same temperature sensation. Then A and B are said to be in thermal equilibrium with each other. The logical and operational test for thermal equilibrium is to use a third object or a test body, such as a thermometer. This is often called zeroeth law of thermodynamics: If A and B are in thermal equilibrium with a third body C, then A and B are in thermal equilibrium with each other.

The above idea suggests that the temperature of a system is a property which eventually attains the same value as that of other systems when all these systems are put in contact.

7.1.3 Measuring Temperature

As our physiological perception of temperature of a substance varies, there are many measurable physical properties that vary in that substance. Examples are the volume of a liquid, the length of a rod, the electrical resistance of a wire, the pressure of a gas kept at constant volume, the volume of a gas kept at constant pressure, and the color of a lamp filament. Any of these properties can be used in the construction of a thermometer.

We cannot, in the strictest sense of the word, measure the temperature, but we can say which of the two bodies is hotter (or at the higher temperature) or which is cooler (or at a lower temperature). In other words, temperatures can be compared. We can observe differences in hotness and the practical measurement of temperature is thus really the comparison of temperature differences.

There are two considerations in the measurement of temperature: (*i*) the selection of a standard temperature difference, and (*ii*) the use of an instrument, called a thermometer, to compare the temperature differences.

7.1.4 Different Scales of Thermometer

The temperatures at which pure substances change their state from solid to liquid and their state from liquid to vapor, at the same pressure, are reproducible. The temperature at which solid and liquid can exist together in equilibrium is called the melting point, and the temperature at which liquid and vapor can exist together in equilibrium is called the boiling point.

The fixed temperatures chosen to define the fundamental interval of temperature are the ice point and the steam point, where the ice point is the equilibrium temperature between ice and water at standard atmospheric pressure, and the steam point is the equilibrium temperature between liquid water and its vapor at a pressure of one standard atmosphere.

The standard atmosphere is defined as the pressure due to a column of mercury 760 mm high, of density 13.5951 gm/c.c. (or 13595.1 Kg/m^3). This pressure is equal to 101.325 kPa or 1,103,250 dynes/cm^2.

The fundamental interval is the difference in temperature between the ice point and the steam point. In the centigrade or the Celsius notation, this interval is subdivided into a hundred equal intervals, called degrees, the ice point being quite arbitrarily named 0°C, and the steam point 100°C. One centigrade degree is thus a temperature interval which is one hundredth of the fundamental interval. When we say that the temperature of the water in the bucket is 20°C, we mean that the difference between the temperature of the water and the ice point is 20/100 of the fundamental interval (temperature difference between the steam point and the ice point.) Comparison of the Kelvin, Celsius and Fahrenheit temperature scales is shown in Fig. 7.1.

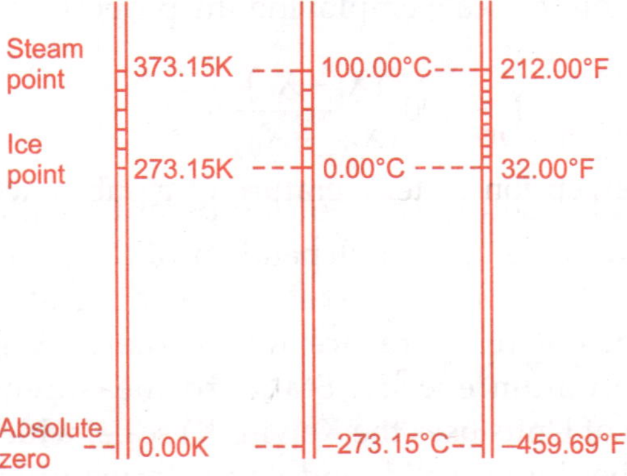

Fig. 7.1 *Comparison of the Kelvin, Celsius and Fahrenheit temperature scales*

In the Fahrenheit notation, which is obsolete now, the fundamental interval is divided into 180 equal degrees. The ice point and the steam point are marked quite arbitrarily 32°F and 212°F, respectively. One Fahrenheit degree is a temperature interval which is 1/180 of the fundamental interval, and is 5/9 of a centigrade degree.

Any property of any suitable substance which varies as the temperature is changed can be used to compare temperature differences with the fundamental interval. For example, the volume of a liquid enclosed in a vessel, the volume of a fixed mass of gas maintained at a constant pressure, the pressure of a fixed mass of gas maintained at constant volume, the electrical resistance of a piece of metal, the saturated vapor pressure of a liquid are among the many measurable physical properties which alter as the temperature changes. Any of these properties can be made the basis of a scale of temperature.

Consider any quantity whose magnitude changes with a change in temperature, and denote this quantity by X. Let X_0 be the value of X at the ice point and X_{100} its value at the steam point. Let X_t be the value at some unknown temperature, $t°C$, which is to be determined. Let equal changes in the value of X denote equal changes in temperature, so that changes in X are proportional to changes in temperature.

A rise in temperature of 100 centigrade degrees (from 0°C to 100°C) causes a change of $(X_{100} - X_0)$, while a rise in temperature of t centigrade degrees (from 0°C to $t°C$) causes a change of $X_t - X_0$. Hence,

$$t = k(X_t - X_0) \tag{7.1}$$

where k is a constant of proportionality, and

$$100 = k(X_{100} - X_0) \tag{7.2}$$

Dividing Eq. (7.1) by Eq. (7.2), we get

$$\frac{t}{100} = \frac{(X_t - X_0)}{(X_{100} - X_0)} \tag{7.3}$$

The temperature $t°C$ on the scale employing the property X is given by the equation

$$t = 100\frac{(X_t - X_0)}{(X_{100} - X_0)}°C. \tag{7.4}$$

There is a theoretical absolute scale of temperature called the Kelvin Absolute Thermodynamic Scale, which is quite independent of the properties of any individual substance. In an absolute temperature scale, a reading of zero coincides with the theoretical absolute zero of temperature where there exists the thermodynamic equilibrium state of minimum energy. Standard measurement of temperature in the International System of Units uses the Kelvin (K) scale, which is based on the unique temperature at which the liquid, solid, and vapor forms of water can be maintained simultaneously. This point is at 0°C in the Celsius scale or at 273.16 Kelvin's (written as 273.16 K without the degree symbol) above the absolute zero and is designated as the "triple point" which is actually at 0.01°C. The Kelvin scale can therefore be considered as the Celsius (°C) temperature scale shifted by 273.15 degrees. We are familiar with the temperature of 0°C as the temperature at which water freezes when the temperature is reduced from above zero, or, coming from the side below zero, ice melts when the temperature is increased towards zero. It is called the triple point because water exists under ideal conditions as both a liquid and a solid under vapor pressure conditions. Vapor pressure or equilibrium vapor pressure is the pressure exerted by a vapor in thermodynamic equilibrium with its condensed phases (solid or liquid) at a given temperature in a closed system. The equilibrium vapor pressure is an indication of a liquid's evaporation rate. It relates to the tendency of particles to escape from the liquid (or a solid). A substance with a high vapor pressure at normal temperatures is often referred to as *volatile*.

7.1.5 Types of Thermometers

(*i*) Mercury-in-Glass Thermometer

Mercury is easily visible, expands considerably, and expands fairly regularly according to the standard scale. It enables a wide range of temperatures to be measured because its melting point is about –39°C, and its normal boiling point about 357°C.

In a mercury-glass thermometer the mercury is filled in a thin walled glass bulb attached to a uniform capillary tube, which is sealed at the other end. The space above the bulb is normally evacuated.

The length of the mercury column above the bulb, say X, is proportional to the temperature of the substance in which the bulb is immersed. Equal changes in length X denote equal changes in temperature on the mercury-in-glass scale. Let l_0 be the length at the ice point, l_{100} the length at the steam point, and l_t the length at t°C. Since X is proportional to l, using Eq. (7.4) we can write

$$t = 100\left(\frac{l_t - l_0}{l_{100} - l_0}\right) \tag{7.5}$$

A thermometer is graduated by marking on the stem the positions of the mercury levels at the ice point and the steam point, and dividing the portion of the stem between these marks into a hundred equal divisions.

The experiment with an ungraduated thermometer, described below, illustrates the basic principle underlying the use of the instrument.

(*a*) The Ice-Point Reading:

A large funnel is filled with chips of pure ice, moistened with distilled water. The thermometer is held upright with the bulb immersed in the mixture. The steady level of the mercury column is marked by means of a piece of adhesive paper. The thermometer is removed, and the distance of the paper from the bottom of the bulb is measured. Let this length be l_0.

(*b*) The Steam-Point Reading:

The following points must be kept in mind while carrying out the steam-point measurement:

(1) The bulb of the thermometer must be immersed in steam and not in boiling water. A thermometer in the vapor, on which some pure liquid has condensed, registers the equilibrium temperature between pure water and its vapor and not of the vapor alone.

(2) The whole thermometer stem is immersed in the steam, and the bulb also must be immersed in the steam.

(3) Extreme care must be taken in reading the barometer since a pressure change of 1 cm of mercury causes about 0.4 centigrade degrees change in the boiling point at 100°C. See Fig. 7.2.

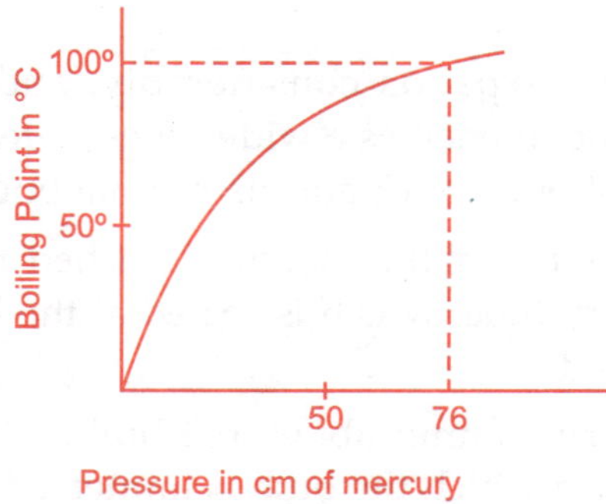

Fig. 7.2 *Boiling point of water against pressure*

The thermometer is then put in a hypsometer, shown in Fig. 7.3, in which the bulb and stem are surrounded with a double jacket C of steam when the water in A is boiled. The pressure gauge at the side shows whether the pressure inside the apparatus is atmospheric or not.

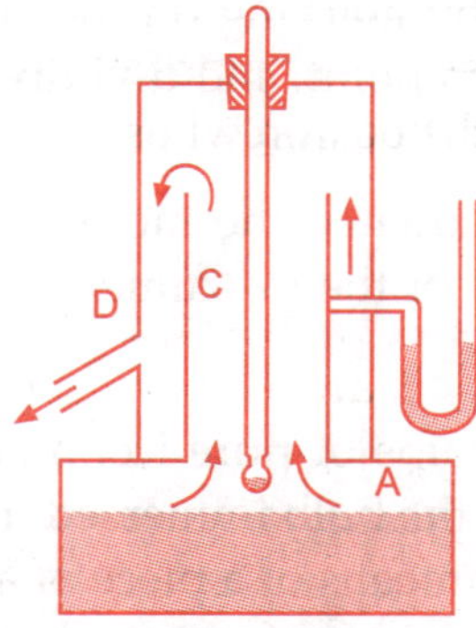

Fig. 7.3 *Hypsometer*

The thermometer is left for about ten minutes in the steam, and the position of the top of the mercury column is marked as before. The length of the mercury column is then measured. The barometer reading is taken which is, say, 750 mm of mercury. Let the corresponding boiling point be 99.6°C. If the thermometer were being permanently graduated, the correct position of the 100°C mark would be found by proportion according to the relation

$$\frac{l_{99.6} - l_0}{l_{100} - l_0} = \frac{99.6}{100} \tag{7.6}$$

Since mercury freezes at about –39°C, and boils at about 357°C, under normal pressure, the range of an ordinary mercury-in-glass thermometer is between its freezing point and about 300°C. If the space above the mercury is filled with nitrogen, or some other inert gas, the pressure of the gas, when both heated and compressed by the rising mercury thread, raises the boiling point, and such a thermometer can be used for temperatures considerably above the normal boiling point of mercury. This way it can be used upto 500°C, or even as high as 800°C for mercury in quartz. Alcohol can be used in thermometers between –130°C and 60°C. Liquid pentane can be used down to –200°C. Alcohol thermometers and others, whose range do not cover both the ice point and the steam point, must be graduated by comparison with mercury or other such thermometers which cover these fixed points.

(ii) Clinical Thermometer:

A clinical thermometer is used to measure the temperature of a human body. The normal temperature of the human body is about 36.9°C, and hence the scale in the clinical thermometer is graduated from about 35°C to about 43°C.

Fig. 7.4 *Clinical thermometer*

As shown in Fig. 7.4, a clinical thermometer has a short and very fine capillary tube, constricted with a slight kink close to the junction of stem and bulb. When the bulb is put in the mouth or the armpit of the person whose body temperature is being measured, the expanding mercury forces its way past the constriction, but when the bulb is removed from the mouth, the contracting mercury does not pull the thread back to the bulb, thus staying in the stem showing the highest temperature reached. It must be reset for another measurement by shaking the thread down vigorously through the kink. The front of the stem is usually shaped to act as a lens all along the graduations, so that the fine thread of mercury can be clearly seen as a wide ribbon when viewed in the right position. The thermometer must be left in the mouth or the armpit for half a minute to two minutes depending on the make, in order to reach the temperature of the body.

(iii) Six's Maximum-and-Minimum Thermometer:

When we require both the maximum and minimum temperatures of a substance, we can use this thermometer. This is shown in Fig. 7.5.

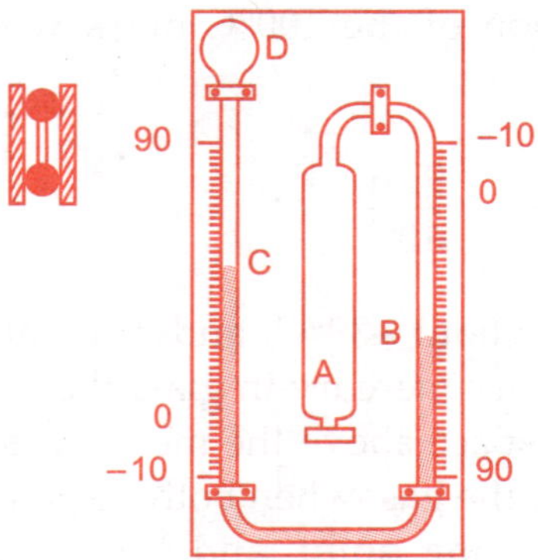

Fig. 7.5 *Maximum – and – Minimum Thermometer*

The thermometer bulb, *A*, contains alcohol or some other suitable liquid. *BC* is a thread of mercury. Above *C* there is more alcohol, which also partly fills the bulb *D*. Above the surface of mercury at both *B* and *C* is a light steel index (shown enlarged at the side of the figure) with springs resting on the top of the mercury levels pushing against the tube walls by their spring action. The springs are pushed up when the mercury level goes up, but will stay in their places when the mercury level in the tube goes down. When the alcohol in *A* expands, the thread *BC* is pushed around, sending the index above *C* upwards, and leaving that above *B* in the same position. When the alcohol in A contracts, the thread *BC* is drawn back, leaving the *C* index recording the maximum temperature reached. The *B* index similarly records the minimum temperature. To reset, each index is drawn down to the mercury surface by a magnet.

(*iv*) Constant Volume Gas Thermometer:

Consider a fixed mass of gas maintained at constant volume, in a vessel fitted with a device for measuring the pressure. Let p_0 and p_{100} be the pressures at the ice point and the steam point, respectively, and p_t the pressure at some unknown temperature $t°C$. Let us assume that equal changes in pressure denote equal changes in temperature. Then following the Eq. (7.5), we have the temperature given by

$$t = 100\left(\frac{p_t - p_0}{p_{100} - p_0}\right)°C \tag{7.7}$$

A constant volume hydrogen thermometer is shown in Fig. 7.6.

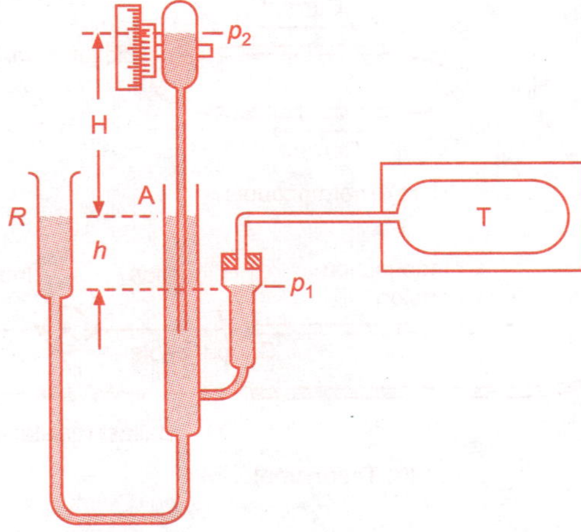

Fig. 7.6 *Constant Volume Hydrogen Thermometer*

(*v*) Platinum Resistance Thermometer:

The resistance of pure metals increases with increase in temperature. We assume that equal changes in resistance denote equal changes in temperature. If R_0 and R_{100} be the resistances at the ice point and the steam point, respectively, and R_t the resistance at $t°C$ then, following Eq. (7.5), we get

$$t = 100\left(\frac{R_t - R_0}{R_{100} - R_0}\right)°C \tag{7.8}$$

This, however, differs appreciably from the gas thermometer reading for the same temperature.

(*vi*) Thermoelectric Thermometers:

When a closed circuit made of two dissimilar metals or alloys has the two junctions maintained at two different temperatures, a current flows in the circuit. This arrangement is called a thermocouple. The size and direction of the current for a given pair of wires depend on the nature of the metals and the temperatures of the junctions. If one junction is maintained at a steady temperature, say 0°C, the temperature of the other may be obtained from a galvanometer G included in the circuit. The sketch of a thermocouple and a practical thermoelectric thermometer is shown in Fig. 7.7.

A more accurate estimate of the temperature can be made by measuring the electromotive force (E.M.F.) developed, instead of the current, because the E.M.F. depends only on the temperatures for two given materials, whereas the current depends on the total resistance of the circuit, which changes with the temperature.

Some of the metal combinations commonly used in thermoelectric thermometers and the maximum temperatures that can be measured using them are:

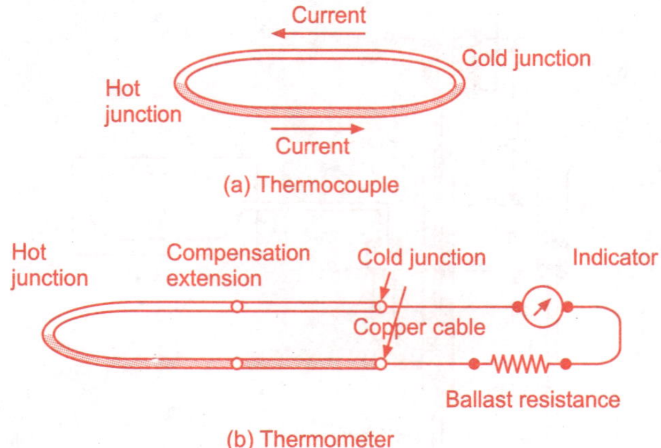

Fig. 7.7 *Practical Thermoelectric Thermocouple for High Temperature*

Platinum with platinum-rhodium (1600°C)

Chromel with alumel (1260°C)

Iron with constantan (1090°C)

Platinum with platinum-iridium (1000°C)

(*vii*) Radiation and Optical Pyrometers:

High temperature thermometers using thermocouples are called pyrometers. They use the accurate quantitative observations of the well known fact that as the temperature of a very hot body rises, it glows more brightly (the basis of radiation pyrometers) and also changes in color, from a dull red at about 600°C, through orange at about 1100°C, to white above 1400°C (the basis of optical pyrometry).

7.1.6 Absolute Zero of Temperature

The pressure in a perfect vacuum is absolute zero. In the same way there is an absolute zero of temperature. The temperature at the absolute zero on the absolute thermodynamic centigrade scale is –273.16°C, accurate to the nearest 1/100°. It is commonly taken as –273°C.

For many purposes it is convenient to measure temperatures upward from absolute zero. The ice point is then 273 K on this scale (without the degree symbol) starting at absolute zero.

7.2 EXPANSION

7.2.1 Expansion of Solids

When we heat a metal rod, the temperature increases causing the rod to expand, i.e.; to increase in length and thickness. The expansion in length is proportional to the length of the rod and the rise in temperature. The expansion also depends on the material of the rod. When equal lengths are heated through the same difference in temperature, it is found that the expansion of iron is about two-thirds that of brass and about half that of zinc.

7.2.2 Types of Expansion

Depending on the shape of the material, we can broadly distinguish three types of expansion, i.e.; linear, superficial, and cubical.

(*i*) **Linear Expansion:** When the material is in the form of a rod, with the length dimension far more than the width and thickness, we deal with only the linear expansion. Let the length of a rod at 0°C and t°C be l_0 and l, respectively. Then

$$l = l_0 (1 + \alpha t) \tag{7.9}$$

where α is a constant for the particular material concerned, called the coefficient of linear expansion of the material.

The coefficient α varies for a given material with temperature, but for all practical purposes this change is so small and we can consider α to be constant with respect to temperature. Hence, Eq. (7.9) can also be used to denote the expansion between any two temperatures, as

$$l_2 = l_1 [1 + \alpha(t_2 - t_1)] \tag{7.10}$$

where l_1 and l_2 are the lengths of the rod at temperatures t_1 and t_2, respectively.

(*ii*) **Superficial Expansion:** When the material has both length and breadth dimensions but negligible thickness, as in the case of a plate, we deal with the superficial expansion. Superficial expansion is the increase in area of a material. The coefficient of superficial expansion, α_s, is the increase in area of a unit area of the material for unit temperature rise.

(*iii*) **Cubical Expansion:** When a material in the form of a solid having length, breadth and thickness dimensions is heated, it expands in all the three dimensions increasing the volume of the material. The coefficient of cubical expansion, α_c, is the increase in volume for unit volume of the material for unit temperature rise.

7.2.3 Relation between Coefficients

Substances which have the same properties in all directions are called isotropic. Many crystals, and substances such as wood, do not have the same properties in different directions, and those are said to be anisotropic.

Consider a cube of unit side dimensions made of an isotropic material and having a coefficient of linear expansion α. If this cube is heated through a rise of t°, the length of one side increases to $(1 + \alpha t)$. The area of one face will be

$$A = (1 + \alpha t)^2 = 1 + 2\alpha t + \alpha^2 t^2 \approx 1 + 2\alpha t \qquad (7.11)$$

because α is normally of the order of 10^{-5} and hence α^2 term can be neglected in comparison with the constant term and the term containing α. Using the coefficient of superficial expansion, the area of the face at the higher temperature is also given by

$$A = (1 + \alpha_s t) \qquad (7.12)$$

Comparing Eqs. (7.11) and (7.12) we get

$$\alpha_s = 2\alpha \qquad (7.13)$$

The volume of the heated cube of an isotropic substance is given by

$$V = (1 + \alpha t)^3 = (1 + 3\alpha t + 3\alpha^2 t^2 + \alpha^3 t^3) \approx (1 + 3\alpha t) \qquad (7.14)$$

where the higher order terms involving a are neglected. Using the coefficient of cubical expansion, the volume is obtained as

$$V = (1 + \alpha_c t) \qquad (7.15)$$

Comparing Eqs. (7.14) and (7.15), we get

$$\alpha_c = 3\alpha \qquad (7.16)$$

For anisotropic solids, the coefficients of expansion may be different in different directions. Consider a unit cube of an anisotropic material to have coefficients of linear expansion α_1, α_2, and α_3 along the three edges at right angles to one another. The volume at $t°C$ is given by

$$V = (1 + \alpha_1 t)(1 + \alpha_2 t)(1 + \alpha_3 t) \qquad (7.17)$$

By multiplying and neglecting the product terms $\alpha_1\alpha_2$, $\alpha_2\alpha_3$, $\alpha_3\alpha_1$, and $\alpha_1\alpha_2\alpha_3$, the volume is obtained as

$$V = [1 + (\alpha_1 + \alpha_2 + \alpha_3)t] \qquad (7.18)$$

Hence the coefficient of cubic expansion for an anisotropic solid is given by

$$\alpha_c = \alpha_1 + \alpha_2 + \alpha_3 \qquad (7.19)$$

Expansion of a Hollow Vessel: Consider a hollow cube of unit sides. Let the coefficient of linear expansion on each side be α. For a rise in temperature of $t°C$, each side becomes $(1 + \alpha t)$, and the volume inside is $(1 + \alpha t)^3 \approx (1 + 3\alpha t)$. But this is just what happens when we consider a solid cube of the same material, hence the volume of the hollow space inside the cube increases just as if the hollow space itself were made of the material of the cube. This applies to vessels of any shape.

7.2.4   Determination of Coefficient of Linear Expansion using Roller Lever Method

The roller lever apparatus is shown in Fig. 7.8. It consists of a hollow tube AB of the metal, whose coefficient of linear expansion is to be measured, firmly clamped at C. The free end of the tube moves over a cylindrical roller R with a thin pointer P attached to it. The pointer moves over a protractor scale graduated in degrees. R rolls freely on a glass plate. Firm contact between the tube and the roller is secured by hanging a weight on a loop of thread over the tube. The distance CB (= l) is measured, the air temperature $t_1°C$, taken, and the initial reading θ_1, of P on the protractor scale is observed. Steam is passed through the tube until there is a steady flow from the end B for some time, and the pointer P is quite steady, when the final reading θ_2 on the protractor scale is taken. The temperature of the steam, $t_2°C$ is noted. The diameter of the roller R is measured as d.

As the rod expands in length, the roller rotates. The roller rotates about its center, and at the same time its center is free to move to the right as well. When the roller rotates through one complete revolution in the clockwise direction, the center of the rod expands through one circumference, πd, of the roller. But if the center of the roller were fixed, an expansion of the rod by a distance πd would rotate the roller by one revolution. Hence the expansion of the tube for one rotation of the roller is $2\pi d$. This is illustrated in Fig. 7.8(*b*). Let points A and A' coincide initially. As the roller moves to the right by a distance πd, point B on the rod will coincide with A'. Hence the total length of the rod OA is now $l + 2\pi d$. The angle turned through by the roller is $(\theta_2 - \theta_1)$ degrees, and this is $(\theta_2 - \theta_1)/360$ revolution.

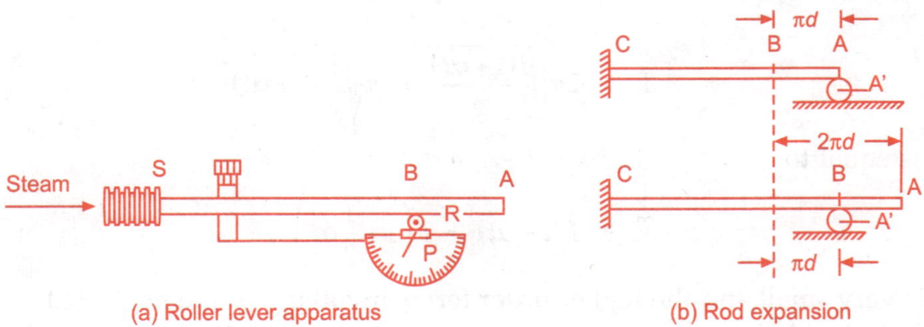

Fig. 7.8 *Simple Expansion Apparatus*

The expansion is then $2\pi d(\theta_2 - \theta_1)/360$ m. The temperature rise is $t_2 - t_1°C$. The coefficient of linear expansion is given by the increase in length per unit temperature rise, and is given by

$$\alpha = \frac{2\pi d(\theta_2 - \theta_1)}{l(t_2 - t_1)360} \tag{7.20}$$

7.2.5 Applications of Expansion of Solids

Allowance for expansion in the laying of railway lines and the erecting of steel bridges, the shrinking of metal tyres on cartwheels are a few of the many applications of expansion of solids. In accurate measurements of length, allowance has to be made for the expansion of the measuring scale, which is therefore accurate at one temperature only.

(*i*) Compensation of Clocks:

The hands of a clock are attached to wheels which come midway in a train of gears between the driving wheel actuated by the spring or weights, and the escapement wheel which controls the rate at which the whole train of wheels run. In the pendulum clock, each oscillation of the pendulum moves a claw-shaped arm so that one tooth of the last wheel of a train of wheels (called the escapement wheel) can pass. If the pendulum oscillates more rapidly, the whole train of wheels will run too fast and the hands connected to the wheels that show the time, will run fast. Consequently, the clock will gain time. If the pendulum oscillates too slowly, the clock will run slow and lose time.

The pendulum has a bob suspended at the end of a wire of negligible mass. For a simple pendulum of length l, the time period T for one complete swing, or to and fro motion, is given by

$$T = 2\pi\sqrt{\frac{l}{g}} \qquad\qquad (7.21)$$

where g is the acceleration due to gravity at the place. When the temperature rises by $t°C$, the new length of the pendulum is given by $l(1 + \alpha t)$. The new period T' is given by

$$T' = 2\pi\sqrt{\frac{l(1+\alpha t)}{g}} = 2\pi\sqrt{\frac{l}{g}}(1+\alpha t)^{\frac{1}{2}} \qquad\qquad (7.22)$$

This is equal to

$$T' = T(l+\alpha t)^{\frac{1}{2}} \approx T\left(1+\frac{1}{2}\alpha t\right) \qquad\qquad (7.23)$$

since αt is very small and the higher order terms involving αt are neglected. Hence, we have the fractional change in period, due to thermal expansion of the pendulum rod, given by

$$\frac{T'-T}{T} = \frac{1}{2}\alpha t. \qquad\qquad (7.24)$$

EXAMPLE 7.1

When the temperature rises by 10°C, find the fractional change in the period of a simple pendulum of brass whose coefficient of linear expansion is $\alpha = 0.000018$.

SOLUTION

From Eq. (7.24), we get

$$\frac{T'-T}{T} = \frac{1}{2}\alpha t = 0.5 \times 0.000018 \times 10 = 0.0009.$$

The simplest method of compensation for temperature error in pendulum type clocks is by using Harrison's Gridiron. The actual construction of such a pendulum is shown in Fig. 7.9.

The distance l between the point of support O and a fixed point supposed to be the centre of gravity G has to remain constant. Two rods, x cm, of a material of coefficient of linear expansion α_1, and y cm, of a material of higher coefficient of linear expansion α_2, are used.

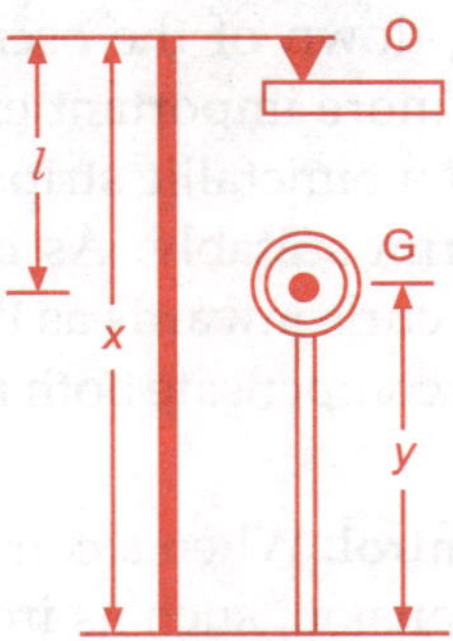

Fig. 7.9 *Principle of Compensated Pendulum*

If $l = x - y$, is to remain the same at all temperatures, the expansion of the first rod downwards for a rise in temperature t must balance the expansion of the second rod upwards. Hence,

$$x\alpha_1 t = y\alpha_2 t \text{ or } \frac{x}{y} = \frac{\alpha_2}{\alpha_1}. \tag{7.25}$$

If x is of iron ($\alpha_1 = 0.000012$) and y is of brass ($\alpha_2 = 0.000018$), then

$$\frac{x}{y} = \frac{0.000018}{0.000012} = \frac{3}{2}.$$

In the actual construction of a gridiron pendulum shown in Fig. 7.10, the downward expanding members are of iron, and the upward expanding members are of brass, the total lengths of the iron and brass being in the ratio 3 : 2.

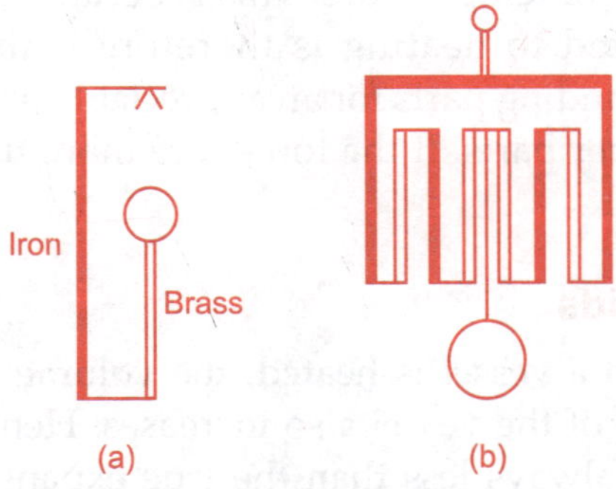

Fig. 7.10 *Grid Iron Pendulum*

(*ii*) **Compensation of Watches:** Similar to pendulum clocks, the hands of the watches are attached to wheels which are midway in a train of gears, between the driving wheel and the escapement wheel. The escapement wheel is actuated by a small flywheel, controlled by a hairspring.

The rate at which the balance wheel oscillates is determined by (*i*) the stiffness of the hairspring, and (*ii*) moment of inertia of the balance wheel. A rise in temperature weakens the hairspring (reduces its stiffness) and increases the diameter of the balance wheel, both resulting in a slowing down of the oscillations of the balance wheel. The weakening of the hairspring is the more important effect. The rim of the balance wheel is made in two or three segments of a bimetallic strip with brass on the outside and steel on the inside, and weighting the rim suitably. As brass has the greater coefficient of expansion, the end of each segment curls inwards as the temperature rises, thus reducing the moment of inertia sufficiently to compensate both for radial expansion and weakening of the spring.

(*iii*) **Bimetallic Thermostat Control:** When a compound strip made of two metals of widely different coefficients of expansion, such as iron and brass, is heated it will bend into an arc with the metal of higher expansion coefficient on the outside of the curve. The principle is shown in Fig. 7.11, where two metal strips form part of an electric circuit when the contact pieces touch.

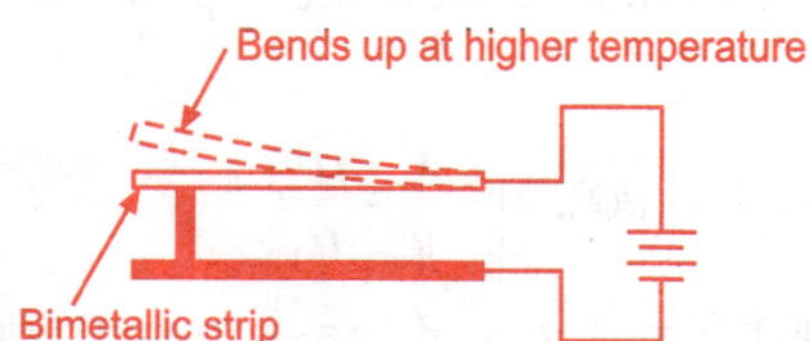

Fig. 7.11 *Bimetallic Thermostat Control*

The upper one is a bimetallic strip which bends up when the temperature rises and breaks the circuit. This thermostat is sensitive to temperature changes of one centigrade degree, provided the rate of change is not too great.

(*iv*) **Fracture of Glassware on Heating:** Thick glass vessels will fracture when subjected to sudden change of temperature. This is because glass is a poor conductor of heat, and the part subjected to heating is therefore maintained at a much higher temperature than the surrounding parts for an appreciable time, during which it expands and exerts forces on the other parts. If the forces are more than the glass can withstand safely, it will fracture.

7.2.6 Expansion of Liquids

When a liquid contained in a vessel is heated, the volume of the liquid increases and simultaneously the volume of the vessel also increases. Hence, the expansion observed when a liquid is heated is always less than the true expansion of the liquid itself. We must thus distinguish between the (*i*) real coefficient of expansion, which is the true increase in volume of unit volume for unit temperature rise, and the (*ii*) apparent coefficient of expansion, which is the observed increase in volume of unit volume for unit temperature rise, where the expansion of the vessel is disregarded.

In experiments conducted over a considerable range of temperature, the mean coefficient of expansion between the initial and final temperatures is found. As many liquids expand irregularly, the value of the mean coefficient of expansion depends on the range of temperature concerned, which should be specified. There are two types of coefficients of expansion that must be distinguished, i.e.;

(*i*) Zero coefficient, where the expansion is considered as a fraction of the volume at 0°C, and

(*ii*) Coefficient of expansion considered as a fraction of the volume at the lower value of the two temperatures.

If V_0 be the volume at 0°C, V_t at t°C, and α_0 the zero coefficient, then we have

$$V_t = V_0 (1 + \alpha_0 t) \tag{7.26}$$

while if V_1 and V_2 be the volumes at t_1°C and t_2°C, respectively, and α_m the mean coefficient between those two temperatures, then we can write,

$$V_2 = V_1[1 + \alpha_m (t_2 - t_1)] \tag{7.27}$$

From Eq. (7.26), we have $V_2 = V_0(1 + \alpha_0 t_2)$ and $V_1 = V_0(1 + \alpha_0 t_1)$, and hence the relation between V_1, V_2 and the zero coefficient is given by

$$\frac{V_2}{V_1} = \frac{1 + \alpha_0 t_2}{1 + \alpha_0 t_1}. \tag{7.28}$$

There is an appreciable difference between α_0 and α_m as a rule, and it is good to specify which coefficient is being used. It is convenient and less confusing to use as a general rule the equation

$$V = V_0 (1 + \alpha t) \tag{7.29}$$

where V_0 stands for the original volume at the lower temperature, which may sometimes be 0°C, V the volume at the higher temperature, t the rise in temperature, and α the mean coefficient of expansion between the temperatures concerned.

Anomalous Expansion of Water

When water at 0°C is heated, it contracts as the temperature rises initially, the volume of a given mass being least at about 4°C, (the density then being the highest), above which point rise in temperature causes expansion. The exact temperature at maximum density, at pressure of one atmosphere, is given as 3.98°C. An accurate knowledge of this temperature is important, since the liter is defined as the volume occupied by one kilogram of pure, air-free water at the temperature of maximum density and 760 mm pressure. The liter (found experimentally to be 1000.028 c.c.) is the generally accepted practical unit of volume.

Most liquids contract when they solidify. The atoms in the liquid form bonds and move closer together. However, when water at 0°C is frozen – it becomes ice – then it expands and becomes less dense. This is why ice floats – it is less dense than water. Pure water at 0°C has a density of 1000 Kg/m^3 but pure ice at 0° has a density of 920 Kg/m^3. Most liquids expand when they are heated. The molecules move faster, jostle each other more and are further apart on average. Water is strange in this respect. If water at 0°C is heated, it contracts and becomes less dense. It keeps contracting when heated until 4°C, then it starts expanding. This is important for life on Earth. It means the bottom of a lake is the last part to freeze, so fish can usually survive the winter.

An accurate determination of the temperature at the maximum density of water can be done by the following experiment due to Joule and Playfair. Two long columns of water are connected by horizontal tube b at the bottom and a horizontal cross channel c at the top as shown in Fig. 7.12. A small glass float is placed in the trough. One column is kept slightly below 4°C, and the other is warmed slightly above 4°C and the temperatures of the two columns at which the float is nearly stationary (showing no convection currents, and hence equal density) are observed. The difference in the temperature between the two columns is of the order of 0.8°C. The temperature of maximum density is determined as 3.95°C.

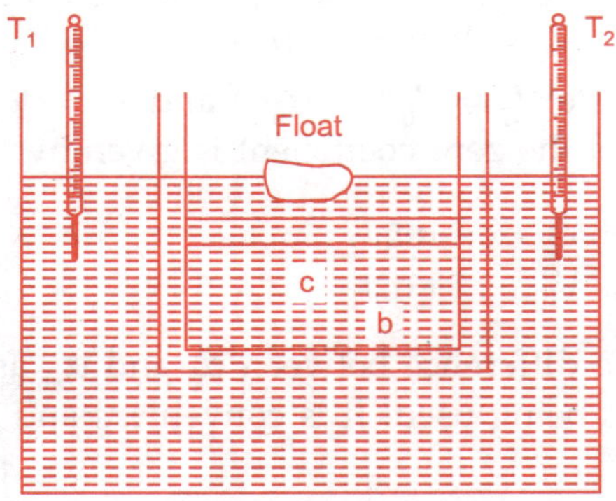

(a) Joule and play fair experiment to demonstrate anomalous expansion of water

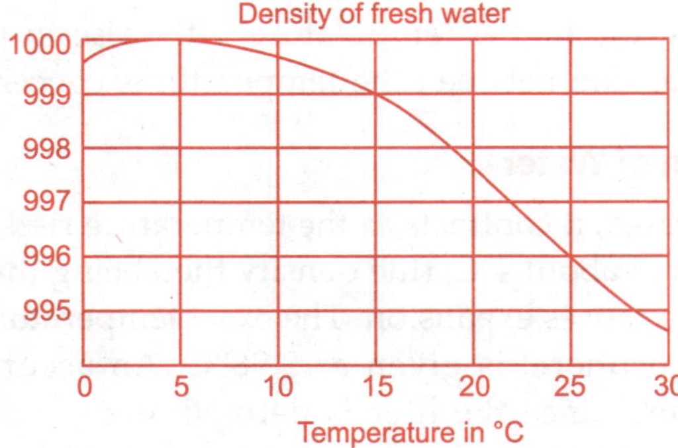

(b) Variation of water density with temperature

Fig. 7.12

Effects of Expansion of Water

(*i*) *Freezing of Ponds:* When the atmospheric temperature is lower than that of the water in the pond, heat escapes from the pond into the atmosphere through the top surface of water. If the air above the pond is, say at –10°C, convection will be maintained normally until all the water in the pond cools down to 4°C. When cooling below this temperature, the coolest liquid stays at the top, and a thin layer of ice is formed. The eventual thickness of the coating of ice is determined by the rate at which heat is lost through the coating of ice by conduction and the length of time for which this operates. The rest of the water in the pond below the layer of ice will be at 4°C, at the start of the freezing. Heat is also lost from the pond through some conduction, and eventually water in the whole of the pond must reach 0°C.

7.2.7 Expansion of Gases

When a gas in a closed container is heated, the pressure inside the container increases. Expansion of gases due to rise in temperature is affected both by its volume and the pressure. This is illustrated by Boyle's Law and Charles's Law as explained below.

Boyle's Law: The volume of a fixed mass of gas maintained at a constant temperature depends on the pressure to which it is subjected. Boyle's Law (1660) states that for a fixed mass of gas at constant temperature the product (pressure × volume) is a constant. All the common gases follow this law almost closely under ordinary conditions; however, it is not strictly true for real gases. Only at extremely low temperatures all gases may obey it. Any substance which does obey Boyle's law under all conditions is called an ideal gas or a perfect gas.

According to Boyle's law, if p is the pressure, and V is the volume occupied by the fixed mass of gas at that pressure

$$pV = k, \text{ a constant.} \tag{7.30}$$

Charle's Law; Expansion of Gas at Constant Pressure: Charle's Law states that for a fixed mass of gas heated at constant pressure, the volume increases by a constant fraction of the volume at 0°C for each centigrade degree rise in temperature.

The volume coefficient a for any gas is defined as the increase in volume of unit volume at 0°C, for each centigrade degree rise in temperature, when a fixed mass of that gas is heated at constant pressure.

Let V_0 denote the volume occupied by the chosen mass at 0°C, and let V be the volume at t°C. Note here that V_0 stands specifically for the volume at 0°C, and not just the original volume at any selected initial temperature; and t stands for the actual temperature t°C, and not for any selected temperature rise. Then

$$\alpha = \frac{V - V_0}{V_0 t} \tag{7.31}$$

and hence, we have

$$V = V_0 (1 + \alpha t). \tag{7.32}$$

The value of α is approximately equal to (1/273) for most gases. The apparatus shown in Fig. 7.13, can be used to study the variation of volume with temperature for a fixed mass of air maintained at constant pressure.

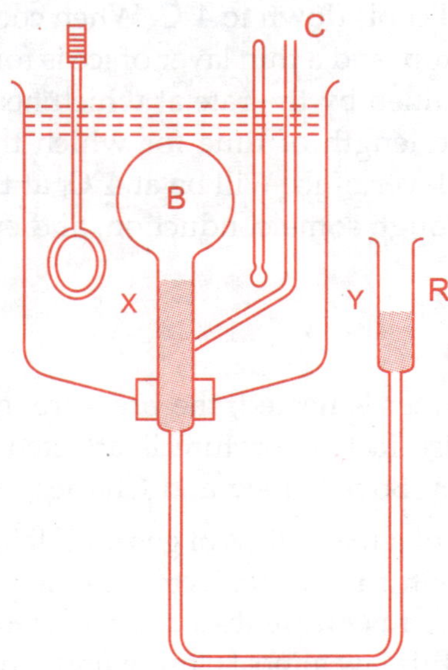

Fig. 7.13 *Constant-Pressure Apparatus*

Bulb B of known volume, contains dry air. The lower part of B is a graduated tube to which is connected a branch C open to the atmosphere. The mercury levels X and Y may be adjusted by raising and lowering the resrvoir R. When X and Y are at the same level, the pressure of the gas in B is equal to the atmospheric pressure. Both B and C are surrounded by a water bath containing an electric immersion heater which also serves as a stirrer. The temperature of the bath is observed by a mercury thermometer. The tube C is needed because equality of height of two levels such as at X and Y only denotes equal pressure if the two columns are at the same temperature.

After the first leveling of X and Y, and the recording of the volume of the enclosed air at the cold temperature, the heater is switched on until a temperature rise of, say, 20°C is obtained. Then the heater is switched off, and the water bath is stirred thoroughly. After levelling X and Y and waiting for them be steady, the levels are adjusted again if necessary. The volume of the air and the temperature of the bath are noted. This procedure is repeated several times, obtaining five or six sets of readings between room temperature and 100°C.

The expansion of the glass bulb in the experiment is a little less than 1 per cent of the observed change in the volume of the gas. It is important to read the barometer at the beginning of the experiment, and also at intervals to ensure that the atmospheric pressure remained steady during the experiment. The results of the experiment are plotted in Fig. 7.14, showing volume against temperature.

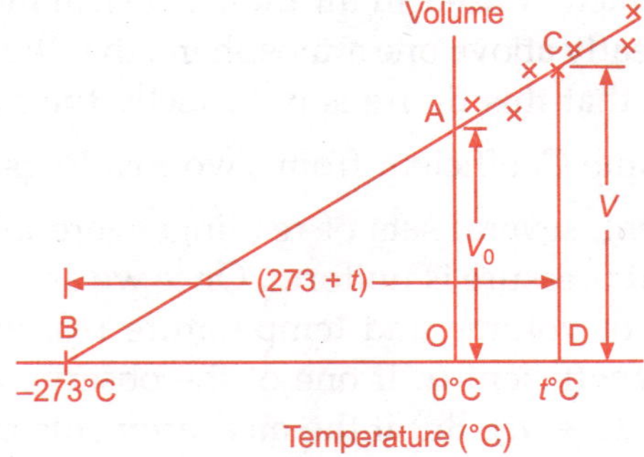

Fig. 7.14 *Variation of Volume against Temperature at Constant Pressure*

The points all lie on a straight line showing that equal changes in temperature lead to equal changes in volume at constant pressure as stated in Charle's law.

By extending the line backwards, the volume V_0 at 0°C is found. Extrapolating further backwards, the line meets the temperature axis at about −273°C. This introduces the idea of an absolute zero of temperature at about −273°C, which can be thought of as a suitable coordinate origin chosen so that the line CA passes through it.

The volume coefficient of expansion can be found from the graph as follows. The increase in volume per degree temperature rise is given by the slope of the graph, which is *AO/BO* and the volume at 0°C, V_0, is represented by *AO*. Hence the increase in volume per degree temperature rise, expressed as a fraction of V_0 is represented by

$$\frac{AO/BO}{AO} = \frac{1}{BO} = \frac{1}{273} \tag{7.33}$$

The volume at any temperature t°C is represented by *CD*. From the similar triangles *AOB* and *CDB* we have *CD/BD = AO/BO*. Consequently, we have

$$\frac{V}{273+t} = \frac{V_0}{273} \tag{7.34}$$

This shows that the volume of a fixed mass of air at constant pressure is directly proportional to the absolute temperature, obtained by adding 273 to the centigrade temperature. Alternately, writing

$T = 273 + t$, and $T_0 = 273$, we can write Eq. (7.34) as

$$\frac{V}{T} = \frac{V_0}{T_0} \tag{7.35}$$

All gases give similar results with values of α fairly close to (1/273). Hence, Charle's law can be restated as

"For a fixed mass of any gas heated at constant pressure, the volume increases by 1/273 of the volume at 0°C for each centigrade degree rise in temperature."

This generalization, which is used in all the gas calculations, holds only when the constant pressure is not greatly above one atmosphere, that the exact value of the fraction is not precisely 1/273, and that this figure is not exactly the same for all gases.

Determination of Volume Coefficient from Two Readings:

In the above experiment, several sets of readings were taken to check whether the expansion of air at constant pressure is uniform. Once we are assured that the expansion is uniform, only two sets of volume and temperature readings are really required in order to find the volume coefficient α. If one of the observations is V_0 at 0°C, then we can use the relation, $V = V_0(1 + \alpha t)$. But if the measurements made are the volume V_1 at t_1°C, and V_2 at t_2°C, then we have $V_1 = V_0(1 + \alpha t_1)$, and $V_2 = V_0(1 + \alpha t_2)$, and hence

$$\frac{V_2}{V_1} = \frac{1 + \alpha t_2}{1 + \alpha t_1} \tag{7.36}$$

Finally, we get after simplification,

$$\alpha = \frac{V_2 - V_1}{V_1 t_2 - V_2 t_1}. \tag{7.37}$$

Pressure Increase at Constant Volume

Consider a fixed mass of gas of volume V_1 at temperature t_1°C, and pressure p_1, and suppose it to be heated to some temperature t_2°C, at which the volume is nV_1, the pressure remaining constant at p_1. Next, let the volume be reduced again to V_1 at the high temperature, by increasing the pressure to p_2. If the gas obeys Boyle's law, then $p_2 V_1 = p_1 n V_1$, or $p_2 = np_1$.

Thus, if the temperature is raised keeping the volume constant at V_1 by suitably increasing the pressure, the law of increase of pressure at constant volume will be the same as the law of increase of volume at constant pressure, provided Boyle's law is obeyed.

That is, the rise in temperature which causes an increase from V_1 to nV_1 if the pressure is constant at p_1, causes an increase of pressure from p_1 to np_1 if the volume is kept constant at V_1. If Boyle's law is not obeyed perfectly, then the theoretical basis of this argument fails. Nevertheless, experiments show that when a fixed mass of gas is heated at constant volume, the pressure increases by a constant fraction of the pressure at 0°C, for each degree centigrade rise in temperature.

The pressure coefficient β for any gas is the increase in pressure, expressed as the fraction of the pressure at 0°C, for one centigrade degree rise in temperature, when a fixed mass of that gas is heated at constant volume. If p_0 is the pressure at 0°C, and p the pressure at t°C, then

$$p = p_0 (1 + \beta t) \tag{7.38}$$

Constant Volume Experiment

The apparatus used is called Jolly's constant volume air thermometer, shown in Fig. 7.15. A bulb V of about 100 c.c. capacity is filled with dry air. V is connected by a glass capillary tube T and rubber pressure tubing to a movable reservoir R. A fixed reference mark X is made on the tube A. The mercury in the tube is always brought to this position, which marks the constant volume maintained, before taking a reading. The bulb V is heated in a water-bath which is well stirred, and temperatures are read with a mercury thermometer.

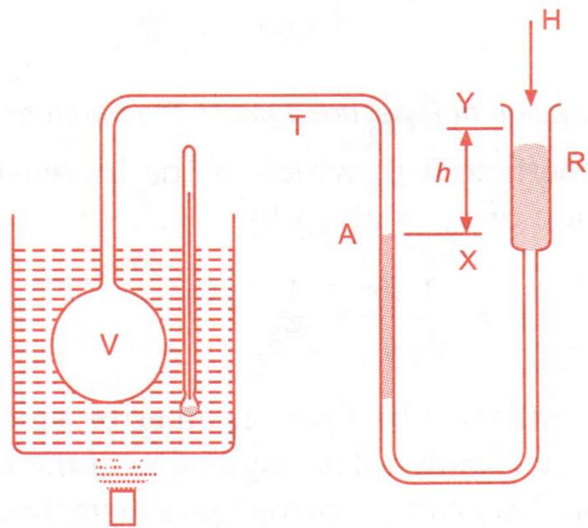

Fig. 7.15 *Constant Volume Air Thermometer*

Reading from the barometer H is first taken. Then starting with cold water in the bath, the temperature of the bath is taken and mercury in A is brought to the fixed mark X and the reading Y of the upper surface of mercury in R is taken. The difference $(Y - X)$ gives the "head" of the mercury h (h is considered negative if Y is below X), and the total pressure is then $p = (H + h)$ cm of mercury. The water bath is then heated through about 20 degrees, and heating is stopped and the bath well stirred. The mercury is brought to X. If there is a slow downward creep of the mercury, it means that the air in the bulb V is still rising to the temperature of the bath, and relevelling is necessary when the mercury is steady. The temperature of the bath is then taken, and the level Y noted as before. Heating is then resumed, and readings are taken. The barometer is checked again at the end of the experiment.

From the results obtained, a graph of pressure against temperature is drawn as shown in Fig. 7.16. The graph is a straight line indicating that the pressure increases uniformly as the temperature rises. When extrapolated backwards towards the negative side of the temperature axis, the line meets the temperature axis at about –273°C. The slope of the graph,

$$\frac{dp}{dt} = \frac{AO}{BO} = \frac{p_0}{273} \tag{7.39}$$

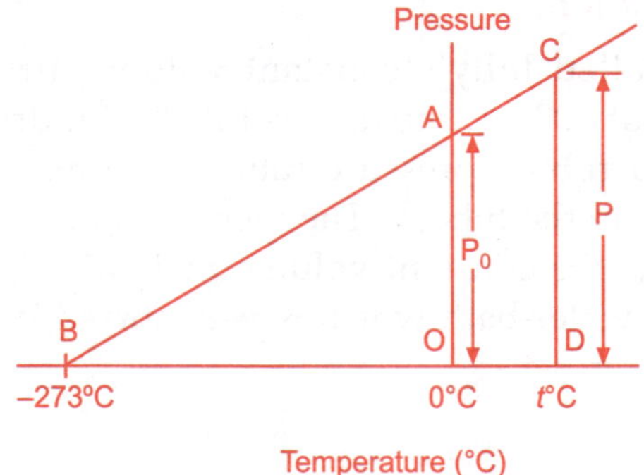

Fig. 7.16 *Variation of Pressure against Temperature at Constant*

Hence, the pressure coefficient β, which is the increase in pressure per degree temperature rise as a fraction of p_0, is given by

$$\beta = \frac{1}{p_0}\frac{dp}{dt} = \frac{1}{273} \tag{7.40}$$

The extra space in the capillary tube T causes an error, since the exact temperature of this part of the apparatus is not known. The expansion of the bulb introduces an error in β of the order of 1 percent. This can be corrected satisfactorily enough by adding the value of g, the coefficient of cubical expansion of glass to the observed value of β. From Fig. 7.16, the slope of the graph can also be written as

$$\frac{dp}{dt} = \frac{CD - AO}{BD - BO} = \frac{p - p_0}{t} \tag{7.41}$$

Hence, we can write

$$\beta = \frac{1}{p_0}\frac{p - p_0}{t} \tag{7.42}$$

This can be rewritten in the form of

$$p = p_0 (1 + \beta t). \tag{7.43}$$

The pressure coefficient can be calculated from two readings of above constant volume experiment, at any two temperatures. Let p_1 be the pressure at $t_1°C$, p_2 be the pressure at $t_2°C$, and p_0 the pressure at 0°C. From Eq. (7.43), we have

$$p_1 = p_0 (1 + \beta t_1), \ p_2 = p_0 (1 + \beta t_2) \tag{7.44}$$

Dividing p_2 by p_1 and rearranging we get β as

$$\beta = \frac{p_2 - p_1}{p_1 t_2 - p_2 t_1} \tag{7.45}$$

Experiments have shown that β is very close to 1/273 for all the permanent gases and the law of increase with pressure at constant volume can be stated as: For a fixed mass of any gas heated at constant volume, the pressure increases by 1/273 of the pressure at 0°C, for each degree rise in temperature. With the symbols T and T_0, standing for absolute temperatures, we can write

$$\frac{p}{T} = \frac{p_0}{T_0} \tag{7.46}$$

7.2.8 Ideal Gas or the Perfect Gas

At extremely low pressures Boyle's law is closely obeyed by all gases. If Boyle's law is obeyed, the values of α and β for the same gas should be equal. This can be verified as follows. Let us start with a fixed mass of gas at 0°C, at pressure p_0 and volume V_0. Let us heat it to t°C, in two different ways: (a) first at constant pressure, (b) then cooling it down again and repeating at constant volume. We see that

(a) at constant pressure p_0, we have

$$V = V_0(1 + \alpha t)$$
$$pV = p_0 V_0 (1 + \alpha t) \tag{7.47}$$

(b) at constant volume V_0, we have

$$p = p_0 (1 + \beta t)$$
$$pV = p_0 V_0(1 + \beta t) \tag{7.48}$$

But if Boyle's law is obeyed, all values of the product pV at the same temperature must be the same and hence from Eqs. (7.47) and (7.48), we get

$$pV = p_0 V_0(1 + \alpha t) = p_0 V_0 (1 + \beta t) \tag{7.49}$$

From Eq. (7.49), we can conclude that $\alpha = \beta$, if the gas obeys Boyle's law.

Real gases behave in this way only at extremely low pressures, where they have $\alpha = \beta = 0.0036608 = (1/273)$ nearly. But a gas which behaves in this way at all pressures is called the ideal or the perfect gas. Substituting this value of $\alpha = \beta = (1/273)$, in Eq. (7.49), we can write

$$pV = p_0 V_0\left(1 + \frac{t}{273}\right) = \frac{p_0 V_0}{273}(273 + t) = R(273 + t) \tag{7.50}$$

where $\qquad\qquad R = (p_0 V_0/273)$.

7.2.9 Absolute Scale of Temperature

From Eq. (7.50), we have $pV = R (273 + t)$. A graph of pV against t can be plotted for this perfect gas. As can be seen from Fig. 7.17, the product pV is zero when $t = -273$°C.

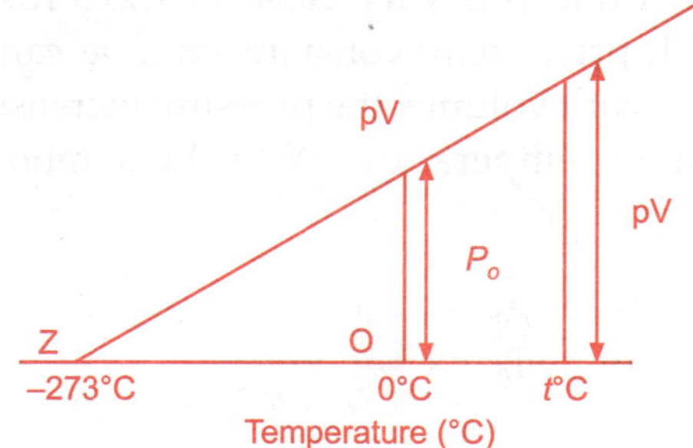

Fig. 7.17 *Variation of pressure against temperature for perfect gas*

This temperature denoted by Z on the graph, is the absolute zero of temperature. If the origin of the temperature is transferred to the point Z, and the temperature measured from this point is denoted T, then we have

$$T = (273 + t) \tag{7.51}$$

Using this notation, we can write Eq. (7.49) as

$$pV = RT. \tag{7.52}$$

7.2.10 Isothermal and Adiabatic Processes

Isothermal Process: If a system undergoes changes in pressure and volume with the temperature remaining constant throughout, then the process is called isothermal.

Consider a gas in a cylinder with a light frictionless piston. Let the cylinder and the piston be made up of a conducting material. If the piston is pushed down very slowly, the heat energy produced will be quickly transmitted to the surroundings. The pressure of the gas, p, increases and its volume, V, decreases. The temperature remains constant, as the heat energy produced is transmitted to the surroundings. This is an isothermal process. During an isothermal process, the condition pV = constant is satisfied.

The condition for a process to be isothermal is (*i*) The process should be slow and (*ii*) perfect heat conduction with the surroundings must be ensured.

Adiabatic Process: A process in which no heat enters or leaves the system is adiabatic.

Consider a gas enclosed in a cylinder provided with a light piston. Let the piston and the cylinder be made up of materials which do not conduct heat, so that no heat can either enter or leave the system. When the piston is pushed down quickly, the heat energy produced increases the temperature of the gas. In this process, the pressure, p, increases and the volume, V, decreases. This is an adiabatic process. Similarly when the gas expands quickly, the temperature of the gas decreases, and hence the process is adiabatic.

Thus, the temperature increases in an adiabatic compression, whereas the temperature decreases in an adiabatic expansion. The equation relating the pressure and the volume in an adiabatic process is given by

$$pV^\gamma = \text{constant} \tag{7.53}$$

where $\gamma = C_p/C_v$ is the ratio of the specific heat capacities of the gas at constant pressure and constant volume, respectively.

Conditions for a process to be adiabatic are that (*i*) the process should be fast, and (*ii*) the process takes place in perfect thermal insulation with the surroundings.

Examples of adiabatic processes are, (*i*) expansion of steam inside the cylinder of a steam engine, and (*ii*) the bursting of a balloon.

The change in pressure and volume of the gas in the above two types of processes can be represented in the form of a graph, as shown in Fig. 7.18.

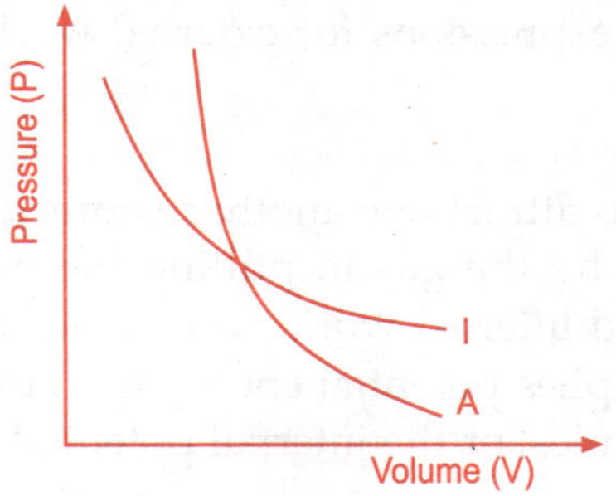

Fig. 7.18 *Pressure vs. volume plot for adiabatic (A) and isothermal (I) processes*

It is seen that the adiabatic curve A is steeper than the isothermal curve I. It can be shown that the slope of an adiabatic curve is γ times the slope of an isothermal curve,

$$\gamma = \frac{\text{Slope of adiabatic curve}}{\text{Slope of isothermal curve}} \tag{7.54}$$

7.2.11 Gas Laws

For a discussion of gas laws, it is necessary to discuss the external work done and the internal work by a gas.

External Work Done by an Expanding Gas: If unit mass of gas at a pressure of p, and volume V, expands by dV, the work done by the gas is pdV. This is called the external work. For a finite expansion from V_1 to V_2, the external work done is

$$W = \int_{V_1}^{V_2} pdV \tag{7.55}$$

This result is true whatever the form of the relation between p and V, and applies to real gases as well as to the ideal gas.

For an ideal gas p can be substituted with RT/V, and the final result can be expressed in several ways. If p_1 and V_1 are the initial conditions, p_2 and V_2 are the final conditions,

T the constant absolute temperature, R the gas constant for unit mass, the external work in absolute units is given by

$$W = \int_{V_1}^{V_2} RT\frac{dV}{V}$$

Integrating, we get

$$W = p_1V_1\ln\frac{V_2}{V_1} = p_2V_2\ln\frac{V_2}{V_1} = RT\ln\frac{V_2}{V_1} \tag{7.56}$$

where $\ln(\ )$ represents logarithm to the base e. In Eq. (7.56), (V_2/V_1) can be replaced with (p_1/p_2), to obtain three more expressions for external work.

Internal Work

In an actual gas the molecules attract one another appreciably. When the volume of the gas increases, work is done by the gas in pulling the molecules away against these attractive forces. This is called internal work. Just as the work done in raising a weight vertically against gravity supplies potential energy to it, the internal work increases the potential energy of the molecules, or the internal potential energy of the gas as a whole.

Law of Conservation of Energy: Suppose a small quantity dQ of heat energy is supplied to unit mass of gas. The molecules move with increased velocities and let the resulting increase in the internal kinetic energy of the molecules be dE, the internal work being done against molecular attractions (increase in internal potential energy) be dI, and the external work performed by the gas in pushing back an external constraint or expanding against an externally applied pressure be dW. Let the total internal energy, which is the sum of the potential and kinetic energy of the molecules, be denoted as U. We have $dU = dE + dI$. Then the law of conservation of energy states that

$$dQ' = dU + dW = dE + dI + dW \tag{7.57}$$

For a perfect gas, since there is no attraction between molecules, $dI = 0$. This is known as Joule's law.

7.2.12 The Perfect Gas Equation

In general, Eq. (7.52) states that the product pV is proportional to the absolute temperature, i.e.; $pV = CT$, where the value of $C = R = p_0V_0/273$. The pressure p_0 and the temperature 273 K, determine the density of the gas. Let us find out the value of the constant C for a unit mass of gas.

The symbol R is used to represent the constant C, when the mass of gas corresponds to one gram-molecular weight or one mole. Gram-molecular weight of oxygen is 32 gm, of helium is 4 gm, of hydrogen 2.016 gm and so on. R has the same value for one mole of all gases and is called the gas constant. The ideal gas equation, referring to one mole of any gas is

$$pV = RT \tag{7.58}$$

It is found experimentally that one mole of any gas under standard temperature and pressure occupies approximately 22.4 liters. The number of molecules in one mole of any gas, denoted by the symbol N, has the value of 6.03×10^{23}, to three significant figures. Hence, in dealing with one mole of any gas, whatever gas is chosen, the number of molecules is the same.

Value of R for one Mole of an Ideal Gas: We know that when p corresponds to 76 cms of mercury column,

$$p = 76(13.6)(9.81) \text{ Pascals (i.e. N/m}^2\text{), and}$$

$$T = 273°K, \text{ then}$$

$$V = 22.4 \text{ liters } (= 0.0224 \text{ m}^3) \text{ for all gases.}$$

Hence,

$$R = \frac{pV}{T} = \frac{76(13.6)(9.81)(0.0224)}{273} \text{ Nm/K}$$

The units may be given in Joules/mole/K also.

7.2.13 Kinetic Theory of Gases

A molecule of any substance is defined as the smallest portion of that substance capable of independent existence. For any given substance, its molecules are all alike in their general properties. The molecules of all bodies are continually moving, and thus possess mechanical energy. The energy of the moving molecules themselves is, in fact, the form of energy known as heat.

Molecules in Solids, Liquids and Gases

In a solid, the molecules are held in position under strong mutual attractions. Each molecule oscillates about its mean position, the amplitude increasing as the temperature rises. This increase in the amplitude of oscillation gives a qualitative explanation for the expansion of the solid when heated. When the melting point is reached, the oscillations are sufficiently violent to overcome the attractive forces, and the rigid structure breaks down, giving a liquid which can flow. The latent heat of fusion absorbed in melting represents the amount of energy needed to breakdown this structure.

In a liquid, contained in a body of fixed volume, the molecules are free to move at random, though they are still held together by the attractions between them. The energies of the molecules are distributed about a certain average value at any temperature, and at any instant some have energies much less than, and others energies much greater than, this value. The faster molecules in the neighborhood of the surface escape, and are free to move about as individuals in the space above. If the most energetic molecules escape, the average energy of those remaining is reduced, so that evaporation cools the liquid.

Now consider what happens when 1 gm of water at 100°C, occupying approximately 1 c.c., is converted into steam at 100°C and under 1 atmosphere pressure. The volume of the steam generated is 1600 c.c. It is clear that the actual bulk of the molecules themselves cannot be more than 1/1600 of the volume occupied by the steam. Also, the molecules must be oscillating violently with large amplitudes, nearly 12 times as far apart in the gaseous state as they were in the liquid state. It can be shown very easily that the attractive forces between molecules diminish greatly as the average distances between the molecules increase. Let us suppose the steam to be maintained at 100°C. Let us expand it and apply Boyle's law. If the volume of the steam is increased ten times, the pressure is reduced to about 1/10 atmosphere, the molecules now fill only 1/16,000 of the total volume occupied, and the average distance apart is increased to 25.2 times the original distance in the liquid state.

Reducing the pressure thus (*i*) reduces the fraction of the occupied volume which is actually occupied by the molecules themselves, and (*ii*) by increasing the average separation of the molecules reduces the effect of their attractions for one another. At extremely low pressures, the fraction of the total volume filled by the molecules themselves is vanishingly small, and so is the effect of attractions. But at extremely low pressures the physical behaviour of an actual gas approaches that of the ideal gas. Hence, low pressure conditions with an actual gas are the conditions with an ideal gas at all pressures.

The difference between a real gas and an ideal gas, therefore, is of the same kind as the difference between a real gas at high pressure and a real gas at low pressure. In one case, the molecules occupy an appreciable fraction of the total volume and attract one another appreciably. In the other case, the molecules fill only an infinitesimally small fraction of the total volume, and attractions between them are entirely negligible. In applying the kinetic theory to an ideal gas, then, we shall neglect the volume of the molecules themselves, and also neglect the effect of their attractions.

7.2.14 Pressure of an Ideal Gas

The following assumptions are made in order to arrive at expressions for the pressure.

1. The molecules behave as if they were hard, smooth and perfectly elastic spheres to which we can apply the laws of motion.

2. The molecules are continually in random motion, colliding with one another, and with the walls of the containing vessel.

3. There are very large number of molecules that we can apply statistics and work in terms of averages.

4. The average kinetic energy of the molecules is proportional to the absolute temperature.

5. The molecules do not exert any appreciable attractions on one another.

6. The space actually filled by the molecules themselves is only an infinitesimal fraction of the total volume occupied.

A perfectly elastic sphere of mass m and velocity v impinging perpendicularly on a fixed rigid wall has its momentum reversed from mV to $-mV$, and the wall receives an impulse equal to the change of momentum of the sphere, $2mV$. If, during one second, a large number of such impacts occur, the average force on the wall, by Newton's second law of motion, is equal in absolute units, to the total change of momentum in that second. The pressure is then obtained by taking the force on unit area of the wall.

Consider a gas in a cubical vessel whose walls are perfectly elastic. Let each edge be of length l. Call the faces normal to the x-axis A_1 and A_2 each of area l^2, as shown in Fig. 7.19.

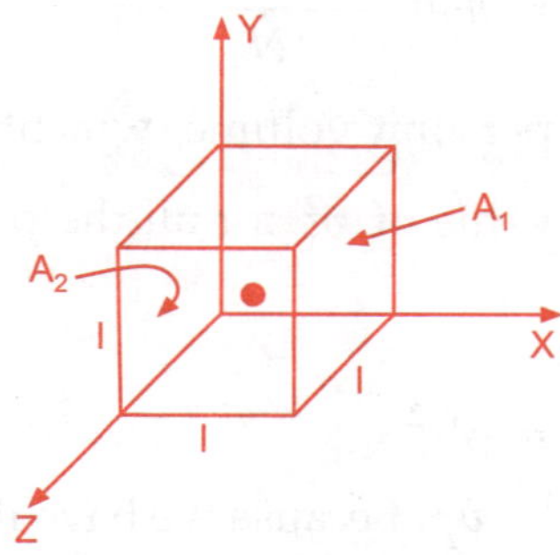

Fig. 7.19 *Cubic box with gas molecules moving inside*

Consider a molecule which has a velocity v. We can resolve v into components v_x, v_y and v_z in the directions of the edges. If this particle collides with A_1, it will rebound with its x-component of velocity reversed. There will be no effect on v_y or v_z, so that the change Δp in the particle's momentum will be

$$\Delta p = p_f - p_i = -mv_x - mv_x = -2\,mv_x \tag{7.59}$$

normal to A_1. Hence the momentum imparted to A_1 will be $2\,mv_x$, since the total momentum is conserved.

Suppose that this particle reaches A_2 without striking any other particle on the way. The time required to cross the cube will be l/v_x. At A_2 it will again have its x-component of velocity reversed and will return to A_1. Assuming no collisions in between, the round trip will take a time $2l/v_x$. Hence the number of collisions per unit time this particle makes with A_1 is $v_x/2l$. This gives the time for one collision, $\Delta t = 2l/v_x$. Consequently, the rate at which transfer of momentum to A_1 takes place is given by

$$\frac{\Delta p}{\Delta t} = 2mv_x \frac{v_x}{2l} = \frac{mv_x^2}{l} \tag{7.60}$$

In order to obtain the total force on A_1, that is, the rate at which momentum is imparted to A_1 by all the gas molecules, we must sum up (mv_x^2/l) for all the particles. Then, to find the pressure, we divide the force by the area of A_1, namely l^2.

If m is the mass of each molecule, the pressure, p, is given by

$$p = \frac{\dfrac{m}{1}\left(v_{x_1}^2 + v_{x_2}^2 + ...\right)}{l^2} \tag{7.61}$$

where v_{x_1} is the x-component of the velocity of particle 1, v_{x_2} is the x-component of velocity of particle 2, etc. If N is the total number of particles in the container and n is the number per unit volume, then $N/l^3 = n$ or $l^3 = N/n$. Hence

$$p = mn\left(\frac{v_{x_1}^2 + v_{x_2}^2 + ...}{N}\right) \tag{7.62}$$

But mn is simply the mass per unit volume, which is the density ρ. The quantity, $\left(v_{x1}^2 + v_{x2}^2 + ...\right)/N$ is the average value of v_x^2 for all the particles in the container. Let us call this $\overline{v_x^2}$. Then

$$p = \rho \overline{V}_x^2 \tag{7.63}$$

For any particle, $v^2 = v_x{}^2 + v_y{}^2 + v_z{}^2$. Because we have many particles and because they are moving entirely at random, the average values of $v_x{}^2$, $v_y{}^2$, and $v_z{}^2$ are equal and the value of each is exactly one third the average value of v^2. There is no preference among the molecules for motion along any one of the three axes. Hence, $\overline{v_x^2} = \frac{1}{3}\overline{v^2}$, so that

$$p = \rho \overline{V}_x^2 = \frac{1}{3}\rho \overline{V}^2 \tag{7.64}$$

Although we derived this result by neglecting collisions between particles, the result is true even when we consider collisions. Because of the exchange of velocities in an elastic collision between identical particles, there will always be some molecule that will collide with A_2 with momentum mv_x corresponding to the one that left A_1 with this same momentum. Also the time spent during collisions is negligible compared to the time spent between collisions. Hence neglecting collisions was convenient in the derivation, but will not affect the result.

7.2.15 Root-Mean-Square Velocity

The quantity $\overline{v^2}$ is called the mean-square-velocity. The square root of $\overline{v^2}$ is called the root-mean-square velocity of the molecules and gives an idea of average molecular speed. Using Eq. (7.63), we can calculate this root-mean-square velocity from measured values of the pressure and density of the gas. Thus,

$$V_{ms} = \sqrt{\overline{v^2}} = \sqrt{\frac{3p}{\rho}} \tag{7.65}$$

EXAMPLE 7.1

Calculate the root-mean-square velocity of hydrogen molecules at 0°C and one atmospheric pressure. Assume hydrogen to be an ideal gas at one atmospheric pressure.

SOLUTION

Under these conditions hydrogen has a density of $\rho = 8.99 \times 10^{-2}$ Kg/m^3. Then since $p = 1$ atmosphere $= 1.01 \times 10^5$ N/m^2, we have

$$V_{ms} = \sqrt{\frac{3p}{\rho}} = 1840 \text{ m/s.}$$

This is equal to 6624 km/hr.

7.2.16 Van der Waal's Equation

In order to account for the difference between the behaviour of a real gas and that of an ideal gas, consider molecular attractions change the pressure p to $(p + x)$ while the finite volume occupied by the molecules reduce from V to $(V - y)$.

If the molecules attract one another, the pressure p observed at the walls of the container is less than the true pressure within the body of the gas, since a molecule within the gas is attracted symmetrically from all sides (and hence unaffected by any resultant force), while a molecule impinging on a wall is retarded by an attractive pull acting towards the body of the gas only. The effect of the attraction on any one individual molecule is assumed to be proportional to the number of molecules per unit volume or, since we are considering a fixed number N of molecules in volume V, to N/V. But the number of impacts thus affected in any interval is also proportional to the density that is, to N/V. The resulting pressure reduction is thus proportional to

$$\Delta p = \frac{N}{V}\frac{N}{V} = \frac{N^2}{V^2} \tag{7.66}$$

We can say that the true pressure inside the gas exceeds the observed pressure at the walls by an amount proportional to $(1/V^2)$, say (a/V^2), where a is a constant of proportionality for unit mass of the particular gas. Hence instead of p, we should write $(p + a/V^2)$ in the equation of state.

The space in which the molecules can move about is less than the observed volume V, since it is diminished by the space actually filled by the volume of the molecules themselves; and the volume should be written $(V - b)$, where b is a constant for unit mass of the gas considered. Hence the equation of state for a real gas is written as

$$\left(p + \frac{a}{V^2}\right)(v - b) = RT \tag{7.67}$$

where R is the gas constant, and a and b are constants for unit mass of any one gas, but different for different gases. This equation is known as Van der Waal's equation.

7.2.17 Kinetic Energy of a Gas

If we multiply each side of Eq. (7.64) by the volume V, we obtain

$$pV = \frac{1}{3}\rho V \overline{v^2} \tag{7.68}$$

where ρV is simply the total mass of gas, since ρ is the density. We can also write the mass of gas as μM, in which μ is the mass in moles (gram molecular weight) and M is the molecular weight. Making this substitution yields

$$pV = \frac{1}{3}\mu M \overline{v^2} \tag{7.69}$$

The translational kinetic energy of the molecules is $\dfrac{1}{3}\mu M \overline{v^2}$, and hence the quantity

$\dfrac{1}{3}\mu M \overline{v^2}$ in Eq. (7.69) is two-thirds the total kinetic energy of the molecules. Hence, we can write

$$pV = \frac{2}{3}\left(\frac{1}{2}\mu M \overline{v^2}\right) \tag{7.70}$$

The equation of state of an ideal gas is

$$pV = \mu RT \tag{7.71}$$

Combining Eqs. (7.69) and (7.70), we obtain

$$\frac{1}{2}\mu M \overline{v^2} = \frac{3}{2}RT. \tag{7.72}$$

That is, the total translational kinetic energy per mole of the molecules of an ideal gas is proportional to the temperature.

7.3 ABSORPTION OF HEAT

When a hot body and a cold body are placed in contact, it is observed that the hot body becomes cooler and the cold body becomes warmer, and that this continues until both are at the same temperature. We say that the heat has passed from the hotter to the cooler body, or the cooler body has absorbed heat from the hotter body.

The unit of heat Q is defined quantitatively in terms of a specified change produced in a body during a specified process. Thus if the temperature of one Kg of water is raised from 21°C to 22°C by heating, we say that one kilocalorie (Kcal) of heat has been added to the system.

Heat is measured in energy units also. The unit of heat is Joules (J). In fact, 1 Kcal/°C = 4.187 × 10³ J/K.

Water Equivalent and Heat Capacity of a Body: The heat capacity, or thermal capacity, of a body is defined as the quantity of heat required to raise the body through one degree. The water equivalent is the mass of water which has the same thermal capacity as the body.

Suppose that we find the thermal capacity of a given body to be w calories per centigrade degree. As w calories raise w gm of water through one degree, the value of the thermal capacity, in calories per centigrade degree, will be the same numerically as the water equivalent (in grams of water).

7.3.1 Specific Heat

The specific heat of a substance is the quantity of heat required to raise the temperature of one unit mass of the substance through one centigrade degree. The unit in which specific heat is expressed is "Joules per kg per Kelvin" or "calories per gram per centigrade degree". Specific heat of a substance is thus the thermal capacity of unit mass of that substance.

The specific heat of all substances varies with temperature. For example, in the case of diamond, the specific heat at −186°C is 0.0025, at 22°C it is 0.122, and at 827°C it is 0.429. The specific heat of copper is 0.91 at 0°C and 0.95 at 100°C.

In many cases, for example, with ferromagnetic materials and alloys such as brass, there is a sharp and very significant rise in specific heat in the neighborhood of temperatures associated with changes in internal structure. Therefore, we must distinguish between the specific heat of a substance at a given temperature, and the mean specific heat over the range between two given temperatures, which is the quantity usually found by experiments.

Let a quantity of heat Q Joules be supplied to a body of mass m kg, to raise its temperature from θ_1°C to θ_2°C. Let s be the mean specific heat between these two temperatures. Then the number of Joules required, on the average, to raise 1 kg of the substance through one centigrade degree is $Q/[m(\theta_2 - \theta_1)]$, and so we have

$$s = \frac{Q}{m(\theta_2 - \theta_1)} \tag{7.73}$$

Rearranging this equation, the heat input, Q, can be obtained as

$$Q = ms(\theta_2 - \theta_1) \tag{7.74}$$

Eq. (7.74) can be expressed in words as

Heat gained = (mass) (specific heat) (temperature rise)

In order to raise the temperature of a body of mass m and specific heat s through one degree requires ms Joules. Thus,

Thermal capacity of a body = mass(specific heat)

Specific Heats of Gases: In dealing with gases, we must specify whether heat is supplied to the gas at constant volume or at constant pressure, because both are possible. Hence a gas will have specific heat at constant volume and a specific heat at constant pressure.

Consider a unit mass of gas contained in a vertical cylinder with a movable piston on which weights may be placed to alter the pressure on the gas. Supply sufficient heat to raise the temperature of the gas by 1°C, at the same time keeping the volume constant by increasing the load on the piston. The quantity of heat supplied is called the specific heat at constant volume.

Next, repeat the experiment, but this time with constant pressure on the gas. Supply sufficient heat to raise the temperature of the gas by 1°C, but this time without altering the weights on the piston. The quantity of heat supplied is called the specific heat at constant pressure. The gas expands, raising the weights, and consequently does work on an external body. Hence, specific heat of gas at constant pressure must be greater than the specific heat at constant volume, for in addition to raising the temperature of the gas, the energy required to do the work to raise the weights must also be furnished.

The specific heat of a gas thus depends on the conditions under which the heat is supplied. Constant volume and constant pressure are only two out of innumerable sets of conditions which may involve combinations of changes in pressure and volume, and hence the specific heat of a gas may have any one of innumerable different values.

7.3.2 Latent Heat

When a solid is melted, the change from solid state to liquid state takes place at a definite temperature under given fixed conditions. Heat must be supplied at that temperature in order to melt the solid, and the addition of heat does not increase the temperature until all the solid has melted. The quantity of heat required to change unit mass of a substance from the solid state to the liquid state without change of temperature is called the latent heat of fusion of the substance. The reverse change from liquid to solid involves the removal of heat without change in temperature.

When a liquid is vaporized, heat must be supplied. When the liquid is boiling, at a fixed temperature under a given pressure, the heat supplied causes the liquid to vaporize at a fixed temperature. The quantity of heat required to change unit mass of a substance from the liquid state into vapor without change of temperature is called the latent heat of vaporization (or latent heat of evaporation) of the substance.

The units in which latent heat is measured are Joules/kg.

The latent heat of fusion supplied to melt a quantity of ice is actually stored in the resulting water. This heat has been used in doing the mechanical work necessary to change a rigid solid body into a fluid. Similarly, the latent heat of evaporation is really expended in doing mechanical work in converting a fluid of definite volume into the gaseous state and also increasing the volume.

An experiment described below will give an idea of the values of the latent heat of fusion and the latent heat of evaporation of water, usually called the latent heat of ice and latent heat of steam, respectively. Two or three cubes of ice are placed in a thin copper can, which is placed over a Bunsen burner shielded carefully from draughts. The temperature is read at intervals of one minute until this reaches 100°C, when it may be read at five minute intervals until the can boils dry. A graph of temperature against time is plotted, and the times t_1, t_2, t_3, for the parts OA (melting of the ice), AB (raising of the water from 0°C to 100°C) and BC (conversion of all the water into steam at 100°C) are obtained from the graph shown in Fig. 7.20.

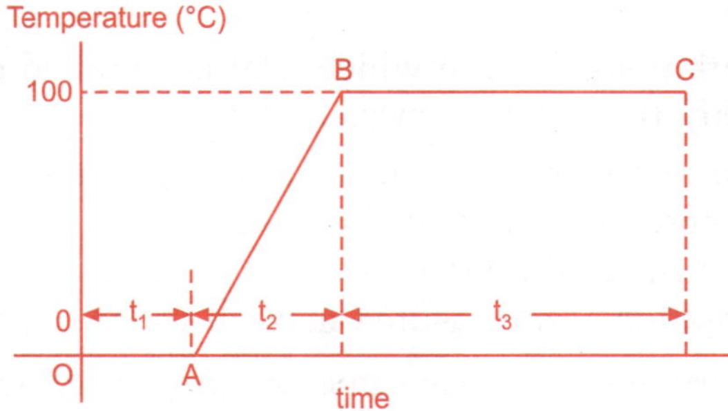

Fig. 7.20 *Graph of temperature against time to determine latent heats*

Then, if heat is supplied at a constant rate

$$\frac{\text{Heat to melt one kg of ice at 0°C}}{\text{Heat to raise one kg of water through 100°C}} = \frac{t_1}{t_2}$$

But 100 calories are needed to raise the temperature of 1 gm of water through 100°C. This is taken as the reference in order to measure the heat in the other two regions based on the method of constant heat supply that will be discussed later. Therefore heat required to change 1 gm of ice at 0°C into water at 0°C is 100 (t_1/t_2) calories. Similarly,

$$\frac{\text{Heat to convert water at 100°C to steam at 100°C}}{\text{Heat to raise same mass of water from 0°C to 100°C}} = \frac{t_3}{t_2} \qquad (7.75)$$

Hence the quantity of heat required to convert 1 gm of water at 100°C into steam at 100°C is 100 (t_3/t_2) calories.

7.3.3 Principle of Calorimeter

Calorimetry deals with the measurement of quantities of heat. Calorimeters are used for such measurements and there are numerous methods of measuring heat. The methods can be classified into two groups.

A. Thermometric Calorimetry, in which a change of temperature is recorded. Five distinct methods are available.

(1) Method of Mixtures, in which the heat to be measured is transferred to a known mass of water. This is the best known method, one of the simplest to use, and an accurate method.

(2) Method of Cooling, which depends on the fact that the rate of loss of heat from a hot body under constant conditions follows a regular law.

(3) Method of constant heat supply, which was mentioned in the determination of latent heats of fusion and latent heat of vaporization in section 7.32.

(4) Electrical method, in which the quantity of energy concerned is measured in electrical units.

(5) Method of continuous flow, in which heat is supplied electrically at a steady rate to a steadily flowing stream of liquid.

B. Latent Heat Calorimetry, in which the quantity of heat to be measured is made to melt a mass of solid whose latent heat of fusion is known, or to evaporate a mass of liquid of known latent heat of evaporation, or in which a known quantity of heat is supplied by the condensation of a measured mass of vapor of known latent heat.

Darling's Fuel Calorimeter: This calorimeter can be used to measure the calorific value of a fuel, such as coal. If m gms of coal completely burned in oxygen raises the temperature of w_1 gm of water and an apparatus of water equivalent w_2 gm from θ_1°C to θ_2°C, then the latent heat supplied is $(w_1 + w_2)(\theta_2 - \theta_1)$ calories, and the calorific value of coal is

$$H_c = (w_1 + w_2)\,(\theta_2 - \theta_1)/m \text{ calories/gm} \qquad (7.76)$$

The fuel calorimeter is shown in Fig. 7.21. The coal is first finely powdered and heated gently to dry it. An amount of m gm is weighed and placed in the crucible C, held in place by clips at the top of the tube A. The bell-jar B is clamped to the brass plate R, and leads P and Q from a 12 volt supply are joined to about 5 cm of Nichrome wire, which dips into the coal and is used to ignite it. A stream of oxygen supplied under pressure from a cylinder enters at O, and leaves via the tube A and the perforated plate below, so that the stream of bubbles impinges on the underside of R.

The oxygen supply is turned on, and the bell-jar lowered gently into the calorimeter, which contains measured quantity of water. The initial temperature is taken. When the oxygen supply has been adjusted to give a steady stream of fine bubbles, the coal is ignited. As it burns, the oxygen supply is adjusted to keep it going steadily. Towards the end, the supply is increased. As soon as all the coal is burned, the oxygen supply is disconnected and the bell-jar is flooded, so that the heated crucible gives up heat to the water. The bell-jar is lifted up and down to stir the water well, and the final temperature is taken. The necessary calculations are carried out in Example 7.2 below.

EXAMPLE 7.2

If 1.2 gm of coal on burning raises the temperature of 1400 gm of water in an apparatus of water equivalent 100 gm through 3.6°C degrees, find the calorific value of coal.

SOLUTION

Referring to Eq. (7.76), the calorific value of coal is obtained as

$$H_c = \frac{(1400+100)\times 3.6}{1.2} = 4,500 \ \text{calories/gm}$$

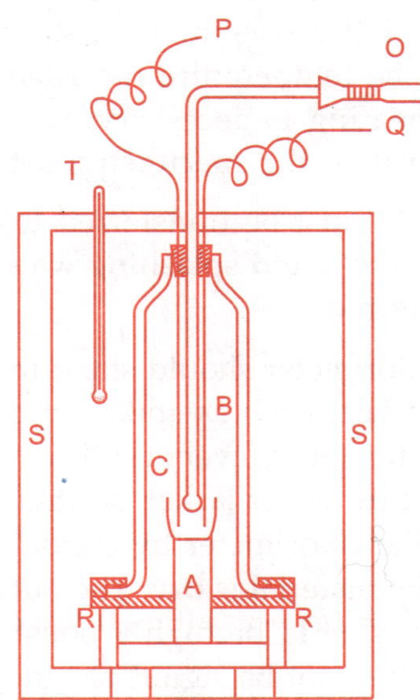

Fig. 7.21 *Darling's Fuel Calorimeter Pressure*

7.3.4 Specific Heat of a Solid by Method of Mixtures

In the method of mixtures, the heat to be measured is transferred to a known mass of water. From the change in the temperature of this mass of water caused by the added heat, we can find out the heat transferred to the water.

A thin copper calorimeter, of mass w and specific heat s, is filled with water of mass W, at $\theta_1°C$. A known mass m of the solid is heated to a suitable high temperature $\theta_2°C$ and dropped into the calorimeter, and the final highest temperature of the mixture $\theta_3°C$ observed after thorough stirring. Then, if no heat is lost to the surroundings, all the heat lost by the hot body in cooling to the final temperature has been gained by the water and calorimeter by raising to the final temperature.

The heat lost by the solid of mass m in cooling from θ_2°C to θ_3°C is given by $ms(\theta_2 - \theta_3)$, where s is the specific heat of the solid to be determined. The heat gained by water of mass W in raising its temperature from θ_1°C to θ_3°C is $W(\theta_3 - \theta_1)$, and the heat gained by the copper calorimeter of mass w and specific heat s_c in rising from θ_1°C to θ_3°C is $ws_c(\theta_3 - \theta_1)$. Since heat lost by the solid is equal to the heat gained by the water in the calorimeter and the calorimeter itself, we have

$$ms\,(\theta_2 - \theta_3) = W\,(\theta_3 - \theta_1) + ws_c\,(\theta_3 - \theta_1) \tag{7.77}$$

The unknown value of the specific heat of the solid, s, can be calculated using this equation.

The following precautions must be taken and corrections done in the course of the experiment:

(a) During the heating of the specimen, it is important to ensure that the specimen does reach the temperature recorded by the thermometer as θ_2°C. For this, prolonged heating is necessary, with the thermometer in contact with the specimen. Further, the specimen must be kept dry.

(b) The hot solid specimen must be transferred to the calorimeter very quickly, and care must be taken to avoid splashing water out of the calorimeter when the specimen is transferred.

(c) During mixture, the calorimeter should stand in an outer container, supported by poor conductors of heat such as spikes of cork, cardboard. The outer can surrounds the calorimeter with a layer of still air, and should itself be surrounded by a constant-temperature water jacket, so that the losses which do occur are minimum. Covering the calorimeter by cotton wool or felt around it should reduce convection if the material is dry. The outside of the calorimeter and the inside of the outer can should be highly polished to reduce radiation losses. The loss by radiation may not be significant in view of the small temperature excesses realized in the experiment. Continuous stirring during the rise in temperature is extremely important. The calorimeter should be fitted with a lid provided with holes for the thermometer and stirrer, to check the evaporation.

When the hot solid is transferred and the temperature of water in the calorimeter is increasing, the calorimeter is also losing heat to the surroundings continuously. It is important to correct for the cooling which unavoidly occurs in spite of every care. Before introducing the solid, allow the water in the calorimeter to attain the temperature of the surroundings, which is noted as θ_0. At the moment when the hot solid is introduced into the calorimeter the temperature is taken, and thermometer readings are observed at half minute intervals, stirring all the while, until the temperature becomes steady. Stirring and thermometer reading are continued for a period of at least as long as the rise in temperature takes place, and a graph of temperature against time is plotted as shown in Fig. 7.22.

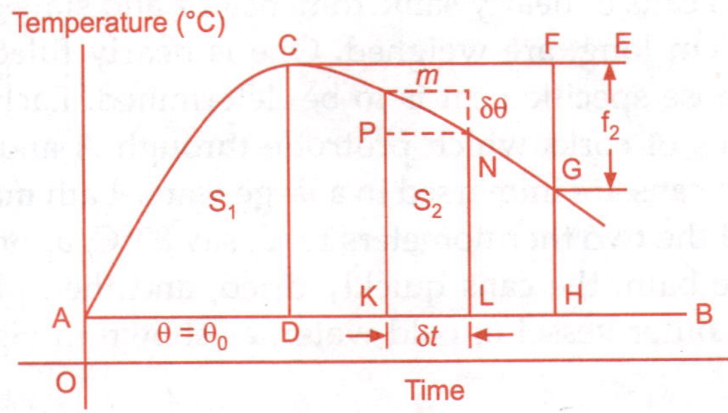

Fig. 7.22 *Cooling correlation graph*

The air temperature line $(\theta = \theta_0)$ AB is drawn. At the highest point C of the curve, a horizontal line CE is drawn, and a perpendicular CD is dropped to AB. From any point F on CE a perpendicular FGH to AB is drawn, cutting the graph at G. The fall f_2 represented by FG gives the cooling in the time represented by DH. It is required to find the fall f_1 which occurs during the time represented by AD.

Assuming that f_2 is the cooling taking place in time DH, the correction, f_1, to be added to the observed maximum temperature to give the true temperature is given by the relation, DH. $f_2 = AD.f_1$. Noting $DH = S_1$ and $AD = S_2$ we get

$$f_1 = f_2 \frac{S_2}{S_1}. \tag{7.78}$$

The method of mixtures may also be used to find the specific heat of a liquid, having the liquid in the calorimeter and introducing a heated solid of known specific heat.

7.3.5 Specific Heat of a Liquid by the Method of Cooling

The rate of loss of heat from a hot body under constant conditions follows a regular law. The factors upon which the rate of loss of heat from the surface of a body can possibly depend, when the body is supported by non-conducting material in an enclosure, are (1) temperature of the surface, (2) temperature of the surroundings, (3) area of the surface, (4) nature of the surface, (5) nature and pressure of the surrounding gas, (6) nature of the surface of the enclosure, and (7) volume and shape of the enclosure, which determine to the same extent the relative importance of conduction and convection through the gas.

It must also depend on the rate at which heat is supplied to the surface from within the body, that is, whether the body is a good conductor or not. For this reason, the method to be described is unsatisfactory for solids. But with a liquid in a thin metal can, either stirring or natural convection currents in the liquid distribute heat to the surface.

If all the above factors are identical for two bodies, then the rate of loss of heat from these two bodies must be exactly the same. This is the principle of the method of cooling.

Two thin aluminum cans of nearly same dimensions and state of the surface, about 2 cm in diameter and 5 cm long, are weighed. One is nearly filled with water and the other with paraffin whose specific heat is to be determined. Each is supported from a large board A by means of corks which protrude through A and carry thermometers. After assembly, the two cans are immersed in a large water-bath maintained at a suitable high temperature, until the two thermometers read, say 80°C, approximately. The board A is removed from the bath, the cans quickly dried, and then placed in an enclosure surrounded by a large outer vessel of cold water, as shown in Fig. 7.23.

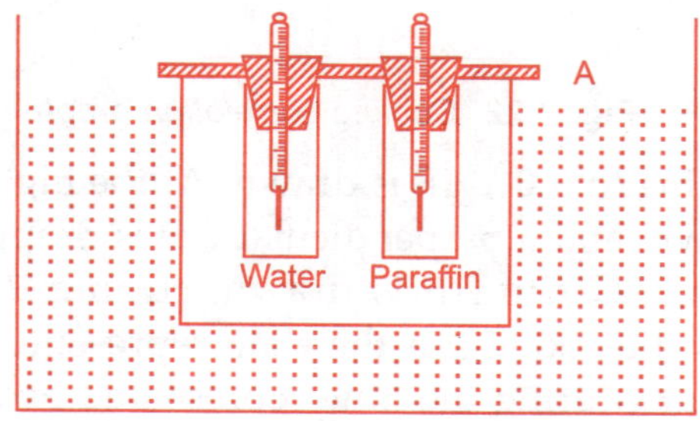

Fig. 7.23 *Method of cooling*

Reading of the two thermometers are started at once, and are continued at intervals of about two minutes until both have fallen to, say, 40°C. The cans are then removed, and weighed to find the mass of water and of paraffin.

A graph of temperature against time is plotted for each can as shown in Fig. 7.24. Choose two suitable temperatures θ_1°C, and θ_2°C, and draw lines parallel to the time axis to cut the curves at A, B, and C, D. Drop perpendiculars AA', etc. on to the time axis. Then $A'C'$ represents the time t_1 for the paraffin to cool from θ_1°C to θ_2°C, and $B'D'$, the time t_2 for the water to cool through the same range. Let m_1 and m_2 be the masses of the cans containing respectively paraffin and water, s the specific heat of the metal, and x the specific heat of the paraffin. Let M be the mass of paraffin and W the mass of water.

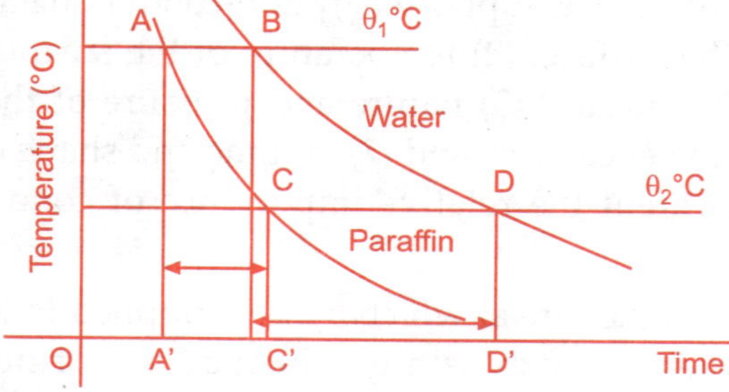

Fig. 7.24 *Cooling curves for water and liquid*

Heat lost by paraffin and container cooling from $\theta_1°C$ to $\theta_2°C$ is given by

$$Q_1 = (m_1s + Mx)\,(\theta_1 - \theta_2) \tag{7.79}$$

Rate of loss of heat is

$$\frac{dQ_1}{dt} = \frac{(m_1s + Mx)(\theta_1 - \theta_2)}{t_1} \tag{7.80}$$

Heat lost by water and container in cooling from $\theta_1°C$ to $\theta_2°C$ is given by

$$Q_2 = (m_2s + W)\,(\theta_1 - \theta_2) \tag{7.81}$$

The corresponding rate of loss of heat is given by

$$\frac{dQ_2}{dt} = \frac{(m_2s + W)(\theta_1 - \theta_2)}{t_2} \tag{7.82}$$

As the two rates of loss of heat are equal, we have $\dfrac{dQ_1}{dt} = \dfrac{dQ_2}{dt}$, and hence

$$\frac{(m_1s + Mx)}{t_1} = \frac{(m_2s + W)}{t_2} \tag{7.83}$$

The mean value of the specific heat of paraffin between $\theta_1°C$ to $\theta_2°C$, x, can be calculated.

With this apparatus, the water and paraffin cannot be stirred and there may be slight variations in the conditions for the two cans. It is better to use a single large calorimeter instead of the two small cans, and to perform two successive cooling experiments, the first with the water and the second with the liquid, when proper stirring is possible. The advantage of the method of cooling is that it avoids transference and mixing.

7.3.6 Specific Heats of Gases

It was mentioned earlier that a gas has two specific heats, (*i*) specific heat at constant pressure, and (*ii*) specific heat at constant volume.

Specific Heat at Constant Pressure, C_p (Regnault's Method): The gas is stored in the reservoir R which is surrounded by a water-bath at constant temperature, monitored by a thermometer T as shown in Fig. 7.25. The volume of reservoir is V, the temperature of gas in R is T_1 and the pressure is P_1. The stream of gas passes through the regulator A, which is adjusted during experiment to keep the reading of the manometer M constant, thus ensuring constant pressure conditions.

The gas passes through a spiral tube immersed in a bath of heated oil, the temperature of which is observed by a thermometer T_2. The heated gas then passes through the tube at C into a metal vessel inside an ordinary calorimeter, escaping finally by the tube D.

The rise in temperature of the thermometer in the calorimeter during the experiment is observed. Let the the initial temperature of the oil bath be T_1, and that of the calorimeter T_2. The mass of water in the calorimeter, and the water equivalents of its metal parts, the fall in temperature of the gas and the total heat supplied to the calorimeter are obtained.

After passing the gas for half an hour, the pressure in R is measured as P_2. Temperature in calorimeter rises to T_3, and the volume is V. We have

The mass of gas, M,

Mass of water in the calorimeter, m,

Water equivalent of calorimeter and its contents, w.

The rise in the temperature in calorimeter and its contents is $(T_3 - T_2)$.

The mean fall in temperature of gas is given by, $\left(T_1 - \dfrac{(T_2 + T_3)}{2} \right)$

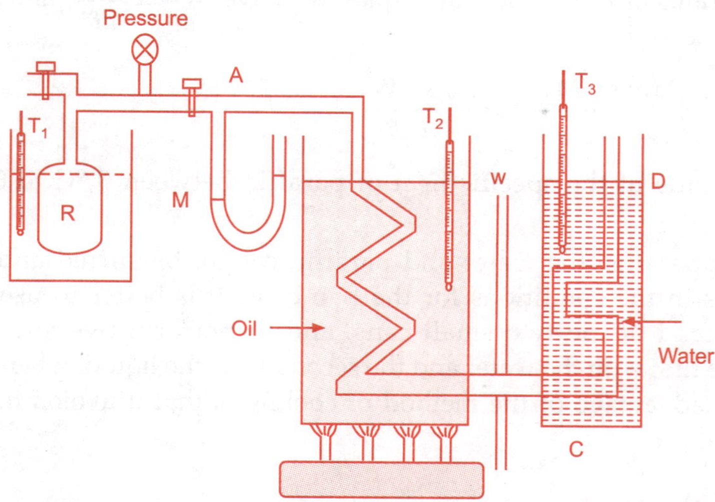

Fig. 7.25 *Specific Heat at constant pressure—Regnault's method*

Expressing the fact that the heat lost by the gas is equal to the heat gained in the calorimeter, we have

$$(m + w)(T_3 - T_2) = C_p M \left(T_1 - \frac{(T_2 + T_3)}{2} \right)$$

Hence the specific heat at constant pressure is obtained as

$$C_p = \frac{(m + w)(T_3 - T_2)}{M \left[T_1 - \dfrac{(T_2 + T_3)}{2} \right]}$$

The mass of gas is found as follows. Let the density of the gas under normal temperature and pressure conditions (NTP) be ρ. During the experiment V amount of gas at pressures $(P_1 - P_2)$ and temperature T has flowed. Reducing these to the NTP conditions, we have

$$\frac{(P_1 - P_2)V}{T} = \frac{76V_0}{273}.$$ Therefore, we have

$$V_0 = \frac{273(P_1 - P_2)V}{76T}.$$ Hence, the mass of gas is obtained as

$$M = \rho V_0 = \frac{273\rho(P_1 - P_2)V}{76T}.$$ This value of M is used in calculating the C_p above.

Specific Heat of a Gas at Constant Volume, C_v (Joly's Differential Steam Calorimeter):

This apparatus, used to determine the specific heat of a gas at constant volume, consists of two identical hollow copper spheres, suspended by fine wires from the arms of a sensitive balance in a chamber into which steam at 100°C can be passed. The wires pass through plugs of plaster of paris in the roof of the chamber, and just above these are two small electric heaters, so that moisture condensing in the opening is either absorbed or evaporated and cannot obstruct the free motion of the wire. Shields to prevent the water condensed on the roof from falling on the spheres are attached to the top of the chamber, and trays of thin metal are fixed below the spheres to catch the water condensed on the spheres. the apparatus is shown in Fig. 7.26.

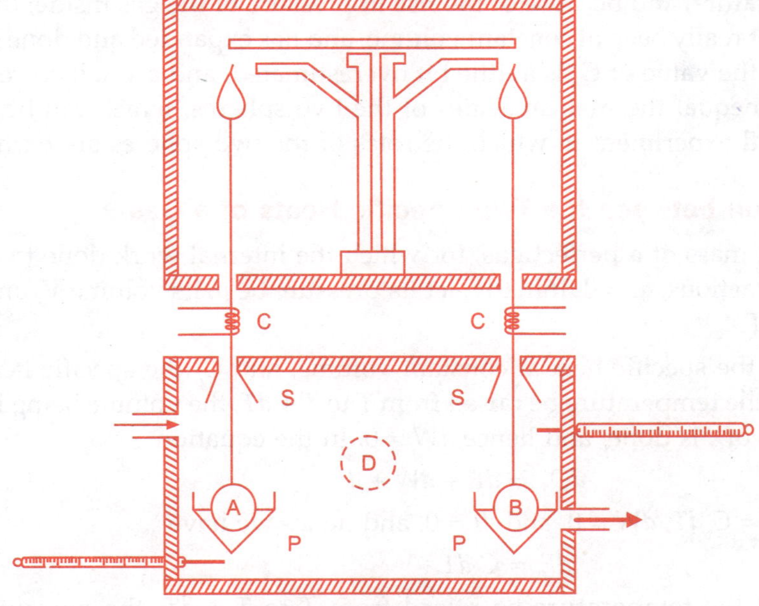

Fig. 7.26 *Jolys differential steam calorimeter*

Both spheres are at first evacuated and counterpoised on the balance. One sphere is then filled with the gas under test at a pressure of about 22 atmospheres, replaced in the steam chamber, and counterpoised again. The difference in weight gives the mass m of the gas enclosed.

The steady temperature θ_1°C of the steam chamber is then observed. Steam is now passed into the chamber for about five minutes until condensation on the spheres is complete. The temperature θ_2°C of the steam is observed by a high-range thermometer. The balance is then counterpoised again, and the further change in weight w is found. The mass of steam which condenses in providing the heat needed to raise the two identical copper spheres themselves from θ_1°C to θ_2°C is the same, and hence w gives the difference between the two masses condensed, that is, the mass of steam, the condensation of which is due to the enclosed gas.

The specific heat of a gas at constant volume is denoted by the symbol C_v. The heat supplied to raise the temperature of mass m gm of gas of specific heat C_v from θ_1°C to θ_2°C is thus wL calories, where L is the latent heat of steam at temperature θ_2°C. Hence

$$mC_v\,(\theta_2 - \theta_1) = wL \tag{7.85}$$

and the specific heat at constant volume is obtained as

$$C_v = \frac{wL}{m(\theta_2 - \theta_1)} \tag{7.86}$$

A test for leakage during the experiment is made at the end, and corrections are made for: (*a*) the expansion of the sphere containing the gas, both on account of its own rise in temperature, and because of the rise in pressure of the gas inside; this means that the gas has not really been at constant volume, and has expanded and done some external work, so that the value of C_v is a little bit overestimated, and a small correction must be applied; (*b*) unequal thermal capacities of the two spheres, which can be compensated for in a second experiment in which the roles of the two spheres are exchanged.

7.3.7 Relation between the Two Specific Heats of a Gas

Consider unit mass of a perfect gas, for which the internal work done to overcome the molecular attractions, $dI = 0$. Initially, let its pressure be p, its volume V, and its absolute temperature T.

Let C_v be the specific heat at constant volume, and C_p the specific heat at constant pressure. Let the temperature be raised from T to $T + dT$, the volume being kept constant. No external work is done, and hence $dW = 0$. In the equation

$$dQ_1 = dE + dW + dI$$

we have, $dQ_1 = C_v dT$, $dW = 0$, and $dI = 0$, and hence we have

$$dE = C_v dT. \tag{7.87}$$

Again, let the temperature be raised from T to $T + dT$, the pressure being kept constant. The volume then increases to $V + dV$, and the external work done is $dW = pdV$.

The heat put in, $dQ_2 = C_p dT$, and as $dQ_2 = dE + dW + dI$ with $dI = 0$, we have

$$dE + dW = C_p dT \tag{7.88}$$

But dE is the same in both cases, and is equal to $dE = C_v dT$, and hence Eq. (7.88) can be written as

$$C_p dT = C_v dT + dW, \text{ or } (C_p - C_v)dT = dW \tag{7.89}$$

If J is the mechanical equivalent of heat in ergs/calorie, then

$$dW = pdV/J \text{ calories} \tag{7.90}$$

Further, if p is constant, for an ideal gas, we have $pV = RT$, and

$$pdV = RdT$$

Hence $dW = RdT/J$. Consequently, we can write Eq. (7.89) as

$$(C_p - C_v) = R/J, \text{ calories per mole per centrigrade degree} \tag{7.91}$$

7.3.8 Ratio of Specific Heats for a Gas

The ratio of specific heat of a gas at constant pressure to that at constant volume, C_p/C_v, is denoted by γ. This ratio of specific heats is an important quantity for a given gas, because the adiabatic change in a gas is governed by this ratio.

When a gas is expanded or compressed in such a way that no heat enters or leaves the gas, the change is said to be adiabatic. When a perfect gas undergoes adiabatic change, the equation $pV = RT$, which is always true under all possible conditions, must of course hold good. But as T is changing as well as p and V, Boyle's law cannot be applied, because Boyle's law holds good for constant temperature conditions. Let us establish a relation between p and V during an adiabatic change.

Consider a unit mass of gas. In general we have

$$dQ = dE + dW + dI \tag{7.92}$$

For a perfect gas $dI = 0$, and since no heat is supplied or removed, $dQ = 0$. Hence

$$dE + dW = 0 \tag{7.93}$$

Suppose the volume changes from V to $V + dV$ at constant pressure p, and the temperature falls from T to $T - dT$. Then we have $dE = C_v dT$, and from Eq. (7.90), we have $dW = pdV/J$, and consequently,

$$C_v dT + \frac{pdV}{J} = 0 \tag{7.94}$$

Since we have $pV = RT$, we can substitute $p = RT/V$ in Eq. (7.94) and hence

$$C_v dT + \frac{RT}{J}\frac{dV}{V} = 0 \tag{7.95}$$

Rewriting this equation we get

$$C_v \frac{dT}{T} + \frac{R}{J}\frac{dV}{V} = 0 \tag{7.96}$$

From Eq. (7.91), we have $R/J = C_p - C_v$, and substituting this in Eq. (7.96) we get

$$C_v \frac{dT}{T} + (C_p - C_v)\frac{dV}{V} = 0 \tag{7.97}$$

Dividing by C_v throughout, we get

$$\frac{dT}{T} + \left(\frac{C_p}{C_v} - 1\right)\frac{dV}{V} = 0. \tag{7.98}$$

Denoting the ratio, $\dfrac{C_p}{C_v} = \gamma$, we get

$$\frac{dT}{T} + (\gamma - 1)\frac{dV}{V} = 0. \tag{7.99}$$

Integrating

$$\ln T + (\gamma - 1)\ln V = \ln T + \ln V^{\gamma-1} = \text{a constant or}$$
$$\ln TV^{\gamma-1} = \text{another constant} \tag{7.100}$$

Taking antilogarithms

$$TV^{\gamma-1} = \text{another constant} \tag{7.101}$$

Again, from the relation $pV = RT$, we can get $T = pV/R$, which is put in Eq. (7.101) to obtain

$$\frac{pV}{R}V^{\gamma-1} = pV^{\gamma} = R \text{ (constant)} = \text{a constant.} \tag{7.102}$$

This is the required relation between p and V during an adiabatic change.

Many processes in nature are adiabatic, because of the rapidity with which they take place. The pressure variations in the atmosphere when a sound wave travels are adiabatic in nature.

7.3.9 Determination of Latent Heat

Experimental determination of the latent heat of fusion and the latent heat of evaporation of water are discussed below.

Latent Heat of Fusion of Ice by the Method of Mixtures: The calorimeter is weighed, about two-thirds filled with water at about 30°C and weighed again. It is placed in its enclosure on a non-conducting stand, and the temperature is taken. Small pieces of dried ice are added one by one, and the water is stirred until each piece has melted before the next piece is added. When the temperature has fallen to about 5°C the thermometer is read. The calorimeter is then weighed to find the mass of ice added.

Let W be the mass of water, w the mass of the calorimeter, s the specific heat of material of calorimeter, m the mass of ice melted. Let θ_1°C and θ_2°C be the initial and final temperatures of the calorimeter, and L cal/gm the latent heat of fusion of ice.

The heat lost by the water in cooling from θ_1°C to θ_2°C is $W(\theta_1 - \theta_2)$ cal. The heat lost by the calorimeter cooling through the same range is $ws(\theta_1 - \theta_2)$ cal.

The heat gained by m gm ice melting at 0°C is mL cal; the heat gained by m gm melted ice in rising from 0°C to θ_2°C is $m\theta_2$ cal. Assuming that no heat has escaped during the experiment, the heat gained will be equal to the heat lost and hence

$$W(\theta_1 - \theta_2) + ws(\theta_1 - \theta_2) = mL + m\theta_2 \tag{7.103}$$

where L can be determined.

Latent Heat of Vaporization of Water by the Method of Mixtures:

A calorimeter is weighed, two-thirds filled with water, and weighed again. It is placed in its outer can, and the temperature is taken. A current of dry steam is blown on the surface of the water until a rise in temperature of about 30° is obtained. The steam passes from the generator through a water trap, as shown in Fig. 7.27, designed to obstruct the passage of water condensing in the delivery tube from the boiler. The final temperature of the calorimeter is taken after stirring, and the calorimeter along with the contents is weighed again.

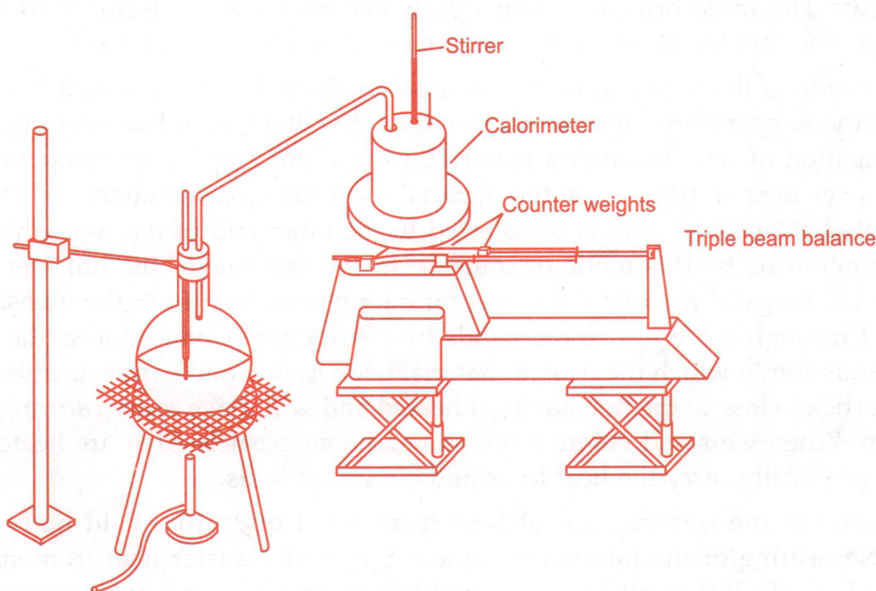

Fig. 7.27 *Latent heat of vaporization by method of mixtures*

Let m be the mass of steam condensed, W the mass of water, w the mass of calorimeter, and s the specific heat of the calorimeter material. Let the initial and final temperatures be θ_1°C and θ_2°C, and the boiling point θ_3°C. Let L be the latent heat of steam at this temperature.

Heat lost by m gm steam condensing at θ_2°C is $Q_1 = mL$.

Heat lost by m gm steam-water cooling from θ_2°C to θ_3°C, $Q_2 = m(\theta_2 - \theta_3)$.

Heat gained by W gm water rising from θ_1°C to θ_3°C, $Q_3 = W(\theta_3 - \theta_1)$

Heat gained by w gm copper rising from θ_1°C to θ_3°C, $Q_4 = ws\,(\theta_3 - \theta_1)$.

Assuming that there is no heat lost to the surroundings, the heat gained will be equal to the heat lost, and hence we have,

$$mL + m(\theta_2 - \theta_3) = W(\theta_3 - \theta_1) + ws\,(\theta_3 - \theta_1) \tag{7.104}$$

In this equation, all quantities except L are observed, and hence L can be calculated.

7.4 TRANSFERENCE OF HEAT

A hot utensil from the stove is taken down using tongs, because, if we touch it with bare hands it will be extremely uncomfortable. We do not like to stand before a hot stove in a poorly ventilated kitchen on a hot afternoon because the air in the vicinity of the stove is very hot and again very uncomfortable. Similarly, we do not stand in the hot sun on a summer afternoon, because sun rays are extremely hot and cause discomfort. The above are examples of different ways in which heat is transferred from a hot body to a colder body. The three principal means by which heat may be transferred from one point to another are by (*i*) conduction, (*ii*) convection, and (*iii*) radiation.

Conduction is the mode of heat transmission through a body which is unequally heated or the temperatures at different points on the body are different. Conduction is the propagation of heat through a material, from a point at higher temperature to a point at lower temperature, while the material itself remains stationary. When a metal rod is heated at one end, heat is transferred to the other end of the rod, which is at a lower temperature, by this mode of transfer. In the example of the hot utensil being held with the tongs, if we hold the vessel for a long time, heat from the utensil may be conducted through the tongs to the hands by conduction. Convection is the mode of heat transmission in which the heated material itself moves and carries the heat with it. The air particles close to the hot stove get heated and will move away carrying the heat with them. When we stand before a hot stove, the air particles that are heated by the hot stove physically carry the heat to us and pass it on to us.

Radiation is the transference of heat from a hot body to a cold body without appreciable heating of the intervening space. Sun rays transfer heat from sun to the earth by this mode. If the path of the rays is blocked then heat will not be transferred by this mode. That is why, if we stand under a thickly foliated tree, or block the sun rays with an umbrella, we can protect ourselves from the hot sun.

7.4.1 Transference of Heat by Conduction

Consider a thin slab of material of uniform thickness as shown in Fig. 7.28. Let the thickness of the slab be δx, and area be A. The face 1 is maintained at a steady temperature of $\theta°C$ and face 2 at $(\theta - \delta\theta)°C$.

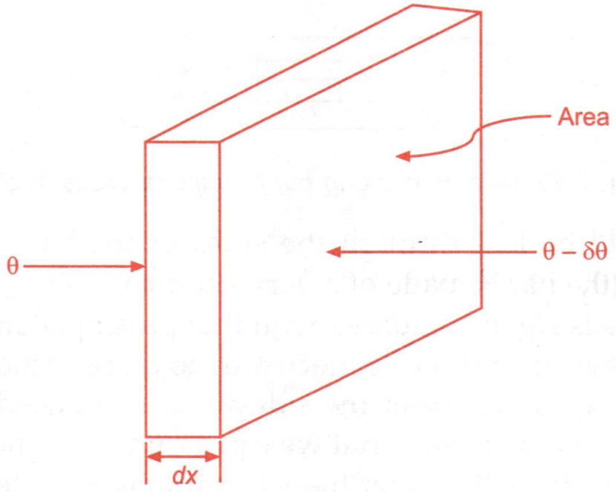

Fig. 7.28 *Transference of heat by conduction*

Under steady state conditions, heat flows at a steady rate from face 1 to face 2, in a direction normal to the faces. The rate of flow of heat is found to be proportional to the (*i*) area A, (*ii*) the temperature fall, $-\delta\theta$, and inversely proportional to the thickness, δx. Hence, writing the rate of flow of heat by $\delta Q/dt$, we have $\delta Q/dt \propto A(-\delta\theta/\delta x)$, which can be put in the form

$$\frac{\delta Q}{\delta t} = -KA\frac{\delta\theta}{\delta x} \tag{7.105}$$

where the constant of proportionality K is a constant for the material, and is called the coefficient of conductivity. In words,

"When heat is flowing normally between the faces of a slab of uniform thickness, under steady state conditions, the rate of flow of heat is equal to the product of the coefficient of thermal conductivity, area of slab, and temperature gradient".

In Eq. (7.105), if $A = 1$, and $\dfrac{\delta\theta}{\delta x} = 1$, then the rate of heat flow is

$$\frac{\delta Q}{\delta t} = -K \tag{7.106}$$

Hence, the coefficient of thermal conductivity, K, of a material is the rate of flow of heat through unit area under unit temperature gradient, normal to the faces of a thin slab of material of uniform thickness, under steady state conditions.

Temperature distribution along a bar of uniform cross section under steady state conditions:

Consider a long cylindrical bar of uniform cross-sectional area A. Let the ends of the rod be maintained at temperatures θ_1 and θ_2, with θ_1 being the greater, as shown in Fig. 7.29.

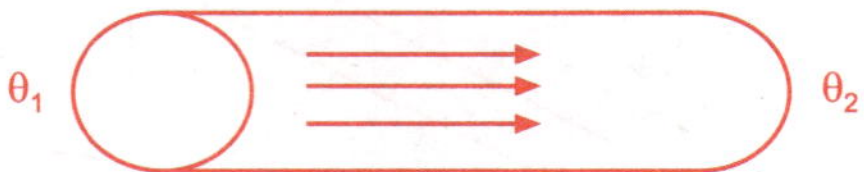

Fig. 7.29 *Heat flow along bar of uniform cross section*

Normally, heat will be lost through the sides of the bar due to convection and radiation. If, however, the bar is made of a very good conductor, the proportion of heat escaping from the sides is small compared with that passing along the bar. If the radius of the bar, r, is large, then the ratio of surface area to cross-section being $2\pi rl/\pi r^2 = 2l/r$, the proportion of heat escaping from the sides is also reduced. Provided we have a thick bar of very good conductor material we can assume that no heat (negligible heat) escapes from the sides of the bar. Under these conditions the rate of flow of heat can be assumed to be same all along the bar.

For any section of the bar, at a distance x from the hot end, we have, $\dfrac{\delta Q}{\delta t} = -KA\dfrac{\delta \theta}{\delta x}$,

where the area A and the value of K are the same for all values of x. That is, the temperature gradient is the same throughout the length of a thick bar of a very good conductor. The variation of θ with the distance x is shown in Fig. 7.30.

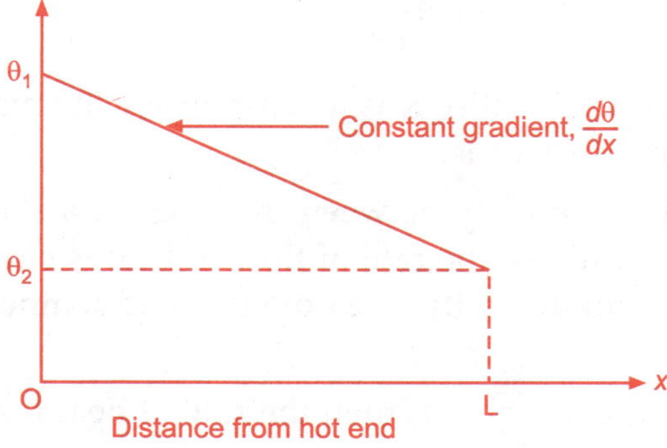

Determination of Conductivity

Searle's Method for the Conductivity of a Very Good Conductor: Let us find the coefficient of thermal conductivity of copper, which is a very good conductor of heat.

The specimen is taken in the form of a thick bar of copper, about 5 cm in diameter, polished, surrounded by dry felt, and enclosed in a wooden box. Steam circulating through a chamber attached to one end of the bar maintains this at 100°C. Water from a constant head apparatus, running through a copper spiral tube soldered on the other end maintains this end at a steady low temperature, and also allows the rate of flow of heat through the bar to be measured. Thermometers T_1 and T_2 are placed in holes in the bar at a known distance d apart, the holes containing mercury to ensure good thermal contact. These holes should be as narrow and shallow as possible, for they disturb the ideal flow conditions, as shown in Fig. 7.31.

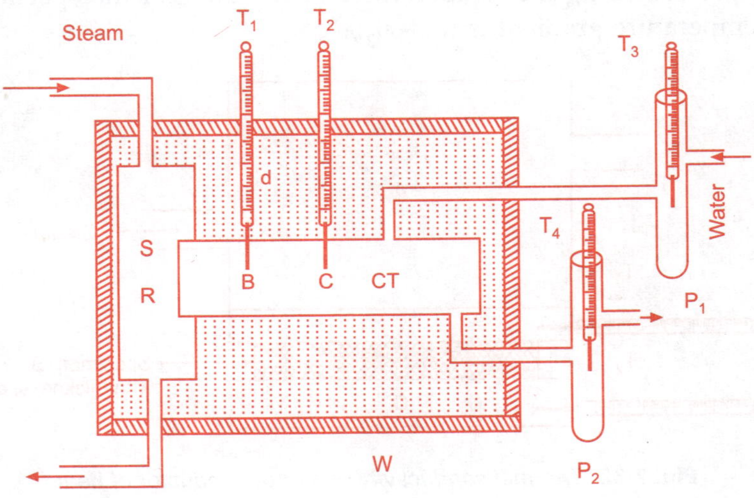

Fig. 7.31 *Searle's method to determine thermal conductivity of a good conductor*

The diameter of the bar, D, is measured in several places, the value of the area $\pi D^2/4$ calculated in each case, and the average taken. The distance d between the centers of the holes in which the thermometers are kept, is also taken. The currents of cold water and steam are started, and the apparatus is left to attain a steady state. When the readings θ_1, θ_2, etc., of all the four thermometers are steady, these are observed, and the water flowing through the copper tube in time t sec is collected and its mass m found. If s be the mean specific heat of water between $\theta_3°$ and $\theta_4°$, the heat supplied in t sec to this water is $ms(\theta_3 - \theta_4)/t$ cal/s. This is the rate of flow of heat down the bar.

The uniform temperature gradient is $\left(\dfrac{\theta_1 - \theta_2}{d}\right)$ °C/cm. Hence, as the area, rate of flow of heat, and temperature gradient are all known, K is calculated by substituting in the expression

$$K = \frac{\text{rate of flow of heat}}{\text{area} \times \text{temperature gradient}} \tag{7.107}$$

Determination of Thermal Conductivity of a Very Bad Conductor

The apparatus is shown in Fig. 7.32 and consists of a steam chest B, the bottom of which is a thick brass block, with a hole to receive a thermometer T_1. The specimen S in the form of a thin circular disc, is sandwiched between the block and a second cylindrical brass block C, holding a second thermometer, T_2. The diameter, D, and the thickness, d, of the specimen are measured. The block C is suspended from a fixed support by threads. Hence, heat can reach the specimen only by conduction.

Steam is passed through the chest until the thermometers T_1 and T_2 register steady temperatures, θ_1 and θ_2, respectively. Since brass is an extremely good conductor, the thermometers are recording the temperatures of the faces of the specimen. Thus, the steady state temperature gradient is $(\theta_1 - \theta_2)/d$.

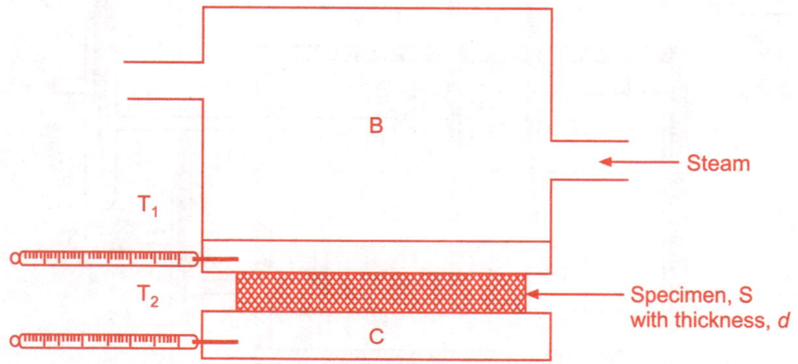

Fig. 7.32 *Thermal conductivity of a poor conductor of heat*

The rate of loss of heat from block C to the surroundings from its side and base equals the rate of flow of heat downwards through the specimen in the steady state. In order to determine the rate of flow of heat through the specimen, the specimen alone is kept on top of block C, and the block is heated gently with a Bunsen burner from $\theta_2°$C to a slightly higher temperature of $\theta_3°$C. The time t taken for the temperature to fall to $\theta_4°$C, which is as far below $\theta_2°$ as $\theta_3°$ was above it, is found. The average loss of heat over this interval is then given by

$$\frac{dQ}{dt} = \frac{ms(\theta_3 - \theta_4)}{t} \tag{7.108}$$

This is the required rate of flow through the specimen in the steady state. The area A of the specimen is given by $A = \pi D^2/4$. Since the area, rate of flow of heat and the temperature gradient are known, the coefficient of thermal conductivity, K, is calculated using Eq. (7.107) as

$$K = \frac{ms(\theta_3 - \theta_4)}{t} \frac{d}{A(\theta_1 - \theta_2)} \tag{7.109}$$

It is assumed that no heat is lost from the curved edges of the specimen in the steady state. Hence the specimen must be quite thin.

Conduction Through Composite Walls in Series

Consider two parallel sided slabs S_1 and S_2 of thicknesses x_1 and x_2, respectively, and coefficients of thermal conductivities k_1 and k_2, respectively. Heat flows from face F_1 of S_1, which is at $\theta_1°C$ to face F_2 of S_2 which is at $\theta_2°C$. Let θ be the temperature of the interface between the two slabs. Let A be the area of cross section of the slab. The rate of flow of heat through the composite slab, dQ/dt, is given by

$$\frac{dQ}{dt} = \frac{k_1 A(\theta_1 - \theta)}{dx_1} = \frac{k_2 A(\theta - \theta_2)}{dx_2} \tag{7.110}$$

From Eq. (7.110), we can get the interface temperature, θ as

$$\theta = \frac{(k_1\theta_1/x_1) + (k_2\theta_2/x_2)}{(k_1/x_1) + (k_2/x_2)} \tag{7.111}$$

Further, we can get the following relations from Eq. (7.110)

$$(\theta_1 - \theta) = \frac{dQ}{dt}\frac{x_1}{k_1 A} \tag{7.112}$$

and

$$(\theta - \theta_2) = \frac{dQ}{dt}\frac{x_2}{k_2 A} \tag{7.113}$$

Subtracting Eq. (7.113) from Eq. (7.112) and solving for dQ/dt, we get

$$\frac{dQ}{dt} = \frac{(\theta_1 - \theta)}{\dfrac{x_1}{k_1 A} + \dfrac{x_2}{k_2 A}} \tag{7.114}$$

Conduction Through Conductors in Parallel

Consider two thin parallel slabs which are connected by two conducting rods, normal to the planes of the slabs, as shown in Fig. 7.33. Let the distance between the slabs be x.

Let the areas of cross section of the two rods be A_1 and A_2, with the coefficients of thermal conductivities k_1 and k_2, respectively. Let the slab S_1 be at a steady temperature of $\theta_1°C$ and S_2 be at a lower temperature of $\theta_2°C$. Heat will flow from slab S_1 towards S_2, through the two rods. The heat flow rate through the first rod is given by

$$\frac{dQ_1}{dt} = \frac{k_1 A_1(\theta_1 - \theta_2)}{x} \tag{7.115}$$

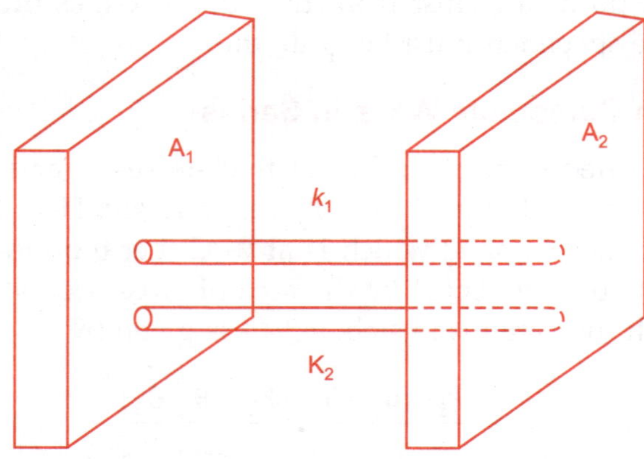

Fig. 7.33 *Conductors in parallel*

and heat flow rate through the second rod is

$$\frac{dQ_2}{dt} = \frac{k_1 A_2 (\theta_1 - \theta_2)}{x} \qquad (7.116)$$

The total heat flow from S_1 to S_2 can be obtained by adding the heat flows through both the rods. Hence, the total heat flow rate is given by

$$\frac{dQ}{dt} = \frac{dQ_1}{dt} + \frac{dQ_2}{dt} = \frac{(\theta_1 - \theta_2)}{x}(k_1 A_1 + k_2 A_2) \qquad (7.117)$$

7.4.2 Transference of Heat by Convection

Mode of transference of heat by a heated material which moves carrying its heat with it is called convection. This mode of heat transference can take place only in fluids (liquids and gases), where the particles can move freely.

Consider the case of a hot body cooling in air. The rate of flow of heat per unit area depends on the temperature difference between the body and the air, and upon the thermal conductivity of the air. The rise in temperature produced in the layer of air in contact with the hot body depends on the specific heat under these conditions. The change in density resulting from this rise in temperature depends on the volume coefficient of the air. Before it was heated, the weight of this layer of air was balanced by the upward thrust exerted on this layer by the lower layer of air. As the layer got heated its density comes down and it becomes lighter. However, the upward thrust is the same as before, and hence, there is a net upward thrust on this layer now. The upward force on this layer of air accelerates it until it moves at such a speed that the viscous resistance to its motion equals the net upward thrust. As heated air rises from the surface of the body a fresh layer of cold air moves up to take its place and thus the cooling process continues.

Convection can be either

(*i*) free or natural convection, or

(*ii*) forced convection.

The example discussed above is a free convection phenomenon. During summer in coastal areas such as Mangalore and Chennai, the sea water gets heated slowly during the day. By night, the air over the water gets sufficiently heated and rises up. A strong breeze will blow from land towards sea at night to take the place of the rising hot air. During daytime, the land gets heated quickly and the air over land becomes hot and rises upwards. A strong breeze will blow from sea towards land during daytime to take the place of the rising hot air over land. These natural phenomena are examples of natural convection.

In forced convection, a steady stream of fluid is sent past the hot body by external means. Ceiling fans used in homes during summer create such forced convection by creating a forced circulation of air. The radiator fan in an automobile cools the radiator by creating a forced convection in the vicinity of the radiator. In cold countries such as Canada, the warm air from furnaces is driven into the cold rooms in homes during winters for home heating.

7.4.3 Transference of Heat by Radiation

Transference of heat from a hot body to a cooler body without appreciable heating of the intervening space is called radiation. It is now known that thermal radiation or "radiant heat" is a means of transferring energy by transverse electromagnetic waves, and is similar in nature to light, radio waves, ultra-violet rays and X-rays. It is not necessary to have an intervening medium for this mode of transference of energy. However, it is difficult to imagine the propagation of waves without a medium, and it is still usual to speak of electromagnetic waves travelling in ether, which is regarded as a medium pervading all space. Scientific proof for the existence of such an ether has never been obtained. If there is no ether in space, then we must suppose that space itself has the necessary electromagnetic properties for wave propagation. All electromagnetic radiations travel in free space with the velocity of light, 3×10^6 m/s.

Suppose that by means of suitably transmitting prisms, a complete spectrum of sunlight can be cast on a screen. It will have colors of a rainbow - red, orange, yellow, green, blue, indigo and violet. If a sensitive thermal detector is moved along the colored portion, it will show that the screen is receiving heat as well as light. If the instrument is moved beyond the red end of the spectrum, it indicates strong reception of heat, showing that there is an invisible extension of the sun's spectrum beyond the red end. This is known as the infra-red region. the same instrument shows that the heating effect extends beyond and above the violet end of the spectrum, into the ultra-violet region. Whenever any radiation, visible or invisible, acts on a body, heat is generated.

Even at the extreme limits of the electromagnetic spectrum, heat transference takes place by radiation, even though only a part of the energy is involved. When radio waves are absorbed by aerials, current is induced in it. Whenever a current flows in a conductor heat is produced. While radiations from all parts of the spectrum may generate heat when absorbed we will deal with only infra-red radiations. This is because, (*i*) radiation emitted from most practical sources of heat is in the infra-red region, and (*ii*) other radiations such as X-rays and radio waves have remarkable distinctive properties which make their heating effects of trivial interest.

Optical Properties of Infra-Red Radiation

Infra-red radiation, like light, travels in straight lines in a homogeneous medium. On a hot summer day, a person is shielded from direct radiation from the sun by standing under a tree with thick foliage, or under an umbrella or a shamiana.

Infra-red radiation travels in empty space with the same velocity as that of light. When the radiation of light from a source is cut off, the radiation of heat is also cut off. For example, when eclipses of the sun occur, the heat is cut off at the same instant as the light.

Infra-red radiation, like light, is absorbed by dark, rough surfaces and is reflected by light and smooth surfaces. Infra-red radiation, or radiant heat, is a form of energy propagation, and has the same nature as that of light. The only difference between the visible light and the infra-red radiation is that the later is below the range of frequencies that are seen by the human eye.

Radiating Powers of Different Surfaces

A dull blackened surface is a good emitter and a good absorber, and a silvered and polished surface is a bad emitter and a bad absorber. In general, for all other types of surfaces, good absorbers are good emitters and bad absorbers are bad emitters.

The different radiating properties of different surfaces can be demonstrated using a cubical tin box which is applied with different surface finishes on its four vertical faces. Face 1 is made a dull black with lamp black, face 2 is coated with black enamel, face 3 is painted a dull white and face 4 is brightly polished. Water is maintained steadily boiling in the cube using a Bunsen burner underneath the cube. A thermocouple which must be screened from the direct radiation of the Bunsen burner under the cube, is held at exactly the same distance from the centre of each face in turn. The observations will show that face 1 is radiating strongly, faces 2 and 3 slightly less strongly, and face 4 very feebly. All the four faces under test are at the same temperature and considering the two extremes 1 and 4, the experiment shows that a dull black surface radiates much more strongly than a polished metal surface. We know that a blackened surface absorbs radiation strongly, and a polished metal surface absorbs only a fraction of the radiation falling on it.

Although it is hard to believe that a polished surface is a bad emitter it is true. A polished surface is a poor absorber of radiation but reflects most of the radiation falling on it. In the above experiment of the cube, the blackened surface absorbs the heat radiated from within the body and simultaneously passes it outwards. A polished surface reflects back into the body the radiation which strikes its inner surface, absorbing little so that there is little available to be passed out.

7.4.4 The Black Body

Black surfaces absorb all the wavelengths in the visible spectrum of light, and hence the surface looks black. In physics, the term black body is used to denote a perfect absorber of all radiations.

A perfectly black body, which would absorb all types of electromagnetic radiations completely, is extremely difficult to imagine, but we will consider only that part of electromagnetic radiations which includes ultraviolet, visible, and infra-red radiation. A cavity in the form of a hollow sphere, with the inside coated with rough black material, and with a small hole in the surface for admission of radiation will absorb almost all of the radiation falling on the hole, and such a cavity may be regarded as a perfectly black body.

Since a perfectly black body is the most perfect absorber of radiation, it must also be the most perfect emitter. The radiation proceeding from such a cavity for any temperature of the walls is called the black body radiation or cavity radiation for that temperature.

Suppose that a perfectly black body is contained in a uniform temperature enclosure. It absorbs all the radiation falling on it, and the intensity and the kind of radiation it emits is exactly the same in intensity and kind of radiation it absorbs. From this we conclude that the radiation within a uniform temperature enclosure itself is exactly the same in intensity and kind as the radiation absorbed and emitted by a perfectly black body at the same temperature. Further, as the radiation within the enclosure depends only on the temperature, the intensity and kind of the radiation from a perfectly black body depends only on its temperature. Hence, the particular intensity and kind of radiation which depends only on the temperature of the radiator, is referred to as the temperature radiation, or full radiation, or, cavity radiation, or black body radiation.

7.4.5 Intensity of Radiation; Emissive and Absorptive Powers

(*i*) Emissive Power:

Emission of energy by radiation from the surface of a body can be considered as

(*a*) the total energy emitted per square centimeter per second, which is the total emissive power e for that particular temperature, or

(*b*) the energy emitted per square centimeter per second within a finite wavelength range between the wavelengths λ and $\lambda + d\lambda$, which depends on the size of the wavelength interval $d\lambda$ itself, and is expressed as $e_\lambda d\lambda$, where e_λ is called the emissive power for wavelength λ.

For unit area and unit time, the total energy of the radiation of all wavelengths is e, while the energy between λ and $\lambda + d\lambda$ is $e_\lambda d\lambda$. Clearly e is the sum of all the terms $e_\lambda d\lambda$ for all the wavelengths concerned so that

$$e = \int e_\lambda d\lambda \tag{7.118}$$

for a continuous range of λ's. The capital letters E and E_λ will be used instead of e and e_λ when referring to a perfectly black body.

(*ii*) Absorptive Power

Consider a body of any type of surface at equilibrium in a uniform temperature enclosure. The energy emitted per square centimeter per second in all directions between wavelengths λ and $\lambda + d\lambda$ is $e_\lambda d\lambda$ where e_λ is the emissive power of the surface for that particular wavelength and temperature.

Let the energy falling on unit area per second within the same wavelength limits be dQ, which depends only on the temperature of the enclosure. Let the fraction of the incident energy absorbed between the wavelengths λ and $\lambda + d\lambda$ be a_λ, This is called the absorptive power of the surface. The total energy absorbed by unit area per second is thus $a_\lambda dQ$.

As the temperature of the body is constant, the rate of emission of energy equals the rate of absorption,

$$e_\lambda d\lambda = a_\lambda dQ \tag{7.119}$$

Rewriting this equation

$$\frac{e_\lambda}{a_\lambda} = \frac{dQ}{d\lambda} \tag{7.120}$$

Since $d\lambda$ is a chosen fixed interval, and dQ depends only on the temperature of the enclosure, the right hand side of this equation must be constant. Therefore, whatever the nature of the body, the ratio e_λ/a_λ is constant.

Kirchhoff's Law: If the body is a perfect black body, we write $E\lambda$ for the emissive power, and the absorptive power is I, and consequently, $e_\lambda/a_\lambda = E_\lambda$. Hence, at any given temperature Kirchhoff's law states that,

"The ratio of the emissive power of any body for any wavelength to its absorbing power for that wavelength at that temperature is constant, and is equal to the emissive power of a perfect black body at that temperature".

The ratio, e_λ/E_λ, is often referred to as the emissivity. Since $e_\lambda/E_\lambda = a_\lambda$, the emissivity as defined here is equal to the absorbing power.

E_λ increases as the temperature rises. Hence, it is important to realize that e_λ/a_λ increases as the temperature rises and is only constant in the sense that it has the same value for all kinds of surface at a fixed temperature and wavelength.

An ideal surface for which al is constant for all wave-lengths (or for which e_λ is exactly the same fraction of E_λ at all wavelengths), is called a grey body.

7.4.6 Stefan's Law for the Total Radiation from a Black Body

The total energy emitted per square centimeter per second, E, from a perfectly black body depends only on the absolute temperature T. The energy emitted per square centimeter per second between wavelengths λ and $\lambda + d\lambda$ by a perfectly black body, given by $E\lambda d\lambda$, depends on the temperature T, the wavelength λ and the size of the interval $d\lambda$. The emissive power, E_λ, depends on the temperature T and the wave-length λ only.

Stefan's law for the total radiation from a black body states that the energy emitted per square centimeter per second is proportional to the fourth power of the absolute temperature

i.e.

$$E = \sigma T^4 \tag{7.121}$$

where σ, a constant, is called Stefan's constant and the value is $\sigma = 5.735 \times 10^{-5}$ erg/cm^2/sec/deg^4. The above equation refers specifically to emission, and not to the net loss. If the black body is surrounded by a perfectly black surface at temperature T_0, the gain from the surroundings is σT^4, and the net loss of energy per square centimeter per second is given by

$$E_{net} = \sigma(T^4 - T_0^4) \tag{7.122}$$

This law is often referred to as the fourth power law, and is also called the Stefan-Boltzmann Law, since Boltzmann later gave a theoretical proof for it on thermodynamical grounds.

7.4.7 Temperature of the Sun

Sun's surface temperature can be found by using Stefan's law, if we know the total energy emitted per square centimeter per second from the sun. Determination of the rate at which the sun's radiation falls on unit area of the earth's surface enables the total energy per square centimeter per second from the sun.

The solar constant, S, is the energy received per square centimeter of the earth's surface in unit time, considering the earth to be at its mean distance from the sun and the radiation falling normally on the absorbing surface. This must be corrected for the absorption of the atmosphere. The mean value of S is about 1.93 calories per square centimeter per minute or about 1.34×10^6 ergs per square centimeters per second. The value of S at any time is obtained by letting the sun's radiation to enter a hollow chamber through an aperture of known area. The inside of the chamber must be black to absorb the radiation. The rate of reception of energy is found by surrounding the chamber with a continuous flow calorimeter. The rate of supply of heat by radiation is measured by reproducing the same temperature rise by supplying heat electrically at a known rate when the apparatus is screened from radiation. Such an instrument is called a Pyroheliometer, shown in Fig. 7.34.

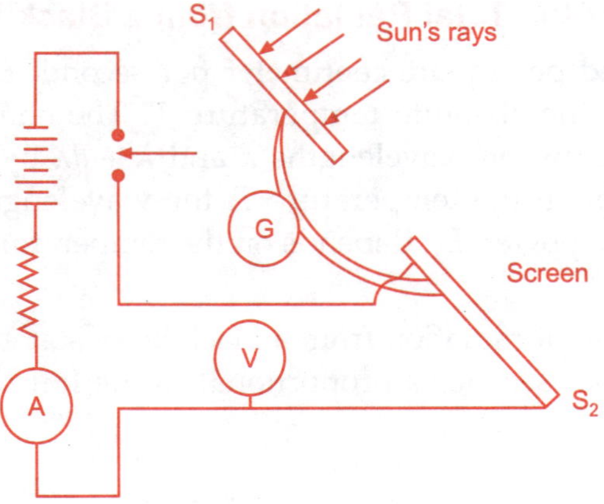

Fig. 7.34 *Pyroheliometer*

If r is the radius of the sun and R is the mean distance of the earth from the sun, then the energy radiated from the sun's surface, given by $4\pi r^2/3$, will be spread over an area of $4\pi R^2/3$ at a distance R from the sun. Let E be the energy emitted per square centimeter per second at the sun's surface. Then the energy received per square centimeter of the earth's surface per second, S, is given by

$$S = E\left(\frac{4\pi r^2/3}{4\pi R^2/3}\right) = E\frac{r^2}{R^2}. \tag{7.123}$$

With $r = 6.955 \times 10^5$ km (4.33×10^5 miles), $R = 1.4848 \times 10^8$ km (9.28×10^7 miles), and $S = 1.34 \times 10^6$ ergs per square centimeter per second we get

$E = 6.16 \times 10^{10}$ ergs per square centimeter per second.

Using Stefan's law, expressed in Eq. (7.121), where σ value is taken as 5.75×10^{-5} units, the temperature of the surface of the sun is obtained as $T = 5,720$ K. This is the temperature of the black body which would have the same value of E as that computed for the sun; that is, the brightness temperature of the sun's surface.

Utilization of Solar Energy

Countries close to the equator such as India, countries in South East Asia, North Africa, South and Central America receive a large amount of solar energy. The summers are very hot with the temperatures sometimes reaching 45°C in some places. If properly utilized, solar energy can supply a large portion of our energy needs.

The many devices which have been developed to utilize solar energy fall into three groups; heating systems, heat engines, and electrical generators.

A flat plate collector fixed on the roof of the house has been used to heat water, which can store heat for several days. The plate may be coated with an oxide film which is a good absorber of visible radiation but a poor emitter.

Steam and hot-air engines have been used to convert solar energy into mechanical energy, however, they are not very efficient.

Direct conversion of solar energy into electricity has been done using semiconductor photovoltaic cells. Solar furnaces, concentrating the radiation onto a small area by means of mirrors have been used. Such devices have also been used in solar cookers.

In some way or other, it is possible to trace the source of all known energy sources today to solar energy.

7.5 THERMODYNAMICS

Thermodynamics is the branch of physics that deals with the concepts of heat and temperature and the inter-conversion of heat and other forms of energy. Thermodynamics is a macroscopic science. It deals with bulk systems and does not go into the molecular constitution of matter. In fact, its concepts and laws were formulated in the nineteenth century before the molecular picture of matter was firmly established. Thermodynamic description involves relatively few macroscopic variables of the system, which are suggested by common sense and can be usually measured directly. A microscopic description of a gas, for example, would involve specifying the coordinates and velocities of the huge number of molecules constituting the gas. The description in kinetic theory of gases is not so detailed but it does involve molecular distribution of velocities. Thermodynamic description of a gas, on the other hand, avoids the molecular description altogether. Instead, the state of a gas in thermodynamics is specified by macroscopic variables such as pressure, volume, temperature, mass and composition that are felt by our sense perceptions and are measurable.

7.5.1 Relation between Heat and Work

Heat is one of the many forms in which energy can exist. The kinetic energy due to motion possessed by a body is converted into heat when friction forces the motion to slow down and stop. Electrical energy is converted into heat when passed through a resistor. Similarly, heat can be converted into mechanical energy in internal combustion engines or steam engines. In this section, we will study the conversion of heat energy into mechanical energy.

First Law of Thermodynamics: When mechanical work is completely converted into heat or heat is completely converted into mechanical work, for each unit of work that is converted into heat a definite quantity of heat is produced, and for each unit of heat converted into work a definite quantity of work is furnished.

This can be put mathematically as

$$W = JH \qquad\qquad (7.124)$$

where W is the work, H is the heat energy and J is the mechanical equivalent of heat, which is approximately 4.2 Joules per calorie.

7.5.2 Heat Engines

A heat engine is a thermo mechanical device that converts heat continuously into mechanical work. Heat engines use a working substance, which performs mechanical work when supplied with heat. All practical heat engines use one of the two working substances, either water (in the form of steam in reciprocating steam engines or steam turbines), or air (in internal combustion engines).

Carnot's Cycle, the Ideal Heat Engine

Consider a cylinder with walls that are non-conducting in nature and with perfectly conducting base, having a piston which is non-conducting and frictionless, as shown in Fig. 7.35. Let a fixed mass of a working substance be inside the cylinder, and let the piston be loaded with weights so as to keep the pressure of the working substance constant. Consider also that the cylinder can be placed on any one of three stands to enable required type of heat conductivity at the cylinder base. The three types of stands are:

   (*i*) a perfectly conducting stand X at a high temperature T_1, called the source,

   (*ii*) a perfectly non-conducting stand Y, and

   (*iii*) a perfectly conducting stand Z at a lower temperature T_2, called the sink.

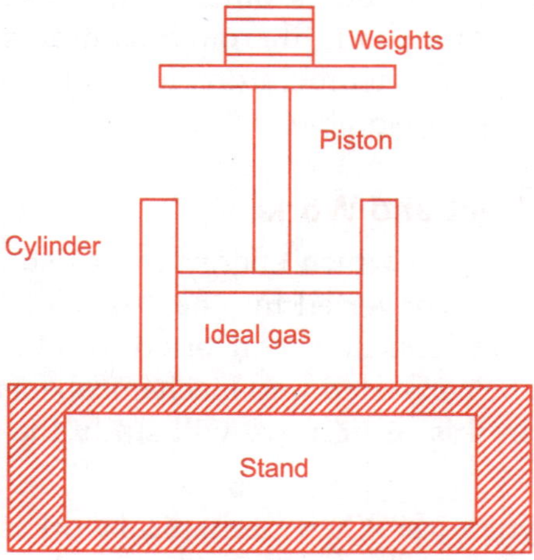

Fig. 7.35 *Carnot's cycle–ideal heat engine*

A sequence of operations are carried out on the working substance, and the resulting pressure and volume changes on it are represented in a *p-V* diagram as shown in Fig. 7.36.

1. At A, where the temperature is T_1, the cylinder is placed on the stand X, and the weights are slowly removed from the top of the piston. The working substance is allowed to expand isothermally to the state represented by point B.

2. At B, the cylinder is placed on stand Y, and further expansion of the substance is allowed. Since the stand Y is non-conducting, the expansion is adiabatic and it is continued until the temperature has fallen to T_2. This corresponds to the point C in the p-V diagram.

3. At C, the cylinder is placed on stand Z and the weights are slowly added to the top of the piston. Consequently, external work is done on the working substance, which is reduced in volume giving out heat isothermally to the sink Z. This is continued until the point D.

4. At D, the cylinder is placed on stand Y, and further weights are added to the top of the piston to compress the substance until the point A is reached. The process from D to A is under adiabatic conditions, since the stand Y is non-conducting.

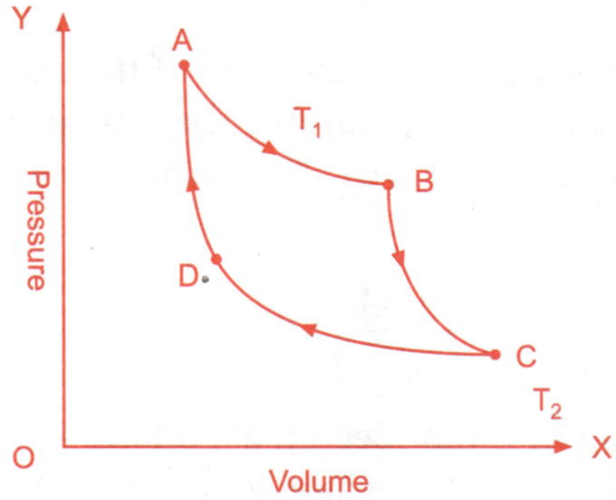

Fig. 7.36 *Pressure-volume diagram for Carnot's cycle*

Let us interpret the p-V diagram shown in Fig. 7.36. The area beneath the lines AB and BC represent the external work done by the substance on the piston in expanding. The area beneath the lines CD and DA is the work done on the substance by the weights placed on the piston. The difference between the work done by the substance and the work done on the substance, represented by the difference between the areas under lines AB, BC and lines CD, DA, is given by the area of the closed figure $ABCDA$. This represents the net amount of external work delivered by the engine during one cycle of its operation.

Considering the heat absorbed and the heat rejected by the substance during one complete cycle of operation, let Q_1 be the quantity of heat absorbed at the temperature T_1, and let a smaller quantity Q_2, be the heat rejected at temperature T_2. By the first law of thermodynamics, the difference $Q_1 - Q_2$, must be equivalent to the mechanical work done by the substance in one cycle.

Since the working substance has been taken through a complete cycle of operations from point A through states B, C, and D, and then back to A, whatever has happened to the working substance during the cycle is quite independent of the internal energy of the substance at A. Hence the absorption and rejection of heat or the mechanical work done during the cycle are independent of the working substance used.

We may choose to carry out the Carnot cycle by starting at any point, such as A in Fig. 7.36, and traversing along each process in a direction against the arrows shown in the figure. Then an amount of heat Q_2 is removed from the lower temperature reservoir at T_2, and an amount of heat Q_1 is delivered to the higher temperature reservoir at T_1. This is possible only if work is done on the system by an outside agency. In this reversed cycle work must be done on the system which extracts heat from the lower temperature reservoir. Any amount of heat can be removed from this reservoir by simply repeating the reverse cycle. Hence, the system acts like a refrigerator, transferring heat from a body at a lower temperature to one at a higher temperature by the help of work supplied to it from outside.

Efficiency of the Engine

The efficiency of the engine is defined as the ratio of the external work delivered to the total energy absorbed at the high temperature. The total energy taken in by the engine is Q_1, and the external work delivered is $Q_1 - Q_2$. Hence the efficiency, e, can be expressed mathematically as

$$e = \frac{Q_1 - Q_2}{Q_1} = 1 - \frac{Q_2}{Q_1} \qquad (7.125)$$

From the above equation, it can be seen that so long as the heat Q_2 delivered to the exhaust is not zero, the efficiency of a heat engine is less than one, i.e.; less than 100%. Every heat engine rejects some heat during the exhaust stroke. This represents the heat absorbed by the engine that is not converted to work in the process.

Reversible and Irreversible Processes

In the cylinder-piston arrangement described above, let the mass of gas have a volume V, at pressure p, and at temperature T. In an equilibrium state these thermodynamic variables remain constant with time. Suppose that the cylinder is placed on a large heat reservoir maintained at this same temperature T. Now, let us change the system to another equilibrium state where the temperature T is the same but the volume V is reduced by one-half. Consider this change done in two extremely different ways.

1. Let us depress the piston, which is assumed to be frictionless, very slowly by adding weights so that the pressure, volume, and temperature of the gas are, at all times, well-defined quantities. The process must be done with only very small changes at each time and allowing sufficient time for the equilibrium to

be established. Initially small weights are added to the piston. This will reduce the cylinder volume a little and the temperature will tend to rise and the system will depart from equilibrium, but only slightly. A small amount of heat will be transferred to the reservoir and in a short time the system will reach a new equilibrium state, its temperature again being that of the reservoir. Then add some more weights on the piston and repeat the procedure. The procedure is repeated until the volume is reduced to one-half. During the entire process the system is never in a state differing much from an equilibrium state. By decreasing the weights added during each repetition and by increasing the number of such repetitions, we arrive at an ideal process in which the system passes through a continuous succession of equilibrium states, which can be plotted as a continuous line on a p-V diagram as shown in Fig. 7.36. During this process a certain amount of heat Q is transferred from the system to the reservoir.

2. Let us depress the piston with a sudden jerk, and then wait until equilibrium with the reservoir to be re-established. During this process the gas is turbulent and its pressure and temperature are not well defined. The process cannot be plotted as a continuous line on a p-V diagram since we do not know the volumes corresponding to the pressure values. The system passes from one equilibrium state i to f through a series of no equilibrium states.

Processes of type 1 are called reversible. A reversible process is one that can be made to retrace its path by a differential change in the environment. As we cause the piston to move downward slowly, the external pressure on the piston exceeds the pressure exerted on it by the gas by only a differential amount dp. At any instant, if we reduce the external pressure on the piston by removing the weights on it, so that it is less than the internal gas pressure by dp, the gas will expand instead of contracting and the system will retrace the equilibrium states through which it had just passed. Process 1 is not only reversible but also isothermal because the temperature of the gas is very closely equal to the temperature of the reservoir, except for a very small change of dT when the weights are added.

Process 2 is irreversible. The volume is not related to pressure at any instant in a predictable way, and it is not possible to retrace the path. Because of turbulence in the gas, there is a possibility of some amount of energy loss, which also makes the process irreversible.

We could also reduce the volume adiabatically by removing the cylinder from the heat reservoir and putting it on a nonconducting stand. In an adiabatic process no heat is allowed to enter or to leave the system. An adiabatic process can either be reversible or irreversible. In a reversible adiabatic process we move the piston exceedingly slowly, whereas in an irreversible adiabatic process we push the piston down suddenly by sudden loading on the piston.

Efficiency of Ideal and Real Engines

Carnot stated a theorem of great practical importance in connection with the efficiency of a heat engine: The efficiency of all reversible engines operating between the same two temperatures is the same, and no irreversible engine working between the same two temperatures can have greater efficiency than this. This theorem also implies that the efficiency of a reversible engine is independent of the working substance and depends only on the temperatures. Fully reversible engine is an ideal engine and all practical or real engines are strictly irreversible even though they may be made to work under close to reversible conditions.

In order to prove the above theorem, consider two reversible engines H and H', operating between temperatures T_1 and T_2, with $T_1 > T_2$. The two engines may have different working substances, or they may have different initial pressures and lengths of stroke. Let the engine H run forward and H' run backward (refrigerator). Engine H receives heat energy Q_1 at T_1 and gives out heat energy Q_2 at T_2. Engine H' receives heat Q_2' at T_2 and gives out heat Q_1' at T_1. Let us connect the two engines mechanically, and adjust the stroke lengths so that the work done per cycle by H is just sufficient to operate H'. Let us suppose that the efficiency e of engine H is greater than the efficiency, e', of H', and check whether this is possible at all. By definition, we have $e > e'$, and in view of Eq. (7.125) this can be put as

$$\frac{Q_1 - Q_2}{Q_1} > \frac{Q_1' - Q_2'}{Q_1'} \tag{7.126}$$

Since the work per cycle done by one engine is equal to the work done per cycle on the other engine, we have $W = W'$, and hence,

$$Q_1 - Q_2 = Q_1' - Q_2' \tag{7.127}$$

Using this relation in Eq. (7.126), we get

$$\frac{1}{Q_1} > \frac{1}{Q_1'} \text{ or } Q_1 < Q_1' \tag{7.128}$$

From Eq. (7.127), we have

$$Q_1' - Q_1 = Q_2' - Q_2 \tag{7.129}$$

Hence from Eqs. (7.129) and (7.128) we see that $Q_2 < Q_2'$. Thus, the hot source gains heat $Q_1' - Q_1$ (positive) and the cool source loses heat $Q_2' - Q_2$ (positive). But no work is done in the process by the combined system $H + H'$, so that heat from a body at lower temperature has been transferred to a body at higher temperature without performing work, in direct contradiction to the second law. Hence, we conclude that e cannot be greater than e'. Likewise, by reversing the engines we can use the same reasoning to prove that e' cannot be greater than e. Hence, $e = e'$.

Now, suppose that H is an irreversible engine and H' is reversible. Following a similar procedure we can prove that e_{ir} cannot be greater than e'. But H cannot be reversed, and hence it is not possible to prove that e' cannot be greater than e_{ir}. Therefore, e_{ir} is either equal to or less than e'. Since $e' = e = e_{ir}$, we have

$$e_{\text{irreversible}} \le e_{\text{reversible}}. \tag{7.130}$$

SUMMARY

1. Heat is a form of energy that flows between a body and its surrounding medium by virtue of temperature difference between them. The degree of hotness of the body is quantitatively represented by temperature.

2. A temperature-measuring device (thermometer) makes use of some measurable property (called thermometric property) which changes with temperature. Different thermometers lead to different temperature scales. To construct a temperature scale, two fixed points are chosen and assigned some arbitrary values of temperature. The two numbers fix the origin of the scale and the size of its unit.

3. The Celsius temperature (t_C) and the Fahrenheit temperature (t_F) are related by

$$t_F = (9/5)t_c + 32$$

4. The ideal gas equation connecting pressure (P), volume (V) and absolute temperature (T) is: $PV = \mu RT$, where μ is the number of moles and R is the universal gas constant.

5. In the absolute temperature scale, the zero of the scale is the absolute zero of temperature– the temperature where every substance in nature has the least possible molecular activity. The Kelvin absolute temperature scale (T) has the same unit size as the Celsius scale (T_c), but differs in the origin:

$$T_c = T - 273.15 \; T_c$$

6. The latent heat of fusion (L_f) is the heat per unit mass required to change a substance from solid into liquid at the same temperature and pressure. The latent heat of vaporisation (L_v) is the heat per unit mass required to change a substance from liquid to the vapour state without change in the temperature and pressure.

7. The three modes of heat transfer are conduction, convection and radiation.

8. The Zeroeth law of thermodynamics states that 'two systems in thermal equilibrium with a third system are in thermal equilibrium with each other'. The Zeroeth Law leads to the concept of temperature.

9. Internal energy of a system is the sum of kinetic energy and potential energy of the molecular constituents of the system. It does not include the over-all kinetic energy of the system. Heat and work are two modes of energy transfer to the system. Heat is the energy transfer arising due to temperature difference between the system and the surroundings. Work is energy transfer brought about by other means, such as moving the piston of a cylinder containing the gas, by raising or lowering some weight connected to it.

10. The first law of thermodynamics is the general law of conservation of energy applied to any system in which energy transfer from or to the surroundings (through heat and work) is taken into account. It states that $\Delta Q = \Delta U + \Delta W$ where ΔQ is the heat supplied to the system, ΔW is the work done by the system and ΔU is the change in internal energy of the system.

11. For an ideal gas, the molar specific heat capacities at constant pressure and volume satisfy the relation, $Cp - Cv = R$, where R is the universal gas constant.

12. Equilibrium states of a thermodynamic system are described by state variables. The value of a state variable depends only on the particular state, not on the path used to arrive at that state. Examples of state variables are pressure (P), volume (V), temperature (T), and mass (m). Heat and work are not state variables. An Equation of State (like the ideal gas equation $PV = (\mu\, RT)$ is a relation connecting different state variables.

13. A quasi-static process is an infinitely slow process such that the system remains in thermal and mechanical equilibrium with the surroundings throughout. In a quasi-static process, the pressure and temperature of the environment can differ from those of the system only infinitesimally.

14. The second law of thermodynamics disallows some processes consistent with the First Law of Thermodynamics. It states:

Kelvin-Planck statement

No process is possible whose sole result is the absorption of heat from a reservoir and complete conversion of the heat into work.

Clausius statement

No process is possible whose sole result is the transfer of heat from a colder object to a hotter object. But simply, the Second Law implies that no heat engine can have efficiency η equal to unity or no refrigerator can have coefficient of performance α equal to infinity.

15 A process is reversible if it can be reversed such that both the system and the surroundings return to their original states, with no other change anywhere else in the universe. Spontaneous processes of nature are irreversible. The idealised reversible process is a quasi-static process with no dissipative factors such as friction, viscosity, etc.

Geometrical Optics

8

8.1 INTRODUCTION

8.1.1 Transparent, Translucent and Opaque Substances

We can see objects which are directly in front of us. However, if there is a brick wall in between the object and our eyes, we will not be able to see the object. If a clear glass sheet is kept in between our eyes and the object, we have no difficulty in seeing the object. In fact, we use specially ground glasses to correct our vision. We call such substances which allow us to see through them without any difficulty as transparent. Atmosphere contains air and air is present between our eyes and the objects, however, that does not prevent us from seeing the objects. Hence air is transparent. So also are some gases, such as hydrogen and helium. In a shallow lake, with crystal clear water without any disturbing waves on its surface, we can see the bottom of the lake very clearly. Clear water is transparent. So also are some liquids such as alcohol.

If we interpose a stained glass sheet or translucent glass sheet in between the object and our eyes, we will not be able to see the object clearly. We may be able to see only a hazy picture of the object from the other side. If a bright bulb is burning on the other side of the translucent glass sheet, we will be able to see the diffused light passing through the glass, but not the actual shape of the bulb. Such a partially transparent glass is a translucent substance. A translucent substance can let the light pass through to some extent, but will not allow us to see clearly the objects on the other side. Glasses with water vapor condensed on the surface, a very thin sheet of white paper are examples of translucent substances.

When the substance does not let the light pass through it at all, the substance is called opaque. A brick wall will not let any quantity of light to pass through it. Wooden panels, a very thick curtain, metal sheets are all examples of opaque substances.

Velocity of Light

Light is a form of energy transmitted in some way through space. The velocity of light in free space is a constant, and its magnitude is so large (3×10^8 m/s) that experimental measurement was a challenge until 1676.

The first attempt to measure the velocity of light was by Galileo in a very primitive fashion. The experiment consisted of two people holding two identical lanterns with shutters, and standing on two hill tops separated by about 1.6 km. The first one opened and closed the lantern instantaneously, and the second one opened his lantern as soon as he noticed the flash. Galileo argued that if light travelled with a finite velocity, the flash from the second lantern will be observed by the person with the first lantern after a certain amount of time delay. This experiment could not establish anything conclusively, because light travels with such a high velocity and the time taken to cover 1.6 km will be so small which cannot be detected at all with that method.

The Danish astronomer, Olaus Roemer, was the first to successfully measure the velocity of light. He was not specifically interested in measuring the velocity of light but his observations on one of Jupiter's moons, called Io, led to the discovery. His method involved observations of the motion of Io. When the moon moves around the planet Jupiter, an observer standing on earth will see the moon periodically disappearing from view as it enters the conical shadow cast by Jupiter, as shown in Fig. 8.1.

The orbital period of Io is known to be about 1.769 Earth days. The moon is eclipsed by Jupiter once in every orbit around Jupiter, as seen from the Earth. Roemer noticed that the time between successive eclipses of the moon became shorter as the Earth moved toward Jupiter and became longer as the Earth moved away from Jupiter. From these observations over a long period of time Roemer estimated that when the Earth was nearest to Jupiter, eclipses of Io would occur about eleven minutes earlier than predicted based on the average orbital period observed over many years. And 6.5 months later, when the Earth was farthest away from Jupiter, the eclipses would occur about eleven minutes later than predicted.

Roemer realized that the time difference must be due to the finite speed of light. That is, light from the Jupiter has to travel farther to reach the Earth when the two planets are on opposite sides of the Sun than when they are on the same side. Romer estimated that light required twenty-two minutes to cross the diameter of the Earth's orbit around the sun. The speed of light could then be found by dividing the diameter of the Earth's orbit by the time difference of 22 minutes.

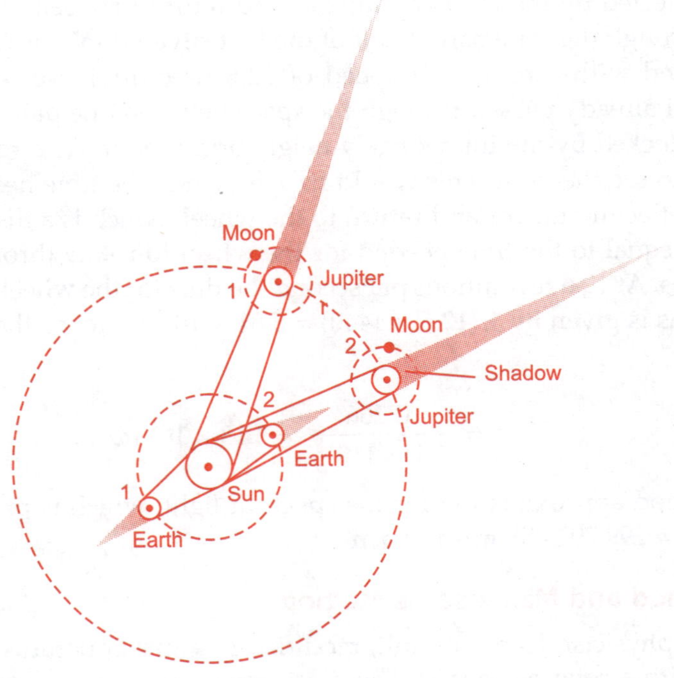

Fig. 8.1 *Observing the eclipses of jupiter's moons from earth*

The velocity of light was measured in the laboratory by the French scientist, H. L. Fizeau in 1849. His apparatus consisted of a wheel with 720 equally spaced teeth, a distant mirror which was kept 8633 m away, and a half silvered mirror which would reflect a beam of light to pass through the gaps between the teeth in the wheel as shown in Fig. 8.2. The width of gaps between the successive teeth in the toothed wheel was equal to the width of the teeth. The silvered half portion of the mirror reflected light and the other half was transparent and allowed the observer to see through it. Light from an arc light was focused on to the half silvered mirror using lenses.

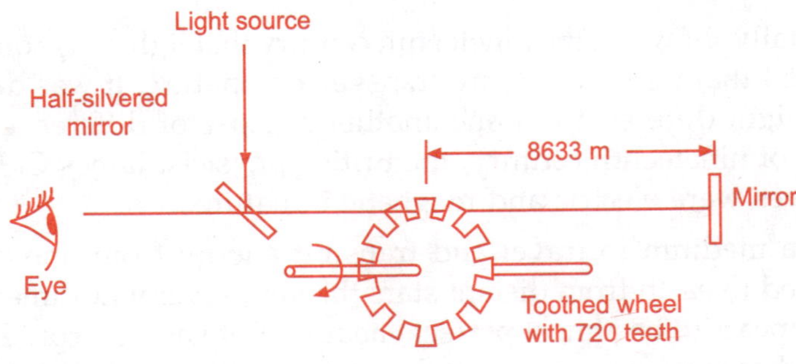

Fig. 8.2 *Fizeau's method to measure velocity of light*

When the toothed wheel was stationary, the light could pass through the gap between the teeth, get reflected by the far away mirror, and return between the same teeth and could be seen through the transparent half of the half silvered mirror. When the toothed wheel was rotated with a rotational speed of 12.6 revolutions/second, the outgoing light ray that had already passed through the space between one pair of teeth will have its return path blocked by the immediately neighboring tooth. As a result the observer will not be able to see the light. This would happen when the time needed for the light to travel to the reflecting mirror and return to the wheel, which is a distance of 2 × 8633 = 17,266 m, was equal to the time needed for the wheel to rotate through 1/(720 × 2) = 1/1440 revolutions. At 12.6 revolutions per second the time for the wheel to rotate through 1/1440 revolutions is given by 1/(12.6 × 1440) = 5.51 × 10^{-5} s. Hence the velocity of light was obtained as

$$c = \frac{17266}{5.41 \times 110^{-5}} = 3.13 \times 10^8 \text{ m/s.} \tag{8.1}$$

This was a good approximation for the speed of light, which is precisely measured presently to be $c = 299{,}792{,}458$ m/s in vacuum.

Foucault's Method and Michelson's Method

In 1850 a French physicist, Jean Foucault, modified Fizeau's apparatus by replacing the toothed wheel with a rotating mirror. The most precise measurements by the Foucault method were done by the American physicist Albert A. Michelson (1852-1931). His first experiments were performed in 1878, while the last, under way at the time of his death, were completed in 1935 by Pease and Pearson.

The most precise determination of the velocity of light to date, jointly by the National Bureau of Standards of USA and the University of Colorado, also in USA, was done using the fact that light is a wave phenomenon. An accurate measurement of frequency and wavelength gives the velocity of light as $c = f.\lambda$. With this approach, the velocity of light in vacuum is obtained to be in the range between 2.99792459×10^8 and 2.99792461×10^8 m/s. For simplicity, we can take the velocity of light, $c = 3 \times 10^8$ m/s.

Optical Medium

It was experimentally shown in the nineteenth century that light was composed of waves. It was also inferred that these waves are transverse in nature. It was demonstrated that the colors of the light differed from one another because of difference in wavelengths. Towards the end of nineteenth century, the British physicist James C. Maxwell showed that the light waves were electric and magnetic in nature.

Waves need a medium to travel and transmit energy from one point to another. Light is transmitted to earth from distant stars through several hexameters (10^{18} meters) of space which appears to be almost perfect vacuum. But vacuum could not support any wave motion and hence it was concluded that space is filled with "ether", a hypothetical medium, through which light waves could propagate.

Since light waves were transverse, it was concluded that ether must be a highly rigid material in order to enable light to travel at the velocity of 3×10^8 m/s. But it was puzzling that if the entire space were filled with this rigid material, how the stars and sun and planets could move through space without any apparent resistance.

Light travels along straight lines so far as it is travelling through a uniform medium, say, air. However, as it enters another medium, it will change direction and follow a different straight path in the new medium. If a beam of light is directed against a pool of water, its direction changes. This is illustrated in Fig. 8.3.

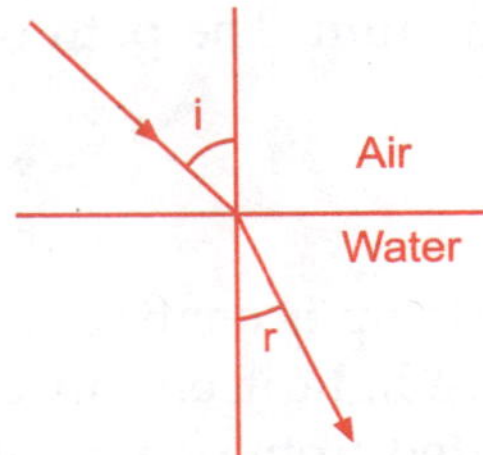

Fig. 8.3 *Refraction phenomenon when light passes to a different medium*

The angle of incidence *i* and the angle of refraction *r* are measured between the ray and the normal to the surface as shown.

In 1621, it was discovered by the Dutch physicist Willebrord Snell that for any ray passing from one medium into another

$$\frac{\sin i}{\sin r} = n \tag{8.2}$$

where *n* is a constant, i.e.; *n* is the same for any pair of media no matter what the angle of incidence is. This ratio *n* is called the refractive index of the second medium with respect to the first.

If the medium through which light is passing is not specified, it is assumed to be empty space.

Rectilinear Propagation of Light

Beams of sun rays coming through narrow gaps between the thick foliage of trees in the morning show that sun rays travel in straight lines. If there are some dust particles or water particles in its path, these rays can be seen very clearly. The light from a torch light or from an automobile headlight will also indicate that light travels in straight lines. In these cases, we do not see the light itself, but the effect of light when it falls on fine particles in its path. The fact that light from a small source such as a torch light, can be intercepted by opaque objects kept in its path, also indicates that light travels in straight lines. When we speak of rays of light we mean the path along which a narrow beam of light travels.

In the above examples, light is travelling in air. If light travels from air to some other transparent medium, such as glass or water, the rays are bent at the boundary. In the new medium, light will continue to move in a different straight path. Such change in direction when light goes from one medium to another is called refraction.

If a ray of light falls on a highly polished surface, it is reflected with a sharp change in direction to a new straight path. If the surface is unpolished, the light is reflected diffusely in all directions. Light travels with a finite velocity, 3×10^8 m/s, in vacuum. The velocity in air is almost the same as in vacuum, but considerably less in other transparent media, such as glass or water. Light travels between any two points in a medium in such a way that the time taken is a minimum. The path of shortest time is also the path of shortest distance—a straight line.

Light is a Form of Energy

Any light emitting object such as a lamp is continuously converting some other form of energy in order to emit light. Hence light itself must be a form of energy, or light is a means by which energy is transmitted from one point to another. Any light source can be subdivided into innumerable number of very small areas, which can be regarded as points. Rays from each such point source enter the eye and are focused by the eye's lens system to form a corresponding image on the retina, the receptive surface of the eye.

8.2   REFLECTION OF LIGHT

When a ray of light is incident on a highly polished surface, such as smooth flat silverware or a plane mirror, it sharply changes its direction and reverses its path. This is called reflection. The reflected ray of light also travels in a straight path. The directions of the incident and reflected rays are related in a definite manner, and these are specified in the form of Laws of Reflection.

Consider a narrow beam of light from a point source falling on a mirror as shown in Fig. 8.4.

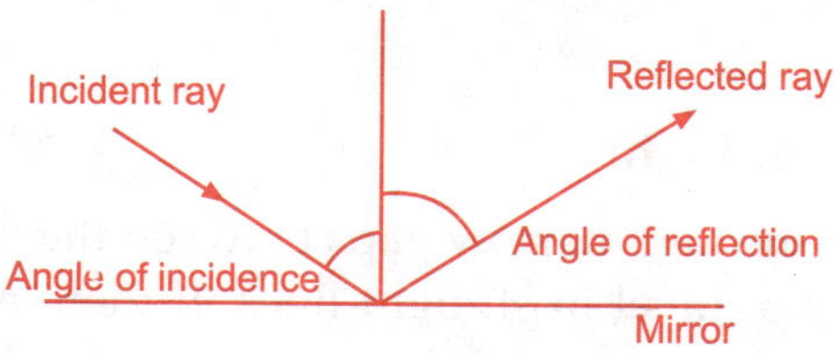

Fig. 8.4 *Reflection in a plane mirror*

Draw a perpendicular to the mirror at the point where the ray meets the mirror. It can be seen that the angle of incidence (angle between the incident ray and the perpendicular to the reflecting surface) is equal to the angle of reflection (angle between the reflected ray and the perpendicular to the surface).

The laws of reflection are given below:

1. The incident ray, the reflected ray and the normal to the surface at the point of incidence all lie in the same plane.

2. The angle of reflection is equal to the angle of incidence for all angles of incidence.

Reflection from Plane Mirrors

The formation of an image in a plane mirror is illustrated in Fig. 8.5.

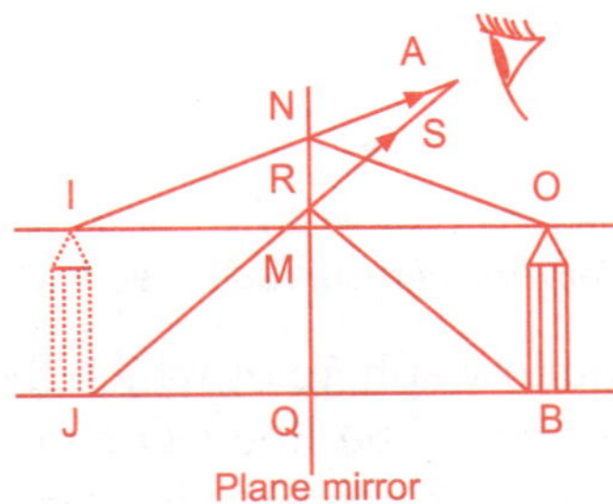

Fig. 8.5 *Virtual image in a plane mirror*

When the pencil is illuminated by a light source, such as the sun or an electric lamp, all points on the pencil send out an infinite number of rays in all directions. Consider the rays going out from O, the tip of the pencil. It will appear to the eye as though the reflected rays were all coming from a point I, which seems to be situated inside the mirror. Of course no light ever reaches or passes through the point I because it is an illusion. Point I may as well be inside a thick wall behind the mirror. The point I is called a virtual image of point O on the object.

Consider another ray OM emanating from point O, which is perpendicular to the mirror. In this case the angle of incidence is zero and hence the reflected ray comes back along the same line as the incident ray. The virtual image of point O is formed where the reflected ray MO and another reflected ray, NA meet at I.

If we want the complete virtual image of the pencil, we must find the virtual image of another point on the pencil, say, the base of the pencil B. This can be shown to be at J, where the rays SJ and BQ meet. Hence the virtual image of the pencil is given by IJ.

8.3 REFLECTION AT CURVED SURFACES

A highly polished and curved outside surface of a stainless steel vessel shows our reflected image which is smaller than normal or attenuated in size. The surface acts as a convex mirror. The curved mirror kept behind a torch light bulb or the headlights of an automobile focus the lights straight ahead, and they are examples of concave mirrors.

The simplest shape of a mirror with a curved surface is that of a small piece of a polished sphere. If the reflecting surface is towards the centre of the sphere, we call it a concave mirror and if the reflecting surface is on the outside of the sphere we call it a convex mirror.

Concave and convex spherical mirrors are shown in Fig. 8.6, where the reflecting surfaces are shown by the full lines.

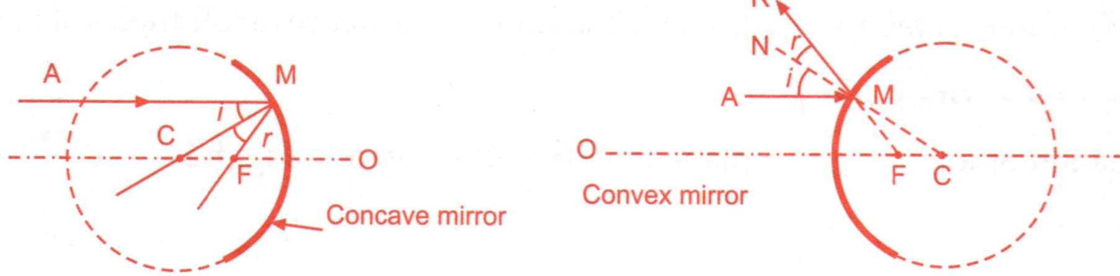

Fig. 8.6 *Mirrors with curved reflecting surfaces*

Point *C* is the centre of the hollow sphere of which the mirror is a part, and is called the centre of curvature of the mirror. The line *CO* connecting the centre of curvature and the centre of the mirror is the principal axis. A ray of light *AM*, parallel to the principal axis, will be reflected by the concave reflecting surface and will pass through the principal axis at point *F* as shown in Fig. 8.6(*a*). This point *F* is called the principal focus of the mirror. In the case of a convex mirror, the ray parallel to the principal axis, *AM*, will be reflected as though coming from *F* as shown in Fig. 8.6(*b*).

8.3.1 Image Formation in a Concave Mirror

A concave mirror is shown in Fig. 8.7, where *OM* is a ray from the tip of the pencil, parallel to the optical axis. The ray *OM* is reflected and passes through the principal focus *F*.

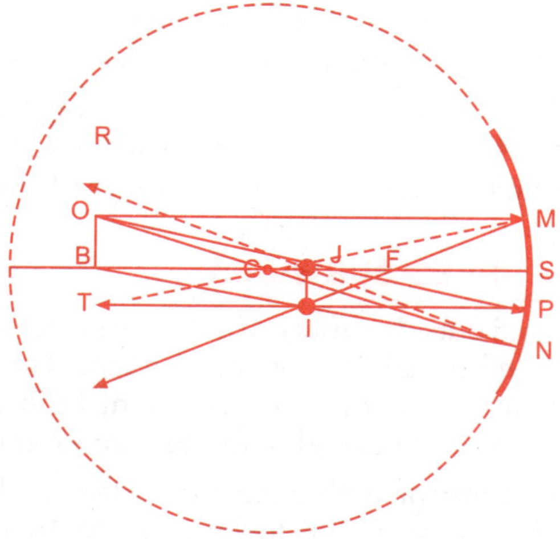

Fig. 8.7 *Image formation in Concave Mirror*

A perpendicular to the mirror at the point of incidence M contains the radius MC passing through the centre of curvature of the mirror C. Hence, the angle of incidence $i = \angle OMC$ and the angle of reflection $r = \angle CMF$ are equal. Since the ray OM is parallel to the principal axis, and the line MC intersects the two parallel lines OM and CS, we have $\angle MCF = i$. Consequently, the triangle MCF is isosceles and we have $MF = CF$. If point M is not too far from S, then $MF = CF = CM/2 = CS/2$, where $CS = \rho$ is the radius of curvature of the mirror. $CF = f$, is the focal length of the mirror. We have the relation, $\rho = 2f$. (Note: $CM = (MF + CF)\cos i = MF + CF$, since $i << 1$).

There will be innumerable rays emanating from the tip of the pencil and going towards the mirror and getting reflected. Let us consider another ray, say, OC. Since it travels along a radius of the sphere, it will strike the mirror at N and be reflected back towards C again. For this ray we have $i = r = 0$. The rays reflected at M and N meet at point I, which is the image of point O, the tip of the pencil. If we follow another ray, which passes through F, say, OF, it will strike the mirror at P and be reflected back parallel to the principal axis passing through the point I, as shown. All the light rays from O do in fact intersect at the point I.

In a similar fashion we can find the image of the base of the pencil B. The ray BC will be reflected back along the same line and any other ray, say, BN will be reflected back along NJ. The point where these two reflected rays intersect is at J and hence J is the image of point B. The image of the pencil OB is given by IJ. If we place a ground glass in this region, an actual image of OB would be focused on the ground glass. Such an image is called a real image.

In order to develop some geometrical relationships between the focal length and the distances from the object to the mirror and that from the mirror to the image, the incident and reflected rays are redrawn in Fig. 8.8.

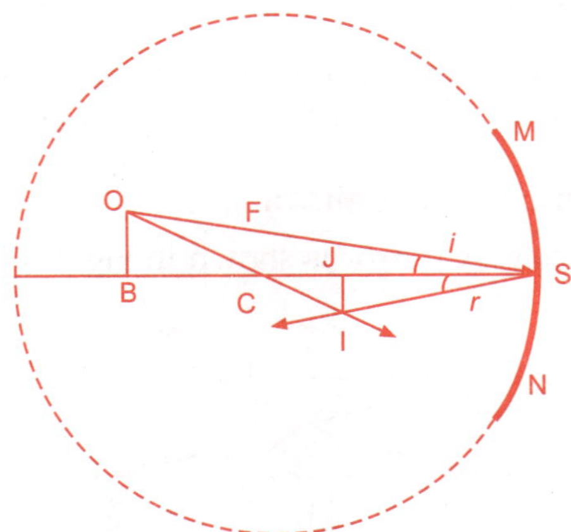

Fig. 8.8 *Geometrical relationship between focal length and distance from the object*

Ray OS is incident and the corresponding reflected ray is SI. The radius at S is SC, which is perpendicular to the mirror at S. Hence, the angle of incidence, $i = \angle OSC$, is equal to the angle of reflection, $r = \angle ISC$. In view of this, the triangles OSB and ISJ are similar. Consequently, we have

$$\frac{OB}{IJ} = \frac{BS}{JS} \tag{8.3}$$

or, in other words,

$$\frac{\text{object size}}{\text{image size}} = \frac{\text{object distance}}{\text{image distance}} = \frac{u}{v} \tag{8.4}$$

In Fig. 8.8, triangles OBC and IJC are also similar. Hence, we have

$$\frac{BC}{JC} = \frac{OB}{IJ} = \frac{u}{v} \tag{8.5}$$

But we have

$$\frac{BC}{JC} = \frac{u-\rho}{\rho-v} = \frac{u}{v} \tag{8.6}$$

where ρ is the radius of curvature of the mirror. Cross multiplying the terms in Eq. (8.6), we get

$$u\,(\rho - v) = v\,(u - \rho). \tag{8.7}$$

Dividing throughout by $uv\rho$, and using the relation that $\rho = 2f$, we get

$$\frac{1}{u} + \frac{1}{v} = \frac{1}{f}. \tag{8.8}$$

In the case of the concave mirror, the object distance u, the image distance v, and the focal length f, are all on the same side of the mirror and are considered positive.

8.3.2 Image Formation in a Convex Mirror

The image formation in a convex mirror is shown in Fig. 8.9.

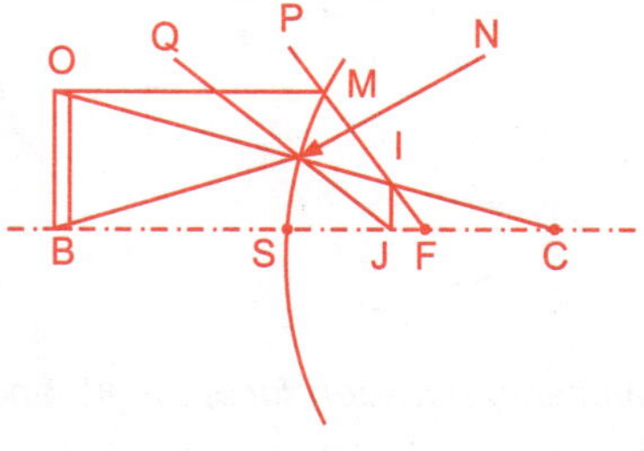

Fig. 8.9 *Image formation in convex mirror*

Ray *OM* from the top of a pillar, parallel to the principal axis, will be reflected along *MP* as though it were coming from the focus *F*, which is behind the mirror. The ray *ON* which is travelling along a radius (point *N*, which is on the mirror surface, is shown with an arrow due to lack of space), will be reflected and will travel back along the same path. Extensions of rays *MP* and *ON* inside the mirror will intersect at I to give the image of point *O*. Similarly we can obtain the image of the base point *B* of the pillar at *J*, which is the intersection of rays *NQ* and *BS* (ray *BN* will reflect back along *NQ*). The image *IJ* of the pillar *BO* is a virtual image because it cannot be projected on a ground glass screen held at *IJ*. The region *IJ* may not be accessible also.

The relation between the object distance, image distance and the focal length given in Eq. (8.8) is applicable in the case of a convex mirror also, if we take care to assign proper signs to the quantities involved. The focal length and the image distance, which are on the opposite side of the object distance, must be considered negative while the object distance is considered positive.

EXAMPLE 8.1

A convex mirror has a radius of curvature of 0.25 m. The object is placed at a distance of 0.3 m, find the image distance.

SOLUTION

Referring to Eq. (8.8) we have $u = 0.3$ m, $\rho = -0.25$ m, and hence the focal length $f = -0.125$ m. Using Eq. (8.8), we get

$$\frac{1}{0.3} + \frac{1}{v} = -\frac{1}{0.125}$$

Hence, $v = -0.088$ m. The negative value for v shows that the image is on the same side as the principal focus.

8.4   REFRACTION AT A PLANE SURFACE

8.4.1 Examples of Refraction

Refraction is the abrupt bending of the path of a ray of light at the boundary separating two transparent media. Some examples of the phenomenon of refraction are:

1. When a straight stick is immersed in a pail of water, the stick seems to be abruptly bent at the surface of water.

2. When we read a meter needle in a measuring instrument such as a volt meter, ammeter etc., the glass cover will introduce an error if we do not look at the meter needle in a perfectly vertical fashion. If we look at the needle from an angle the readings are subject to parallax errors.

3. When light is passed through a triangular prism, the incident ray will be bent at the two boundaries, as shown in Fig. 8.10. The bending angles differ for different frequencies and this causes dispersion and we can see a spectrum. We will study these phenomena in more detail later in the section on prisms.

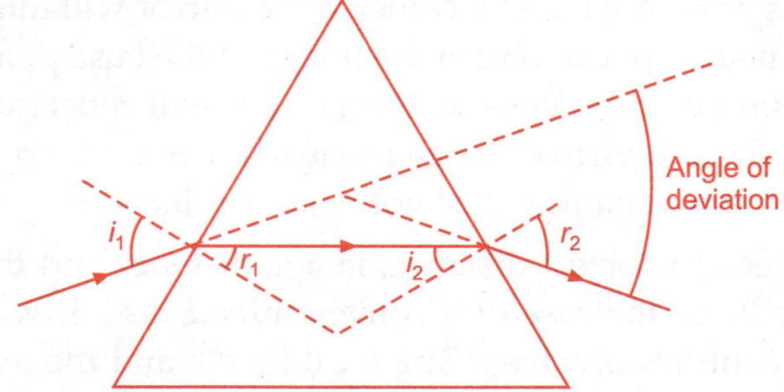

Fig. 8.10 *Refractions in triangular prisms*

8.4.2 Laws of Refraction

A ray of light travelling through a medium reaches the boundary separating that medium with another medium, as shown in Fig. 8.11.

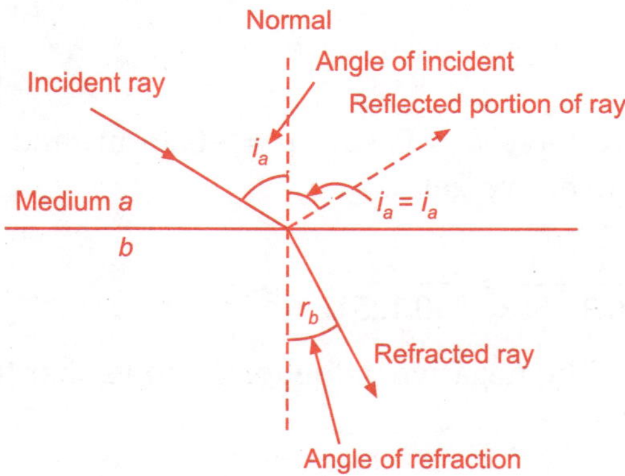

Fig. 8.11 *Refraction of light*

The directions of the beam incident on the boundary, and the refracted beam are specified in terms of the angles they make with the normal to the boundary at the point of incidence. A part of the beam is also reflected, which follows the laws of reflection that we have already discussed earlier.

The laws of refraction are:

Law 1. The incident ray, the refracted ray, and the normal at the point of incidence all lie in one plane.

Law 2 (Snell's Law): For the light of a given color, striking the boundary between two given media, the ratio of the sine of the angle of incidence to the sine of the angle of refraction is a constant for all angles of incidence.

The discovery of this law is due to Willebrord Snell (1591-1676). The ratio

$$n = \frac{\sin i}{\sin r} \tag{8.9}$$

is called the refractive index for the boundary concerned, and is a property of the two media separated by the boundary. This is also called relative refractive index, because the index of one medium is expressed relative to that of the other medium.

If refraction is considered when light ray travels from a vacuum to a medium, then the corresponding refractive index is termed as absolute refractive index. Consider a beam of monochromatic light (light of a single frequency) travelling in vacuum, making an angle of incidence i_0, with the normal to the surface of a substance a, and let r_a be the angle of refraction. The corresponding refractive index is given by

$$n_a = \frac{\sin i_0}{\sin r_a} \tag{8.10}$$

which depends only on the medium a and hence is referred to as the absolute refractive index of medium a.

The absolute refractive indices of most of the common glasses used in optical instruments fall between 1.46 and 1.96. Only a very few substances have refractive indices larger than 1.96. For example, diamond has a refractive index of 2.42.

The refractive index of air under standard conditions is about 1.0003, and hence it is assumed as unity for all practical purposes. The refractive indices of gases increase uniformly as their densities increase.

When a ray is passing from vacuum into a medium a, the angle of refraction, r_a, is always less than the angle of incidence i_0. Hence the ray is always bent towards the normal in the medium a. If the light is passing from the denser medium into the vacuum, the reverse is true and the ray is bent away from the normal.

The relative refractive index and the absolute refractive index can be related as follows:

Consider uniform flat rectangular plates of materials a and b placed parallel to each other with an arbitrary space between them, as shown in Fig. 8.12.

Let the space all around the plates be vacuum. A ray incident on the plate a from vacuum will have an angle of incidence of i_0. The angle between the ray and the normal at the point of incidence in plate a is r_a. From plate a, light enters the vacuum with angle of incidence of r_a and angle of refraction of i_0. When the light ray enters the plate b, the angle of incidence is i_0 and the angle of refraction is r_b.

Applying Snell's law to the refractions that have taken place, we obtain

$$n_a = \frac{\sin i_0}{\sin r_a}, n_b = \frac{\sin i_0}{\sin r_b} \tag{8.11}$$

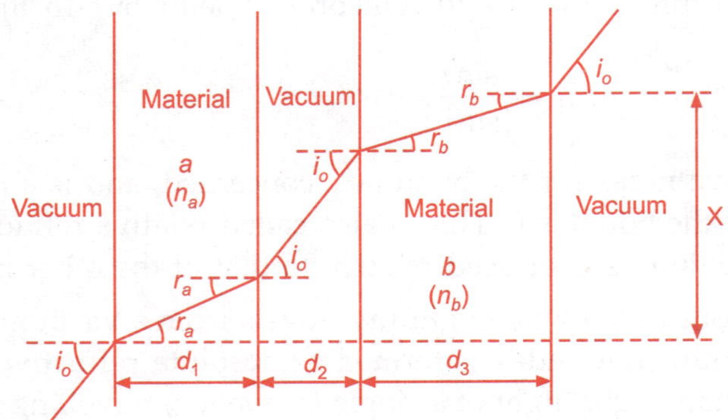

Fig. 8.12 *Refraction in two media separated by vacuum*

Dividing the second equation by the first we obtain n_a/n_b as

$$\frac{n_b}{n_a} = \frac{\sin r_a}{\sin r_b} \qquad\qquad (8.12)$$

This shows that the relative refractive index is obtained as the ratio of absolute refractive indices of the two mediums concerned.

8.4.3 Refraction through Multiple Refracting Media

Consider a ray of light passing through media m_1, m_2 and m_3 without any gap between each medium, as shown in Fig. 8.13.

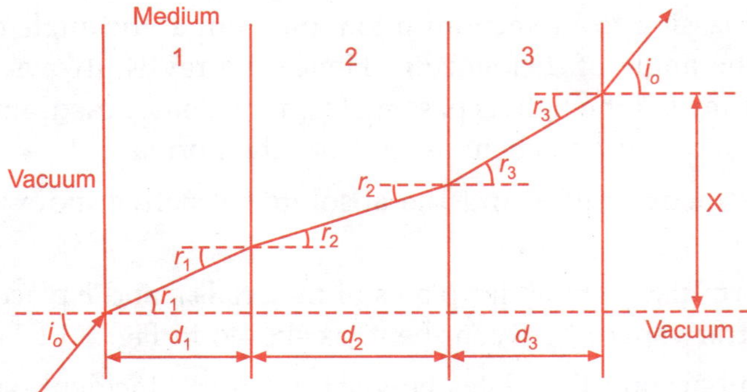

Fig. 8.13 *Refraction through multiple media*

Light travelling in vacuum, adjacent to medium m_1, is incident on m_1 with an angle of incidence of i_0. The angle of refraction when the ray enters a is r_1. The ray continues in medium a and hits the medium m_2 with the angle of incidence being r_1. The angle of refraction after the ray enters medium m_2 is r_2. Similarly, the angle of incidence when the ray hits medium m_3 is r_2 and the angle of refraction after entering medium m_3 is r_3. We have the relations

$$\frac{n_3}{n_2} = \frac{\sin r_2}{\sin r_3}, \frac{n_2}{n_1} = \frac{\sin r_1}{\sin r_2} \tag{8.13}$$

If we multiply the two relations in Eq. (8.13), we get

$$\frac{n_3}{n_1} = \frac{\sin r_1}{\sin r_3} \tag{8.14}$$

This indicates that the medium m_2, in between the media m_1 and m_3, does not influence the final outcome at all. The point where the ray emerges out of medium m_3 is displaced from the point where the ray is incident on medium m_1 by a distance x given by

$$x = d_1 \sin r_1 + d_2 \sin r_2 + d_3 \sin r_3$$

When there are multiple media $m_1, m_2, m_3, \ldots\ldots, m_N$, placed side by side, the ratio of the refractive index (absolute) of the last medium, m_N, to that of medium m_1, is given by

$$\frac{n_N}{n_1} = \frac{\sin r_1}{\sin r_N} \tag{8.15}$$

and the incident ray on m_1 is displaced from the emerging ray from the last medium m_N by an amount x given by

$$x = \sum_{i=1}^{N} d_i \sin r_i \tag{8.16}$$

8.4.4 Refractive Index in Terms of Real and Apparent Depths

When the water is clear and the surface is free of waves we are able to see the bottom of a shallow lake. We have seen that an object at the bottom of the lake appears to be nearer to the surface creating an illusion that the water is not very deep. However, if we put our hand and try to grab the object, we realize that the object is not exactly where we expected it to be. This is because of the effect of refraction. The apparent depth where we think the object is and the real depth where it is are different. This fact is used to find the refractive index of the liquid.

The liquid whose refractive index is to be found is placed in a tall jar. A pin O is kept at the bottom of the jar. A plane mirror M is kept over the top of the jar as shown in Fig. 8.14.

A second pin P is then moved vertically above the mirror until its image Q in M appears to coincide with the image I of O formed by refraction at the liquid surface. Since $MQ = MP$, Q can be located. The apparent depth of I is obtained as

$$SI = MQ - MS = MP - MS$$

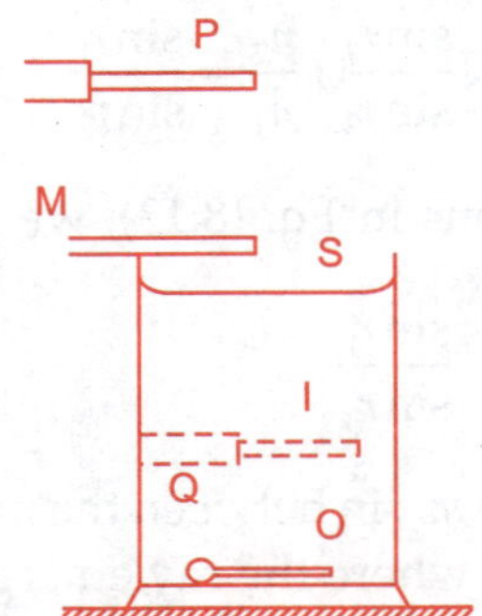

Fig. 8.14 *Refractive index of a liquid by real and apparent depth method*

where the letters P, M, S, Q and O identify the positions of the top pin, mirror, water surface, image of top pin P, bottom of the jar, respectively. Also Q and I are at the same height. The refractive index of air-liquid combination is obtained as

$$_a\mu_l = SO/SI$$

8.5 TOTAL INTERNAL REFLECTION

8.5.1 Critical Angle

A number of rays emanating from a point source P in a medium a and striking the surface of a second medium b are shown in Fig. 8.15. Let the refractive indices of a and b be given by n_a and n_b, respectively, with $n_a > n_b$. From Snell's law, we have

$$\frac{n_b}{n_a} = \frac{\sin i_a}{\sin r_b} \tag{8.17}$$

Since $n_a > n_b$, we have $\sin r_b > \sin i_a$. Thus, there must be some value of i_a less than 90° for which $\sin r_b = 1$, and consequently $r_b = 90°$. This is illustrated by ray 3 in the diagram (counting from left), which emerges just grazing the surface at an angle of refraction of 90°.

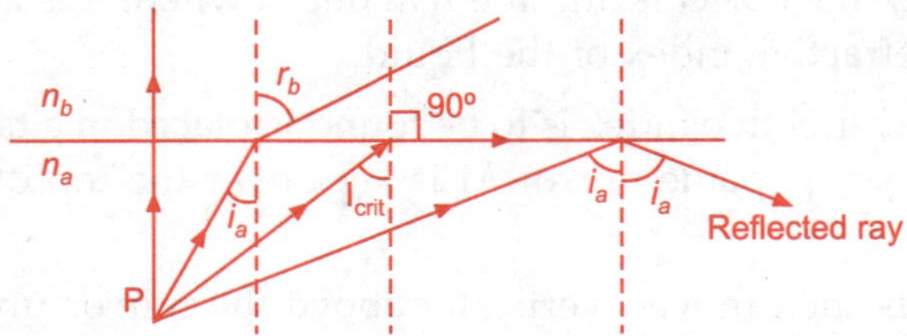

Fig. 8.15 *Total internal reflection*

The angle of incidence for which the refracted ray emerges parallel to the surface is called the critical angle and is designated by i_{crit} in the diagram. If the angle of incidence is greater than the critical angle, the ray does not pass into the medium b, but is totally reflected at the boundary back into medium a. Total internal reflection is possible only when a ray is incident on the surface of a medium whose refractive index is smaller than that of the medium in which the ray is travelling.

In Eq. (8.17), if $r_b = 90°$, then $\sin r_b = 1$ and hence we have

$$\sin i_{\text{crit}} = \frac{n_b}{n_a}. \tag{8.18}$$

EXAMPLE 8.2

Find the critical angle for a glass-air interface, if $n_{air} = 1.0003$ and $n_{glass} = 1.5$.

SOLUTION

Using Eq. (8.18), we have

$$\sin i_{\text{crit}} = \frac{1.0003}{1.5} = 0.6669$$

Hence, $\qquad i_{\text{crit}} = 41.83°.$

8.5.2 Totally Reflecting Prisms

Since i_{crit} is less than 45° in the glass-air example given above, it is possible to use a prism of angles 45°-45°-90° as a totally reflecting surface.

Sometimes totally reflecting prisms are preferred over other reflecting surfaces such as metallic plane reflectors. This is because, no metallic surface will reflect 100% of the incident light as in a totally reflecting prism, and the metallic surface may tarnish after sometime unlike a glass prism. However, some loss of light energy will occur in glass prisms due to the reflections, which can be minimized by coating the surfaces with some nonreflecting films.

A totally reflecting 45°-45°-90° glass prism is shown in Fig. 8.16.

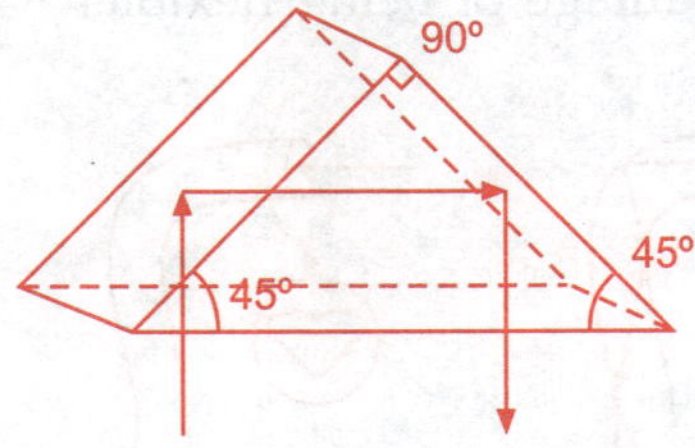

Fig. 8.16 *A porro prism*

It is called a Porro prism. When a ray is incident on the 45° surface, it is totally reflected onto the other 45° surface where it is totally reflected again. The final reflected ray emerges out at 180° to the first incident ray.

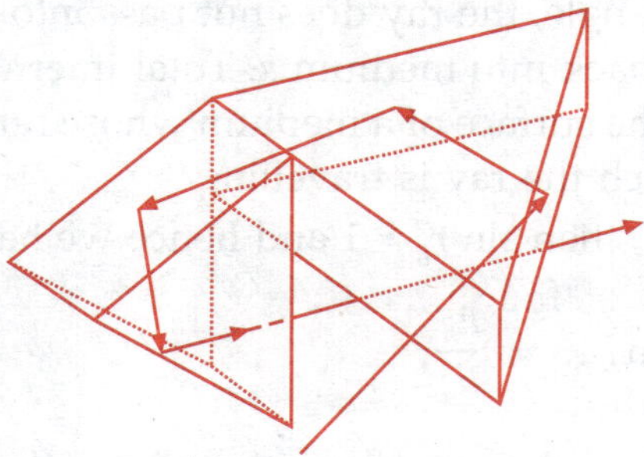

Fig. 8.17 *Combination of porro prisms*

Two Porro prisms are sometimes combined and as shown in Fig. 8.17 for use in binoculars.

8.5.3 Total Reflection used in Endoscopy Communication

A transparent curved rod is shown in Fig. 8.18. If a beam of light enters one end of this rod it gets totally reflected internally whenever it hits the sides. Hence the light is trapped within the rod and acts as a light pipe provided the curvature is not too much. The light ray that enters at one end of the glass rod will travel to the other end although the glass tube is curved.

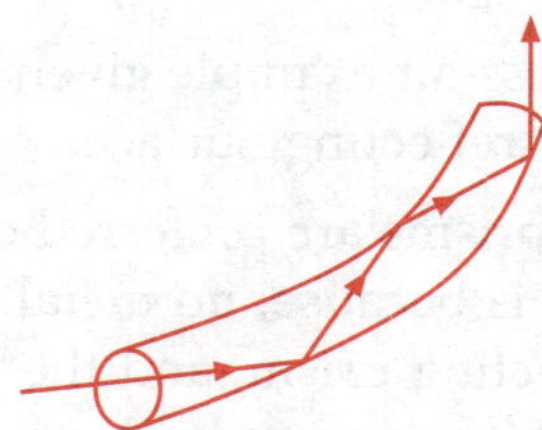

Fig. 8.18 *Light trapped by internal reflections in curved glass rod*

A bundle of such glass rods, or a bundle of glass fibers, will behave in the same way and send the light rays through multiple total internal reflections. The bundle of glass fibers has the additional advantage of being flexible.

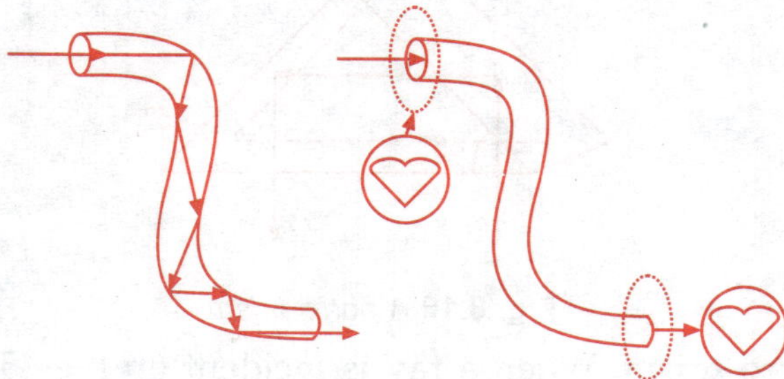

Fig. 8.19 *Endoscopic image transmission in bundle of glass fibers*

Such fiber optic bundles may consist of thousands of fibers, with diameters varying anywhere between 0.002 and 0.01 mm. The fibers in the bundle are so arranged as to have same relative positions at both ends which will enable it to transmit an image as shown in Fig. 8.19.

Fiber optic bundles of a few meters in length have been made for use in endoscopic examination of the interior of the human body. Such a bundle can be enclosed in a hypodermic needle called an endoscope, for the study of tissues and blood vessels far beneath the skin.

8.6 REFRACTION THROUGH A PRISM

8.6.1 Angle of the Prism and Angle of Minimum Deviation

A triangular glass prism is shown in Fig. 8.20 (*a*). The angle of the prism *A* is the vertex angle of the triangle as shown. A ray of light entering one of the faces adjacent to the angle of the prism, is refracted inside the prism once and is again refracted when emerging out of the prism on through the other face, as shown in Fig. 8.20(*b*). Consider a ray *PQ* entering the face *AB* as shown. The angle of incidence of ray *PQ* must as a rule be greater than about 30°, to avoid total internal reflection on face *AC*.

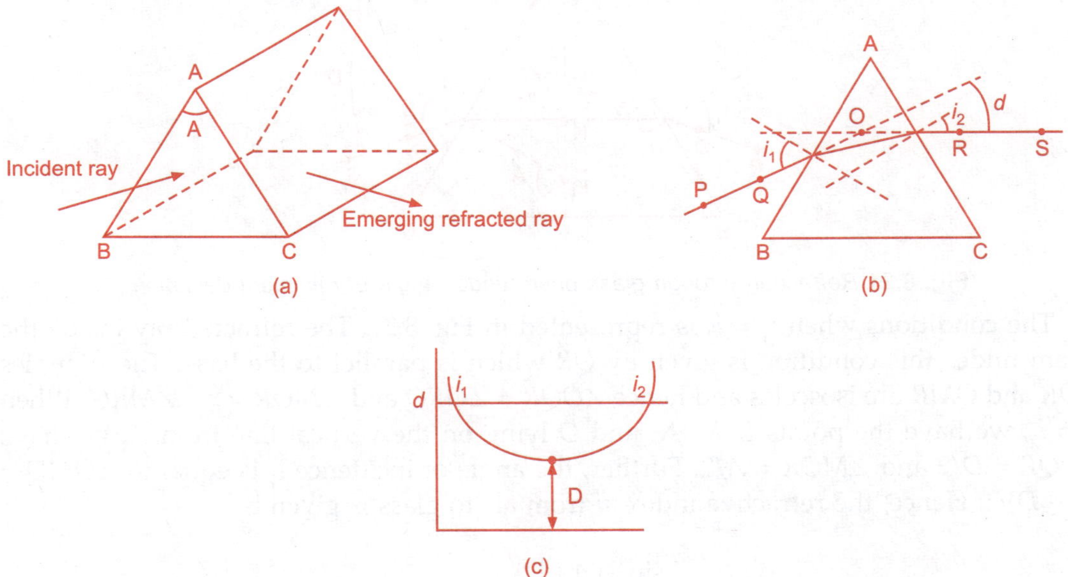

Fig. 8.20 *Refraction through a glass prism*

The prism is kept on a paper with its triangular base resting on the paper. Two pins are stuck on the paper representing the incident ray on the face *AB*. Viewing from the face *AC* two more pins are stuck on that side to align with the pins *PQ* seen through the prism face. Lines are drawn around the prism showing the triangular outline of the prism on the paper. The prism is removed and the lines *PQ* and *RS* are extended to meet

O inside the triangle representing the prism. The angle i_1 between the line PQ and the normal to the face AB, and the angle i_2 between the line RS and the normal to the face AC are measured. The angle of deviation d is the angle between PQ, which is the incident ray into the prism, and RS, which is the ray emerging out of the prism. The experiment can be repeated for several values of the incident angle i_1 and a graph of the deviation d against the incident angle i_1 can be drawn as shown in Fig. 8.20 (c). This graph will show that there is an angle of minimum deviation, say, D, which corresponds to the minimum point on the curve in Fig. 8.20 (c).

If we assume that the incident ray is PQ then angle of incidence is i_1 and the path of the ray is given by $PQRS$ and the angle of deviation is d. For the same angle of deviation d, it is also possible to assume that SR is the incident ray in which case the angle of incidence is i_2 and the path of the ray is given by $SRQP$. Hence, for the same angle of deviation there are two possible angles of incidence, i_1 and i_2. This can be seen from Fig. 8.20 (c) also. For the same value of d there are two possible values of angles of incidence. However, at the angle of minimum deviation, D, there is only one value of angle of incidence. Hence, at this point i_1 must be equal to i_2.

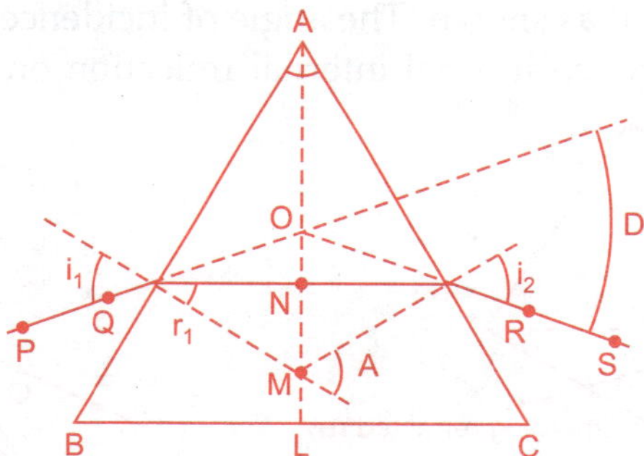

Fig. 8.21 *Refraction through glass prism under angle of minimum deviation*

The conditions when $i_1 = i_2$ is represented in Fig. 8.21. The refracted ray inside the prism under this condition is given by QR which is parallel to the base. The triangles QOR and QMR are isosceles and hence $\angle OQR = \angle ORQ$ and $\angle MQR = r = \angle MRQ$. When $i_1 = i_2$, we have the points L, M, N, and O lying on the vertical line from A. We have $\angle OQR = D/2$ and $\angle MQR = A/2$. Further, the angle of incidence i_1 is equal to $\angle OMQ = (A + D)/2$. Hence, the refractive index m from air to glass is given by

$$\mu = \frac{\sin i_1}{\sin r_1} = \frac{\sin \frac{1}{2}(A+D)}{\sin \frac{1}{2}A} \tag{8.19}$$

8.6.2 Dispersion through a Prism

Newton showed that when a beam of sunlight is passed through a glass prism onto a white screen, a band of color pattern or a color spectrum is formed on the screen. The pattern of colors is formed in a definite order with red, orange, yellow, green, blue, indigo and violet going from the vertex of the prism downwards, as shown in Fig. 8.22.

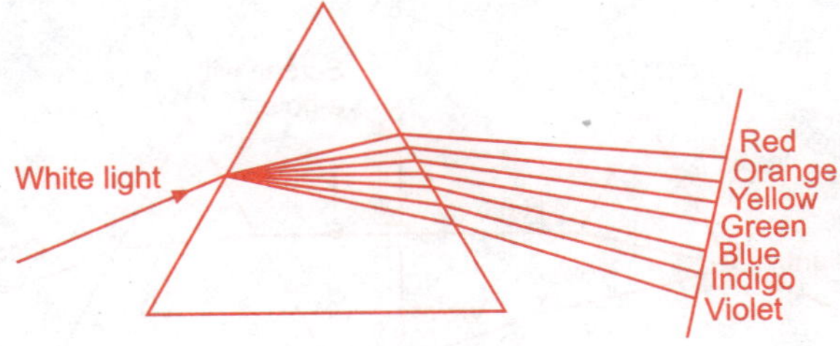

Fig. 8.22 *Dispersion of white light*

Using two similar prisms with the vertex of one aligned with the base of the other, as shown in Fig. 8.23, the colors are recombined again to show only white light.

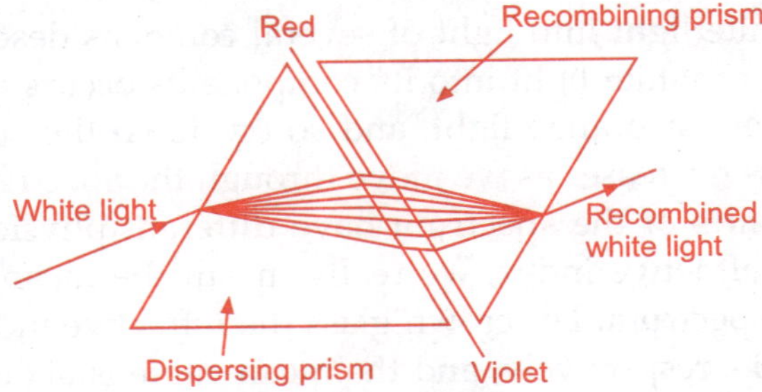

Fig. 8.23 *Two prism combination which recombines the dispersed white light*

Further, if two prisms are kept at right angles to each other, the displacements of the colors will be the resultant of the displacements given by the individual prisms and the color pattern will appear in an inclined fashion, as shown in Fig. 8.24.

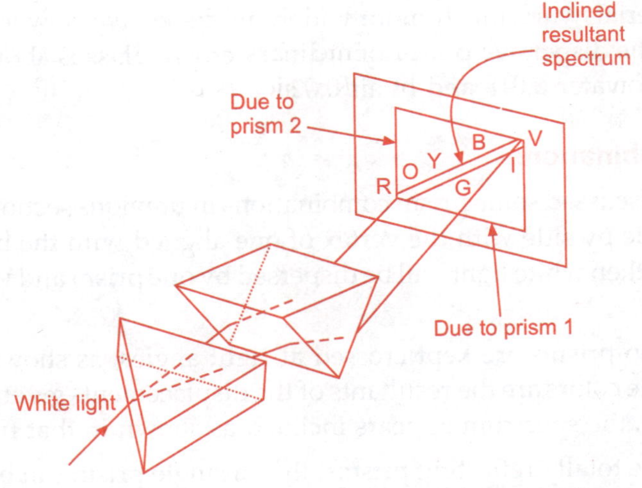

Fig. 8.24 *Twp prism combination with a resultant dispersion*

If any one color emerging out of the prism is isolated by passing it through a hole in a screen and is made to fall on a second prism, it will not split up any further. This is shown in Fig. 8.25.

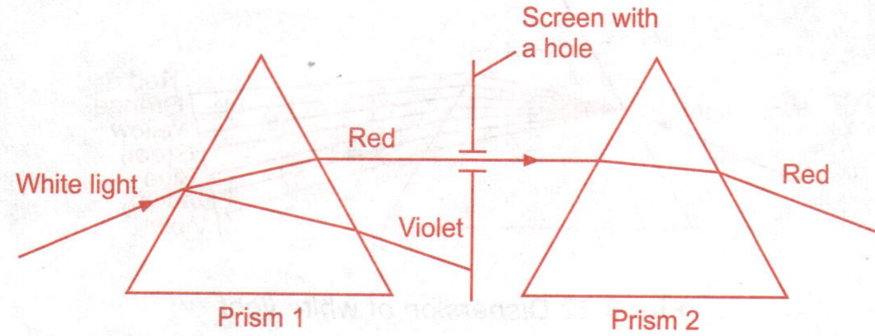

Fig. 8.25 *Two prism combination illustrating non-dispersion of ray of single color light deviation*

The splitting of white light into light of several colors as described above is called dispersion. Dispersion of white light into its components occurs because the refractive index of glass is greater for orange light, and so on down the spectrum. Further, the velocity of light in glass decreases as we move through the spectrum from red to violet. Hence, the different parts of the spectrum have different physical behaviors. So far, when we mentioned refractive index, we really meant the mean refractive index, for the yellow part of the spectrum. For crown glass the refractive indices for red and blue light are 1.515 and 1.530, respectively, and the mean value is about 1.52.

In dispersion due to refraction, the angle between the extreme ends of the spectrum is called the angle of dispersion. For a given prism, the spread of the dispersed colors depends on the difference between m for the ends of the spectrum, while the mean deviation of the middle of the spectrum depends on the mean refractive index. The ratio of the dispersion to the mean deviation is called the dispersive power of the material of the prism. Materials vary much more widely in dispersive power than in the mean refractive index. The dispersive power of ordinary crown glass is about 0.0165, of dense flint glass 0.030, of water 0.018 and of air 0.028.

8.6.3 Prism Combinations

We have already discussed some prism combinations in previous sections. If two identical prisms are kept side by side with the vertex of one aligned with the base of another, as shown in Fig. 8.23, then white light will be dispersed by one prism and will be recombined by the other prism.

Similarly, if two prisms are kept crossed at right angles, as shown in Fig. 8.24, the displacements of the colors are the resultants of the displacements given by the individual prisms. As a result, the spectrum appears inclined as shown in that figure.

If the prisms are totally reflecting prisms, then a single prism can be used to turn the ray by 180° or by 90°. By using two totally reflecting prisms, it is possible to offset the ray without any change in the direction. This is shown in Fig. 8.17 as two Porro prism combination.

Combinations of totally reflecting 45° prisms are used in prismatic binoculars to invert the images. The reinversion of the image is done in two stages, by two lateral inversions.

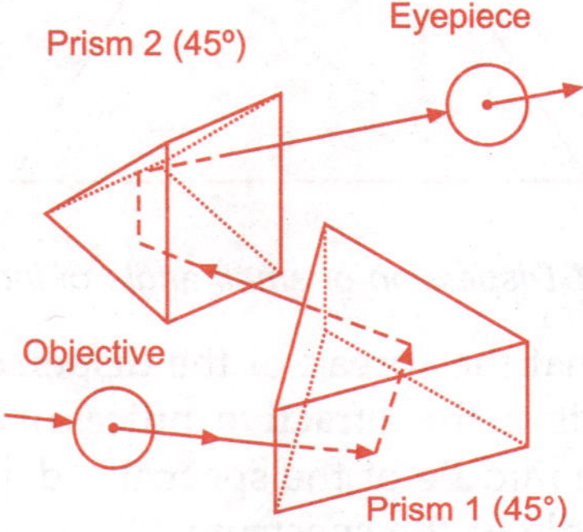

Fig. 8.26 *Prismatic binoculars*

As shown in Fig. 8.26, in addition to laterally inverting as a mirror would do, the two prism arrangement also displaces the rays sideways so that they return along the next limb. The two prisms are set so that the right angle edges are horizontal and vertical in the figure - not parallel to one another. The main advantage of prism type binoculars is the wide field of view they can give with a reasonably high magnification.

8.6.4　Dispersion for Small Angles of Incidence

For small angles, we can approximate $\sin \theta \cong \theta$, provided θ is expressed in radians. This can be very easily verified if we refer to Trigonometric Tables, or with a calculator that can compute trigonometric quantities. For small angles of incidence angles of refraction also will be small, and hence we can write Snell's law as

$$\mu = \frac{\sin i}{\sin r} \approx \frac{i}{r} \tag{8.20}$$

or, we can write $i = \mu r$.

Referring to Fig. 8.27, we can write

$$\angle A = \angle r_1 + \angle r_2 \tag{8.21}$$

$$\angle d = (i_1 - r_1) + (i_2 - r_2) \tag{8.22}$$

Substituting $i_1 = \mu r_1$ and $i_2 = \mu r_2$ in Eq. (8.22), we obtain the deviation as

$$d = (\mu - 1)A \tag{8.23}$$

This expression does not contain i_1 and i_2 at all, and hence for small angles of incidence the deviation is constant.

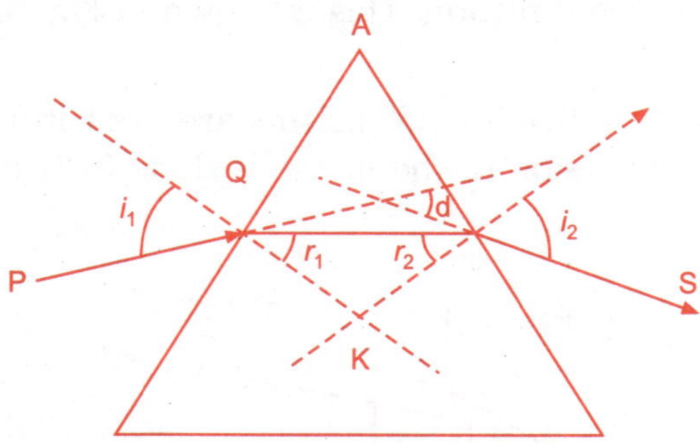

Fig. 8.27 *Dispersion of small angle of incidence*

It was mentioned earlier that the spread of the dispersed colors when white light is passed through a prism depends on the refractive indices of the end colors of the spectrum and the mean deviation of the middle of the spectrum depends on the mean refractive index for the color in the middle of the spectrum.

Consider the colors violet and red. Let d_v and d_r be the deviations of the violet and red colors. If μ_v and μ_r are the corresponding refractive indices of violet and red colors, respectively, then for small angles of incidence we have from Eq. (8.22)

$$d_v - d_r = (\mu_v - 1)A - (\mu_r - 1)A = (\mu_v - \mu_r)A \tag{8.24}$$

If μ is the mean refractive index corresponding to the middle of the spectrum between violet and red, then $d = (\mu - 1)A$ is the mean deviation. The dispersive power of the prism, which is the ratio of the dispersion to the mean deviation is given by

$$\frac{d_v - d_r}{d} = \frac{\mu_v - \mu_r}{\mu - 1}. \tag{8.25}$$

This is a property of the material of the prism only and does not depend on the value of prism angle A.

8.6.5 Deviation without Dispersion

When white light is passed through an optical system, if the path of the light ray is deviated from its original path without any dispersion, then we have the phenomenon of deviation without dispersion. We know that light passing through a prism undergoes deviation as well as dispersion. Light passing through a single prism, therefore cannot have deviation without dispersion.

We can combine two prisms of different materials in such a way that the deviation introduced by one is cancelled by the other without completely cancelling the dispersion. Consider again the colors violet and red. Let d_v and d_r be the deviations of the violet and red colors. Using a two prism combination as shown in Fig. 8.28,

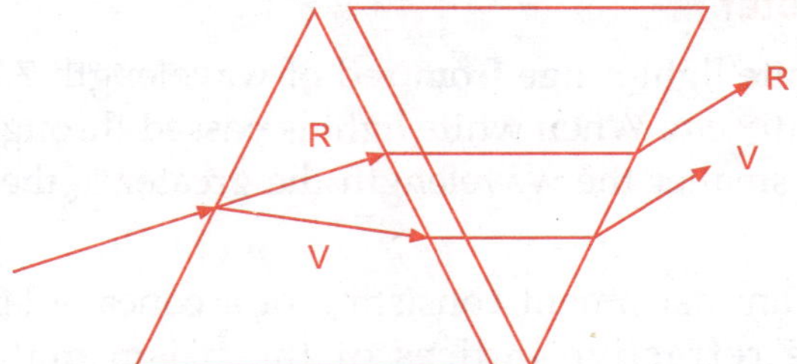

Fig. 8.28 *Two prism combination illustrating non-deviation*

we have from Eq. (8.25)

$$\left(\frac{d_v - d_r}{d}\right) - \left(\frac{d_v - d_r}{d}\right) = 0. \tag{8.26}$$

This shows that the dispersion is cancelled without affecting the deviation d. From Eq. (8.25), we can see that in order for the quantity $(d_v - d_r)$ to be the same in both the prisms we must have $(\mu_v - \mu_r)A$ for both the prisms.

Dispersion without Deviation

We can invert Eq. (8.25) and write

$$\frac{d}{d_v - d_r} = \frac{\mu - 1}{\mu_v - \mu_r} \tag{8.27}$$

We can combine two prisms in such a way that $(\mu - 1)A$ is same for both, then we can cancel the deviation d, but retain the dispersion.

8.6.6 The Rainbow

When sun rays fall on water particles in the atmosphere, an observer with his or her back to the sun can see one or sometimes two rainbows. Rainbows are parts of concentric circles with different colors corresponding to the light spectrum. The centers of these concentric circles have their centre on the line joining the sun to the observer's eye. If two rainbows appear, the inner one is called the primary rainbow which subtends an angle of about 41° at the eye, and the outer one is called the secondary rainbow which subtends an angle of about 53°. Both of them show the spectrum of colors. The primary bow shows violet on the inner edge and red on the outer edge, whereas the secondary bow shows the same colors in a reverse order. Other bows are sometimes visible near the inner edge of the primary bow and the outer edge of the secondary bow. The portions of the sky within the primary bow and just outside the secondary bow are brighter than other parts of the sky. The primary bow is more intense than the secondary bow and other bows.

8.6.7 The Spectrometer

The constituents of white light range from red of wavelength 7.7×10^{-5} cm, to violet of wavelength about 4×10^{-5} cm. When white light is passed through a prism the colors are dispersed because the smaller the wavelength the greater is the refractive index of the glass.

A spectrometer is an instrument consisting of a concave lens and a prism, and is used to measure the refractive indices of the prism material and to measure the wavelength of the light. The working principle of a spectrometer is shown in Fig. 8.29.

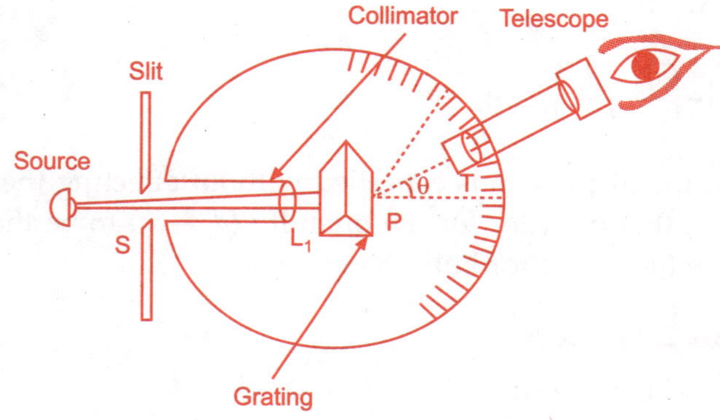

Fig. 8.29 *The spectrometer*

It consists of a collimator, which is a tube with one end having a slit S and the other end having an achromatic lens L_1. Achromatic lenses are corrected to bring two wavelengths (typically red and blue) into focus in the same plane. The prism P is mounted on a table, which can rotate about a vertical axis at right angles to the plane containing the optic axes of the telescope and the collimator. The telescope, fitted with cross wires in the position of T, can be rotated about the same axis as the prism table, and carries a vernier, which moves over an angular scale engraved on the frame.

Focusing is done in three stages. First focus the eyepiece on the cross wires. Next take the instrument out of doors, and adjust the telescope by moving the eyepiece and the cross wires as one unit until the image of a distant object coincides without parallax with the crosswire. Then set telescope and collimator in line, and adjust the distance between S and L_1 until a sharp image of the slit coincides without parallax with crosswire.

The second adjustment consists of setting the telescope to receive parallel light. Thirdly, the collimator is made to give parallel light. Instead of taking the instrument out of doors it may be convenient to use a parallel light projector set up for the purpose—a master collimator kept permanently in adjustment.

The Spectroscope: When used merely to examine the spectra, the instrument is a spectroscope.

8.7 REFRACTION AT A SPHERICAL SURFACE

When a ray of light is incident on a plane surface separating two optical media, if the angle of incidence is greater than the angle of total internal reflection, then the ray is subjected to refraction. Even if the surface separating the two media is curved, a ray of light will be refracted if the angle of incidence is greater than certain value. Let us discuss the phenomenon of refraction at spherical surfaces.

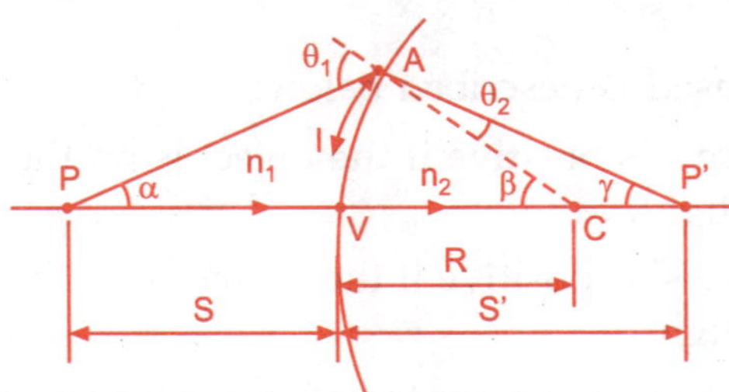

Fig. 8.30 *Refraction at a spherical surface*

Consider Fig. 8.30, which shows a curved surface separating two media. C is the centre of curvature of the curved surface and R is the radius of curvature.

Consider the two rays, PA and PV, leaving the point P. The ray going towards A will be refracted at the surface and bend down (towards the normal to the surface, AC) to travel towards point P', while the ray going along the principal axis hits the interface normally and hence proceeds straight without refraction meeting the ray from A at P'. Hence the object at P has its image formed at P'. If the rays are "paraxial", then the angles α, β, γ, θ_1 and θ_2 are all small. Employing Snell's law, we have $n_1 \sin \theta_1 = n_2 \sin \theta_2$. Since the angles are all small,

$$n_1\theta_1 = n_2\theta_2 \tag{8.28}$$

From plane geometry, the exterior angle of a triangle is equal to the sum of the two opposite angles. Consequently, in triangle PAC

$$\theta_1 = \alpha + \beta \tag{8.29}$$

and from triangle $P'AC$

$$\beta = \theta_2 + \gamma \tag{8.30}$$

We can eliminate θ_1 and θ_2 between these two equations. Substituting for θ_2 from Eq. (8.28) in Eq. (8.30), we get β as

$$\beta = \frac{n_1}{n_2}\theta_1 + \gamma \tag{8.31}$$

Substituting for θ_1 from Eq. (8.29) in Eq. (8.31), we get

$$\beta = \frac{n_1}{n_2}(\alpha+\beta) + \gamma \tag{8.32}$$

After simplification we have

$$n_1\alpha + n_2\gamma = \beta\,(n_2 - n_1)\tag{8.33}$$

Denoting the object distance as s and the image distance as s', we have

$\beta = \ell/R$, $\alpha = \ell/s$, and $\gamma = \ell/s'$, and we can write Eq. (8.33) as

$$\frac{n_1}{s} + \frac{n_2}{s'} = \frac{n_2 - n_1}{R}.\tag{8.34}$$

The sign convention used is described below:

1. An object distance s is positive if the object is on the same side of the surface as the incoming light.

2. An image distance s' is positive if the image is on the same side of the surface as the outgoing light.

3. Radius R is positive if the centre of curvature is on the same side of the surface as the outgoing light.

EXAMPLE 8.3

An object under clear water seems to be at a depth of 1.5 m when seen from directly above it. The index of refraction of air is 1 and of water is 1.333. What is the actual depth of the object?

SOLUTION

As shown in Fig. 8.31, the image is on the incoming side of the surface and hence s' = –1.5 *m*. The surface is flat and hence $R = \infty$. From Eq. (8.34), we have

$$\frac{1.333}{s} + \frac{1.000}{-1.5} = \frac{1.000 - 1.333}{\infty} = 0.$$

Consequently, we get $s = 2.0$ *m*. Hence, the object is actually deeper than it appears to be.

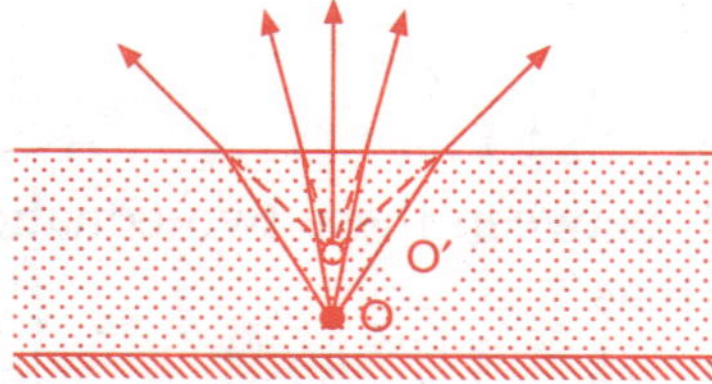

Fig. 8.31 *Object under water*

8.8   REFRACTION THROUGH A THIN LENS

A lens is made of some dense transparent material such as glass, which has either two spherical surfaces or one spherical and one plane face. Lenses can be convex type or concave type. Convex lenses, which are also called as converging lenses, may be biconvex, Plano-convex or convex meniscus lenses, all of which are thicker in the middle as shown in Fig. 8.32(*a*). Concave lenses, also called as diverging lenses, may be biconcave, Plano-concave or concave meniscus lenses which are thinner in the middle as shown in Fig. 8.32(*b*).

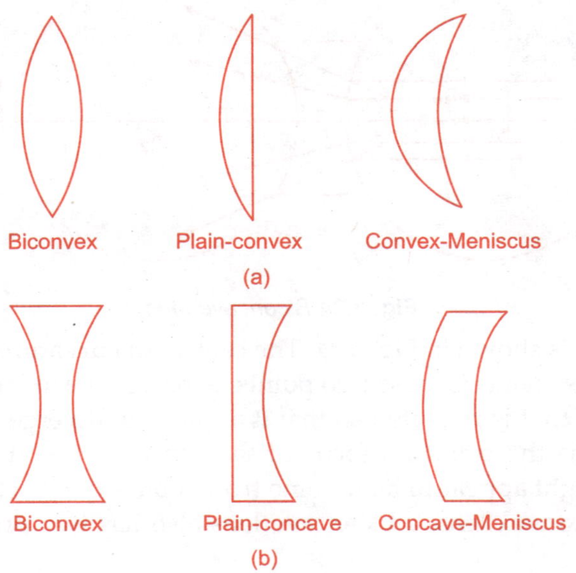

Fig. 8.32 *Lenses*

A biconvex lens is shown in Fig. 8.33. The centers of the spheres of which the surfaces are parts, are called the centers of curvature of these surfaces and are shown by points K_1 and K_2 in the figure.

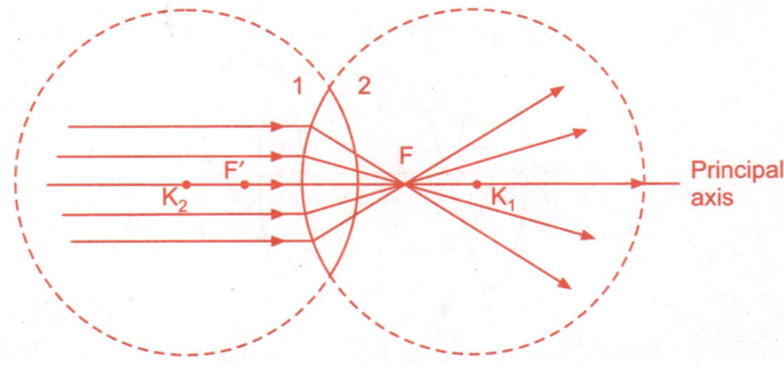

Fig. 8.33 *Biconvex lens*

The line joining the two centres of curvature, $K_1 K_2$, is called the principal axis of the lens. When rays parallel to the principal axis are incident on the convex surface of the lens, they are refracted and meet at a point F on the principal axis on the opposite side.

Point *F* is called the principal focus of the lens. If parallel rays of light were travelling from right to left, they would be focused on the left side of the lens at *F'*, which is another principal focus of the lens. Both *F* and *F'* are real principal foci of the lens since rays of light actually pass through them.

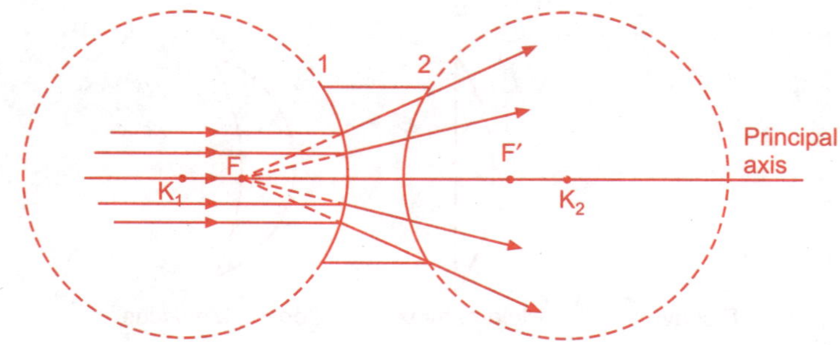

Fig. 8.34 *Biconcave lens*

A biconcave lens is shown in Fig. 8.34. The centres of curvature K_1 and K_2 are shown, and the principal axis connects these two points. A narrow beam of parallel rays parallel to the axis and close to it is refracted so that it appears to diverge from a point *F* on the principal axis. This is the principal focus of the concave lens. It is a virtual principal focus, since rays of light appear to have come from it but do not actually pass through it. As in the case of biconvex lens, a biconcave lens also has two principal foci, on either side of the lens.

Optical Centre of a Lens

Consider the biconvex lens shown in Fig. 8.35. When a ray *OP* is incident at point *P* on the lens surface as shown, this ray is refracted along *PQ* inside the lens, where *Q* is a point on the opposite surface of the lens.

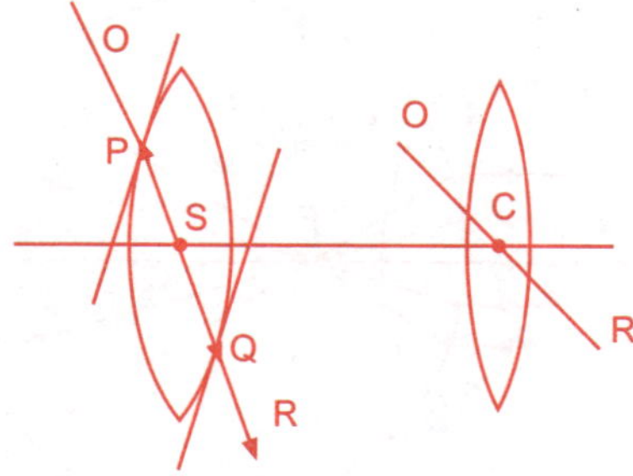

Fig. 8.35 *Optical centre of a biconvex lens*

At *Q* the ray emerges out of the lens when another refraction takes place and the ray moves along the line *QR*. The tangents at the points *P* and *Q* are parallel to each other. Consequently the incident ray *OP* and the out coming ray *QR* are parallel to each other. The ray *PQ* inside the lens intersects the principal axis of the lens at *S*.

When the thickness of the lens is significant, the line *PQ* is not parallel to the rays *OP* and *QR*. However, if the lens is thin, then we can assume that *PQ* is a continuation of the ray *OP* and *QR* is a continuation of *PQ*. In this case we can assume that points *P* and *Q* come closer to point *S* and points *P*, *Q*, and *S* coincide on a point *C*, which is called the optical centre of the lens. A ray of light striking the optical centre of a thin lens continues its path without deviation.

The thickness of a thin lens is not more than about 2 per cent. Practical measurements are usually made from the middle of the lens. The points where the two faces of the lens cut the axis are always supposed to coincide with the optical centre.

8.8.1 Power of a Lens

The distance from the optical centre of the lens to its principal focus is called the focal length of the lens, denoted by f. The focal length of a convex lens is considered positive and that of a concave lens is considered negative.

The power, P, of a lens is the reciprocal of its focal length, f. The power of the lens is expressed in dioptre (D) units. The power of a lens with one meter focal length is taken as one dioptre (D). Hence, we have

$$P = 1/f \text{ dioptres or } D. \tag{8.35}$$

For example, the power of a convex lens having a focal length of + 50 cm, i.e.; 0.5 m, is given by

$$P = 1/0.5 = 2D$$

Opticians commonly give the power of the lens in dioptres.

Focal Plane

Any narrow beam of parallel rays inclined at a small angle to the axis can be focused at a point G, which lies on a plane passing through the focus *F* and normal to the principal axis. Such a plane, normal to the axis and at the focus *F* is called the focal plane, which is shown in Fig. 8.36.

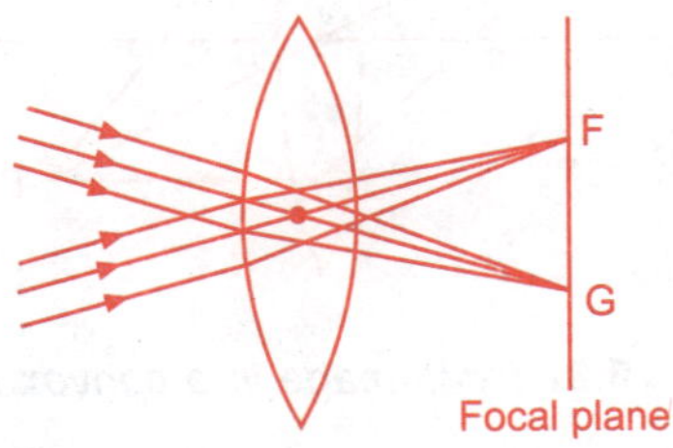

Fig. 8.36 *Focal plane*

The position of *G* can be located by finding where the undeviated ray passing through the optical centre cuts the focal plane. Rays of light from a very distant object are almost parallel, so that a very distant object can be considered as a collection of points each of which sends a beam of parallel rays to the lens. Corresponding to each of these object points there is an image point G in the focal plane, and consequently, the convex lens forms a real image of a distant object in the focal plane.

This property can be used to determine the focal length of a thin lens. A screen is placed to coincide with the focal plane of convex lens. The real image of a distant object on the other side of the lens is focused on the screen. When the image of the distant object is properly focused on the screen, the distance between the optical centre of the lens and the screen gives the focal length. If the object is not very distant, the image distance becomes slightly greater than f.

8.8.2 Image Formation due to a Lens

Real Image

When a small object is placed at a distance greater than f from the convex lens on the principal axis, a real image of the object is formed on a screen kept on the focal plane. The image is real because the rays of light forming the image points actually pass through these image points. A real image can be held on a screen or on a photographic plate. A real image can also serve as an object for a second lens, and can actually be seen if one is suitably placed so that rays after forming the image points can continue their journey to the eye.

In Fig. 8.37, the rays from the point A on a long object OA, reaching the lens are made to converge by it and are brought to a focus at B. Parallel rays from all the points on the object pass through the principal focus so that an image of the whole object is formed in a plane perpendicular to the axis which includes B. The size of the image in comparison with that of the object depends on the distances of the object and the image. The image appears inverted always.

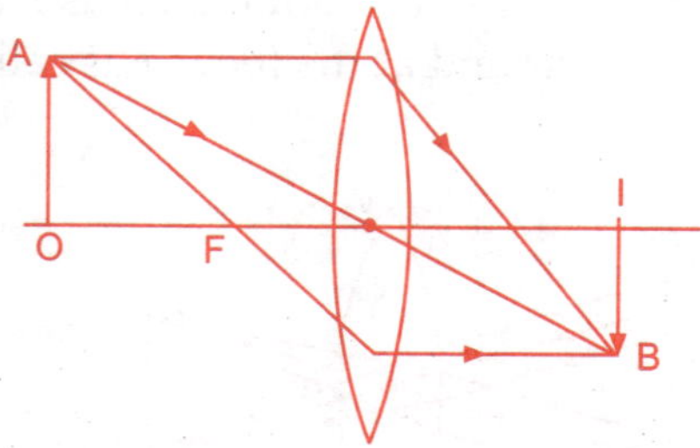

Fig. 8.37 *Real Image in a convex lens*

The convex lens converts a diverging beam into a converging beam in order to form a real image. The lens can achieve only a certain amount of converging. When the object distance is less than f, the object is very close to the lens and the rays from the object will be diverging considerably. In such a case a real image cannot be formed and the image is virtual.

Virtual Image

Unlike the real images, virtual images are not formed by the actual rays intersecting, but are formed when the rays are traced back and intersect. As a consequence, virtual images cannot be shown on a screen, or they cannot serve as objects for a second lens. Both convex and concave lenses can form virtual images.

(*a*) **Convex Lens:** As mentioned earlier, when the object distance is less than the focal length of the lens, the image formed is virtual. Referring to Fig. 8.38(a), rays from *A* appear to come from *B*. As a consequence, the rays diverge less. The point *B* is the virtual image of *A*. The rays do not actually intersect at *A* and the image is formed when rays which are traced back intersect at *B*. While the real image is upside down, the virtual image is always erect.

(*b*) **Concave Lens:** A biconcave lens is shown in Fig. 8.38(*b*). Rays from *A* strike the lens and when they cross to the other side of the lens they diverge more and will never meet. However, they appear to have come from point *B* although they have not really passed through it. The virtual image formed by a concave lens is always smaller than the object and are not inverted.

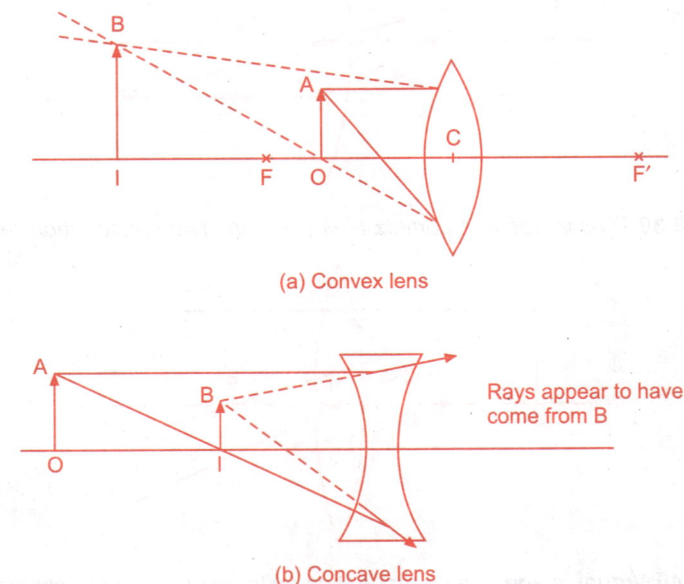

(a) Convex lens

(b) Concave lens

Fig. 8.38 *Virtual images*

Image Construction for a Thin Lens

Innumerable rays emanate from each point of an object which pass through the lens. By following the paths of these rays and identifying the images corresponding to each point on the object it is possible to obtain the optical image of the object.

If we can locate one point on the image we can then locate the whole image. By tracing just two rays from a point A on the object, it is possible to locate the corresponding point B on the image. These two rays can be chosen from (*i*) a ray parallel to the principal axis, which is refracted from the principal focus F, (*ii*) a ray through the optical centre C, which continues undeviated, and (*iii*) a ray through F, the principal focus on the object side (or, if the object is nearer to the lens than F, a ray directed as if from (F), which emerges parallel to the principal axis; this is the counterpart of (*i*).

If these rays intersect at a point B on the far side of the lens, then B is the real image of A, and the whole image is real. If they appear to diverge from a point B on the same side of the lens as the object, then B is the virtual image of A and the whole image is virtual.

A converging lens forming a real image, when the object is further from the lens than F is illustrated in Fig. 8.39, whereas when the object is nearer than F it forms a virtual image which is illustrated in Fig. 8.40.

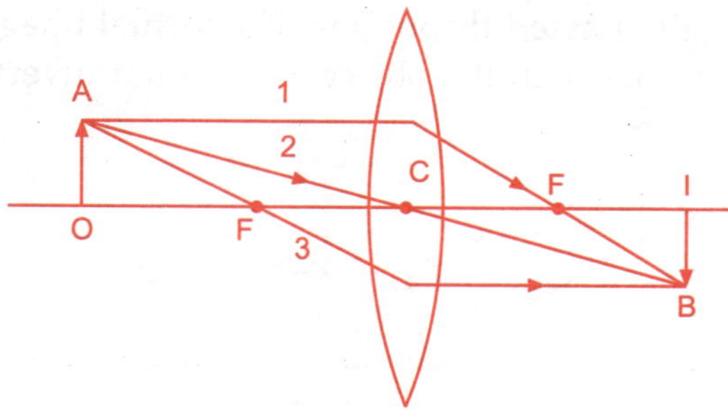

Fig. 8.39 *Real image in a convex lens (with any two construction rays)*

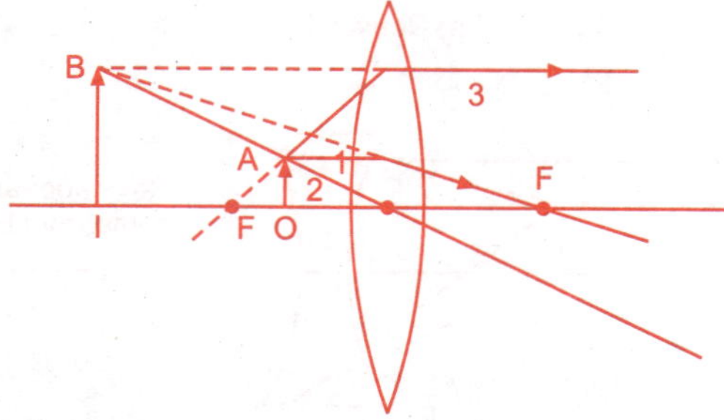

Fig. 8.40 *Virtual image in a concave lens (with any two construction rays)*

A diverging lens forming a virtual image is illustrated in Fig. 8.41. In this diagram, ray 3 is going towards the focus on the other side of the lens until it falls on the lens when it goes parallel to the principal axis. When the parallel portion of the ray is traced back, it passes through the point B, which is the virtual image of A. Since the path of the rays is reversible A and B are interchangeable and are called conjugate foci. In these figures A is drawn as the tip of the object OA standing upright on the vertical axis and the image BI drawn in by dropping a perpendicular on the axis from B.

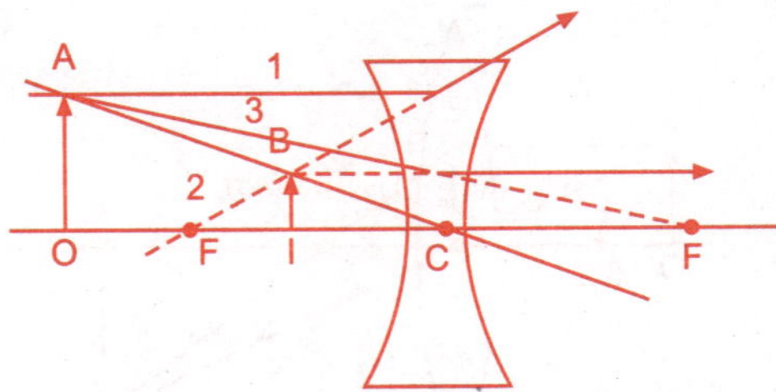

Fig. 8.41 *Virtual image in a convex lens (with any two construction rays)*

Magnification of the Image:

The magnification of the image is given by

$$\text{Magnification, } m = \frac{\text{Height of image}}{\text{Height of object}} \tag{8.36}$$

The relation between object and image distances, focal length and magnification can be expressed as

$$\frac{1}{v} - \frac{1}{u} = \frac{1}{f}$$

$$m = \frac{v}{u} \tag{8.37}$$

In the above equation, New cartesian sign convention (*NC*) is used where

u is the object distance, positive to the left of C,

v is the image distance, positive to the right of C, and

f is the focal length, positive for a convex lens (F is to the right of C), and negative for a concave lens (F is to the left of C)

Distances above the axis are positive and below the axis are negative.

EXAMPLE 8.4

A cylindrical rod OA is placed vertically at a distance of 0.3 m from a convex lens of focal length 0.2 m. Calculate the position, nature and size of the image.

SOLUTION

The object and the image relative to the lens are shown in Fig. 8.42. Employing the sign convention described earlier we have $u = -0.3$ m, (left of C) and $f = 0.2$ *m*, (F is to the right of C).

Using Eq. (8.37) and substituting the known quantities we get

$$\frac{1}{v} - \frac{1}{(-0.3)} = \frac{1}{0.2}$$

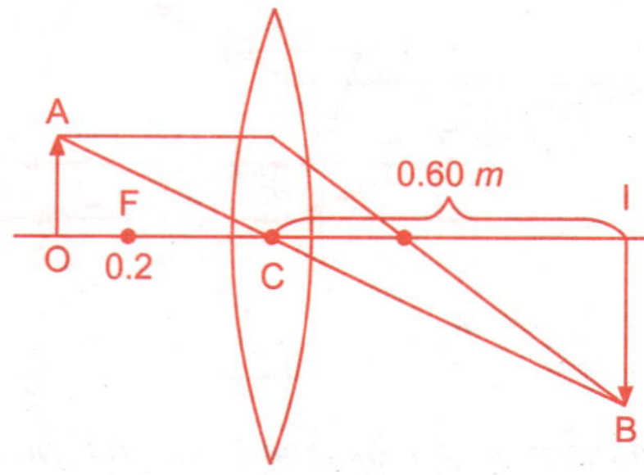

Fig. 8.42 *Real image in a convex lens*

From this we get $v = 0.6$ m. Further, the magnification is given by

$$m = 0.6/(-0.3) = -2m$$

This shows that the image is inverted (and real). The image is twice the size of the object. We can also conclude that the image is real from the fact that the object distance is more than the focal length of the lens.

8.8.3 Combination of Lenses

When two thin lenses are in contact, they will act as a single lens with an effective focal length. A combination of a concave and a convex lens is shown in Fig. 8.43.

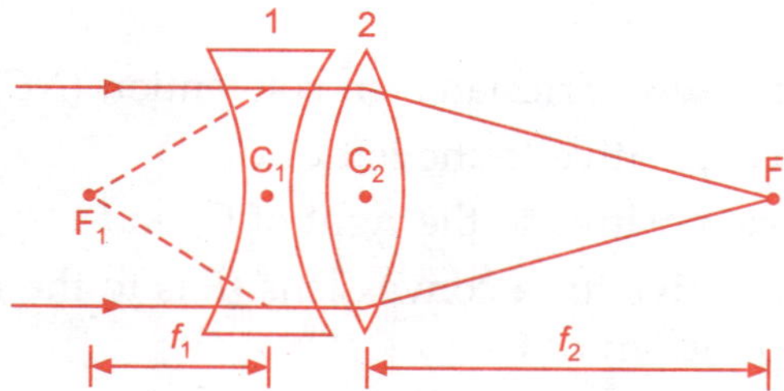

Fig. 8.43 *Combination of thin lenses*

Let f_1 and f_2 be their individual focal lengths and let f be the effective focal length. A parallel beam which is parallel to the axis falling on lens 1 (concave), comes to a focus at a distance F beyond lens 2 (convex). Referring to Fig. 8.43, the parallel beam is made to diverge from F_1 and the light hits lens 2 as though it were coming from a real object at F_1 of which a real image is formed at F.

For lens 2, we have $u = f_1$, $v = f_2$. Referring to Eq. (8.37), we get

$$\frac{1}{f_2} - \frac{1}{f_1} = \frac{1}{f}. \tag{8.38}$$

The sign of the power is same as that of the focal length. Hence from Eq. (8.35), we have

$P = + 5D$ for $f = 0.2m$, and $P = -2D$ for $f = -0.5$ m. Consequently, from Eq. (8.38), we get

$$P = P_1 + P_2 \qquad (8.39)$$

Hence, lenses of powers $5D$ and $-2D$ give a combination power of $3D$.

8.9 HUMAN EYE

Eye is the most versatile optical instrument. All other optical instruments come as aids to the eye, whereas the eye itself is so sophisticated that it cannot be replaced by any artificial means.

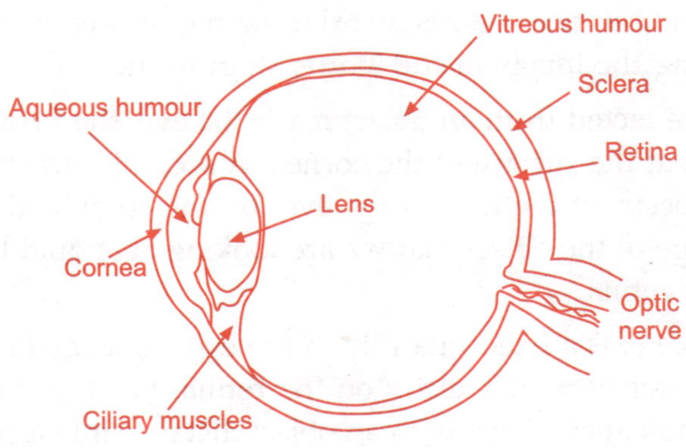

Fig. 8.44 *Schematic diagram of a human eye*

A schematic diagram of the eye is shown in Fig. 8.44. At the very front of the eye is the sharply curved cornea. Behind the cornea there is the lens ($n = 1.396$), and the space between the cornea and the lens is filled with a liquid called aqueous humour ($n = 1.336$). Most of the refraction of the incoming light takes place when the light passes from air into this liquid. The index of the lens is different at the center and the edges. The curvature of the lens can be adjusted by the Ciliary muscles which are attached to the lens in order to form images of objects at different distances. Behind the lens the vast spherical area is filled with a fluid called vitreous humour, which is similar to the aqueous humour.

The back wall of the spherical interior is a light-sensitive layer called the retina. The retina contains photosensitive cells, which produce nerve signals when light falls on them, and nerve cells that process these signals and send them to the brain for interpretation.

The muscles that hold and move the eye are fastened to the tough outer membrane of the eye called sclera. The front part of the sclera is conspicuously curved and part of it is transparent. The transparent part is called the cornea. The middle layer of the eyeball sandwiched between the sclera (white of the eye) and the retina is called the choroid. This contains the connective tissues and the blood vessels that supply nutrients to the inner portion of the eye. The choroid layer inside the sclera contains a dark pigment which ensures that the light entering the retina are absorbed and are not scattered back into the central chamber.

The details of a small object that we can distinguish depend on the size of the image of that object on the retina. This depends on the angle subtended by the object at our eye. In order to enlarge the apparent size of the object, it must be brought closer to the eye so that the angle subtended by it at the eye is larger. Eventually, if the body is brought too much closer, eye can no longer focus the image of the object on the retina. The closest distance at which the eye can comfortably focus on the object is called the near point. We can find our near points by bringing the printed words of a book close to the eyes while keeping the image of the words sharply focused.

When light rays reflected from an object reach the eye, the first and more important refraction takes place at the surface of the cornea. A second refraction and the resulting bending of the rays occur at the lens in the eye. By the combined action of the cornea and the lens, the image of the object that we are looking at should be formed accurately on the light sensitive retina.

The image distance v, is a fixed quantity in the eye. A normal relaxed eye forms the image of a distant object ($u = \infty$), exactly on the retina. We know $(1/u)+(1/v)=(1/f)$, and hence if we want to look at an object with an object distance much smaller, then we have to reduce the focal length, f. This change in the focal length of the lens is accomplished by the ciliary muscles, which act to squeeze the lens and make it more convex.

An infant can form clear images of objects when they are as close as 70 mm to the eye. A young adult person can focus sharply at objects which are as close as 100 mm to the eye. A 60 year old person may be able to focus sharply only at a distance of 2 m. Old people will not be able to see objects which are closer than this distance without the aid of glasses. The ability of the eye to focus on objects at different distances is called accommodation. Accommodation is possible by tightening the muscles around the eye lens. When these muscles are relaxed and not tightened or strained, a normal eye is focused at infinity.

8.9.1 Common Defects for the Eye

Focusing in both relaxed position and in accommodating (lens muscle squeezed) position by a normal eye are shown in Fig. 8.45. In the relaxed position, far away objects are focused on the retina and in order to focus nearer objects the lens muscles must squeeze the lens and make it more convex.

Myopia, or the shortsightedness, and farsightedness are the two common defects of the eye. In both cases opticians can help by prescribing glasses with diverging lenses for nearsighted persons, and converging lenses for the farsighted.

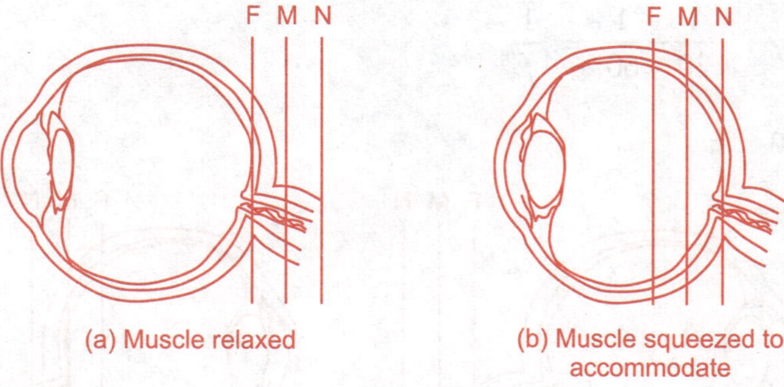

Fig. 8.45 *Focusing of normal eye*

The lens of a myopic or a shortsighted eye is too strong and the distant objects are focused in front of the retina, instead of on the retina, when the eye is relaxed. Since the accommodation that can be effected with the aid of the Ciliary muscles is to make the lens stronger or to make it more convex, far off objects cannot be accommodated through any muscular effort. This situation is illustrated in Fig. 8.46. In order to see objects that are far off, a nearsighted person can make use of glasses with appropriate focal length. Consider a myopic person who cannot focus clearly on objects more than 1 m away. The focal length of the lens of the glasses must produce a virtual image 1 m to the front, where the eye can use it as a virtual object. This is obtained from the relation

$$\frac{1}{\infty} + \frac{1}{(-1)} = \frac{1}{f}$$

from which we can get the focal length to be −1 m.

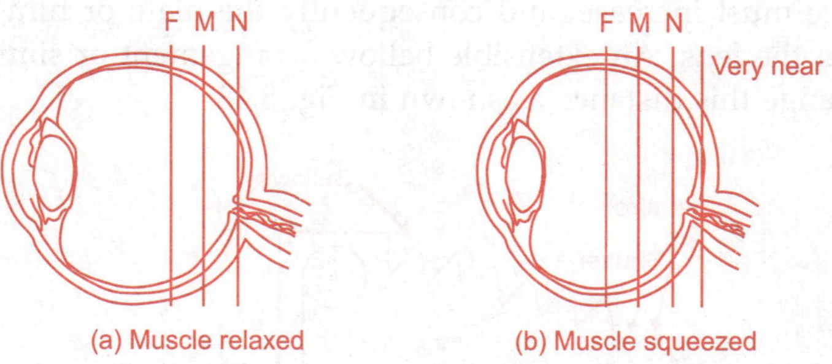

Fig. 8.46 *Focusing of near-sighted eye*

The lens of a farsighted eye is not strong enough to focus even distant objects when it is relaxed. Thus a farsighted eye must exert a constant squeeze on the lens muscles, and consequently the eye gets tired very easily.

This is shown in Fig. 8.47. Again, it is possible to use corrective glasses in order to focus closer objects on the retina. Consider a farsighted eye that cannot focus on objects closer than 60 cm. If an object held at a distance of 20 cm must be made visible, the focal length of the lens of the glasses is given by the relation

$$\frac{1}{20} - \frac{1}{60} = \frac{1}{f}$$

and hence $f = 30$ cm.

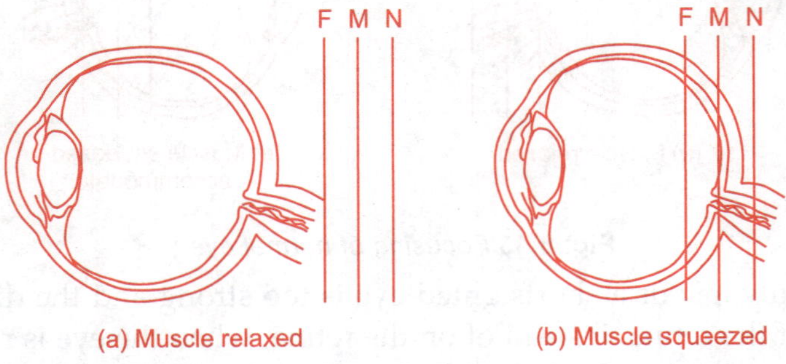

Fig. 8.47 *Focusing of far-sighted eye*

8.9.2 Photographic Cameras

A photographic camera is essentially similar to the human eye in its operation. It consists of a converging lens, a variable area aperture and a stop mechanism that controls the duration of the aperture opening which allows the passage of light rays through the aperture onto a light sensitive plate or film that captures the real, inverted and usually diminished image of the object which is being photographed.

Assuming that the thin lens theory applies to the camera lens, an object at a very large distance forms its image in its focal plane. As the object moves closer to the lens, the image distance must increase, and consequently the plate or film must be moved farther away from the lens. An extensible bellow arrangement or similar arrangement is provided to change this distance as shown in Fig. 8.48.

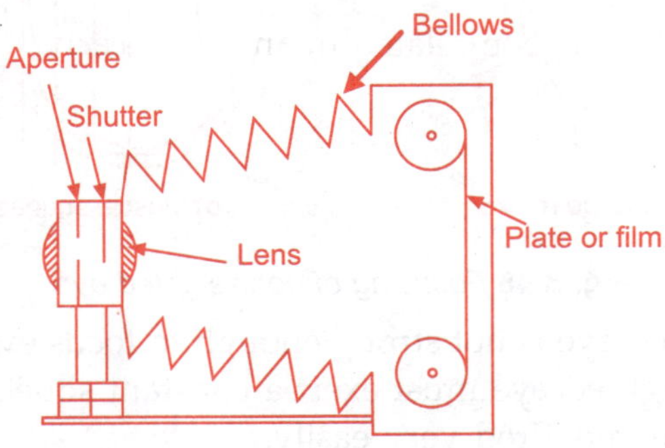

Fig. 8.48 *Photographic camera*

The interior of the camera is blackened in order to absorb the light reflected from the plate. The aperture and the aperture stop are provided very close to the lens.

The light entering the camera through a diameter d is proportional to the area of the stop, and hence to d^2. The illumination of the plate, at a distance nearly f from the lens, should be proportional to $1/f^2$ according to the law of inverse squares. Hence, the illumination of the plate is proportional to d^2/f^2. For a given object illumination, lens used at the same relative aperture d/f should give the same plate illumination. The lens mount is usually marked with "f-numbers", representing the diameter of the stop for each adjustment. If $d/f = 1/4.5$, then $d = f/4.5$. While continuous adjustment is usually possible, the graduations on the mount, $f/4.5$, $f/6.3$, $f/7.7$,... etc., form a series in which the denominator numbers increase successively by a factor of $\sqrt{2}$. Each gives an illumination of the plate half that due to the one before, presumably requiring twice the exposure.

The plate or film is coated with a fine grained emulsion of silver bromide in gelatin. Light activates the grains, so that when the exposed plate is immersed in a suitable reducing solution (developer), the activated grains are reduced to opaque metallic silver. After some time, the plate is removed from the developer, washed in water to remove the remaining developer, and the undeveloped bromide dissolved out from the grains using sodium thiosulphate (a process called fixing). The resulting plate is a "negative", in which the lighter parts of the object appear with deeper darkening. The negative is printed by passing light through it onto a light sensitive paper held in contact with it. The darker parts of the negative give the lighter parts of the print, thereby reproducing the original object in the right way.

8.10 PHOTOMETRY

Photometry deals with the measurement of light illumination and the units used for such measurements. Proper illumination is very important from the point of visibility in a particular environment. The instruments with the help of which illumination measurements are made are called Photometers. In order to measure illumination, we need to define illumination and the related quantities clearly.

Just as sound is a form of energy felt by the human or animal ears, illumination is the radiant form of energy sensed by the human or animal eyes. Illumination is of considerable practical importance. Its definitions and measurement methods apply to the transfer of any kind of radiant energy. The basic unit of illumination, the candela, is a fundamental unit in the SI system.

Following the wave theory of light, light travels in the form of waves, with a velocity of 300×10^6 m/s. The corresponding wavelength of visible light ranges from about 360 to 830 nanometers (1000 nm $= 1$ μm $= 10^{-6}$m). Typical photometric units include lumens, lux, and candelas.

Light from a point source illuminates the space around it. In order to define the area illuminated by that source, it is necessary to understand the concept of a solid angle.

8.10.1 Projected Area and Solid Angle

The concept of solid angle is explained by extending the concept from the plane angle. The angle on a plane surface between two intersecting straight lines is measured in radians or **rad.** One radian is the angle subtended at the center of a circle by a portion of the circumference equal in length to one radius. Since the circumference of a circle is given by $2\pi r$, there are 2π radians in a circle, and the conversion between degrees and radians is 1 rad = $(180/\pi)$ degrees.

A solid angle extends this concept to three dimensional fields. The solid angle is measured in steradians or sr. One steradian (sr) is the solid angle formed by a pyramid with its vertex at the centre of a sphere, and its base covering an area of R^2 on the surface of the sphere, where R is the radius of the sphere.

The total surface area of a sphere is given by $4\pi R^2$. Therefore the sphere subtends a solid angle of $4\pi R^2/R^2 = 4\pi$ steradians at the centre of the sphere. The solid angle subtended by a hemisphere at the centre of the sphere is 2π steradians. Imagine a square sheet of paper with side dimension R glued on to the surface of a sphere. The sheet forms a surface area of R^2 on the surface of the sphere. That area subtends a solid angle of 1 sr at the centre of the sphere.

The solid angle is denoted either by ω, the lowercase Greek letter omega, or Ω, the uppercase omega. Both plane angles and solid angles are dimensionless quantities.

8.10.2 Quantities used in Photometry and their Units

The Table 8.1 gives the units normally used in photometry. They are basically weighted for the spectral response of the human eye.

TABLE 8.1 Units used in Photometry

Quantity	Photometric
Power	Lumen (lm)
Power per unit area	lm/m^2 = lux (lx)
Power per unit solid angle	lm/sr = candela (cd)
Power per area per solid angle	lm/m^2-sr = cd/m^2 = nit

8.10.3 Luminous intensity

Luminous intensity of a source is a measure of luminous energy (visible light) emitted by the source. Luminous intensity of a point source is defined as the total amount of luminous energy given out per second per sq. cm area by the source in all directions. Symbol L (or C) is used for luminous intensity of a source.

Units of Luminous Intensity

The **candela** is one of the seven base units of the SI system. It is defined as follows:

The candela is the luminous intensity, in a given direction, of a source that is emitting monochromatic (single frequency) radiation of frequency 540×10^{12} Hertz having a radiant intensity of 1/683 Watt per steradian in that direction.

The candela is abbreviated as *cd* and its symbol is I_v.

As a historical note, the original unit used for luminous intensity was a standard candle, which was a candle of sperm wax, seven eighths of an inch in diameter, weighing one-sixth of a pound and burning at the rate of 120 grains per hour. This unit was called candle, commonly expressed as candle power. The next international standard was introduced in 1948, which was the international candle or candela (Latin word for candle). Candela was defined as one-sixtieth of the luminous energy coming out per second from one sq. cm area of a black body maintained at the freezing point of platinum (1773°C). The candela was also defined as the luminous intensity, in the perpendicular direction, of a surface of 1/600000 square meter of a black body at the temperature of freezing platinum under a pressure of 101325 Newtons per square meter.

Luminous Flux

Luminous flux across any surface is the rate of flow of luminous energy through the surface. In the absence of absorption, the luminous flux across any imaginary closed surface surrounding a source is equal to the amount of luminous energy emitted per second by the source, i.e., luminous intensity of the source.

The unit of luminous flux is called Lumen. Lumen is defined as the luminous flux through unit area of the surface of an imaginary sphere of unit radius when a point source of light of one candela is placed at its centre (alternatively, lumen is the luminous flux per unit solid angle due to a point source of light of one candela).

As the surface area of a sphere of unit radius is 4π, the total luminous flux across any closed surface surrounding a point source of one candela is 4π lumens. Therefore, total luminous flux due to a source of luminous intensity L is $4\pi L$ lumens.

The **lumen** is the unit for luminous flux. The abbreviation is lm and the symbol is Φ_v. The lumen is the luminous flux emitted by an isotropic point source into unit solid angle (1 sr) having a luminous intensity of 1 candela. The lumen is the product of luminous intensity and solid angle, cd · sr. If a light source is isotropic, the relationship between lumens and candelas is 1 cd = 4π lm. In other words, an isotropic source having a luminous intensity of 1 candela emits 4π lumens into space, which just happens to be 4 steradians. We can also state that 1 cd = 1 lm/sr.

A fundamental method used to determine the total flux (lumens) is to measure the luminous intensity (candelas) in many directions, and then numerically integrate over the entire sphere. Later on, we can use this "calibrated" lamp as a reference in an integrating sphere for routine measurements of luminous flux.

In a light bulb we want a high number of lumens with a minimum of power consumption and a reasonable lifetime. Projection devices are also characterized by lumens to indicate how much luminous flux they can deliver to a screen.

8.10.4 Intensity of Illumination or Illumination

Intensity of illumination at a point on a surface is defined as the luminous flux per unit area, across a small element of area, surrounding the point, the rays being perpendicular to the area. If dF is the luminous flux through an element of area dS, surrounding a point on a surface, the flux being perpendicular to dS, then intensity of illumination I, is given as

$$I = \frac{dF}{dS} \tag{8.40}$$

The units used for the measurement of illumination are lux, phot and foot candle. Lux or meter candle is the illumination on a square meter area of a spherical surface of radius one meter, due to a point source of one candela at the centre of the sphere. Also one lux = one lumen per sq. meter. Phot is equal to one lumen per sq. cm. Therefore, 1 phot = 10^4 lux. One foot candle is equal to one lumen per sq. ft.

It must be noted that the brightness of a surface is not the same as the illumination of the surface. Illumination depends on the luminous flux on the surface, while the brightness depends on the reflecting power of the surface. This is the reason why the brightness of white chalk on a blackboard is very high compared to that of black polish on the board.

When light is incident on a surface, a portion is reflected, a portion is transmitted and the remaining portion is absorbed. If I is the illumination on the surface, R the portion reflected, T the portion transmitted and A the portion absorbed, then

$$I = R + T + A$$

or, $$l = \frac{R}{I} + \frac{T}{I} + \frac{A}{I} = r + t + a \tag{8.41}$$

where r, t and a are, respectively, the reflection coefficient, the transmission coefficient and the absorption coefficient of the surface.

Illuminance denotes luminous flux density. It has a special name, **lux**, and is lumens per square meter, or lm/m^2. The symbol is E_v. Most light meters measure this quantity, as it is of great importance in illumination engineering.

Luminance is the luminous flux per unit area. It also has a special name, **nit**, and is cd/m^2 or lm/m^2-sr. The symbol is L_v. It is most often used to characterize the "brightness" of flat emitting or reflecting surfaces. A typical use would be the luminance of a laptop computer screen. They have between 100 and 250 nits, and those readable in sunlight have more than 1000 nits. Typical CRT monitors have between 50 and 125 nits.

8.10.5 Inverse Square Law

The total luminous flux due to a point source of luminous intensity, L, is $4\pi L$ lumens. Therefore, the illumination on (or luminous flux through unit area of) a spherical surface of radius r, with the source at its centre is given by

$$I = \frac{4\pi L}{4\pi r^2}, \text{ or } I = \frac{L}{r^2} \tag{8.42}$$

If the radius of the spherical surface is doubled, the surface area is increased four times. As the luminous flux through the spherical surface remains the same, equal to $4\pi L$, illumination becomes one fourth. If the radius is trebled, the illumination becomes one-ninth and so on.

Hence, for a given source, $I \propto \dfrac{1}{r^2}$.

Thus the illumination at a point on a surface due to a source varies inversely as the square of the distance of the point from the source. This is the inverse square law in photometry. This law holds well when the source is a point source. It is also true when the size of the source is very small compared to the distance of the point from the source.

Oblique Incidence

The expression for illumination at a point due to a source given by Eq. (8.42) holds good for normal incidence, i.e., when the luminous flux is normal to the surface. For oblique incidence, as shown in Fig. 8.49, illumination I is given by the relation

$$I = I_0 \cos \theta \tag{8.43}$$

where $I = \dfrac{L}{d^2}$ is the illumination for normal incidence and θ is the angle of incidence.

Hence,

$$I = \frac{L \cos \theta}{d^2} \tag{8.44}$$

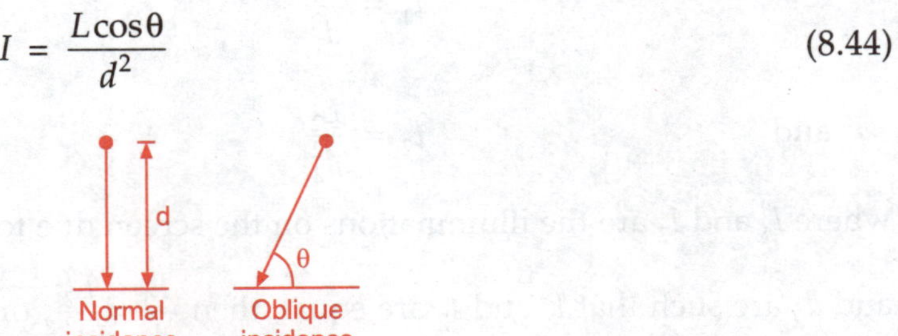

Fig. 8.49 *Normal and oblique incidence*

In Fig. 8.50, AB represents area on which illumination is I and angle of incidence is θ. AC represents the area on which illumination is I_0 for normal incidence.

$$I \propto \frac{1}{AB}$$

and

$$I \propto \frac{1}{AC}$$

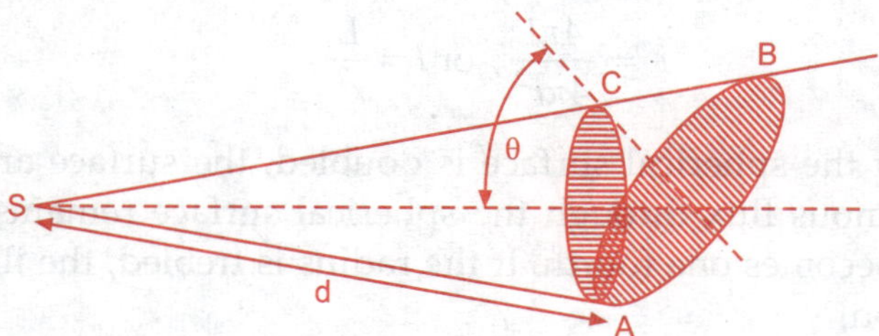

Fig. 8.50 *Oblique incidence*

$$\frac{I}{I_0} = \frac{AC}{AB} = \cos\theta$$

Hence, $I = I_0 \cos\theta$.

From Eq. (8.43), it follows that the illumination at a point on a surface due to a source is directly proportional to the cosine of the angle of incidence. This is known as Lambert's cosine law.

It follows from Eq. (8.44), that the illumination at a point due to a source is (*i*) directly proportional to the luminous intensity of the source, (*ii*) directly proportional to the cosine of the angle of incidence, and (*iii*) inversely proportional to the square of the distance of the point from the source.

8.10.6 Principle of Photometry

If two sources S_1 and S_2 of luminous intensities L_1 and L_2, respectively, are placed at distances d_1 and d_2 from a screen, then we have

$$I_1 = \frac{L_1}{d_1^2}$$

and

$$I_2 = \frac{L_2}{d_2^2}$$

where I_1 and I_2 are the illuminations on the screen due to the sources. If the distances d_1 and d_2 are such that I_1 and I_2 are equal, then $\dfrac{L_1}{d_1^2} = \dfrac{L_2}{d_2^2}$, or

$$\frac{L_1}{L_2} = \frac{d_1^2}{d_2^2}. \tag{8.45}$$

Hence, the luminous intensities of two sources are directly proportional to the squares of the distances at which they produce equal illumination. Equation (8.45) is the condition for photometric balance. Two sources are said to be matched, when condition (8.45) is satisfied. The condition holds good only when the sources are small compared to the distances.

EXAMPLE 8.5

At what distance from a light source of 64 candela will the intensity of illumination become 40 lux?

SOLUTION

$$I = \frac{L}{d^2}$$

I = Illumination = 40 lux

L = Luminous intensity = 64 cd

$$40 = \frac{64}{d^2} \text{ or } d^2 = 1.6$$

i.e., $\qquad d = 1.265$ m

Hence, the distance is 1.265 m.

EXAMPLE 8.6

A source of light is at a height of 2 m above the centre of a circular table of diameter 3m. Compare the intensities of illumination at the centre and at the edge.

SOLUTION

From Lambert's cosine law

$$I = \frac{L\cos\theta}{d^2}$$

At the centre, $\qquad I_C = \frac{L}{2^2}$

At the edge, $\qquad I_E = \frac{L\cos\theta}{2.5^2}$ $\qquad$ (Note that at E, $d^2 = 2^2 + 1.5^2 = 2.5^2$)

$$\frac{I_C}{I_E} = \frac{2.5^3}{2^3} = \frac{15.625}{8}$$

i.e., $\qquad I_C : I_E = 15.625 : 8.$

EXAMPLE 8.7

A 16 cd lamp is placed at a distance of 2 meters from a 4 cd lamp. Find the position of the point at which the two lamps produce the same intensity of illumination.

SOLUTION

For equal illumination, $\dfrac{L_1}{d_1^2} = \dfrac{L_2}{d_2^2}$

Writing $L_1 = 16$ cd and $L_2 = 4$ cd and the distance of the point of equal illumination from the 16 cd as $x\,m$,

$$\frac{16}{x^2} = \frac{4}{(2-x)^2}$$

or, $\qquad\qquad\qquad x^2 = 4\,(2-x)^2$

i.e. $\qquad\quad 3x^2 - 16x + 16 = 0 \text{ or } (3x - 4)\,(x - 4) = 0$

Hence x can have two values, $x_1 = 4/3$ m and $x_2 = 4$ m.

One position is between the lamps at a distance of 4/3 meters from the 16 cd lamp. The other position is at a distance of 4 meters from the 16 cd lamp and 2 m from the other lamp.

EXAMPLE 8.8

Find the distance at which a 100 cd source is to be placed from a screen to produce the same illumination as that produced by a 25 cd source kept on the same side at a distance of 1 meter.

SOLUTION

For equal illumination, $\qquad \dfrac{L_1}{d_1^2} = \dfrac{L_2}{d_2^2}$

Writing $\qquad\qquad\qquad L_1 = 100$ cd, $L_2 = 25$ cd, $d_2 = 1$ m,

$$\frac{100}{d_1^2} = \frac{25}{1^2}$$

i.e. $d_1^2 = 4$ or $d_1 = 2$m.

100 source has to be placed at a distance of 2 m from the screen.

EXAMPLE 8.9

A photographic print can be made in 16 sec, when held at a distance of 50 cm from a lamp. Find the time of exposure when it is held at a distance of 75 cm from the same source.

Amount of light failing normally on unit area in one sec is illumination.

SOLUTION

For same amount of light, illumination × time = Constant.

i.e.
$$I_1 t_1 = I_2 t_2$$

$$\frac{L}{d_1^2} t_1 = \frac{L}{d_2^2} t_2$$

Substituting the values,

$$\frac{16}{50^2} = \frac{t_2}{75^2} \text{ or } t_2 = \frac{16}{20^2} \times 75^2$$

Time of exposure required = 36 sec.

EXAMPLE 8.10

A lamp with a dirty chimney at certain distance from a screen produces the same illumination as a standard candle distant 10 cm from the screen. When the chimney is cleaned the candle has to be shifted by 2 cm to restore balance. Calculate the percentage of light absorbed by the dirty chimney.

SOLUTION

If L_1 and L_2 represent the luminous intensities of the lamp with dirty chimney and cleaned chimney respectively, then

$$\frac{L_1}{d^2} = \frac{L_1}{10^2} \text{ and } \frac{L_2}{d^2} = \frac{1}{8^2}$$

$$\frac{L_1}{L_2} = \frac{64}{100}$$

Amount absored = $(L_2 - L_1)$

Fraction absorbed $= \dfrac{L_2 - L_1}{L_2} = 1 - \dfrac{L_1}{L_2} = 1 - \dfrac{64}{100} = \dfrac{36}{100}$

Percentage of absorption = 36.

SUMMARY

1. Reflection is governed by the equation $\angle i = \angle r$ and refraction by the

   Snell's law, sin i/sin $r = n$, where the incident ray, reflected ray, refracted ray and normal lie in the same plane. Angles of incidence, reflection and refraction are i, r and r', respectively.

2. The *critical angle of incidence* i_c for a ray incident from a denser to rarer medium, is that angle for which the angle of refraction is 90°. For $i > i_c$, total internal reflection occurs. Multiple internal reflections in diamond ($i_c = 24.4°$), totally reflecting prisms and mirage, are some examples of total internal reflection. Optical fibres consist of glass fibres coated with a thin layer of material of *lower* refractive index. Light incident at an angle at one end comes out at the other, after multiple internal reflections, even if the fibre is bent.

3. *Cartesian sign convention*: Distances measured in the same direction as the incident light are positive; those measured in the opposite direction are negative. All distances are measured from the pole/optic centre of the mirror/lens on the principal axis. The heights measured upwards above *x*-axis and normal to the principal axis of the mirror/lenses are taken as positive. The heights measured downwards are taken as negative.

4. *Dispersion* is the splitting of light into its constituent colours.

5. Natural light, *e.g.,* from the sun is unpolarized. This means the electric vector takes all possible directions in the transverse plane, rapidly and randomly, during a measurement. A polaroid transmits only one component (parallel to a special axis). The resulting light is called linearly polarised or plane polarised. When this kind of light is viewed through a second polaroid whose axis turns through 2π, two maxima and minima of intensity are seen. Polarised light can also be produced by reflection at a special angle (called the Brewster angle) and by scattering through $\pi/2$ in the earth's atmosphere.

6. *The Eye*: The eye has a convex lens of focal length about 2.5 cm. This focal length can be varied somewhat so that the image is always formed on the retina. This ability of the eye is called *accommodation*. In a defective eye, if the image is focussed before the retina (myopia), a diverging corrective lens is needed; if the image is focussed beyond the retina (hypermetropia), a converging corrective lens is needed. Astigmatism is corrected by using cylindrical lenses.

Magnetism   9

9.1 INTRODUCTION TO MAGNETISM

9.1.1 Magnets and Magnetism

A magnet can attract an iron or steel piece when it is brought close to it. A magnetic compass needle always points toward the northerly direction. It has long been known that a particular specimen of iron ore (Fe_3O_4) exhibits the property of attracting small pieces of iron. The name magnetite might have been given to this ore because of its abundance in the province of Magnesia in Asia Minor. A piece of magnetite when freely suspended is found to assume approximately north-south direction. Because of this property it is popularly known as lode stone or leading stone. The property of magnetite that can attract a piece of iron or steel is called magnetism. Lode stone is a natural magnet. Artificial magnets can be made from an iron or steel bar by

(*i*) rubbing the iron bar with a magnet or

(*ii*) passing an electric current through a wire wound round it.

9.1.2 Poles of a magnet

Consider a bar shaped magnet. The magnetic property of a magnet can be explained by assuming that the entire magnetism is concentrated at the ends of the magnet. The two end points are called the poles of the magnet. When a bar magnet is suspended freely in a horizontal plane, its ends will align themselves approximately north-south. The pole pointing to the north is called the N-pole or positive pole of the magnet. The south pointing pole is called the S-pole or negative pole of the magnet. The poles always occur in pairs and they are of equal strength.

9.1.3 Coulomb's Law

A detailed study of the force between two magnetic poles was done by C.A. Coulomb, a French scientist, and the law relating to the force is called Coulomb's law. Coulomb's law states that the force of attraction or repulsion between any two point magnetic poles is directly proportional to the product of their pole strengths and is inversely proportional to the square of the distance between them.

If m_1 and m_2 are the strengths of two point poles situated at a distance d from each other, and f is the force between them, then Coulomb's law can be expressed as

$$f \propto \frac{m_1 m_2}{d^2}$$

or, it may be expressed in the form

$$f = \frac{m_1 m_2}{Kd^2} \tag{9.1}$$

where K is a constant of proportionality depending on the nature of the medium and the system of units employed.

Unit Pole:

In *SI* system, the constant K is written as $4\pi\mu$ where μ is called permeability (absolute) of the medium. Further,

$$\mu = \mu_0 \mu_r$$

where μ_0 = permeability of free space (vacuum, or air for all practical purposes) and μ_r = relative permeability of the medium. Value of μ_0 is $4\pi \times 10^{-7}$ weber^2/Nm^2 or $4\pi \times 10^{-7}$ henry/meter.

Unit pole can be defined as that pole which when placed in vacuum (or air) at a distance of one meter from an identical pole repels it with a force equal to $\dfrac{1}{4\pi\mu_0}$ newton. This unit pole is called weber.

The pole strength of a magnetic pole is equal to $4\pi\mu_0$ times the force experienced by an isolated unit north pole placed in vacuum (or air) near the pole at a distance of one meter.

9.2 MAGNETIC FIELD

The space surrounding a magnet where influence of the magnet is experienced is called a magnetic field. In general, space where magnetic effect is observed is called a magnetic field.

9.2.1 Lines of force

The concept of lines of force helps to study the nature of magnetic fields. An isolated north pole placed at any point in a magnetic field, is acted upon by a force. If the isolated pole is free to move, it moves along the direction of the force. The path or track in space along which an imaginary isolated north pole moves, is called a line of force. A line of force is such that the tangent at any point on it gives the direction of the magnetic field at that point. The lines of force due to a magnet are closed curves starting from N-pole and ending on S-pole. No two lines of force intersect each other; because if they do, there would be two directions of magnetic field at a point, which is not possible. Lines of force in a magnetic field can be drawn with the help of a compass needle.

9.2.2 Neutral points

When two or more magnetic fields are combined there may be certain points at which the resultant field is zero. Such points of zero magnetic fields are called neutral points or null points. If a compass needle is placed at such a point, the compass needle will be undisturbed and may be set in any direction. If a bar magnet is placed along the magnetic meridian with S-pole pointing north, neutral points are obtained on the axial line as shown by cross marks in Fig. 9.1(*a*). For the case of N-pole pointing north, the neutral points are obtained on the equatorial line as shown in Fig. 9.1(*b*).

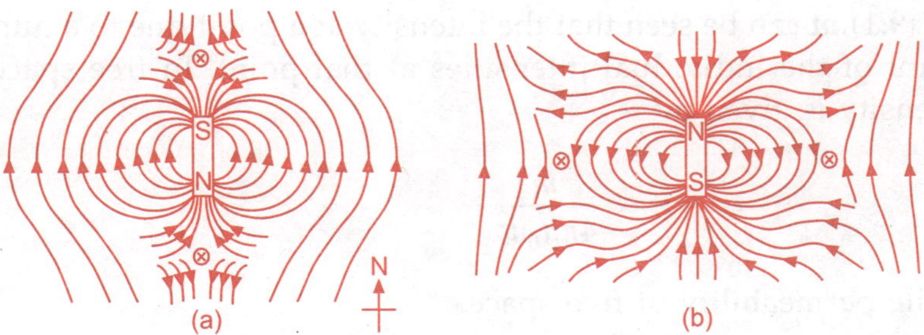

Fig. 9.1 *Lines of forces and neutral points*

We must be able to measure and express quantitatively the various characteristics of magnetism. Unfortunately, a number of unit conventions are used, as shown in the Table 9.1. We will use SI units in our discussions. The advantage of using SI units is that they are traceable back to an agreed set of four base units-meter, kilogram, second, and Ampere.

TABLE 9.1 Quantifying Magnetic Properties

Quantity	Symbol	SI Units (Sommerfeld)	SI Units (Kenelly)	CGS Units (Gaussian)
Field	H	A/m	A/m	Oersteds
Flux Density (Magnetic Induction)	B	Tesla	Tesla	Gauss
Flux	F	Weber	Weber	Maxwell
Magnetization	M	A/m	-	erg/Oe-cm^3

9.3 MAGNETIC INTENSITY AND MAGNETIC POTENTIAL

9.3.1 Magnetic Intensity

The intensity of a magnetic field at a point is defined as the force experienced by an imaginary isolated unit north pole placed at that point. The magnetic intensity has both magnitude and direction at a point and hence it is a vector quantity.

Magnetic intensity at a point is said to be one unit if an isolated unit north pole placed at that point experiences a force of one newton. This is expressed in newtons/weber (or ampere/meter). The intensity at a point is said to be H if a unit north pole placed at that point experiences a force H. A magnetic pole of strength m, kept in a magnetic field having intensity H at the point, experiences a force $f = mH$.

The magnetic intensity F at a distance d from a pole of strength m is given by

$$F = \frac{m \times 1}{4\pi\mu d^2} = \frac{m}{4\pi\mu d^2}$$

From Eq. (9.1), it can be seen that the intensity at a point due to a number of poles is the resultant of the individual intensities at that point. In free space (or air) the magnetic intensity is given by

$$F = \frac{m}{4\pi\mu_0 d^2} \tag{9.2}$$

where μ_0 is the permeability of free space.

Oersted is a unit of magnetic intensity in the CGS system. The conversion between the CGS system unit and the SI unit is: 1 newton/weber = $4\pi \times 10^{-3}$ oersteds.

9.3.2 Uniform Magnetic Field

A uniform magnetic field has the same magnetic intensity at every point in the field, in magnitude and direction. A uniform magnetic field may be represented by parallel lines.

9.3.3 Couple Acting on a Bar Magnet in a Uniform Magnetic Field

If a bar magnet of pole strength m and magnetic length $2l$ is suspended freely in a uniform magnetic field of intensity H, the magnet comes to rest in the direction of the field.

If the magnet is deflected through an angle θ from its initial position, the N-pole will be acted upon by a force mH in the direction of the field and S-pole by an equal force mH in the opposite direction as shown in Fig. 9.2. The forces acting on the poles of the magnet are equal and opposite and are parallel to each other. Therefore they constitute a couple of moment C, given by

$$C = (mH)x = (mH)(2l \sin \theta) = MH \sin \theta \tag{9.3}$$

where mH is force on each pole of the magnet, and $x = 2l \sin \theta$ is the moment arm of the couple. Further, $M = m(2l)$, is a constant for the magnet, called the magnetic moment of the magnet. The couple tends to restore the magnet to its initial position.

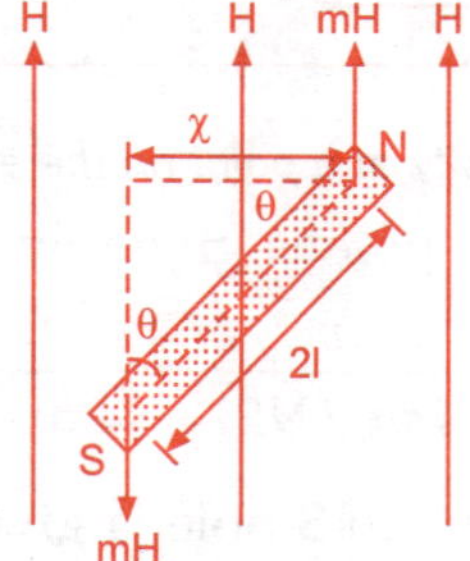

Fig. 9.2 *Couple acting on a bar magnet*

9.3.4 Magnetic Moment

When a magnet of pole strength m and magnetic length $2l$ is placed in a uniform magnetic field of unit intensity, with the magnet at right angles to the field, the moment of the couple experienced by the magnet is given by

$$M \times 1 \times \sin 90° = M \tag{9.4}$$

where $M = 2l$, and $H = 1$. This is a constant for a given magnet and it is called magnetic moment or moment of the magnet.

Thus, the magnetic moment of a magnet is equal to the moment of the mechanical couple required to hold the magnet at right angles to a uniform magnetic field of unit strength. Magnetic moment has both magnitude and direction and hence it is a vector quantity. It is directed from the south pole to the north pole along the magnetic axis.

H is a vector and C is a vector perpendicular to H. This is possible if M is also a vector such that C is perpendicular to both M and H. In other words, C is a vector product of M and H given by

$$\vec{C} = \vec{M} \times \vec{H} \tag{9.5}$$

Magnetic moment is an important quantity and applies not only to permanent magnets but also to electromagnets. A study of magnetic moments of atoms and sub-atomic particles throws light on understanding the nature of matter.

Magnetic moment is expressed in weber-meter, and is related to the CGS units by

$$1 \text{ weber-metre} = \frac{10^{10}}{4\pi} \text{ CGS units (cm-dyne/oersted).}$$

9.3.5 Magnetic Intensity at a Point on the Axial Line of a Bar Magnet

A point on the axis of the bar magnet is said to be in the "A position". In Fig. 9.3, P represents a point on the axial line of a bar magnet NS of moment M, pole strength m and magnetic length $2l$. The point P is at a distance d from the centre O of the magnet.

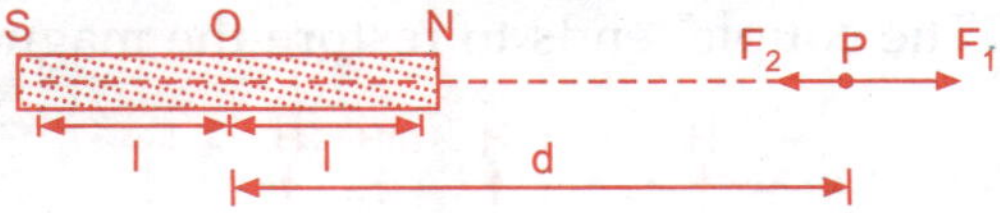

Fig. 9.3 *Magnetic intensity at a point on the axial line of a bar magnet*

The intensity F_1 at P due to N-pole is given by

$$F_1 = \frac{m}{4\pi\mu_0 (NP)^2} = \frac{m}{4\pi\mu_0 (d-l)^2} \tag{9.6}$$

along NP. The intensity F_2 at P due to S-pole is given by

$$F_2 = \frac{m}{4\pi\mu_0 (SP)^2} = \frac{m}{4\pi\mu_0 (d+l)^2} \tag{9.7}$$

along PS. Therefore, the resultant magnetic intensity F_A at P due to the magnet is given by

$$F_A = F_1 - F_2 \qquad F_1 > F_2 \tag{9.8}$$

along NP. Substituting for F_1 and F_2 from Eqs. (9.6) and (9.7) we get

$$F_A = \frac{m}{4\pi\mu_0 (d-l)^2} - \frac{m}{4\pi u_0 (d+l)^2}$$

$$= \frac{m\left[(d+l)^2 - (d-l)^2\right]}{4\pi\mu_0 \left(d^2 - l^2\right)^2} = \frac{m4dl}{4\pi\mu_0 \left(d^2 - l^2\right)^2}$$

$$= \frac{2Md}{4\pi\mu_0 \left(d^2 - l^2\right)^2} \tag{9.9}$$

The relation $(m \times 2l = M)$ has been used in the above.

Finally, we get

$$F_A = \frac{2Md}{4\pi\mu_0\left(d^2 - l^2\right)^2} \qquad (9.10)$$

This field acts along the direction SN.

9.3.6 Magnetic Intensity at a Point on the Equatorial line of a Bar Magnet

A point on the equatorial line or right bisector of a magnet is said to be in the broadside on position or "B position" with respect to the magnet. P is a point on the equatorial line of a bar magnet NS of pole strength m and magnetic length $2l$. The point P is at a distance d from the center O of the magnet as shown in Fig. 9.4.

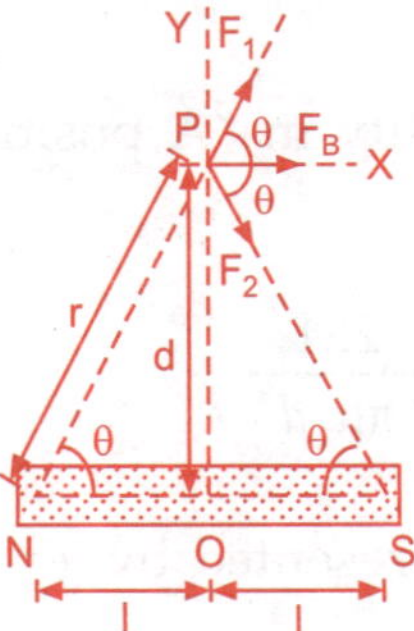

Fig. 9.4 *Magnetic intensity at a point on the equatorial line of a bar magnet*

The point P, being on the right bisector, is equidistant from both N and S. Therefore, we have $NP = SP = r$. Intensity F_1 at P due to N-pole is given by

$$F_1 = \frac{m}{4\pi\mu_0 r^2} \qquad (9.11)$$

along NP. Intensity F_2 at P due to S-pole is given by

$$F_2 = \frac{m}{4\pi\mu_0 r^2} \qquad (9.12)$$

along PS. Since F_1 and F_2 are equal in magnitude, the resultant magnetic intensity F_B at P due to the magnet is along PX, the bisector of the angle between F_1 and F_2 (It must be noted from the figure that PX is parallel to NS. Therefore, the resultant, F_B is given by

$$F_B = F_1 \cos\theta + F_2 \cos\theta = 2F_1 \cos\theta = 2\frac{m}{4\pi\mu_0 r^2}\frac{l}{r}$$

$$F_B = \frac{2ml}{4\pi\mu_0 r^3} = \frac{M}{4\pi\mu_0 r^3} = \frac{M}{4\pi\mu_0\left(d^2 + l^2\right)^{\frac{3}{2}}}$$

$$r^2 = d^2 + l^2$$

$$F_B = \frac{M}{4\pi\mu_0 \left(d^2 + l^2\right)^{\frac{3}{2}}} \tag{9.13}$$

It is evident that F_B acts along PX, in a direction parallel to NS.

9.3.7 Short Magnet

The semi magnetic length of a bar magnet is defined as the distance from the midpoint to one of the poles. A short magnet has a very small semi magnetic length compared to the distance d of the point from the magnet, such that $\frac{1}{d} \ll 1$. Consequently, $r \approx d$.

For a short magnet, the intensity in "A position" represented by F_A^S is obtained from Eq. (9.10) as

$$F_A^S = \frac{2M}{4\pi\mu_0 d^3} \tag{9.14}$$

The intensity in B position represented by F_A^S is given by Eq. (9.13),

$$F_B^S = \frac{M}{4\pi\mu_0 d^3} \tag{9.15}$$

since $\frac{l}{d} \ll 1$, and is negligible. Therefore, in the case of short magnet, intensity at a point in A position is twice the intensity at a point in B position at the same distance.

9.3.8 Magnetic Potential

If a N-pole is brought from infinity to a point in a magnetic field, the work done against the field is the magnetic potential energy of the pole at that point. It is assumed that at infinity, the magnetic intensity is zero, which is taken as reference. The amount of work done in bringing unit N-pole from infinity to the point against the direction of the field is called the magnetic potential of the field at that point.

The concept of magnetic potential is quite similar to the work done in lifting a body against gravity which is the gravitational potential energy of the body. In this case, the work done is mechanical work and therefore easily visualized. In the case of mechanical work, the surface of the earth is taken as reference for measurements of the gravitational potential energy.

9.3.9 Relation between Magnetic Intensity and Potential

The potential difference dV between two points in a magnetic field is the amount of work done in moving a unit N-pole from one point at a lower potential to the second point at a higher potential. If H is the field strength at a point, the force acting on unit N-pole at that point is H. The work done, dW for a small displacement dx of a unit N-pole against the field is given by

$$dW = -Hdx$$

$$dV = -Hdx \tag{9.16}$$

The negative sign indicates that the displacement and the field are in opposite directions. From Eq. (9.16), we have

$$H = -\frac{dV}{dx} \tag{9.17}$$

Thus the intensity of the magnetic field at a point is equal to the negative rate of change of potential with respect to distance, or is equal to the negative potential gradient at the point.

9.3.10 Potential at a Point due to a Magnetic Pole

If H is the magnetic intensity at a point with distance x from a pole of strength m, then

$$H = \frac{m}{4\pi\mu_0 x^2} \tag{9.18}$$

From Eq. (9.16), the work done, dW, in moving unit N-pole against the field is given by

$$dW = -Hdx = -\frac{m}{4\pi\mu_0 x^2}dx \tag{9.19}$$

Total work done, W, in bringing a unit N-pole from infinity to the point against the field is given by the following integral

$$W = \int_{\infty}^{x} -\frac{m}{4\pi\mu_0 x^2}dx = \frac{m}{4\pi\mu_0}\left[\frac{1}{x}\right]_{\infty}^{x} = \frac{m}{4\pi\mu_0 x} \tag{9.20}$$

This corresponds to the potential V at the point

$$V = \frac{m}{4\pi\mu_0 x} \tag{9.21}$$

It must be noted that the potential at a point due to a north pole is positive and that due to a south pole is negative.

9.3.11 Potential at a Point on the Axial Line of a Bar Magnet

From Fig. 9.3, potential at P due to N-pole is given by

$$V_{PN} = \frac{m}{4\pi\mu_0 \, (NP)} = \frac{m}{4\pi\mu_0 \, (d-l)} \tag{9.22}$$

Similarly, potential at P due to S-pole is given by

$$V_{PS} = \frac{-m}{4\pi\mu_0 \, (SP)} = \frac{-m}{4\pi\mu_0 \, (d+l)} \tag{9.23}$$

Therefore, potential V_A due to the magnet at P is given by

$$V_A = \left[\frac{m}{4\pi\mu_0 \, (d+l)}\right] - \left[\frac{m}{4\pi\mu_0 \, (d+l)}\right] = \frac{m(2l)}{4\pi\mu_0 \, (d^2 - l^2)}$$

$$V_A = \frac{M}{4\pi\mu_0 \, (d^2 - l^2)} \tag{9.24}$$

For a short magnet, the potential V_A due to the magnet at P is given by

$$V_A^S = \frac{M}{4\pi\mu_0 d^2} \tag{9.25}$$

where it is assumed that $\dfrac{l}{d} << 1$ for a short magnet.

9.3.12 Potential at a Point on the Equatorial line of a Bar Magnet

Any point on the equatorial line is equidistant from both the poles of the magnet as can be seen from Fig. 9.4. Therefore, the positive potential due to N-pole cancels with the negative potential due to S-pole. This result in

$$V_B = 0 \tag{9.26}$$

Therefore, the equatorial line of a magnet is an equipotential line.

9.3.13 Potential at any Point due to a Magnet

In Fig. 9.5, P is a point at a distance d from the centre O of a bar magnet SN of moment M and semi magnetic length l. Further, the line OP makes an angle θ with SN. The magnetic moment being a vector can be resolved into two components $M \cos\theta$ along OP and $M \sin\theta$ at right angles to OP as shown in Fig. 9.5. Effectively, magnet NS is equivalent to two magnets of moments $M \cos\theta$, denoted as $S_1 N_1$ and $M \sin\theta$ which is denoted as $S_2 N_2$. The point P is in the end on position, with respect to the magnet $S_1 N_1$ of magnetic moment $M \cos\theta$. Therefore, the potential at P is

$$V_{A_1} = \frac{M\cos\theta}{4\pi\mu_0 \left(d^2 - l^2\right)} \tag{9.27}$$

because the point P is in the equatorial line with respect to the magnet S_2N_2 of magnetic moment $M \sin \theta$, its contribution to the potential at P is zero. Hence, the potential V at P due to the magnet is given by

$$V = \frac{M \cos\theta}{4\pi\mu_0 \left(d^2 - l^2\right)} \tag{9.28}$$

For a short magnet the potential at P is given by

$$V^S = \frac{M \cos\theta}{d^2} \tag{9.29}$$

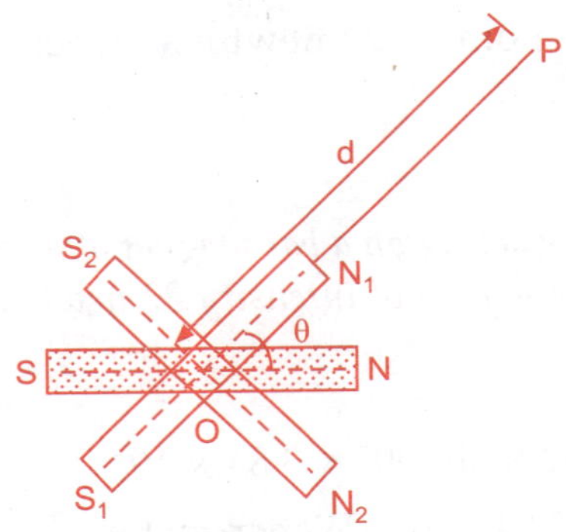

Fig. 9.5 *Potential at any point due to a magnet*

There are similarities between the magnetostatic quantities and the electrodynamic quantities. The relation between the magnetostatic and electrodynamic units are given in Table 9.2.

TABLE 9.2 Relation between Magnetostatic and Electrodynamic Units

Physical quantity	Symbol	Magnetostatic unit (as in this book)	Units used elsewhere
Pole strength	M	Wb	A-m
Flux density	B	Wb/m^2	T; N/A-m; G
Magnetic moment	M	Wb-m	A-m^2
Magnetic intensity	H	N/Wb	N/A-m; A/m
Intensity of magnetisation	I	Wb/m^2	A/m
Permeability of free space	μ_0	Wb/N-m^2	T-m/A; H/m; N/A^2

Wb → weber,   A → ampere,   m → meter,   T → tesla,
G → gauss,   N → Newton,   H → henry,   1G = 10^{-4}T

EXAMPLE 9.1

The force acting on a magnetic pole of strength 7.5×10^{-6} weber is 1.5×10^{-4} N. Find the magnetic intensity at a point where the pole is placed.

SOLUTION

$$F = mH$$

$$1.5 \times 10^{-4} = 7.5 \times 10^{-6}\, H$$

$$H = \frac{1.5 \times 10^{-4}}{7.5 \times 10^{-6}} = 20$$

Magnetic intensity at the point = 20 newtons/weber.

EXAMPLE 9.2

What is the moment of the couple acting on a bar magnet of moment 1.2×10^{-6} Wb-m placed at right angles to a uniform magnetic field of intensity 32 newtons/weber?

SOLUTION

$$C = MH \sin \theta = 1.2 \times 10^{-6} \times 32 \times \sin 90° = 3.84 \times 10^{-5}$$

Moment of the couple = 3.84×10^{-5} Weber-meter.

EXAMPLE 9.3

The moment of a short magnet is $4p \times 10^{-7}$ Wb.m. At what point on its axial line does the intensity of the field become $\dfrac{10^3}{4\pi}$ newton/weber. ($\mu_o = 4\pi \times 10^{-7}$ H/m)

SOLUTION

Short magnet A position

$$F_A^S = \frac{2M}{4\pi\mu_0 d^3}$$

$$d^3 = \frac{2M}{4\pi\mu_0 F_A^s} = \frac{2 \times 4\pi \times 10^{-7} \times 4\pi}{4\pi \times 4\pi \times 10^{-7} \times 10^3} = 2 \times 10^{-3}$$

or,

$$d = \frac{2^{\frac{1}{3}}}{10} = 0.126 \text{ m} = 12.6 \text{ cm.}$$

The intensity of the field is $\dfrac{10^3}{4\pi}$ Newton/weber at a distance of 12.6 cm from the centre of the magnet.

EXAMPLE 9.4

A bar magnet of moment $16\pi \times 10^{-8}$ Wb.m and length 10 cm from the base of an equilateral triangle, the poles being at the extremities of the base. Find the magnitude and direction of the field at the vertex of the triangle.

SOLUTION

The vertex is on the equatorial line of the bar magnet (Fig. 9.6)

$$F_B = \frac{M}{4\pi\mu_0\left(d^2 + l^2\right)^{\frac{3}{2}}}$$

$$M = 16\pi \times 10^{-8}$$

$$\mu_0 = 4\pi \times 10^{-7}$$

$$\left(d^2 + l^2\right)^{\frac{1}{2}} = 0.1 \text{ m}$$

$$F_B = \frac{16\pi \times 10^{-8}}{4\pi \times 4\pi \times 10^{-7} \times 0.1^3} = \frac{100}{\pi}$$

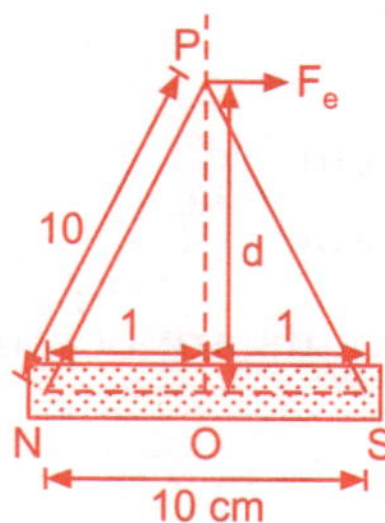

Fig. 9.6 *Bar magnet*

The intensity at the vertex is $100/\pi$ Newton/weber (or 0.4 oersted) and is in a direction parallel to *NS*.

EXAMPLE 9.5

Calculate the pole strength of a bar magnet of magnetic length 10 cm which produces a magnetic field of intensity $250/4\pi$ Newton/weber at a distance of 20 cm from the nearer pole.

SOLUTION

$$l = 5 \text{ cm}$$

distance = 20 cm from the nearer pole

$$d = 0.25 \text{ m}$$

$$\mu_0 = 4\pi \times 10^{-7} \text{ H/m}$$

$$F_A = \frac{2Md}{4\pi\mu_0 \left(d^2 - l^2\right)^2}$$

$$\frac{250}{4\pi} = \frac{2m \times 0.1 \times 0.25}{4\pi\mu_0 \left(0.25^2 - 0.05^2\right)^2}$$

or

$$m = \frac{250 \times 0.2^2 \times 0.3^2 \times 4\pi \times 10^{-7}}{2 \times 0.1 \times 0.25} = 7.2\pi \times 10^{-6}$$

For second face, $\mu = \dfrac{x}{4}$

Pole strength of the magnet = $7.2\pi \times 10^{-6}$ Weber.

EXAMPLE 9.6

A magnet of pole strength $48 \times 4\pi \times 10^{-8}$ weber and of length 10 cm is placed on the magnetic meridian with its S-pole pointing north. If the distance between the neutral points is 30 cm, then find the horizontal intensity of the earth's field.

SOLUTION

$$2D = 30 \text{ cm}$$

$$2t = 10 \text{ cm}$$

S-Pole pointing north; neutral points are obtained on the axial line.

At the neutral point,

$$H = F_A = \frac{2Md}{4\pi\mu_0 \left(d^2 - l^2\right)^2}$$

$$H = \frac{2 \times 48 \times 4\pi \times 10^{-8} \times 0.1 \times 0.15}{4\pi \times 4\pi \times 10^{-7} \times 0.2^2 \times 0.1^2} = \frac{360}{4\pi}$$

Horizontal intensity of the earth's magnetic field = $\dfrac{360}{4\pi}$ Newton/weber or 0.36 oersted.

EXAMPLE 9.7

The field intensities at two points on the axis of a bar magnet, at distance of 10 cm and 20 cm respectively from its centre are in the ratio 12.5 : 1. Find the length of the magnet.

SOLUTION

$$d_1 = 0.1 \text{ m}$$
$$d_2 = 0.2 \text{ m}$$
$$F_1 : F_2 = 12.5 : 1$$

Axial line

$$F_1 = \frac{2Md_1}{4\pi\mu_0 \left(d_1^2 - l^2\right)^2} \text{ and}$$

$$F_2 = \frac{2Md_2}{4\pi\mu_0 \left(d_2^2 - l^2\right)^2}$$

$$\frac{F_1}{F_2} = \frac{d_1}{\left(d_1^2 - l^2\right)^2} \times \frac{\left(d_2^2 - l^2\right)^2}{d_2} = \frac{0.1}{\left(0.1^2 - l^2\right)^2} \times \frac{\left(0.2^2 - l^2\right)^2}{0.2}$$

$$\frac{12.5}{1} = \frac{\left(0.2^2 - l^2\right)^2}{2\left(0.1^2 - l^2\right)^2}$$

or,
$$25 = \frac{\left(0.2^2 - l^2\right)^2}{\left(0.1^2 - l^2\right)^2}$$

$$5(0.1^2 - l^2) = (0.2^2 - l^2)$$

$$4l^2 = 5 \times 0.1^2 - 0.2^2 = 0.01$$

or,
$$2l = 0.1$$

Length of the magnet = 0.1 m or 10 cm.

EXAMPLE 9.8

Two bar magnets of lengths 6 cm and 8 cm respectively, are placed with their north poles turned towards each other, so that their axes are along the same straight line and their centres are 21 cm apart. A magnetic needle placed at a point between the two magnets, 4 cm from the north pole of the shorter magnet on their common axis shows no deflection. Compare their pole strengths. Ignore the earth's magnetic field.

SOLUTION

The point is 7 cm from the centre of the shorter magnet and 14 cm from the centre of the other magnet (Fig. 9.7).

If the pole strengths are m_1 and m_2, the intensities due to the magnets are equal and opposite at the point.

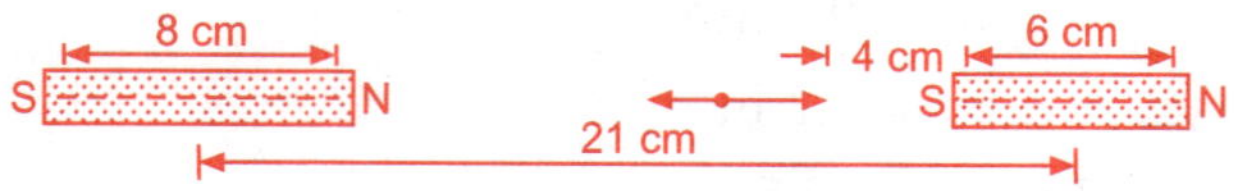

Fig. 9.7 *Two bar magnets with their N-poles facing each other*

$$\frac{(2m_1 \times 2l_1)d_1}{4\pi\mu_0\left(d_1^2 - l_1^2\right)^2} = \frac{(2m_2 \times 2l_2)d_2}{4\pi\mu_0\left(d_2^2 - l_2^2\right)^2}$$

$$\frac{m_1 \times 6 \times 7}{\left(7^2 - 3^2\right)^2} = \frac{m_2 \times 8 \times 14}{\left(14^2 - 4^2\right)^2}$$

or,

$$\frac{m_1}{m_2} = \frac{8 \times 14}{6 \times 9} \times \frac{10^2 \times 4^2}{18^2 \times 10^2} = \frac{32}{243}$$

The pole strengths are in the ratio 32 : 243.

EXAMPLE 9.9

Find the potential at a point situated on a line passing through the middle points of a bar magnet of length 10 cm and moment $600 \times 4\pi\, 10^{-10}$ Wb.m making an angle of 60° with the axis of magnet. The point is 15 cm away from the middle point.

SOLUTION

$$\text{Potential } V = \frac{M\cos\theta}{4\pi\mu_0\left(d^2 - l^2\right)}$$

$$= \frac{600 \times 4\pi \times 10^{-10}\cos 60°}{4\pi \times 4\pi \times 10^{-7}\left(0.15^2 - 0.5^2\right)}$$

$$= \frac{600 \times 10^{-3}}{4\pi} \times \frac{1}{2} \times \frac{1}{0.2 \times 0.1}$$

$$= \frac{15}{4\pi} \text{ Joule/weber.}$$

9.4 MAGNETOMETERS

9.4.1 Deflection Magnetometer

When there is only one uniform magnetic field of intensity H, the magnet sets itself in the direction of H. If another uniform magnetic field of intensity F, is applied perpendicular to H, the magnet deflects through an angle θ and sets itself in the direction of the resultant as shown in Fig. 9.8(a). The deflection θ is given by the relation

$$\tan \theta = \frac{F}{H} \tag{9.30}$$

Rearranging Eq. (9.30), we get

$$F = H \tan \theta \tag{9.31}$$

This is known as tangent law in magnetism.

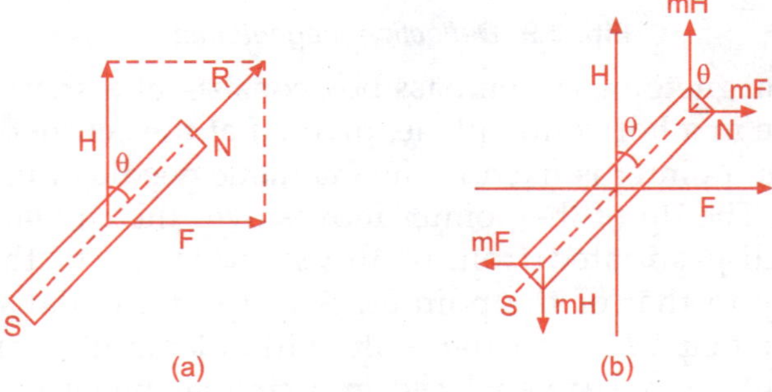

Fig. 9.8 *Deflection magnetometer*

9.4.2 Tangent Law (Principle of Deflection Magnetometer)

A magnet pivoted in a horizontal plane acted upon by two uniform fields that are perpendicular to each other, will set itself in the direction of the resultant of the two fields. The Eq. (9.31) may be obtained, more explicitly, in the following manner.

When there are two fields of intensity F and H, the magnet is acted upon by two couples. The couple due to H is given by

$$C_H = MH \sin \theta \tag{9.32}$$

and the other due to F is given by

$$C_F = MF \sin (90 - \theta) = MF \cos \theta \tag{9.33}$$

These two couples are in opposite directions because couple due to F tends to deflect it away from the original position and the couple due to H tends to restore it to its initial position. In the equilibrium position, the two couples balance each other, and hence $MF \cos \theta = MH \sin \theta$ or we get the tangent law,

$$F = H \tan \theta \tag{9.34}$$

This principle is used in the construction of deflection magnetometer.

In practice, H is the horizontal component of the earth's magnetic intensity and F is due to a bar magnet or a coil carrying current. If F is called the deflecting field then H is the controlling field. The equation, $F = H \tan \theta$, holds good, if and only if both F and H are uniform in the region in which the magnet is suspended. This is why a short magnetic needle is used in a magnetic compass box.

9.4.3 Deflection Magnetometer

Deflection magnetometer consists of a compass box mounted at the middle of a one meter long wooden base board, as shown in Fig. 9.9. Two half-meter scales are fixed to the wooden base board on either side such that the zero ends are located at the centre of the board.

Fig. 9.9 *Deflection magnetometer*

The deflection magnetometer compass box consists of a short powerful magnetic needle, free to rotate in a horizontal plane, pivoted at the centre of a circular scale. A long light aluminium pointer is fixed to the magnetic needle at its centre and at right angles to its length. The tip of the pointer moves over the circular scale divided into four quadrants each graduated from 0° to 90°. Rotation of the magnetic needle corresponds exactly to that of the pointer. A strip of circular plane mirror fitted beneath the pointer helps to read the scale without parallax error. The magnetic needle and the circular scale are enclosed in a circular metal or ebonite box with a glass top.

Tan A position

When a deflection magnetometer is arranged such that the pointer is parallel to the board, the board will be oriented along east-west direction. If a bar magnet is placed on the board, towards one end of the board, as shown in Fig. 9.10, with its axis parallel to the board and passing through the centre of the needle, the needle will be in A position with respect to the magnet. The magnetometer is said to be in Tan A position. In this case, the magnetic intensity F due to the magnet is at right angles to H, the horizontal component of earth's magnetic intensity. Therefore, tangent law can be applied.

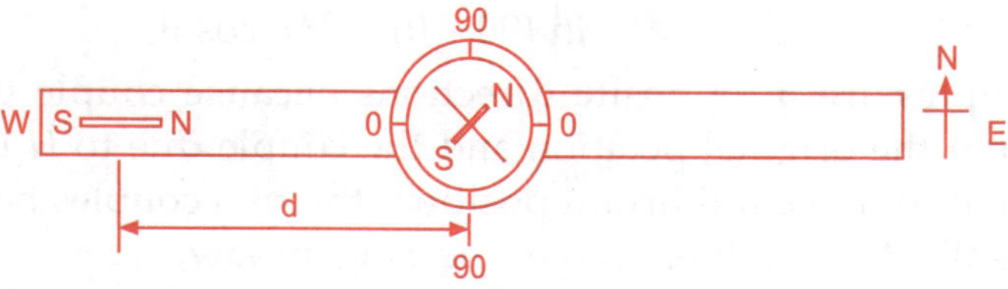

Fig. 9.10 *Deflection magnetometer in Tan A position*

Intensity F is given by

$$F = \frac{2Md}{4\pi\mu_0\left(d^2 - l^2\right)^2} \tag{9.35}$$

where M = magnetic moment, l = semi-magnetic length, d = distance of the centre of the needle from the centre of the magnet, and μ_o = permeability of free space.

Substitution for F in Eq. (9.31) gives

$$\frac{2Md}{4\pi\mu_0\left(d^2 - l^2\right)^2} = H\tan\theta \tag{9.36}$$

where θ is the deflection of the needle or pointer.

Hence,

$$\frac{M}{H} = \frac{4\pi\mu_0\left(d^2 - l^2\right)^2 \tan\theta}{2d} \tag{9.37}$$

For a short magnet,

$$\frac{M}{H} = \frac{4\pi\mu_0 d^3 \tan\theta}{2} \tag{9.38}$$

Before taking measurements in Tan A position, any magnets and magnetic substances not used in the experiments must be removed from the work table. The magnetometer must be placed on the work table, such that the magnetometer is approximately along the east-west direction. The compass box is rotated till the $0-0$ line is parallel to the magnetometer board. The board is then bodily rotated till the pointer reads $0° - 0°$. The magnetometer will now be along east-west with the pointer reading $0° - 0°$, and is set for Tan A position.

Tan B position

When a deflection magnetometer is set-up such that the pointer is perpendicular to the board, the board will be oriented along north-south direction which is along the magnetic meridian. A bar magnet is placed on the board, on one side of the compass box as shown in Fig. 9.11, with its axis at right angles to the magnetometer, and the right bisector of the magnet passing through the center of the magnetic needle. With this arrangement the needle will be in B position with respect to the magnet. Then the magnetometer is said to be in Tan B position. As the field of intensity F due to the magnet is at right angles to H, tangent law can be applied, and the intensity F is given by

$$F = \frac{M}{4\pi\mu_0\left(d^2 + l^2\right)^{\frac{3}{2}}} \tag{9.39}$$

Substitution for F in Eq. (9.2) gives

$$\frac{M}{4\pi\mu_0\left(d^2-l^2\right)^{\frac{3}{2}}} = H\tan\theta \qquad (9.40)$$

where θ is the deflection of the needle or pointer.

From Eq. (9.40), we get

$$\frac{M}{H} = 4\pi\mu_0\left(d^2-l^2\right)^{\frac{3}{2}}\tan\theta \qquad (9.41)$$

For a short magnet, we have $\dfrac{l}{d}\ll1$, and hence

$$\frac{M}{H} = 4\pi\mu_0 d^3\tan\theta \qquad (9.42)$$

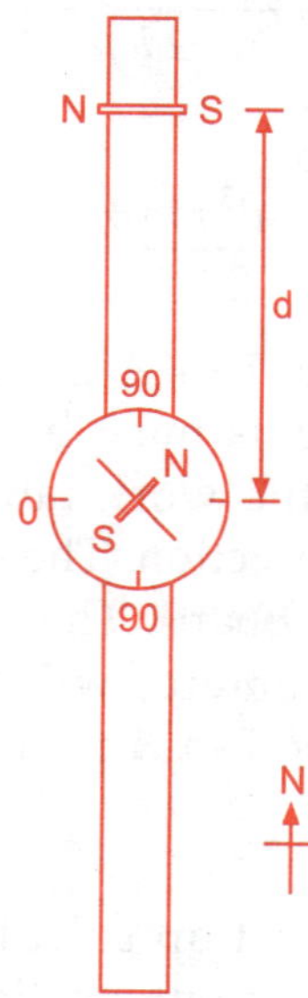

Fig. 9.11 *Deflection magnetometer in Tan B position*

9.4.4 Experimental Determination of M/H Ratio

The determination of the M/H ratio, where the ratio of M, the magnetic moment of a bar magnet to H, the horizontal intensity of the earth's field is done as follows.

A deflection magnetometer must be set either in Tan A position or Tan B position in order to conduct any experiments. A number of readings must be taken in order to eliminate several errors. In order to avoid errors and to have accurate measurements, the following precautions must be taken.

1. The magnetic needle may not pivoted exactly at the centre of the circular scale. The resulting error is eliminated by taking the readings of both the ends of the pointer (θ_1 and θ_2).

2. The magnet may not be symmetrically magnetised. The consequent error is eliminated by repeating the readings after reversing the magnet pole per pole, in the same position (θ_3 and θ_4).

3. The centre of the needle may not be at the centre of the magnetometer board. The error due to this is eliminated by repeating the above readings as in (1) and (2), with the magnet at the same distance, on the other side (θ_5 and θ_6, θ_7 and θ_8).

The average of the eight readings is taken as the true deflection (θ).

The deflection magnetometer is set-up in Tan A position as already described. The given magnet is placed on one arm of the magnetometer board with its axis parallel to the board and passing through the center of the needle. The distance of the magnet is adjusted such that the deflection is in the range 30° to 60°. The distance d of the centre of the magnet from the centre of the needle is noted. The deflections θ_1 and θ_2 of the two ends of the pointer are noted. The magnet is reversed pole per pole in the same position and the deflections θ_3 and θ_4 are noted. In a similar manner, four more readings θ_5, θ_6, θ_7 and θ_8 are taken with the magnet on the other arm, at the same distance from the needle. The average deflection θ is found. The experiment is repeated for different distances. The magnetic length $2l$ of the magnet is found by locating its poles using a compass needle as shown in Fig. 9.12. When the bar magnet is placed close to the compass needle, the compass needle will point at the pole. The direction of the compass needle is marked on the magnet. Repeating this for 2 or 3 positions, the pole can be determined as the meeting point of such lines.

The ratio M/H is calculated in each case, using Eq. (9.37)

$$\frac{M}{H} = \frac{4\pi\mu_0 \left(d^2 - l^2\right)^2 \tan\theta}{2d} \tag{9.43}$$

From Eq. (9.43), the M/H ratio for a short bar magnet is,

$$\frac{M}{H} = \frac{4\pi\mu_0 d^3 \tan\theta}{2} \tag{9.44}$$

The experiments must be repeated several times and the mean value of all the trials will determine the M/H ratio.

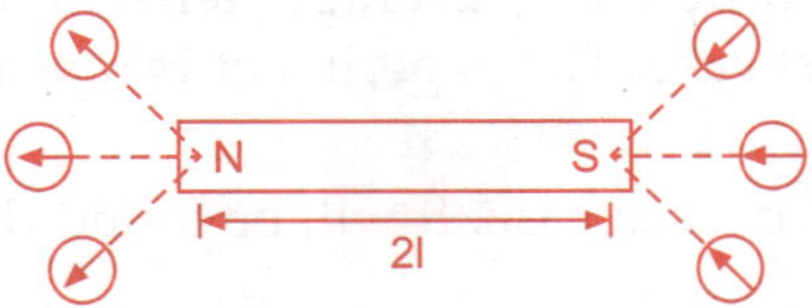

Fig. 9.12 *Finding the magnetic length*

Experiments are conducted with the deflection magnetometer set from Tan B position, and the M/H ratio is obtained as

$$\frac{M}{H} = 4\pi\mu_0 \left(d^2 + l^2\right)^{\frac{3}{2}} \tan\theta \tag{9.45}$$

For a short bar magnet, the M/H ratio is given by

$$\frac{M}{H} = 4\pi\mu_0 d^3 \tan\theta \tag{9.46}$$

The magnetic moment M, and hence pole strength m of a bar magnet can be determined from the mean value of (M/H), and by assuming the value of H at the place.

9.4.5 Comparison of Moments of two Magnets

Using one of the magnets with the deflection magnetometer, the mean value of (M/H) is found as explained above. The experiment is repeated for the other magnet. If M_1 and M_2 are the moments of the magnets, X and Y the mean values of (M/H) for the magnets, respectively, then

$$\frac{M_1}{M_2} = \frac{X}{Y}. \tag{9.47}$$

9.4.6 Verification of Inverse Square Law

The intensity at a point along the axial line of a short bar magnet is double that due to the same magnet at an equal distance in the equatorial position. If F_A^S and F_B^S are the intensities at points on the axial line and the equatorial positions of a short bar magnet, for the same distance, then $F_A^S = 2F_B^S$. This result is obtained on the assumption that the inverse square law holds good. Hence, inverse square law can be indirectly verified by showing that the ratio $\left(\dfrac{F_A^S}{F_B^S}\right)$ is equal to 2. This method is called Gaussian method.

The average deflection θ_A is determined for a short bar magnet with a deflection magnetometer set for Tan A position. For the same magnet with the deflection magnetometer set for Tan B position, average deflection θ_B is found, the distance being the same as in the first case. The experiment is repeated for different distances.

The ratio $\left(\dfrac{\tan\theta_A}{\tan\theta_B}\right)$ is found in each case. It will be found that the ratio is nearly equal to 2. This verifies inverse square law.

Alternative method: The magnetometer is set in Tan *A* position with the board along east-west and the pointer reading 0° – 0°. A long bar magnet or a ball ended magnet is fixed to a wooden stand and is arranged such that one end is vertically above the centre and the other end on the axis of the board in the same plane as the needle as shown in Fig. 9.13. The distance *d* is noted.

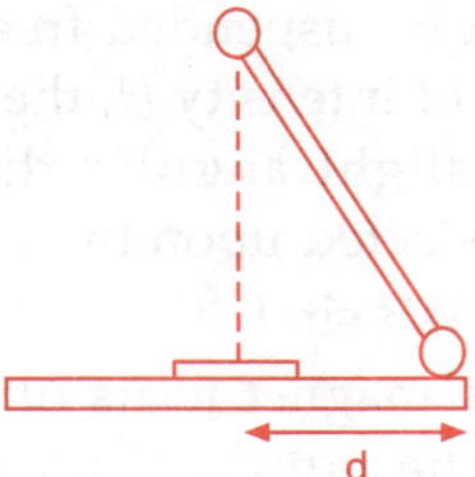

Fig. 9.13 *Verification of inverse square law*

The two readings of the pointer are noted. The magnet is reversed pole per pole at the same distance. Two more readings are taken. Four more readings are taken with the magnet on the other side at the same distance. The average deflection θ is found.

The experiment is repeated for different distances and the value of $d^2 \tan \theta$ is calculated in each case. It will be found that $d^2 \tan \theta$ is nearly a constant. This verifies inverse square law.

The magnetic intensity at the needle due to the pole of the magnet vertically above the needle, is in the vertical direction and hence does not affect the motion of the needle in the horizontal plane. In this case, *F*, in the equation $F = H \tan \theta$, represents the magnetic intensity due to the pole on the magnetometer board. Therefore, we have

$$F = \frac{m}{4\pi\mu_0 d^2} = H \tan \theta \tag{9.48}$$

Since $d^2 \tan \theta$ is constant,

$\tan \theta \propto \dfrac{1}{d^2}$ and hence $F \propto \dfrac{1}{d^2}$.

This alternative method can be conveniently used to determine pole strength of a magnet. The mean value of $d^2 \tan \theta$ is found for different values of *d* as explained above. Pole strength *m* is calculated using the relation

$$m = 4\pi\mu_0 H (d^3 \tan \theta) \tag{9.49}$$

where the value of *H* at the place is used.

9.5 VIBRATION MAGNETOMETER

Vibration magnetometer works on the principle that, whenever a freely suspended magnet in a uniform magnetic field is disturbed from its equilibrium position, it is set into vibrations about the mean position. Vibration magnetometer is used for comparison of magnetic moments and magnetic fields.

When a magnet of moment M is suspended freely, with its axis horizontal in the plane of a uniform magnetic field of intensity H, the magnet sets itself in the direction of H. If the magnet is given a slight angular displacement θ about the axis of suspension initially, the magnet is acted upon by a couple,

$$C = MH \sin \theta \tag{9.50}$$

The couple tries to restore the magnet to its original position. The magnet gains momentum and goes beyond the equilibrium position through almost equal displacement on the other side. The motion repeats and the magnet executes oscillations about the axis of suspension. The period of oscillation of the magnet depends on (*i*) M, the moment of the magnet, (*ii*) I, the moment of inertia of the magnet about the axis of suspension, and (*iii*) H, the intensity of magnetic field in which the magnet is. suspended. The period of oscillation T, is given by the formula,

$$T = 2\pi\sqrt{\frac{I}{MH}} \tag{9.51}$$

9.5.1 Searle's Vibration Magnetometer

Searle's vibration magnetometer consists of a small, cylindrical magnet NS fixed in a brass bob B as shown in Fig. 9.14. The brass bob is used to increase the moment of inertia and hence the period of oscillations. An aluminium pointer PP is fixed below the magnet, for observing the oscillations. The system is suspended by means of an unspun silk thread. The complete system may be arranged in a narrow enclosure fixed to a suitable support.

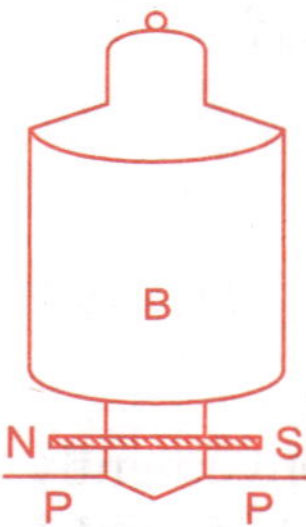

Fig. 9.14 *Searle's vibration magnetometer*

9.5.2 Box type Vibration Magnetometer

The simplest type of vibration magnetometer can be made with a magnet placed in a paper wrap suspended by means of an unspun silk thread within a deep glass bowl as shown in Fig. 9.15. The glass bowl is covered with a cardboard. The thread passes through a hole in the cardboard cover. A small stick is tied to the end of the thread for support.

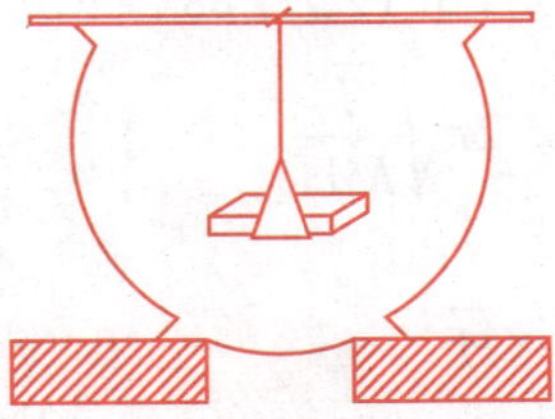

Fig. 9.15 *Simple vibration magnetometer*

The box type vibration magnetometer consists of a box enclosure with glass sides as shown in Fig. 9.16. A non-magnetic stirrup made of stiff copper wire or paper is suspended by means of an unspun silk thread, attached to a torsion head T at the other end. The torsion head can be slowly rotated to adjust the position of the magnet. A strip of a plane mirror is fixed at the base of the box.

A line is marked on the mirror, parallel to the edge of the box at the middle of the base as shown. The slots at the top help to observe and count the oscillations. This type of magnetometer is useful in the study of vibrations of different magnets.

A compass needle is placed on the line marked on the mirror and the box is rotated until the magnetic needle is parallel to the line. Thus the line and hence the magnetometer is set along the magnetic meridian. A rod of non-magnetic substance is now placed in the stirrup and the torsion head is rotated until the rod comes to rest in a direction parallel to the line on the mirror. A magnet placed in the stirrup, sets itself along the magnetic meridian without any twist in the suspension.

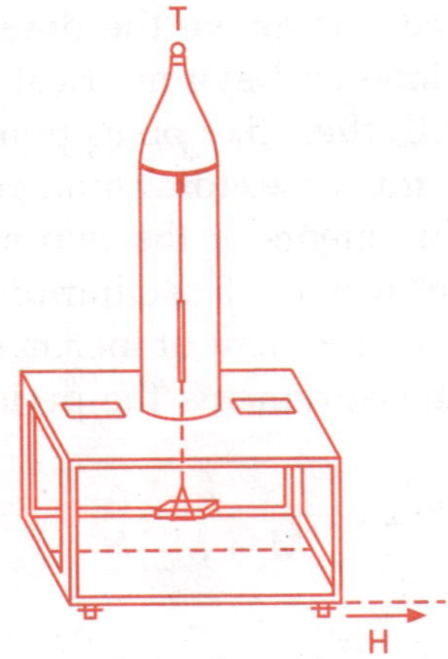

Fig. 9.16 *Box type vibration magnetometer*

9.5.3 Experiments with a Vibration Magnetometer

(*i*) **Comparison of the horizontal intensities of the earth's field at two places:** Searle's vibration magnetometer can be conveniently used for this purpose. The period of oscillations is given by Eq. (9.51) as

$$T = 2\pi\sqrt{\frac{I}{MH}}$$

$$T^2 \propto \frac{1}{H}.$$

If n is the number of oscillations per second or the frequency of oscillations, then $n = \dfrac{1}{T}$, and hence,

$$n^2 \propto H. \tag{9.52}$$

If n_1 and n_2 are the frequencies of oscillations of a magnetometer needle, at the two places then, $n_1^2 \propto H_1$ and $n_2^2 \propto H_2$, where H_1 and H_2 are the horizontal intensities of the earth's field at the two places.

Hence,

$$\frac{H_1}{H_2} = \frac{n_1^2}{n_2^2} \tag{9.53}$$

The periods of oscillation at two places are found by determining the times for a known number of oscillations. From this, the frequencies of oscillations, n_1 and n_2 can be computed. Also, the values of H_1 and H_2 at the two places are compared using the above relation.

(*ii*) **Comparison of magnetic moments of two magnets – The sum and difference method:** Box type vibration magnetometer can be conveniently used for this purpose. The magnets are placed one above the other in the stirrup such that their axes are in the same vertical plane and symmetrical to the axis of suspension. The system is allowed to oscillate with their like poles pointing in the same direction and the time period T_1 is found. In this case, total moment of the combination is $(M_1 + M_2)$, where M_1 and M_2 are the moments of the two magnets. One of the magnets is reversed pole per pole and the period of oscillation T_2 is found. In this case, the resultant moment is $(M_1 - M_2)$. The moment of inertia of the system remains the same in both the cases. For the two measurements the periods are given by

$$T_1 = 2\pi\sqrt{\frac{I}{(M_1 + M_2)H}} \tag{9.54}$$

and

$$T_2 = 2\pi\sqrt{\frac{I}{(M_1 - M_2)H}} \tag{9.55}$$

Hence, we have

$$\frac{T_1^2}{T_2^2} = \frac{M_1 - M_2}{M_1 + M_2}$$

or,

$$\frac{T_2^2}{T_1^2} = \frac{M_1 + M_2}{M_1 - M_2}$$

Rewrite in the form

$$\frac{M_1}{M_2} = \frac{T_2^2 + T_1^2}{T_2^2 - T_1^2} \tag{9.56}$$

The moments of the two magnets are compared using the above equation. Alternatively, if n_1 and n_2 are the corresponding numbers of oscillations during the same interval of time, then,

$$\frac{M_1}{M_2} = \frac{n_1^2 + n_2^2}{n_1^2 - n_2^2} \tag{9.57}$$

(iii) Determination of MH, the product of M, the magnetic moment of a magnet and H the horizontal component of earth's field at a place: The magnet is placed in the stirrup of a box type vibration magnetometer, such that it is symmetrical about the axis of the suspension. The system is allowed to oscillate in the earth's field and the period of oscillation T is found, by finding the time for a known number of oscillations.

The moment of inertia, I, of the magnet about the axis of suspension is calculated using the expression,

$$I = \frac{m\left(L^2 + B^2\right)}{12}$$

where m is the mass of the magnet and L and B are the length and breadth of the magnet, in the plane perpendicular to the axis of suspension. The value of MH is calculated using the relation

$$MH = \frac{4\pi^2 I}{T^2}. \tag{9.58}$$

It may be noted that (*i*) M, the magnetic moment of the magnet can be found by assuming the value of H at the place, and (*ii*) The moments of two magnets can be compared by finding the values of MH for the two magnets. The magnetic moments M_1 and M_2 of the magnets are compared using the relation

$$\frac{M_1}{M_2} = \frac{x}{y}$$

where x and y are the values of MH for the two magnets.

9.5.4 Determination of *M* and *H* by Deflection and Vibration

The determination of M, the moment of a magnet, and H, the horizontal intensity of earth's magnetic field consists in finding

(*a*) M/H for the magnet using a deflection magnetometer and

(*b*) MH for the magnet by vibration.

The mean value of (M/H) for the magnet is determined using a deflection magnetometer as already described. The value of MH for the magnet is found by vibration magnetometer as described above. M and H are calculated using the relation

$$M = \sqrt{PQ} \text{ and } H = \sqrt{\frac{Q}{P}} \tag{9.59}$$

where P and Q represent the values of (M/H) and MH, respectively.

SOLVED PROBLEMS

1. A magnetic needle in A position with respect to a magnet undergoes a deflection of 45°. Find the intensity at the point due to the magnet, given H at the place = 30 N/Wb?

   The intensity at the point due to the magnet is equal to

   $$H = 30 \text{ N/Wb}$$

   The magnetic needle is in A position with respect to the magnet. Therefore, $F = H \tan\theta$, where F is the intensity at the point due to the magnet, θ being 45°. Hence, $F = H = 30$ N/Wb.

2. A short magnet placed in Tan A position in a deflection magneto meter shows a deflection of 45°. If the magnetic moment is $48\pi \times 10^{-8}$ Wb-m, find the distance of the magnet from the compass needle given $H = (75/\pi)$ newton/weber.

   Short magnet, Tan A position

   $$F = \frac{2M}{4\pi\mu_0 d^3} = H\tan\theta$$

   $$d^3 = \frac{2M}{4\pi\mu_0 H\tan\theta} = \frac{2\times 48\pi\times 10^{-8}\times\pi}{4\pi\times 4\pi\times 10^{-7}\times 75\times\tan 45°} d^3 = 8\times 10^{-3}$$

$$d = 0.2 \text{ m}$$

Distance of the magnet from the compass needle is 0.2 m.

3. A magnet of length 0.04 m is placed with the axis east-west at a distance of 0.1 m to the north of a compass needle. If the deflection is 45° obtain the moment of the bar magnet and its pole strength given $H = 25$ newtons/weber.

The magnet placed east-west is north of the compass needle i.e., Tan B position.

$$F = \frac{M}{4\pi\mu_0\left(d^2+1^2\right)^{\frac{3}{2}}} = H\tan\theta$$

$$M = 4\pi\mu_0\left(d^2+1^2\right)^{\frac{3}{2}} H\tan\theta$$

$$= 4\pi\times 4\pi\times 10^{-7}\left(0.1^2+0.02^2\right)^{\frac{3}{2}}\times 25\tan 45°$$

$$= 3.95\times 10^{-7} \text{ Wb-m}$$

Pole strength $= \dfrac{M}{21} = \dfrac{3.95}{4}\times 10^{-7} = 9.9\times 10^{-8}$ Weber

Magnet moment = 3.95 × 10⁻⁷ Wb-m

Hence the pole strength = 9.9 × 10⁻⁸ Wb.

4. A magnet in Tan A position in a deflection magnetometer gives a deflection of 60°. A second magnet in the same position gives a deflection of 30°. If the moment of the first magnet is 2.25 × 10⁻⁶ Wb-m, find the moment of the second magnet.

As the lengths of the magnets are not given, the magnets have to be considered to be short.

Tan A position, $\theta_1 = 60°$, $\theta_2 = 30°$,

$$M_1 = 2.25\times 10^{-6} \text{ Wb-m}, \quad M_2 = ?$$

$$F_1 = \frac{2M_1}{4\pi\mu_0 d^3} = H\tan\theta_1 \text{ and}$$

$$F_2 = \frac{2M_2}{4\pi\mu_0 d^3} = \tan\theta_2$$

$$\frac{M_2}{M_1} = \frac{\tan\theta_2}{\tan\theta_1} \text{ or}$$

$$M_2 = 2.25\times 10^{-6}\times\frac{\tan 30°}{\tan 60°} = 7.5\times 10^{-7}$$

Moment of the second magnet = 7.5 × 10⁻⁷ Wb-m.

5. A deflection magnetometer shows no deflection when two short magnets are placed on either arm with their centers 0.225 m and 0.15 m away from the center of the needle. Compare their magnetic moments.

As the needle shows no deflection, the magnetic intensities due to the two magnets at the needle, are equal and opposite. Therefore,

$$\frac{2M_1}{4\pi\mu_0 d_1^3} = \frac{2M_2}{4\pi\mu_0 d_2^3}$$

or, $$\frac{M_1}{M_2} = \frac{d_1^3}{d_2^3} = \frac{0.225^3}{0.15^3} = 3.3.75$$

The magnetic moments are in the ratio 3.375 : 1.

6. A bar magnet is placed with its N-pole 0.16 m due east of the needle of a deflection magnetometer so that the pole of the magnet and the magnetic needle are in the same horizontal plane, the other pole of the magnet being vertically above the center of the needle. If the deflection of the needle is 30°, find the pole strength of the magnet. [$H = (433/4\pi)$ newton/weber and $\sqrt{3} = 1.732$].

According to the set-up, the deflection field F is due to N-pole of the magnet due east of the needle. Therefore,

$$F = \frac{m}{4\pi\mu_0 d^2} = H\tan\theta$$

$$m = 4\pi\mu_0 Hd^2\tan\theta$$

$$= 4\pi \times 4\pi \times 10^{-7} \times \frac{433}{4\pi} \times 0.16^2 \times \frac{1}{1.732}$$

$$= 8.05 \times 10^{-8} \text{ Wb.}$$

Pole strength = 8.05×10^{-8} Wb.

7. The period of oscillation of a magnet swinging in a horizontal plane at a given place is 31.4 sec. If its moment of inertia is 5×10^{-4} kgm^2 and its magnetic moment is 1.25×10^{-6} Wb-m, calculate earth's horizontal intensity at the place.

$$T = 2\pi\sqrt{\frac{I}{MH}} \quad \text{or}$$

$$H = \frac{4\pi^2 I}{MT^2} = \frac{4(3.14)^2 \times 5 \times 10^{-4}}{1.25 \times 10^{-6} \times (3.14)^2} = 16 \text{ newton/weber}$$

Earth's horizontal intensity at the place = 16 newton/weber

8. A magnet makes 30 oscillations per minute at a place where $H = 32$ newton/weber. At another place, it takes 1.6 sec to complete one oscillation. What is the value of earth's horizontal intensity at the second place?

Same magnet $T \propto \dfrac{1}{\sqrt{H}}$ or $T^2 \propto \dfrac{1}{H}$

$$T_1 = \frac{60}{30} = 2 \text{ sec, } T_2 = 1.6 \text{ sec, } H_1 = 32 \text{ newton/weber, } H_2 = ?$$

$$\frac{H_2}{H_1} = \frac{T_1^2}{T_2^2} \text{ or } H_2 = \frac{32 \times 2^2}{1.6^2} = 50$$

Earth's horizontal intensity at the second place = 50 N/Wb.

9. Two bar magnets are placed in the stirrup of a vibration magnetometer. They make 30 oscillations per minute when their like poles are together and 20 oscillations per minute when their unlike poles are together. Compare their magnetic moments.

With like poles together, $n_1 = 30$ osc./min and with unlike poles together, $n_2 = 20$ osc./min; using the formula,

$$\frac{M_1}{M_2} = \frac{n_1^2 + n_2^2}{n_1^2 - n_2^2} = \frac{30^2 + 20^2}{30^2 - 20^2} = \frac{13}{5}$$

The magnetic moments are in the ratio 13:5.

10. A compass needle makes 30 oscillations per minute in the earth's field. When a bar magnet is placed near it so as not to alter the direction of the needle, it makes 40 oscillations per minute. How many times will the needle oscillate per minute if the magnet is reversed?

In the first case $n_1^2 \propto H$ with $n_1 = 30$ osc./min   (1)

In the second case, direction of the needle is not altered. Further $n_2 > n_1$, therefore the field is $H + F$ (and not $H - F$)

Hence $n_2^2 \propto (H + F)$ where $n_2 = 40$ osc./min   (2)

From (1) and (2)

$$\frac{F + H}{H} = \frac{n_2^2}{n_1^2} \text{ or}$$

$$\frac{F}{H} = \frac{n_2^2 - n_1^2}{n_1^2} = \frac{40^2 - 30^2}{30^2} = \frac{7}{9}$$

$F < H$; therefore, when the magnet is reversed, field is $(H - F)$. In this case,

$$n_3^2 \propto (H - F) \qquad\qquad (3)$$

$$\frac{n_3^2}{n_1^2} = \frac{(H-F)}{F} = 1 - \frac{7}{9} = \frac{2}{9} \text{ or}$$

$$n_3^2 = \frac{2 \times 30^2}{9} = 200$$

$$n_3 = 10\sqrt{2}$$

The magnet makes $10\sqrt{2}$ oscillations per minute.

9.6   TERRESTRIAL MAGNET

9.6.1   The Earth as a Magnet

The earth behaves like a huge magnet, with its poles near the geographic poles. As a result, a freely suspended magnet or a magnetic needle in the earth's magnetic field sets itself approximately along the geographic north-south direction.

The magnetic pole of the earth near the geographic north was referred to as blue pole (S-pole) and the other pole of the earth near the geographic south as red pole (N-pole) as shown in Fig. 9.17. The south magnetic pole is in Boothia Felix, the northern most tip of the North American mainland, in central Nunavut, Canada, and the north magnetic pole is in South Victoria land, in the Antarctic island. Now-a-days, the magnetic north pole is considered close to the geographic north pole and vice versa.

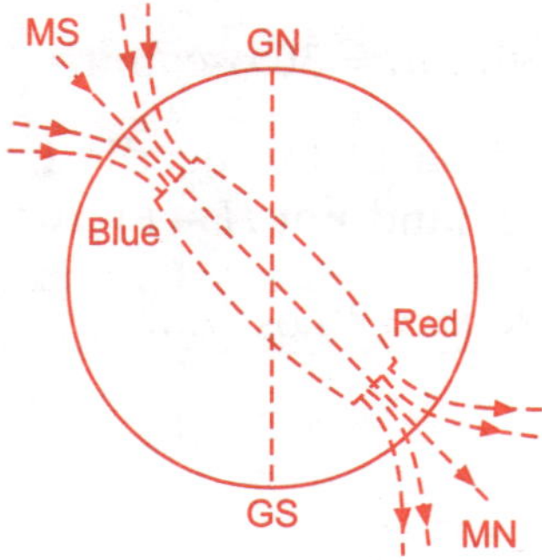

Fig. 9.17 *Terrestrial magnet*

9.6.2 Magnetic Elements

The magnetic elements are those quantities which determine the earth's magnetic field. The magnetic meridian at a place is the vertical plane containing the direction of the magnetic field at that place. It is the vertical plane containing the magnetic axis of a freely suspended magnetic needle.

The geographic meridian at a place is the vertical plane passing through the given place and the geographic axis. The angle between the geographic meridian and the magnetic meridian at a place, δ, is the declination at that place.

When a magnetic needle is free to rotate in the magnetic meridian about a horizontal axis through its center of gravity, it sets itself along the direction of the intensity of the earth's magnetic field at that place. The angle through which the needle dips below the horizontal, is called the magnetic dip at the place or inclination as shown in Fig. 9.18. This angle is between the horizontal intensity and the total intensity of the earth's field at a place and is the magnetic dip, θ, at the place.

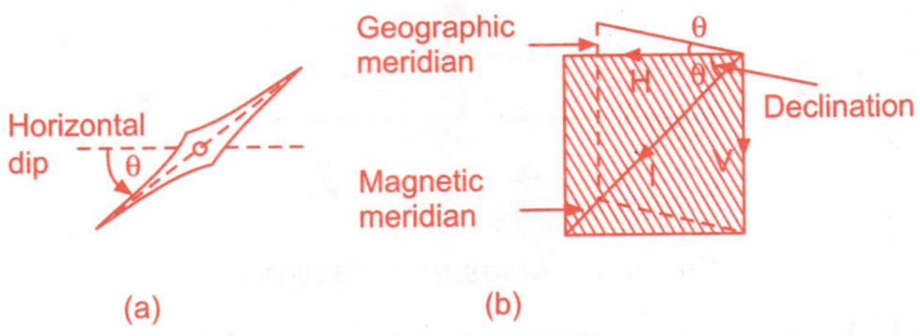

Fig. 9.18 *Magnetic elements*

At the earth's magnetic poles, a freely suspended magnetic needle stands vertical with the dip at the poles being 90°. At points on the magnetic equator, a freely suspended magnet remains horizontal and hence the dip at the magnetic equator is 0°.

The horizontal component of the total intensity I, is earth's horizontal intensity H and the component of I along the vertical is called vertical component V, and are indicated in Fig. 9.18(b). The total intensity, I, is not determined directly. The horizontal component H and the dip θ are measured. By definition, we have

$$H = I \cos \theta \qquad (9.60)$$

and the vertical component V is calculated from the relation

$$V = H \tan \theta \qquad (9.61)$$

H, V and I are related by the relation

$$I^2 = H^2 + V^2 \qquad (9.62)$$

The quantities I, H, V, θ and δ are the magnetic elements at a place. Among them δ, θ and H are the principal magnetic elements, because they are sufficient to describe the earth's magnetic field at a place.

9.6.3 Measurement of Magnetic Elements

1. Measurement of horizontal intensity of the earth's field (H)

The horizontal intensity, H, can be determined by deflection and vibration as explained earlier. H can also be determined using a tangent galvanometer.

2. Measurement of declination

In order to determine the declination, first it is necessary to determine the geographic and magnetic meridians. The simplest method of finding the geographic meridian is as follows. A rod is fixed vertically on the ground. A circle of suitable radius is drawn with the foot of the rod O as the centre as shown in Fig. 9.19. When the tip of the rod's shadow just touches the circle before noon, point P_1 is marked on the circle as shown in Fig. 9.19. The point P_2 is marked on the circle when the tip of the shadow touches the circle in the afternoon. The bisector of the angle $P_1 O P_2$ gives the direction of the geographic meridian.

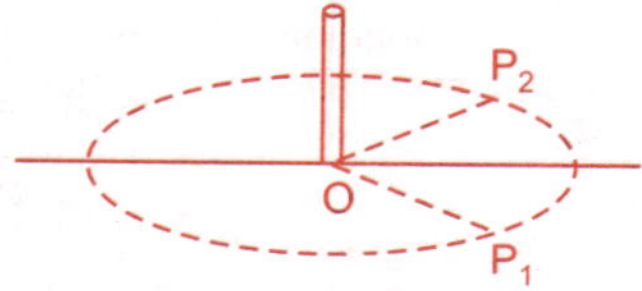

Fig. 9.19 *Measuring declination*

In order to find the magnetic meridian, a magnet is freely suspended and two pins P_1 and P_2 are fixed vertically at its ends as shown in Fig. 9.20. The line a_1b_1 is drawn joining the tips of the pins P_1 and P_2. The magnet is turned upside down and the line a_2b_2 is drawn joining the tips of the pins P_1 and P_2. The bisector of the angle $a_1O a_2$ gives the direction of the magnetic meridian. The declination at a place can be measured from the directions of the magnetic meridian and the geographic meridian.

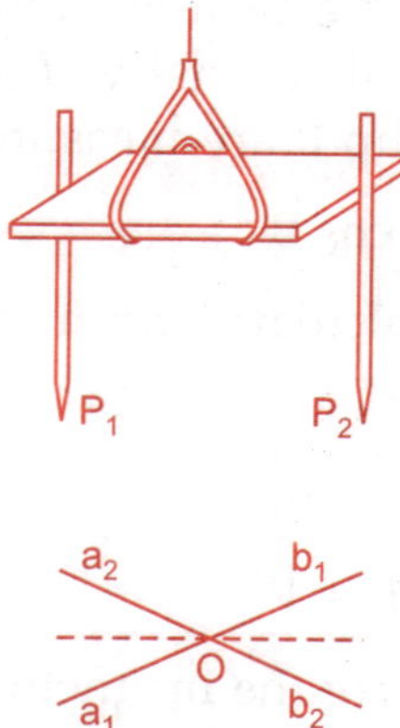

Fig. 9.20 *Magnetic meridian*

3. Measurement of dip

Angle of dip at a place is measured with the help of a dip-circle as shown in Fig. 9.21. It has a magnetic needle resting on two horizontal knife edges, by means of a short horizontal axis of steel, passing through the center of gravity of the needle. The needle can rotate in a vertical plane about the horizontal axis. The ends of the needle move over a vertical circular scale and the axis of the needle passes through the center of the scale. The scale, graduated in degrees, is divided into four quadrants. The zero-zero line of the scale is horizontal. The magnetic needle and the vertical scale are mounted in a wooden box, with glass sides. The box can be rotated about a vertical axis passing through the point of support of the needle. Its position can be read on a horizontal circular scale fitted to the base of the instrument. The base of the instrument is provided with three levelling screws. The instrument is usually provided with a spirit level at the top, in order to ensure proper levelling.

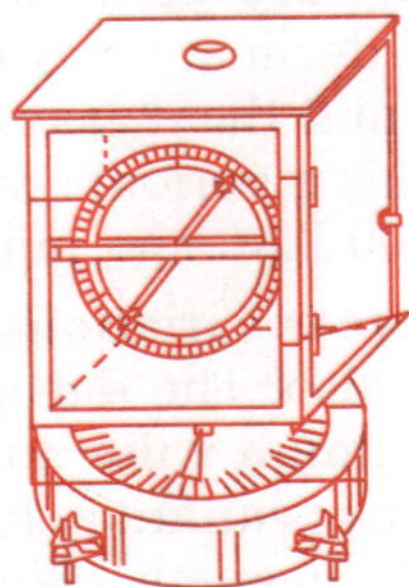

Fig. 9.21 *Measuring dip using dip circle*

Initially the instrument is levelled with the help of levelling screws, using a spirit level. The box is rotated about the vertical axis, until the needle reads 90°-90° on the vertical scale. The reading on the horizontal scale is noted. The box is then rotated through 90°. The needle will now be in the magnetic meridian and the reading of the needle gives the angle of dip at the place.

In order to eliminate errors, the following precautions must be taken:

The pivot of the needle may not pass through the centre of the vertical scale, and hence both the ends of the needle θ_1 and θ_2 are read to eliminate error due to eccentricity of the needle.

In order to eliminate error due to non-horizontality of zero-zero line, the box is rotated through 180° and two more readings θ_3 and θ_4 are taken. The needle is reversed on its bearings and the four readings θ_5 to θ_8 are noted. These readings are taken to eliminate error due to non-coincidence of magnetic and geometric axes of the needle.

The needle is remagnetized in the opposite direction and the eight readings θ_9 to θ_{16} are noted. This eliminates error due to the fact that the pivot of the needle may not pass through the centre of gravity of the needle. Average of the sixteen readings gives corrected value of dip at the place.

True value of dip at a place can be obtained from the apparent values of dip measured in two mutually perpendicular planes. If θ_1 and θ_2 are the apparent dips in two mutually perpendicular planes, then true dip θ is given by the relation,

$$\cot^2\theta = \cot^2\theta_1 + \cot^2\theta_2 \qquad\qquad (9.63)$$

9.6.4 Variation of Magnetic Elements

The values of magnetic elements vary from place to place. We have $H = 0$ and $\theta = 90°$ at the magnetic poles, and $V = 0$ and $\theta = 0°$ at the magnetic equator. A systematic recording of the magnetic elements at a place shows that they are not constant quantities and that they undergo variations. Such variations can be classified as follows:

(*i*) **Daily changes or diurnal variations:** Variations in magnetic elements during different times of the day are grouped as daily variations or diurnal variations. For example, if declination is measured at intervals of one hour, over a period of 24 hours, and the measurement is repeated for a number of days, it is found that there is a periodic daily variation. Declination is found to be maximum at about 1.00 P.M. and minimum at about 8.00 A.M.

(*ii*) **Annual, lunar, and multi-year variations:** Magentic elements may undergo periodic annual variations also. The declination is found to have maximum value in February and minimum value in August. There is a small periodic monthly variation called lunar variation, having the same time interval as the tides due to the moon. The magnetic elements also undergo changes which coincide with the eleven year cycle of sunspot activity.

(*iii*) **Secular changes:** Secular changes are gradual changes in all magnetic elements extending over a long period. The period is about 960 years. These secular changes are probably due to the slow movement of the earth's magnetic poles about the geographic poles.

(*iv*) **Irregular changes:** Sometimes, sudden, irregular and large disturbances are found to occur in the magnetic elements. Such disturbances occur more or less simultaneously, all over the world. These are referred to as magnetic storms. The magnetic storms usually accompany volcanic eruptions, earthquakes, appearance of large sunspots and Aurora Borealis (sudden appearance of intense bright light in the sky at the polar regions).

9.6.5 Magnetic Maps

Magnetic maps are contours having the same value for a certain magnetic element. For example, a map with equal dip contours can be drawn by joining all the places having equal dip. A line drawn through points of equal declination is called Isogonal line. The line passing through points having zero declination is called Agonal line. A line drawn through points having equal values for the angle of dip is referred to as Isoclinic line. The magnetic equator where the angle of dip is zero is called Aclinic line. A line through points at which the horizontal component of the earth's magnetic field is same, is called Isodynamic line.

EXAMPLE 9.10

The horizontal intensity of the earth's field is $(380/4\pi)$ N/Wb and the vertical intensity is $(134/4\pi)$ N/Wb. Find the total intensity and dip.

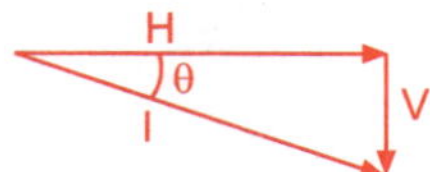

Fig. 9.22 *Total intensity*

SOLUTION

Referring to Fig. 9.22, the total intensity I is given by

$$I^2 = H^2 + V^2$$

$$I^2 = \frac{380^2}{(4\pi)^2} + \frac{134^2}{(4\pi)^2} = \frac{162400}{(4\pi)^2}$$

$$I = \frac{400}{4\pi} = \frac{100}{\pi} \ \text{Newton/Weber}$$

Dip θ is given by

$$\tan\theta = \frac{V}{H} = \frac{134}{380} = 0.3526$$

Hence, the dip angle is $\theta = 19°24'$.

EXAMPLE 9.11

The true value of dip at a place is 45°. If the plane of the dip circle is turned through 60° from the magnetic meridian, determine the apparent dip.

SOLUTION

Apparent dip θ' in a plane inclined at an angle ϕ with the magnetic meridian is given by the relation

$$\tan\theta' = \frac{\tan\theta}{\cos\phi}$$

where θ = true value of dip, and is 45°

$$\phi = 60°, \ \tan\theta' = \frac{\tan 45°}{\cos 60°} = 2$$

$$\theta' = 62°27'$$

Apparent dip = 63°27'.

EXAMPLE 9.12

Calculate the value of the horizontal intensity at a place where the angle of dip is 45° and the total intensity is 28 Newton/Weber.

SOLUTION

$$H = I\cos\theta = 28 \times \cos 45° = \frac{28}{\sqrt{2}} = 20$$

Horizontal intensity = 20 Newton/Weber.

EXAMPLE 9.13

At a place A, the total intensity is 40 Newton/Weber and the angle of dip is 68° while at another place B the total intensity is 44 Newton/Weber and the angle of dip is 72°. Compare earth's horizontal intensities at the two places.

SOLUTION

$$H_1 = I_1 \cos\theta_1 \text{ and } H_2 = I_2 \cos\theta_2$$

$$\frac{H_1}{H_2} = \frac{I_1\cos\theta_1}{I_2\cos\theta_2} = \frac{40 \times \cos 68°}{44 \times \cos 72°} = \frac{40 \times 0.3746}{44 \times 0.3090} = \frac{1.103}{1}$$

Horizontal intensities are in the ratio 1.103:1.

The apparent dips at a place in two mutually perpendicular vertical planes are 30° and 20° respectively. Calculate the true dip at the place.

SOLUTION

True dip θ is given by the equation

$$\cot^2\theta = \cot^2\theta_1 + \cot^2\theta_2$$

θ_1 and θ_2 being the apparent dips

$$\cot^2\theta = \cot^2 30° + \cot^2 20° = \left(\sqrt{3}\right)^2 + (2.748)^2$$

$$\cot^2\theta = 10.55$$

$$\cot\theta = \sqrt{10.55} = 3.25$$

$$\theta = 17°$$

True dip at the place = 17°.

EXAMPLE 9.15

At a place A, the total intensity of the earth's field is 80 N/Wb and the dip is 30°. At another place B, the total intensity is 34.64 N/Wb and the dip is 60°. The time period of a magnet vibrating horizontally at A is 3 sec. What is the period at B?

SOLUTION

$$T \, \alpha \, \frac{1}{\sqrt{H}} \text{ and } H = I \cos \theta$$

$$\frac{T_2}{T_1} = \left(\frac{I_1 \cos\theta_1}{I_2 \cos\theta_2} \right)^{\frac{1}{2}} = \left(\frac{80}{34.64} \times \frac{\cos 30°}{\cos 60°} \right)^{\frac{1}{2}} = 2$$

$$T_2 = 2T_1 = 2 \times 3 = 6$$

Period at B = 6 seconds.

9.7 MAGNETIC MATERIALS

9.7.1 Introduction

The strength of a magnetic field at any point in space is referred to as the magnetic field intensity H and is the force experienced by a unit north pole at that point. When a magnetic substance is placed in a uniform magnetic field there will be a redistribution of the magnetic lines of force as shown in Fig. 9.23. There is crowding of the magnetic lines of force through the specimen placed in the magnetic field. The resultant field inside the specimen, is indicated by B and is called magnetic induction or magnetic flux density.

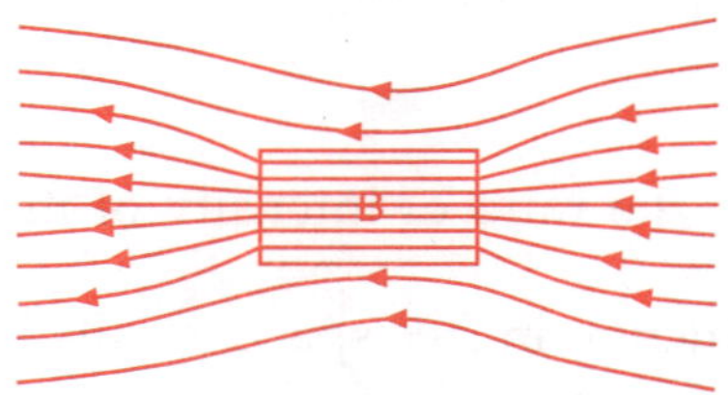

Fig. 9.23 *Redistribution of lines of force*

Magnetic induction or flux density is expressed in weber/meter^2; the unit of B in *SI* system is also called tesla. 1 tesla = 1 weber/meter^2.

The pole strength acquired by the specimen depends on the intensity of the magnetizing field H. The pole strength acquired per unit area of the specimen in a direction perpendicular to H is called intensity of magnetization and is also defined as the magnetic moment per unit volume. If a is the area of cross section of the specimen, perpendicular to the direction of the magnetic field, m the pole strength and $2l$ the length of the specimen in the direction of the field, then the intensity is given by

$$I = \frac{m}{a} = \frac{m \times 2l}{a \times 2l} = \frac{M}{V} \tag{9.64}$$

where $m \times 2l = M$, the magnetic moment; $a \times 2l = V$, the volume of the specimen. It must be noted that I is measured in the same units as B.

Magnetic induction B, magnetizing field H, and the intensity of magnetization I, are related to each other. In *SI* system, the relation connecting B, H and I is given by the equation

$$B = \mu_0 H + I \tag{9.65}$$

If H and I are in A/m, then

$$B = \mu_0 (H + I) \tag{9.66}$$

The fact that the lines of force crowd inside a bar of soft iron placed in a uniform magnetic field as shown in Fig. 9.23, shows that the lines of force find it easier to pass through iron than through air. In other words, iron is more permeable or has greater permeability than air. Therefore, the permeability of a medium may be defined as its conducting power for magnetic lines of force. Permeability μ, is measured by the ratio between magnetic induction B, and the intensity of magnetizing field, H,

$$\mu = \frac{B}{H} \tag{9.67}$$

From Eqs. (9.66) and (9.67), we have

$$\mu = \mu_0 + \frac{I}{H} = \mu_0 \left(1 + \frac{I}{\mu_0 H} \right). \tag{9.68}$$

The ratio $\dfrac{I}{\mu_0 H}$, which is the ratio of intensity of magnetization to the magnetic flux density or magnetic induction in free space is called magnetic susceptibility, and is represented by χ which is a nondimensional number. The susceptibility of a substance is a measure of the ease with which the substance can be magnetized. Soft iron can be easily magnetized as compared to steel, and therefore susceptibility of iron is higher than that of steel. Equation (9.68) can be written as

$$\mu = \mu_0 (1 + \chi)$$

As $\mu = \mu_0 \mu_r$, $\mu_r = (1 + \chi)$; For free space $\chi = 0$, $\mu_r = 1$.

9.7.2 Properties of Magnetic Materials

For quite a long time scientists believed that only substances such as iron, steel, cobalt, nickel and their alloys exhibited magnetic properties. However, in 1845, Michael Faraday, an English scientist, discovered that all substances exhibit magnetic properties, but to different extents. Based on their response to a magnetic field, different materials can, in general, be classified into three kinds: diamagnetic, paramagnetic and ferromagnetic substances.

For many substances, the force experienced in magnetic fields is very small, even when the fields are very strong. Some substances are always pulled in the direction of increasing field intensity and others in the direction of decreasing field intensity, irrespective of the direction of the magnetic field. This is shown in Fig. 9.24, where the N and S are pole pieces of a powerful electromagnet that has one pole flat and the other sharply pointed. The field is stronger near the pointed pole than that near the flat pole. If a small piece of a substance is suspended by means of a thin string between the poles, a small force acts on the material resulting in a small displacement of the piece. If the substance is bismuth, the piece is found to move towards the flat pole because, bismuth is repelled from high field region. If the substance is aluminium, it is found to move towards the pointed pole because, aluminium is attracted towards high field region.

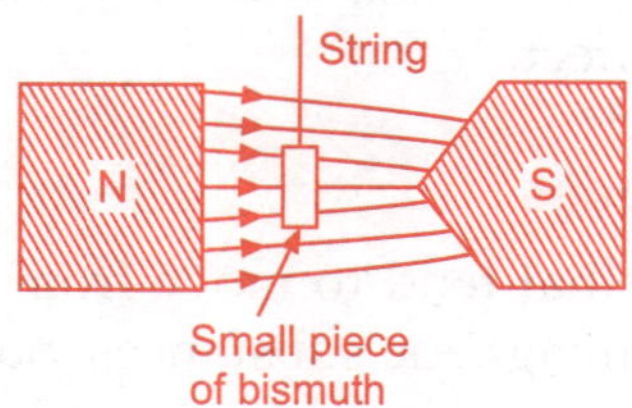

Fig. 9.24 *Pole pieces with flat and pointed poles*

9.7.3 Diamagnetic Substances

Substances such as bismuth that tend to move from stronger to weaker regions in a magnetic field are called diamagnetic substances. Some examples of diamagnetic substances are phosphorus, antimony, sodium chloride, quartz, copper, mercury, water, alcohol, air, hydrogen.

When a diamagnetic substance, in the form of a thin bar, is suspended between the pole pieces of a powerful magnet, it sets itself with its longer axis perpendicular to the direction of the field as shown in Fig. 9.25(*a*). If a diamagnetic liquid taken in a watch glass is placed over two very near pole pieces of a powerful electromagnet, a depression is produced in the middle. If one limb of a *U* tube, containing a diamagnetic liquid, is placed in between the pole pieces of a powerful magnet, the level of the liquid falls as shown in Fig. 9.25(*b*). If a gas is allowed to bubble through a liquid placed in between the pole pieces of a powerful magnet, the gas spreads across the field.

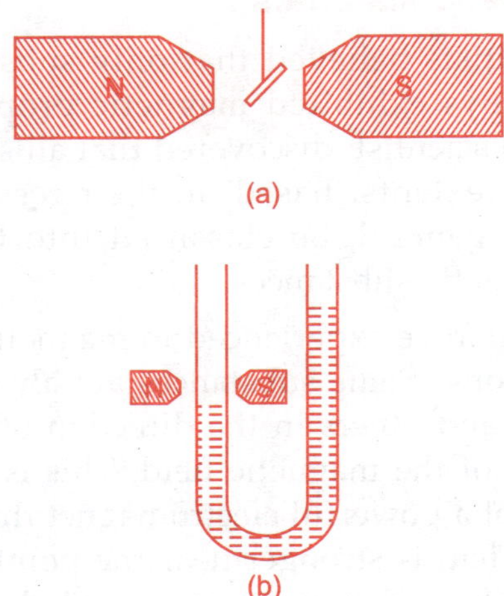

Fig. 9.25 *Diamagnetic substance in a magnetic field*

The relative permeability (μ_r) of a diamagnetic material is slightly less than unity. Susceptibility is constant and has a low negative value (-10^{-6} to -10^{-7}) about -1.5×10^{-6} for bismuth. It is independent of temperature. Diamagnetism is a universal effect and all substances show this effect.

9.7.4 Paramagnetic Substances

Substances such as aluminium that tend to move from weaker to stronger regions in a magnetic field are called paramagnetic substances. Some examples of paramagnetic substances are platinum, chromium, manganese, sodium, crown glass, solutions of salts of nickel, iron, and oxygen.

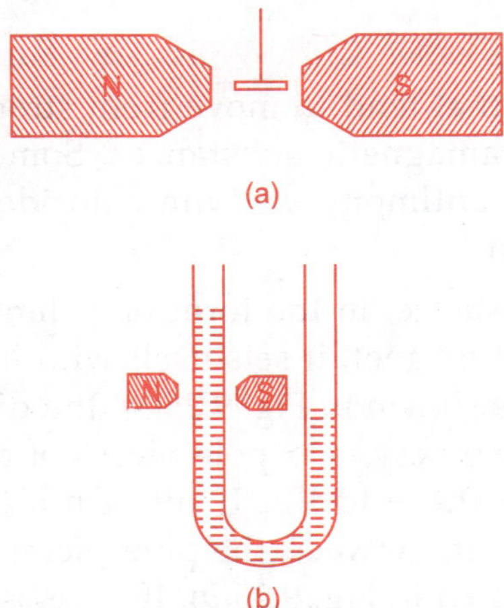

Fig. 9.26 *Paramagnetic substances*

When a paramagnetic substance, in the form of a thin bar, is suspended between the pole pieces of a powerful magnet, it sets itself with its longer axis parallel to the field as shown in Fig. 9.26(*a*). If a paramagnetic liquid taken in a watch glass, is placed over two very near pole pieces of a strong electromagnet, a bulge is produced in the middle. If one limb of a U tube containing a paramagnetic liquid is placed in between the pole pieces of a powerful electromagnet, the level of the liquid rises as shown in Fig. 9.26(*b*). If a paramagnetic gas is allowed to bubble through a liquid placed in between the pole pieces of a powerful magnet, the gas passes straight across the gap.

The relative permeability (μ_r) of a paramagnetic material is slightly greater than unity. Susceptibility has a low positive value (10^{-4} to 10^{-6}). Paramagnetism increases as the temperature is reduced, the effect being large at temperatures near absolute zero.

9.7.5 Ferromagnetic Substances

Substances which can be strongly magnetized are called ferromagnetic substances. Iron, steel, nickel, and cobalt are ferromagnetic. In addition to these metals, a number of alloys and crystalline compounds are known to be ferromagnetic. Ferromagnetic substances exhibit all the properties of a paramagnetic substance in a very strong fashion. Relative permeability is very large and magnetization is about 10^6 times more than that for paramagnetic and diamagnetic substances. The intensity of magnetization of a ferromagnetic substance is not directly proportional to the magnetizing field. In other words, μ varies considerably with the magnetizing field. This is the distinguishing characteristic of a ferromagnetic material.

Ferromagnetic substances have very large susceptibility. Susceptibility varies inversely as the absolute temperature. This is known as Curie law. At a certain temperature called Curie point or Curie temperature, ferromagnetism disappears and the substance becomes paramagnetic. The susceptibility of a ferromagnetic substance, above its Curie point, is inversely proportional to the excess of temperature above the Curie point. This is called Curie-Weiss law. Further, μ decreases suddenly at the Curie point. The Curie point is about 1100°C for cobalt, 400°C for nickel and 770°C for iron.

Ferromagnetic substances exhibit change in the dimensions when subjected to a magnetic field. This phenomenon is known as magnetostriction.

For ferromagnetic materials, magnetization does not have a definite value for a given H. Magnetization is found to depend on the previous history (in terms of its magnetic, thermal and mechanical treatment) of the specimen. Ferromagnetics are said to exhibit a phenomenon known as hysteresis, which is explained in the next section.

There are two other types of magnetism, anti-ferromagnetism and ferrimagnetism which are closely related to ferromagnetism. Anti-ferromagnetic substances exhibit very little external magnetism. Manganese-dioxide is an example of this type. If an anti-ferromagnetic substance is heated to a certain temperature called Neel temperature, the substance becomes paramagnetic. Ferrimagnetic substances are those in which external magnetism is in between ferromagnetism and anti-ferromagnetism.

9.7.6 Cycle of Magnetization, Hysteresis Cycle

Let an unmagnetized bar of ferromagnetic substance, such as iron, be subjected to a slowly increasing magnetising field, H. The intensity of magnetization, I, corresponding to the values of H is plotted as shown in Fig. 9.27. The curve OAB is called initial magnetization curve. When H is further increased, I remains constant and the substance is said to be saturated.

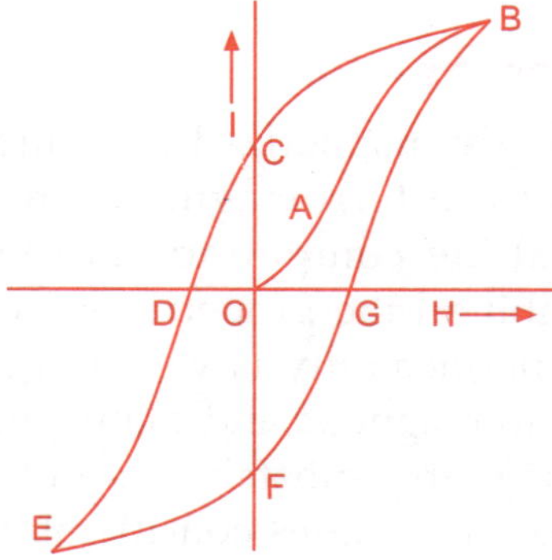

Fig. 9.27 Hysteresis cycle

Let H be gradually decreased from the maximum value H_m to zero. The corresponding values of I will follow the path BC without following the initial curve OAB. Reversing the direction of H and decreasing H gradually from zero to $-H_m$, will follow the path CDE. At this stage, H is slowly increased to zero, and finally increased in its original direction to the maximum value. The curve $EFGB$ corresponds to the change from $-H_m$ to $+H_m$. If the set of operations corresponding to curve $BCDEFGB$ is repeated, same curve is obtained again. The set of operations is called cycle of magnetization. The descending branch, $BCDE$ of the closed curve lies always above the ascending branch $EFGB$, which implies that the magnetization lags behind the magnetizing field. This lagging of the magnetization behind the magnetizing field, is called hysteresis. The closed figure $BCDEFGB$, or the $I - H$ loop, is called hysteresis loop. The shape of the hysteresis loop is characteristic of a magnetic substance.

The process of taking a ferromagnetic substance through a cycle of magnetization involves a loss of energy, which is not recovered. This loss of energy is referred to as hysteresis loss, and appears in the form of heat in the specimen. The area of the hysteresis loop represents the hysteresis loss per cycle per cc of the specimen.

9.7.7 Retentivity and Coercivity

It can be seen from the $I - H$ curve shown in Fig. 9.27, that magnetization has a certain positive or negative value, when the magnetizing field H is zero. The capacity of a substance to retain magnetization is called retentivity. The amount of magnetization which remains even when the magnetizing field is removed is known as residual magnetism or remanence. In Fig. 9.27, *OC* or *OF* represents residual magnetism.

Residual magnetism in a specimen can be reduced to zero by applying a particular value of reversed magnetizing field. The magnitude of this field is called coercive force. The property of the substance is called coercivity. The capacity of a substance to retain magnetization in spite of any subsequent treatment is called coercivity. In Fig. 9.27, *OD* or *OG* represents coercive force.

9.7.8 Uses of Magnetic Materials

Hysteresis curves for ferromagnetic materials afford valuable information regarding the suitability of materials for different purposes. Figure 9.28 shows the shape of hysteresis curve for soft iron and steel. Hysteresis loss is greater for steel than for soft iron.

In the cores of transformers, the material goes through cycles of magnetization many times per second. Hence materials with low hysteresis loss and high permeability have to be used in such applications. For example, Permalloy (78.5% Ni & 21.5% Fe) is used in the cores of communication transformers.

For permanent magnets, hysteresis loss is not a major consideration. The material should have large remanence and high coercivity. Tungsten steel (6% tungsten), cobalt steel (35% Co, 6% Cr, 4% tungsten) and Alnico (10% Al, 18% Ni, 12% Co, 6% Cu) are largely used. Now-a-days, semiconducting materials (oxides of trivalent metal, iron or copper), which are proved to be ferromagnetic, called ferrites, are used in the electronic industry.

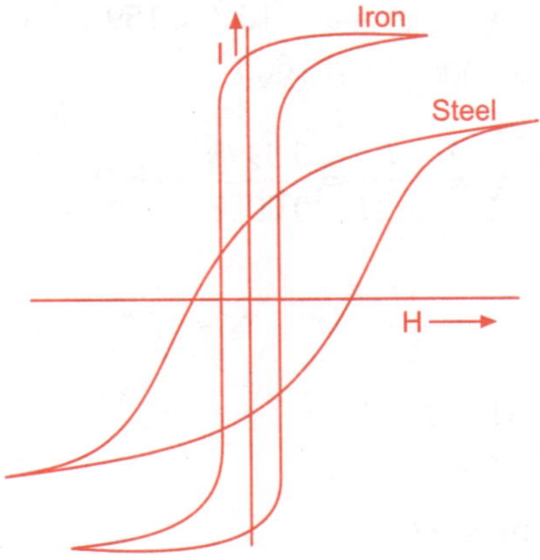

Fig. 9.28 *Hysteresis curves for soft iron and steel*

Paramagnetic substances are used in the measurement of low temperatures. The phenomenon of magnetostriction exhibited by ferromagnetic substances may be used in the production of ultrasonic waves. The study of magnetic properties, susceptibility in particular, of single crystals of organic compounds, helps to deduce useful information about the shape of the molecules in the crystal and the nature of the electron orbits around the atoms of the molecules.

EXAMPLE 9.16

An iron rod of length 60 cm has a cross sectional area of 5 sq. cm. If the flux of magnetic induction through the bar is 2×10^{-4} and the relative permeability of iron is 2000, calculate (i) the strength of the magnetising field (ii) the intensity of magnetisation (iii) the susceptibility and (iv) the magnetic moment of the rod.

SOLUTION

Flux of induction $\phi = 2 \times 10^{-4}$ Weber;

Relative permeability $\mu_r = 2000$; area $A = 5$ sq. cm $= 5 \times 10^{-4}$ m^2

Flux density
$$B = \frac{\phi}{A} = \frac{2 \times 10^{-4}}{5 \times 10^{-4}} = 0.4 \text{ Weber/m}^2$$
$$= 0.4 \ T$$

Strength of magnetizing field,

$$H = \frac{B}{\mu} = \frac{B}{\mu_0 \mu_r} = \frac{0.4}{4 \times 10^{-7} \times 200} = 159.1$$

$$B = \mu_0 H + I$$

Intensity of magnetisation, $I = B - \mu_0 H$

$$I = 0.4 - 4 \times 10^{-7} \times 159.1 = 0.3998 \text{ Weber/m}^3$$

Susceptibility $\chi = \mu_r - 1 = 2000 - 1 = 1999$

or,
$$\chi = \frac{I}{\mu_0 H} = \frac{0.3998}{0.0002} = 1999$$

$$I = \frac{M}{V}$$

Magnetic moment $M = I \times V$

$$I = 0.3998$$

Length of the rod $= 60$ cm $= 0.6$ m;

Area of cross section $= 5$ sq.cm $= 5 \times 10^{-4}$ m^2

$M = 0.3998 \times (0.6 \times 5 \times 10^{-4}) = 1.199 \times 10^{-4}$ Wb-m

Strength of magnetising field $H = 159.1$ Newton/Weber

Intensity of magnetisation $I = 0.3998$ Weber/meter^2

Susceptibility $\chi = 1999$

Magnetic moment of the rod $M = 1.199 \times 10^{-4}$ Wb-m.

SUMMARY

1. North and South poles are two ends of the magnet and they are equal in strength.

2. The space surrounding a magnet where influence of the magnet is experienced is called a magnetic field.

3. **Line of force:** The path or track in space along which an imaginary isolated North pole moves. The lines of force due to a magnet are closed curves starting from N-pole and ending on S-pole.

4. **Neutral points:** The points at which the resultant magnetic field will be zero when, two or more magnetic fields are combined.

5. **Magnetic intensity:** It is defined as the force experienced by an imaginary isolated unit North pole placed at that point.

6. A uniform magnetic field has the same magnetic intensity at every point in the field, in magnitude and direction and it is represented by parallel lines.

7. Magnetic moment of a magnet is equal to the moment of the mechanical couple required to hold the magnet at right angles to a uniform magnetic field of unit strength.

8. A short magnet has a very small semi-magnetic length compared to the distance d of the point from the magnet, such that $\dfrac{l}{d} \ll 1$.

9. If an N-pole is brought from infinity to a point in a magnetic field, the work done against the field is the magnetic potential energy of the pole at that point.

10. The potential difference dV between two points in a magnetic field is the amount of work done in moving a unit N-pole from one point at a lower potential to the second point at a higher potential.

11. **Tangent Law:** When there are two fields of intensity F and H then, $F = H \tan \theta$.

12. Vibration magnetometer works on the principle that, whenever a freely suspended magnet in a uniform magnetic field is disturbed from its equilibrium position, it is set into vibrations about the mean position. It is used for comparison of magnetic moments and magnetic fields.

13. The magnetic pole of the earth near the geographic north was referred to as blue pole (S-pole) and the other pole of the earth near the geographic south as red pole (N-pole).

14. The magnetic meridian at a place is the vertical plane containing the direction of the magnetic field at that place.

15. The strength of a magnetic field at any point in space is referred to as the magnetic field intensity H and is the force experienced by a unit north pole at that point.

16. The pole strength acquired per unit area of the specimen in a direction perpendicular to H is called intensity of magnetization and is also defined as the magnetic moment per unit volume.

17. The magnetic materials are classified into diamagnetic, paramagnetic and ferromagnetic substances.

18. **Diamagnetic substances:** Bismuth, phosphorus, antimony, sodium chloride, quartz, copper, mercury, water, alcohol, air and hydrogen.

19. **Paramagnetic substances:** Aluminium, platinum, chromium, manganese, sodium, crown glass, solutions of salts of nickel, iron, and oxygen.

20. **Ferromagnetic substances:** Iron, steel, nickel, and cobalt.

21. Ferromagnetic substances exhibit change in the dimensions when subjected to a magnetic field which is called as Magnetostriction.

22. The process of taking a ferromagnetic substance through a cycle of magnetization involves a loss of energy which referred to as hysteresis loss.

23. The capacity of a substance to retain magnetization is called retentivity.

24. Residual magnetism in a specimen can be reduced to zero by applying a particular value of reversed magnetizing field. The magnitude of this field is called coercive force.

Electrostatics

10.1 ELECTRIFICATION

The electric current is flow of electrons in materials. The electrons are pushed along a wire with the help of electric forces. These electric forces can attract small particles. Some of the examples are: when small bits of paper are brought close to a comb after prolonged combing of the hair, the paper bits are attracted to the comb and stick to it. When a glass rod is rubbed against a silk cloth, the rod can attract small pieces of paper. Similarly, a piece of amber, rubbed with fur, attracts small bits of straw or paper. A systematic study of the phenomenon was made by Sir William Gilbert, court physician to Queen Elizabeth in 1600 A.D. When some substances are rubbed against some specific matching substances, they attract light matter such as small bits of paper, straw, cork, pith, etc. Sealing wax with flannel, vulcanite with fur, polythene rod with duster, plastic strip with silk etc., exhibit the property. The phenomenon is called electrification. This name comes from the Greek word electron for amber. A substance possessing this property is said to be electrified or has acquired a charge of electricity. Electrostatics is the branch of electricity which deals with the study of stationary electric charges.

10.1.1 The Phenomenon of Electrification

Every atom of matter consists of a central positively charged particle called nucleus surrounded by very minute negatively charged particles called electrons. These electrons revolve around the nucleus in different orbits. In a normal atom, the positive charge of the nucleus is equal in magnitude to the total charge of the electrons. Since there are no unbalanced charges in a normal atom as a whole, it is electrically neutral.

If a body acquires a surplus of electrons, the positive charges in the nucleus will not be able to neutralize their negative charges, and the body becomes negatively charged. When it loses some of its electrons, it becomes positively charged. When a glass rod is rubbed with silk, electrons are stripped from the glass rod and stick to the silk; the glass rod gets positively charged and the silk gets negatively charged by equal amounts. Thus electrification is the separation of the positive and negative charges which are already present in the body.

The parameter of basic importance in electrostatics is 'charge'. It gives rise to the electric fields generating forces which can produce high voltages in low capacitance systems. They may ignite flammable gases, cause shocks to personnel, damage semiconductor devices and upset the operation of microelectronic equipment.

The following conclusions are drawn based on simple observations:

1. There are two kinds of charges – positive and negative.

2. Like charges repel each other and unlike charges attract.

3. Certain substances such as copper, iron, silver, etc., allow charges to flow through them easily. They are known as conductors. Substances such as glass, vulcanite, rubber etc., do not allow the flow of charges through them. They are known as insulators or dielectrics.

4. A conductor can be charged by induction. When a glass rod rubbed with silk is brought close to a conductor, negative charge collects at the near end and positive charge at the far end. If the far end is momentarily connected to earth, positive charge is neutralised by electrons from the earth. As a result the conductor is charged negatively.

5. During electrification, equal and opposite charges are produced and there is no net charge.

6. Charge is always collected on the outer surface of the conductor.

10.1.2 Gold-leaf Electroscope

An electroscope is a simple device that can be used for the detection and measurement of charge. Gold-leaf electroscope consists of two thin gold leaves or aluminium foil L fastened to the lower end of a brass rod R as shown in Fig. 10.1. The rod passes through a hole in the stopper S made of a highly insulating material such as vulcanite fitting into the neck of a metal case with glass sides. The top end of the rod is provided with a metal disc D. A small basin containing fused calcium chloride is kept inside the case in order to keep the air dry. The case is electrically earthed.

The leaf electroscope can detect if a body is electrified, and also it can identify the kind of charge possessed by the body. When a charged body is brought near the disc of an electroscope, the leaves diverge. The increase or decrease in the divergence of the leaves helps to determine the nature of the charge. If the terminal is grounded by touching it with a finger, the charge is transferred through the human body to the earth and the leaves close together.

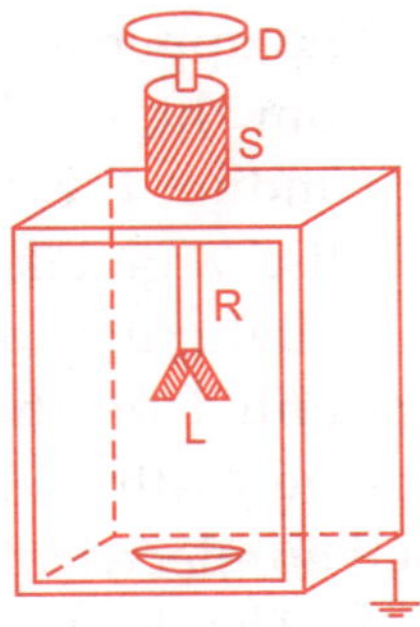

Fig. 10.1 *Leaf electroscope*

The divergence of the leaves of the electroscope is only due to the potential difference between the leaves and the metal base, and not because of repulsion between the leaves. The divergence is a measure of the potential difference.

10.1.3 Charge Distribution on Conductors

When an insulator is charged, the charge is found to remain in the region where it is produced. However, when a conductor is charged, charge is distributed throughout its surface by the free electrons. It is observed that charges normally reside on the outer surface of conductors.

The charge distribution on the surface of a conductor is measured in terms of the surface density of charge. Surface density of charge at any point on a conductor is the amount of charge per unit area around the point. The surface density of charge depends on the shape of the conductor. Surfaces of different shapes can be characterized in terms of curvature of the surface. The curvature of a surface is the reciprocal of the radius of curvature. The surface density of charge is directly proportional to the square of the curvature or inversely proportional to the square of the radius of curvature. Therefore, greater the curvature at a point on the surface of a conductor, greater will be the surface density of charge and hence, greater will be the accumulation of charge at the point.

10.1.4 Discharge Action of Points of charged Conductor

The charge density is maximum at the pointed ends of a charged conductor. Consequently, there will be great repulsion between charges of the same kind. As a result, the charge starts leaking into the surrounding air. Air and dust particles passing across pointed ends of a charged conductor, remove charges gradually from the conductor. This is called the discharging action of points. In view of this property, sharp points should be avoided in conductors in order to retain charge for a long time.

Discharging action of points has application in lightning conductors. Lightning conductor, or lightning Rod, is an apparatus that protects buildings or towers from the destructive effects of lightning. The upper region of the atmosphere is normally at a different electrical potential from the earth. The thick dense clouds which precede a thunderstorm serve to conduct the electricity of the upper air down towards the earth, and an electrical discharge takes place across the air space when the pressure is sufficient resulting in lightning followed by thunder. The lightning rod or the lightning conductor mounted on the top of a building, protects the building by providing a safe passage to this electricity to the ground. Pointed conductors called 'combs' are used in electrical machines such as Van de Graff generator for the collection of charges. The discharging effect of points is used in the electrical smoke and the dust precipitator.

10.1.5 Coulomb's law

As mentioned earlier, two like charges repel each other and two unlike charges attract. In 1785, Coulomb, a French physicist, made quantitative studies of these forces between two charges. He showed that the electrical force between two point charges is directly proportional to the product of the charges and inversely proportional to the square of the distance between them. This relation is known as Coulomb's law.

Coulomb's law holds good for point charges and charged bodies with dimensions that are small compared to the distance between the charges.

Let f be the force between two point charges q_1 and q_2, which are at a distance d apart as shown in Fig. 10.2. Then we have

$$f \propto \frac{q_1 q_2}{d^2},$$

or,
$$f = \frac{q_1 q_2}{K d^2} \qquad (10.1)$$

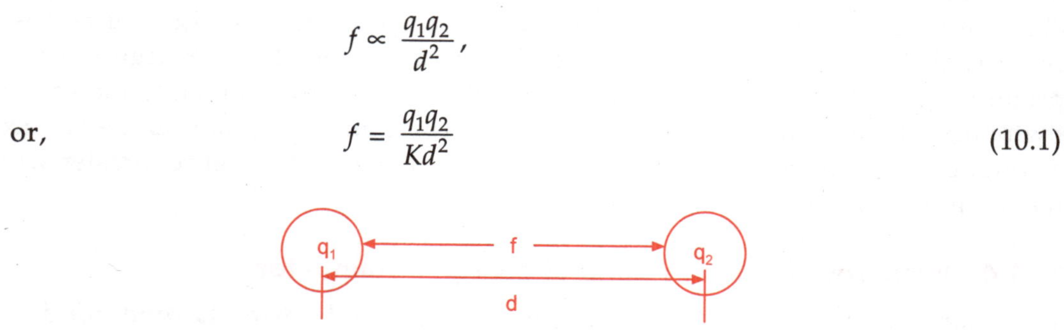

Fig. 10.2 *Electrostatic force*

where K is a constant of proportionality which depends on the units of measurement and the nature of the medium containing the two charges.

Unit of charge

A constant of proportionality is needed to give the strength of the force in newtons when q_1 and q_2 are in coulombs and d in metres. In SI system, the constant K is written as $4\pi\varepsilon$, where ε is called the permittivity (absolute) of the medium; further $\varepsilon = \varepsilon_o \varepsilon_r$, where ε_o = permittivity of free space (vacuum or air for all practical purposes) and ε_r = relative permittivity or dielectric constant of the medium.

The value of ε_o can be taken as 8.85×10^{-12} C^2/Nm^2 and $1/4\pi\varepsilon_o = 9.0 \times 10^9$ Nm^2/C^2 = 9.0×10^9 farad/meter.

Hence, Eq. (10.1) can be written as

$$f = \frac{q_1 q_2}{4\pi\varepsilon\, d^2}. \tag{10.2}$$

From Eq. (10.2), it is seen that the constant of proportionality is equal to $4\pi\varepsilon_o = 9.0 \times 10^9$ Nm^2/C^2. Therefore, the charge that is repelled by a force of 9×10^9 newtons, when placed in free space at a distance one meter from an equal and similar charge, is defined as the unit of charge. This unit of charge is called coulomb.

Although the *CGS* system of units is out of date, originally the unit of charge was defined with the constant of proportionality, K, taken as unity for free space in that system. Therefore, for information, the electrostatic unit of charge in the CGS system is also given here. It is defined as that charge which is repelled by a force of one dyne, when placed in free space at a distance of one cm from an identical charge. 1 coulomb = 3×10^9 esu of charge (the electrostatic unit of charge, esu, is also called as stat coulomb).

10.2 STRENGTH OF ELECTRIC FIELD OR THE ELECTRIC INTENSITY

The electric intensity at any point is defined as the strength of the electric field at that point. Electric intensity at a point is the force experienced by a unit positive charge placed at that point and it is a vector quantity.

In SI system, electric intensity at a point is said to be one unit when one coulomb of charge placed at that point experiences a force of one newton. It is expressed in newton/coulomb, (N/C).

If we have two charges, namely, source charge q_1 and a test charge q_2, then the electric intensity can be defined with the help of the test charge as follows.

If the test charge q_2 kept at a point experience an electrical force f, then the electric intensity E at that point is

$$E = \frac{f}{q_2}. \tag{10.3}$$

The direction of E is the same as the direction of motion of a positive charge placed at that point.

The electric intensity E at a point which is at a distance d from a test charge q_2 is given by

$$E = \frac{q_1 q_2}{4\pi\varepsilon_0\varepsilon_r d^2 q_2} = \frac{q_1}{4\pi\varepsilon_0\varepsilon_r d^2} \tag{10.4}$$

The field strength at a point due to a number of charges is the sum of the individual fields at that point.

10.2.1   Effect of dielectric

The force between two charged particles is influenced by the medium in which they are placed. This is easily demonstrated by suspending two similarly charged balls from the same point. They repel each other and finally come to rest at a certain distance (Fig. 10.3(a)).

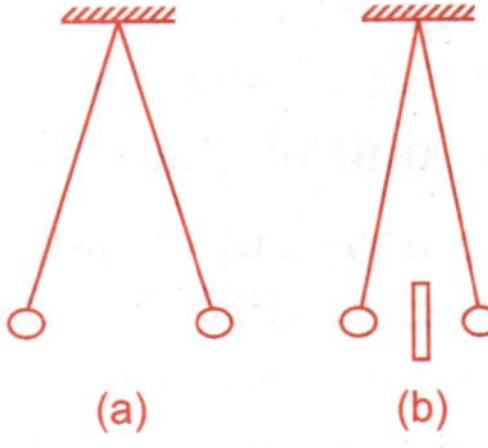

Fig. 10.3 *Suspension of two charged balls (a) without dielectric (b) with dielectric*

If a plate made of an insulator material (dielectric), such as a glass, is inserted between them as shown in Fig. 10.3 (*b*), the final distance between the two charges decreases. This distance will be different for different media. The force is larger in air and is smaller in any other medium (see Eq. 10.4). This is because ε_r for air is 1 and ε_r for any other medium is greater than unity. Hence, the intensity, E, is reduced when air is replaced by a dielectric. Similarly, potential at a point is reduced when air is replaced by an insulator or a dielectric. However, it should be noted that the capacity of a capacitor increases when air is replaced by a dielectric.

An insulator has no free electrons unlike in an electrical conductor. Normally, in the absence of an external field, the positive and negative charge distributions in an atom cancel each other. However, when a dielectric is placed in an electric field, the electrons in the orbits become slightly displaced in one direction. The atoms or molecules in the dielectric are said to be polarized. The phenomenon is referred to as electric polarization. The displaced equal positive and negative charges constitute an electric dipole. The product of charge and distance between them is called the dipole moment. The dipole moment per unit volume is called dielectric polarization and is represented by the letter P which is a vector quantity. A dielectric slab placed in a uniform electric field between two parallel metal plates, to which a potential difference is applied, is shown in Fig. 10.4.

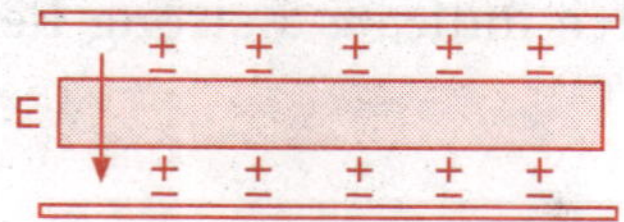

Fig. 10.4 *Dielectric slab subjected to electric field*

The dielectric as a whole is polarized and the molecules are oriented in such way that the positive charges on one face and the negative charges on the other face are aligned perpendicular to the field as shown in Fig. 10.4. These charges are known as polarization charges. The polarization charges always oppose the external field. In other words, when a dielectric is placed in an electric field, the original field is weakened within the dielectric.

Let E be the resultant electric field within the dielectric and P the polarization. Then the quantity $(\varepsilon E + P)$, is found to be independent of the medium: This quantity is represented by D and is called electric displacement or electric induction.

The relation, $D = \varepsilon E + P$, is similar to the relation $B = \mu_0 H + I$ in magnetism. The dielectric polarization P is proportional to E; therefore, $D = \varepsilon E$, where ε is a constant depending on the medium and $\varepsilon = \varepsilon_0 \varepsilon_r$ where ε_r is called dielectric constant of the medium.

10.2.2 Lines of Force

The influence of an electric field is represented by imaginary lines of force as shown in Fig. 10.5. The idea of using lines of force to represent the electric field was introduced by Faraday. The properties of lines of force are:

1. Lines of force originate from a positive charge and terminate on a negative charge.

2. They are perpendicular to the surface of a conductor.

3. The direction of a line of force is that along which a free positive charge tends to move.

4. No lines of force exist within a conductor.

5. Lines of force never intersect one another.

6. The number of lines of force crossing unit area normally around a point represents the field strength E at the point.

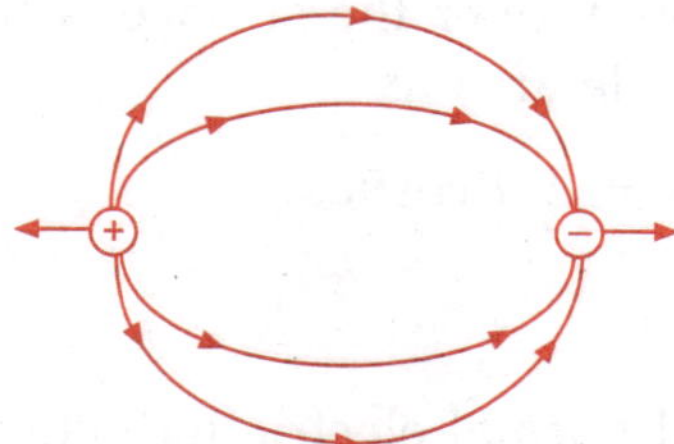

Fig. 10.5 *Electric field lines of force*

Closely spaced lines of force indicate a strong field and widely spaced lines of force indicate a weak field.

10.2.3 Electric Induction

The electric intensity at a distance d from a point charge q is given by the equation

$$E = \frac{q}{4\pi\varepsilon d^2}$$

$$\varepsilon E = \frac{q}{4\pi d^2} \tag{10.5}$$

The product εE, denoted by D, is called electric displacement or electric induction. It can be seen from Eq. (10.5), that the electric induction does not depend upon the medium. Consequently, D is considered to be of a more fundamental nature than E in a dielectric medium.

In a dielectric, E represents the number of lines of force per unit normal area while D represents the number of lines of induction per unit normal area. The total number of lines of induction passing normally through any surface is called the flux of electric induction or total normal electric induction (TNEI) through the surface.

In Fig. 10.6, ds is an element of area of a surface. The electric intensity E at the element makes an angle θ with the normal to the element.

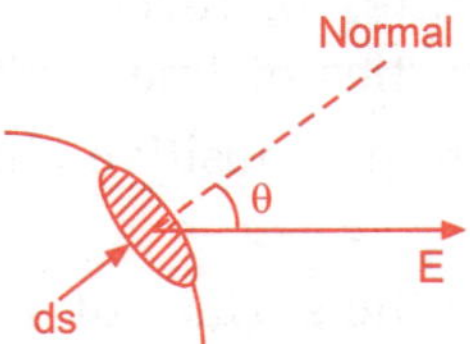

Fig. 10.6 *Electric induction on a surface*

The component of E along the normal is $E \cos\theta$, and that of induction is $\varepsilon(E \cos\theta)$. Therefore, the normal induction through the element of area ds is given by

$$dN = (D \cos\theta)ds$$
$$dN = (\varepsilon E \cos\theta)ds.$$

The total normal induction N over the surface is therefore given by adding the normal induction over the whole area as

$$N = \Sigma \varepsilon E \cos\theta \, ds. \tag{10.6}$$

10.2.4 Gauss' Theorem

The relation between the total normal electric induction across a closed surface and the charge enclosed was discovered by Karl Freidrich Gauss, a German scientist. The relation is named after him as Gauss' theorem. Gauss' theorem states that the total normal electric induction across any closed surface in an electric field is numerically equal to the total charge enclosed by the surface. If q is the charge enclosed by a closed surface, then

$$q = \Sigma \varepsilon E \cos \theta \, ds \qquad (10.7)$$

where the summation extends over the entire closed surface.

If q is positive, the induction or flux is directed outwards and if q is negative, the induction is inwards. If q is outside the closed surface, there is no contribution to the induction and the induction through the surface is zero.

For a closed spherical surface of radius r, enclosing a point charge q at its centre as shown in Fig. 10.7, the electric intensity E, at any point on its surface is given by

$$E = \frac{q}{4\pi\varepsilon r^2} \qquad (10.8)$$

The area of the spherical surface $= 4\pi r^2$.

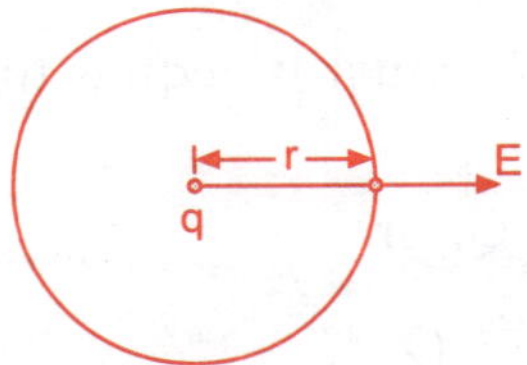

Fig. 10.7 *Closed spherical surface*

In this case E is same at all points on the surface and $\theta = 0$. Total normal electric induction across the surface is given by

$$\Sigma \varepsilon E \cos \theta \, ds = \varepsilon E \Sigma \, ds \qquad (10.9)$$

As, $\cos 0° = 1$, we have $\varepsilon\left(\dfrac{q}{4\pi\varepsilon r^2}\right)4\pi r^2 = q$ indicating that the TNEI (total normal electric induction)over the closed surface is equal to the charge enclosed.

10.2.5 Applications of Gauss Theorem

1. Electric intensity at a point due to a charged spherical conductor

P is a point outside a spherical conductor having a charge Q. The distance of P from the centre of the sphere is d as shown in Fig. 10.8. If a concentric spherical surface of radius d passing through P is imagined, then the total normal electric induction, N, through the surface is given by

(*a*) $\qquad\qquad\qquad N = \Sigma D \cos \theta \, ds \qquad (10.10)$

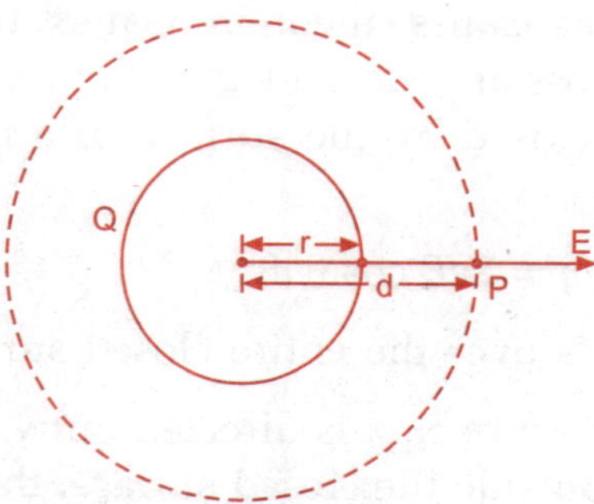

Fig. 10.8 *Spherical conductor*

Here $D = \varepsilon E$ is radial and has the same value at all points on the spherical surface on account of symmetry about $\theta = 0$.

$$N = D\Sigma ds = \varepsilon E 4\pi d^2 \tag{10.11}$$

where $E \to$ Electric intensity, at the point P, ε permittivity of the medium surrounding the spherical conductor and $4\pi d^2 \to$ surface area of the imaginary surface.

By Gauss' theorem this must be equal to the charge Q on the spherical conductor.

$$\varepsilon E 4\pi d^2 = Q, \text{ or}$$

$$E = \frac{Q}{4\pi\varepsilon d^2} \tag{10.12}$$

(b) For a point just outside the sphere very near the outer surface we have

$$E = \frac{Q}{4\pi\varepsilon R^2} = \frac{4\pi R^2 \sigma}{4\pi\varepsilon R^2} = \frac{\sigma}{\varepsilon} \tag{10.13}$$

where R is the radius of the spherical conductor, σ is the surface density of charge or charge per unit area. This is a special case of Coulomb's law.

(c) If the point P is inside the conductor, the charge enclosed by the imaginary spherical surface is zero, because charge resides on the outer surface of the conductor. Hence electric intensity inside a conductor is zero.

2. Coulomb's law

If E is the intensity at a point, at a distance r from a point charge q, then E can be obtained using Gauss' theorem. If an imaginary spherical surface of radius r, with centre at the point charge is considered, electric intensity is radial and is same at all points on the surface. The TNEI over the spherical surface is $\varepsilon E 4\pi r^2$. By Gauss' theorem this must be equal to q, the charge enclosed.

$$\varepsilon E 4\pi r^2 = q.$$

Hence,
$$E = \frac{q}{4\varepsilon\pi r^2} \qquad (10.14)$$

The force f on a point charge q placed on this surface, i.e., at a distance r from the charge q is given by

$$f = q'E = \frac{q'q}{4\pi\varepsilon r^2} \qquad (10.15)$$

3. Electric intensity at a point just outside the surface of a charged conductor. The expression for E can be found using a small cylindrical box (a flat, cylindrical shaped figure)through a small area around a point P on the surface as a Gaussian surface, with generating lines of the cylinder normal to ds as shown in Fig. 10.9. E is zero everywhere inside the conductor.

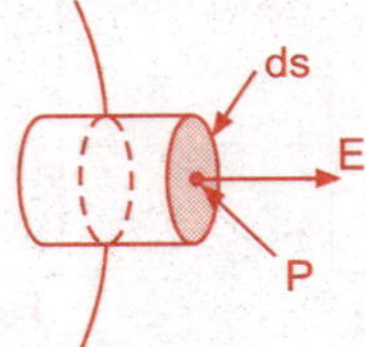

Fig. 10.9 *Pill box as a conductor*

The only contribution to induction is from the end face of area ds, outside the conductor. Charge, q, on the area ds is given by $\sigma\,ds$, where σ is the surface density of charge on ds. Therefore by Gauss' theorem

$$\Sigma\varepsilon E ds \cos\theta = q$$
$$\varepsilon E ds = \sigma ds.$$

Hence,
$$E = \frac{\sigma}{\varepsilon} \qquad (10.16)$$

In a similar manner it can be shown that the electric intensity at a point outside a large charged plane conductor is given by

$$E = \frac{\sigma}{\varepsilon} \qquad (10.17)$$

where σ is the surface density of charge.

4. Electric intensity at a point in between two parallel plane charged conductors

P_1 and P_2 are two parallel plane charged conductors having surface charge densities σ and $-\sigma$, respectively, as shown in Fig. 10.10. The field between two such oppositely charged conductors is uniform. Electric field E, at any point between them can be obtained by considering a pill box (cylindrical box) as the Gaussian surface, whose end faces of area ds are perpendicular to the field; one end face is within the field and the other inside the plate P_1. The only contribution to induction is from the end face within the field; charge enclosed by the pill box is $\sigma\, ds$, the charge on area ds of the place P_1.

Therefore by Gauss theorem, we have

$$\varepsilon E ds = \sigma ds,\ \text{or}$$

$$E = \frac{\sigma}{\varepsilon} \tag{10.18}$$

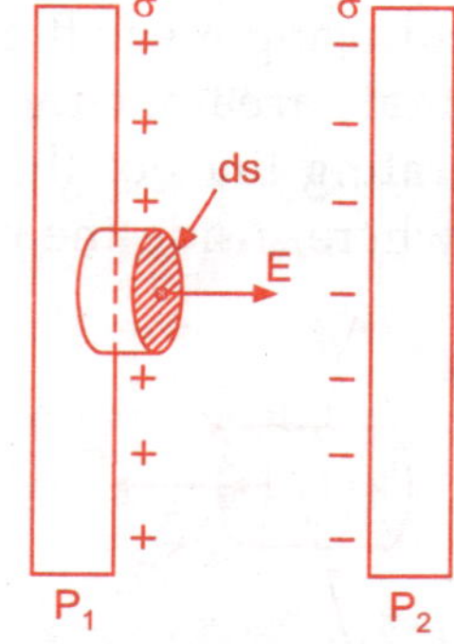

Fig. 10.10 *Electric intensity between two parallel planes*

5. Electric intensity at a point just outside a long uniformly charged cylindrical conductor P is a point outside a uniformly charged cylindrical conductor, the distance P from the axis of the cylinder being r as shown in Fig. 10.11. A Gaussian surface in the form of a cylindrical surface of radius r coaxial with the charged cylinder is considered. TNEI over this surface is $\varepsilon E 2\pi r$ per unit length. By Gauss' theorem

$$\varepsilon E 2\pi r = \lambda$$

where λ = charge per unit length of the cylinder.

$$E = \frac{\lambda}{2\pi\varepsilon r} \tag{10.19}$$

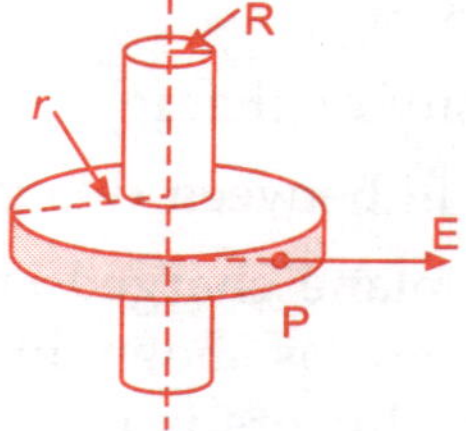

Fig. 10.11 *Electric intensity in a long cylindrical conductor*

If the point P is very close to the charged cylinder, then $r = R$, the radius of the charged cylinder. If σ is the surface density of charge, $\lambda = 2\pi R\sigma$. Substituting for λ in Eq. (10.19), we get

$$E = \frac{\sigma}{\varepsilon}. \tag{10.20}$$

10.2.6 Electric Potential

When two charged bodies are connected by a metal wire or placed in contact, there is a flow of charge from one body to another. The direction of flow of charge is determined by the difference in the electrical levels or the electric potential. It is not determined by the amount of charge on either body.

The water level in hydrostatics or the temperature in heat can be considered analogous to electric potential in electricity. For example, when two vessels containing water at two different levels are connected by a pipe, water will flow from the vessel having water at a higher level to the vessel having water at a lower level. The flow of water will continue until the water levels in both vessels are equalized. The difference in level of water in the two vessels will decide the direction of flow. Heat also flows from a body at a higher temperature to a body at a lower temperature, when the two bodies are brought in contact with each other. Heat will continue to flow until the temperature in the two bodies become equal. The difference in temperature determines the direction of flow of heat.

The potential of a body may be considered as the electrical condition of the body, which determines whether it can deliver positive charge to or receive positive charge from another body. The potential at any point in an electrostatic field can be considered as the condition or the ability of the field, which determines the direction of flow of positive charge placed at that point. A reference level is necessary to measure electric potential. The mean sea level is taken as the reference level or zero level in order to specify altitudes of places or the heights of mountains. In the study of electrostatics, potential of the earth is taken to be at zero potential. The earth is so large such that, it can gain or lose electricity but it does not affect the potential. Hence, it is assumed that earth is at zero potential and thus any body connected to the earth is at zero potential. A body is said to be at a positive potential if electrons flow from the earth to the body, when the body is connected to the earth. A body is said to be at a negative potential, if the electrons flow from the body to the earth.

Electric potential at any point in an electric field is measured by the amount of work done in bringing a unit positive charge from infinity to that point against the direction of the field. It is the potential energy possessed by a unit positive charge placed at that point.

Units of Electric Potential

When we refer to two charged bodies which differ in their charge value, we express the difference in their levels as potential difference. The unit of potential or potential difference is called volt. The potential at a point is said to be one volt, if one joule (one newton.meter) of work needs to be done in bringing one coulomb of charge from infinity to the point.

Electric Potential and Electric Intensity

Let A and B be two points on the same line at distances x and $(x + dx)$, respectively, from a point charge at O, as shown in Fig. 10.12. If E is the electric intensity at A, the work done against the field in moving unit positive charge from B to A is equal to $-Edx$. The negative sign indicates that the field and the displacement are in opposite directions. The work done, by definition, must be equal to the potential difference, dV, between the two points.

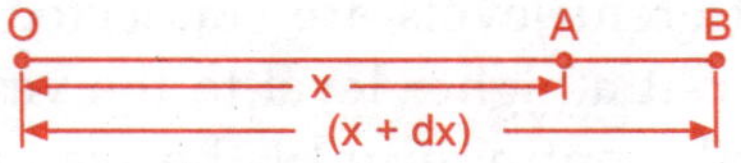

Fig. 10.12 *Potential difference in a body*

Hence, $-Edx = dV$ (10.21)

$$-E = \frac{dV}{dx}$$

The quantity $\dfrac{dV}{dx}$ is called the potential gradient at that point. Therefore, the negative gradient of the electrostatic potential in any direction gives the strength of the field in that direction. Potential at a point is a scalar quantity whereas the potential gradient or the electric field is a vector quantity.

Electrostatic Potential at any point due to a point charge

It is evident from Eq. (10.21), that the potential difference between two points A and B in Fig. 10.12, is given by

$$dV = -Edx$$

Potential at A is numerically equal to the total work done in moving a unit charge from infinity to that point and is given by

$$V = \int_{\infty}^{x} -Edx = \int_{\infty}^{x} -\frac{q}{4\pi\varepsilon x^2}dx = \frac{q}{4\pi\varepsilon x}$$

Hence,

$$V = \frac{q}{4\pi\varepsilon x}$$ (10.22)

10.2.7 Potential of a Conductor

Potential of a conductor is the amount of work done in bringing unit positive charges from infinity to any point infinitesimally close to the surface of the conductor.

The following important electrostatic results for a conductor should be kept in mind:

1. Inside a conductor, electrostatic field is zero.
2. At the surface of a charged conductor, electrostatic field is perpendicular to the surface at every point.
3. interior of a charged conductor cannot have excess charge, and the excess charges reside only on the surface.
4. Electrostatic potential is a constant throughout the volume of a conductor, and has the same value inside and on the surface.

10.2.8 Equipotential Surfaces

Any line or surface passing through points of equal potential are referred to as equipotential lines or equipotential surfaces. In electrostatics, the work done to move a charge from any point on the equipotential surface to any other point on the same surface is zero since they are at the same potential.

Surface with Constant Electric Potential

Equipotential surfaces are always perpendicular to the lines of force. Surface of a charged conductor is equipotential. This is because, if, two points exist on the surface of a charged conductor having a potential difference, then, the embedded electrons of the conductor will flow from the point of lower potential to the point of higher potential until equipotential conditions are established.

Potential inside a charged hollow conductor is uniform and is the same as that on the surface of the conductor. Since the potential inside a conductor is uniform, the electric intensity inside a charged conductor is zero. Therefore, there can be no charge inside the conductor. Consequently, all the charges must reside on the surface of a charged conductor.

10.2.9 Electric Screening or Electrostatic Shielding

The field within a closed conductor is zero. This property can be conveniently used for electrical or electrostatic shielding. Electrical or electrostatic shielding is used to protect electronic instruments and devices from being affected by external charges. The goal of electrostatic shielding is the creation of a region of space in which the electric field is independent of the conditions outside this region.

If a gold leaf electroscope is enclosed inside a cage of wiremesh, the electroscope is unaffected even if the cage is highly charged. This is why it is safer to stay inside a tin shed than in a building during lightning, because, tin shed behaves like an earth connected hollow conductor.

10.2.10 Electrostatic Capacity

Electrostatic capacity, or the capacity for electricity, of a body is the quantity of electricity to be supplied to the body to raise its potential by one unit. This is analogous to the thermal capacity of a body which is the amount of heat necessary to raise the temperature of the body by one degree.

If V is the potential of a conductor due to a charge Q, then the capacity C is given by

$$C = \frac{Q}{V}. \tag{10.23}$$

The capacity of a conductor depends on (*i*) the area of the surface of the conductor, (*ii*) the surrounding medium and (*iii*) the presence of neighbouring conductors.

The unit of capacity is called farad. A conductor is said to have a capacity of one farad, if one coulomb of charge is required to raise its potential by one volt.

$$\text{Capacity in farads} = \frac{\text{charge in coulombs}}{\text{potentials in volts}}$$

Since farad is a large unit, microfarad (mF) and picofarad (rF) are commonly used ($1\ \text{mF} = 10^{-6}\text{F};\ 1\ \rho\text{F} = 10^{-12}\text{F}$).

10.2.11 Capacity of a Spherical Conductor

If V is the potential of a spherical conductor of radius, r, due to a charge Q, then the potential is given by

$$V = \frac{Q}{4\pi\varepsilon r}$$

where ε is the permittivity of the surrounding medium. In the case of a spherical conductor, the intensity and potential at a point outside the conductor can be calculated by considering the charge to be concentrated at the centre of the sphere.

$$\text{Capacity,}\quad C = \frac{Q}{V} = \frac{Q}{\dfrac{Q}{4\pi\varepsilon r}} = 4\pi\varepsilon r. \tag{10.24}$$

10.3 CAPACITOR

A capacitor or condenser is used for storing charges. It is an insulated conductor whose capacity is artificially increased. It consists of two closely held parallel metal plates with an insulating medium (known as dielectric) between them. Capacitors are widely used in power systems, radio, television and telecommunication circuits.

They are used to eliminate sparking in electrical circuits of induction coils, ignition system of automobiles, etc.

The capacity of a capacitor is the ratio of the charge on either conductor or the potential difference between the two conductors. Capacitors are used to store electric charge.

10.3.1 Principle of a Capacitor

Consider a conductor A charged to a potential V with positive charge Q as shown in Fig. 10.13. When an uncharged insulated conductor B is brought near A, conductor A will induce a negative charge on the inner surface of B and a positive charge on its outer surface. As the negative charge on B is closer to A, there will be a slight change in the potential of A.

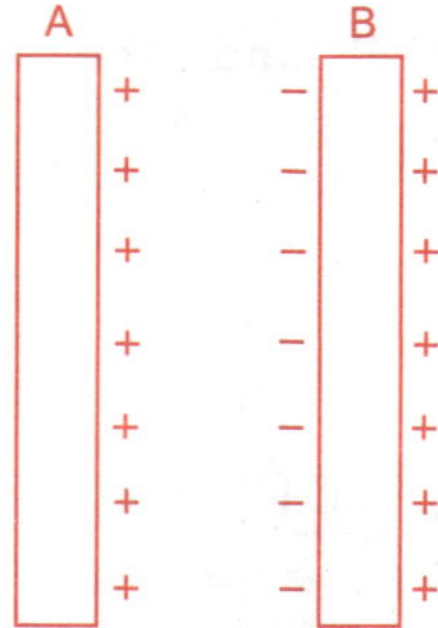

Fig. 10.13 *Two conducting plates as a capacitor*

However, if B is connected to earth, the positive charge on B is removed; the negative charge on B produces a negative potential on A and hence the potential of A decreases. Thus, more charge can be stored in A to raise it to the original potential. Thus, capacity of A is artificially increased by placing an earth connected conductor near A. Such an arrangement constitutes a condenser.

10.3.2 Capacity of a Parallel Plate Capacitor

A parallel plate capacitor has two parallel metal plates, P_1 and P_2, separated by a dielectric as shown in Fig. 10.14. If P_1 is charged and P_2 is earthed, then, the field between the plates is uniform at all points, except at the two ends.

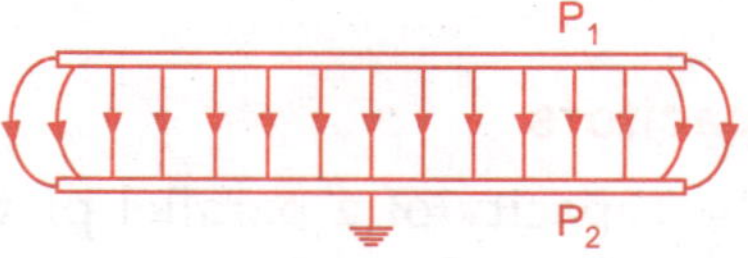

Fig. 10.14 *Two parallel conducting plates as a capacitor*

The electric field, E, between the plates is given by

$$E = \frac{\sigma}{\varepsilon}$$

If Q is the charge on the plate P of area A, then

$$\sigma = \frac{Q}{A}$$

$$E = \frac{Q}{\varepsilon A} \tag{10.25}$$

Since the electric field is equal to potential gradient and the field between the plates is uniform, we have

$$E = \frac{V}{t} \tag{10.26}$$

where V is the potential of the plate P_1, and t the separation between the plates. From Eqs. (10.25) and (10.26), we have,

$$\frac{Q}{\varepsilon A} = \frac{V}{t}$$

Therefore, capacity is given by

$$C = \frac{Q}{V} = \frac{\varepsilon A}{t} \tag{10.27}$$

10.3.3 Dielectric Constant

The capacity, C_d, of a condenser with a dielectric other than air is ε_r times the capacity, C_a, of the same condenser with air as dielectric.

Therefore,
$$\varepsilon_r = \frac{C_d}{C_a}. \tag{10.28}$$

Hence, dielectric constant of a medium is the ratio of the capacity of a condenser with that medium as dielectric to the capacity of the same condenser with air as dielectric.

It should be noted that the dielectric constant is different from dielectric strength. The dielectric strength refers to the breakdown voltage, of a dielectric. It is the maximum voltage below which it behaves as a dielectric and above that voltage it starts conducting.

10.3.4 Guard Ring for Capacitors

In deriving Eq. (10.27) for the capacity of a parallel plate capacitor, it was assumed that the electric field is uniform throughout the region between the plates. However, the field is not uniform near the ends as already mentioned earlier.

In order to overcome the end effects, Lord Kelvin used a circular plate P_1 surrounded by a guard ring G, in plane of the plate P_1, separated by a narrow gap as shown in Fig. 10.15. The area of the earthed plate P_2 is equal to that of P_1 and G together. The plate P_1 and the ring G are kept at the same potential by connecting G and P_1 by a wire. Consequently, the field between the plates will be uniform.

The effective area A = area of plate P_1 + ½(area of the gap).

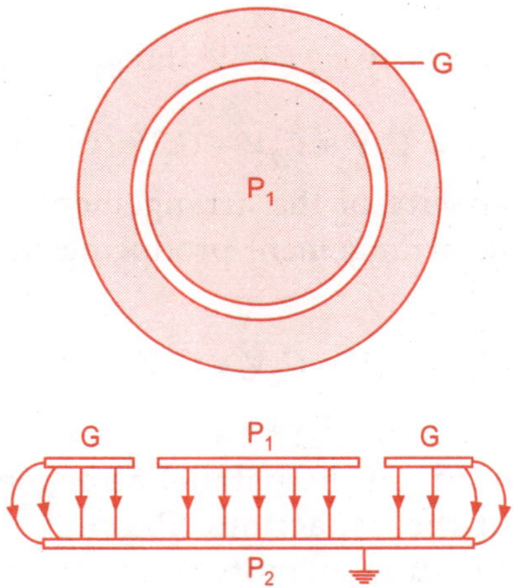

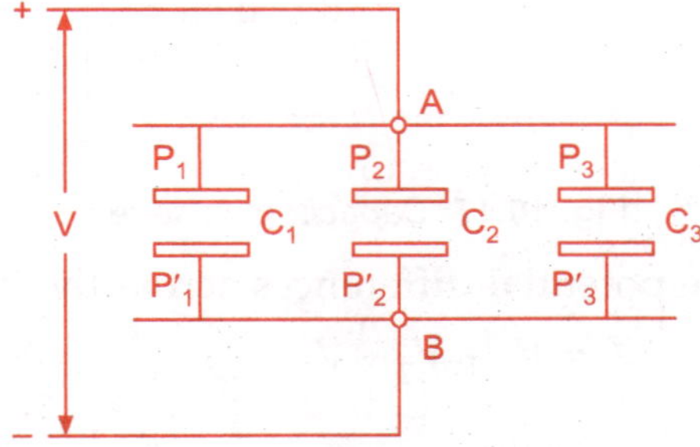

Fig. 10.15 *A plate surrounded by guard ring*

10.3.5 Different Arrangements of Capacitors

Capacitors may be connected in parallel configuration or in series.

1. Capacitors connected in parallel

The parallel arrangement of capacitors C_1, C_2 and C_3 is shown in Fig. 10.16. When the capacitors are connected in parallel (*i*) the potential difference across each capacitoris the same and (*ii*) the total charge of the system is equal to the sum of the charges of all the capacitors together. In the Fig. 10.16, the plates P_1, P_2 and P_3 are connected

Fig. 10.16 *Capacitors in parallel*

together to the common terminal A and the other set of plates P_1', P_2' and P_3' are connected to the terminal B. The capacitors are charged to the same potential V by connecting A and B, respectively, to the positive and negative terminals of a battery.

If the capacitors share respectively charges q_1, q_2 and q_3, Q being the total charge, then

$$Q = q_1 + q_2 + q_3$$

But $C_1 = \dfrac{q_1}{V}$ or $q_1 = C_1 V$. Similarly, $q_2 = C_2 V$ and $q_3 = C_3 V$.

Therefore, $$Q = C_1 V + C_2 V + C_3 V$$

If C is the equivalent capacity of the arrangement, i.e. the capacity of a single condenserthat can replace the arrangement producing the same effect, then we have

$$Q = CV$$
$$CV = C_1 V + C_2 V + C_3 V$$
$$C = C_1 + C_2 + C_3 \tag{10.29}$$

Thus, the equivalent capacity of the parallel arrangement is equal to the sum of the individual capacities of the capacitors. Parallel connection of a number of capacitors can be usedto store large quantity of electricity.

2. Capacitors Connected in Series

Capacitors are sometimes connected end to end in a series arrangement, as shown in Fig. 10.17. In a series arrangementof identical capacitors (*i*) the charge on each plate of the capacitor is same and (*ii*) the total potential difference across the system is equal to the sum of the potential differences across the capacitors.

Three capacitors of capacities C_1, C_2 and C_3 are connected in series as shown in Fig. 10.17. When a potential V, is applied across the system, the charge q taken by each capacitor is the same.

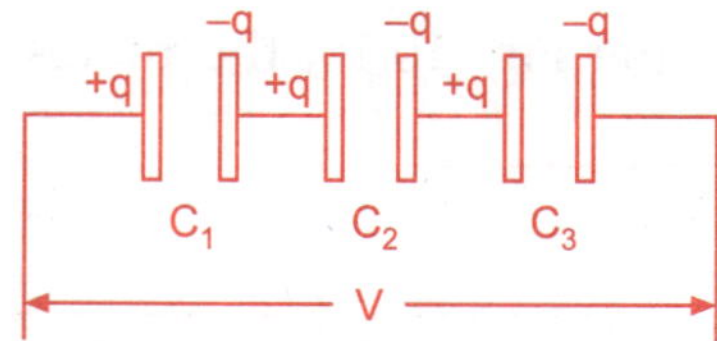

Fig. 10.17 *Capacitors in series*

If V_1, V_2 and V_3 are the potential differences across the individual capacitors then

$$V = V_1 + V_2 + V_3$$

But $V_1 = \dfrac{q}{C_1}$, $V_2 = \dfrac{q}{C_2}$ and $V_3 = \dfrac{q}{C_3}$

Therefore, we have

$$V = \frac{q}{C_1} + \frac{q}{C_2} + \frac{q}{C_3}$$

If C is the equivalent capacity of the system, $V = \frac{q}{C}$, then

$$\frac{q}{C} = \frac{q}{C_1} + \frac{q}{C_2} + \frac{q}{C_3},$$

or,
$$\frac{1}{C} = \frac{1}{C_1} + \frac{1}{C_2} + \frac{1}{C_3} \tag{10.30}$$

Hence, the reciprocal of the equivalent capacity is equal to the sum of the reciprocals of the individual capacities of the capacitors in series. The series connection of a number of capacitors can be used to serve as a potential divider when the source of electric charge is at a very high potential. The arrangement also provides a means of storing charge at a high potential.

10.3.6 Types of Capacitors

1. Mica capacitor

This is a multiplate capacitor as shown in Fig. 10.18. It consists of two sets of metal foils, insulated from each other with mica as the dielectric between them.

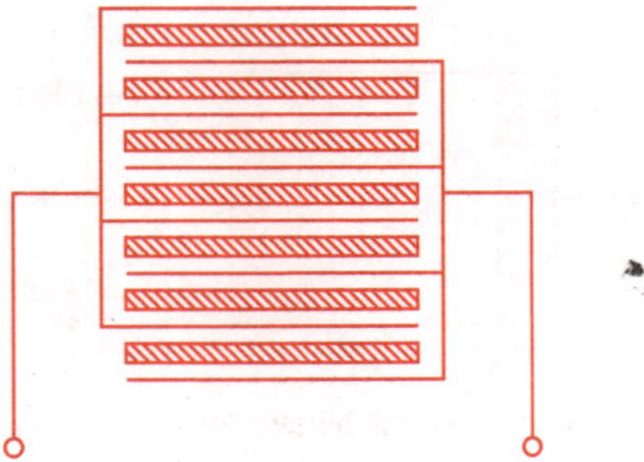

Fig. 10.18 *Multi-plate capacitor*

The capacity of such a capacitor is given by

$$C = \frac{n\varepsilon A}{t} \tag{10.31}$$

where n is the number of dielectric sheets, ε is permittivity, A is common area, and t is the thickness of each dielectric sheet. A thin layer of silver is deposited on the opposite faces of mica in order to establish a good contact with the metal plates.

2. Paper capacitor

This is a parallel plate capacitor consisting of long strips of aluminium or tin foil as electrodes with paraffin waxed paper in between as shown in Fig. 10.19. The foils with the dielectric are rolled up occupying a very small space and the capacitor is compact. Paper capacitors are very easy to make and are with low cost.

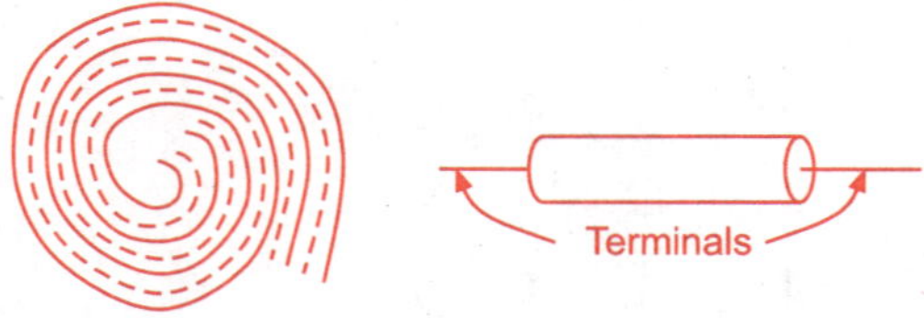

Fig. 10.19 *Paper capacitor*

3. Electrolytic Capacitor

When a direct current is passed through an electrolyte of aluminium borate with aluminium plates as electrodes, a very thin film of molecular dimensions of aluminium oxide is formed on the anode plate. The oxide film acts as an insulator. The aluminium plates with the film in between acting as the dielectric, is rolled into a cylindrical shape as shown in Fig. 10.20. An electrolytic capacitor has a high capacity (in the order of hundreds of μF) compared to multiplate capacitor (a few μF). While using electrolytic capacitor, the anode plate is usually indicated by + signs, which must always be positive relative to the other plate. If this condition is not satisfied, the oxide film breaksdown.

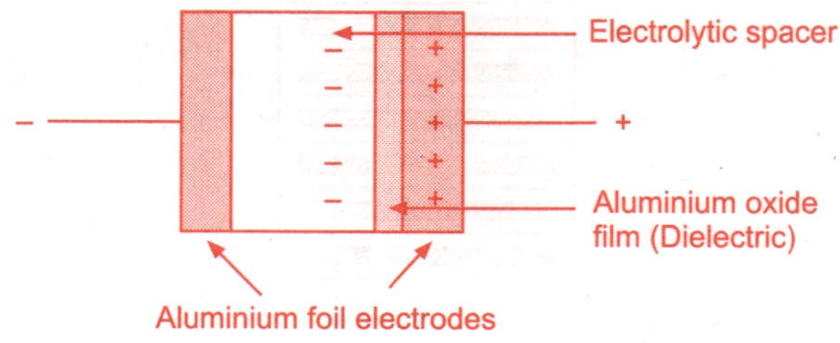

Fig. 10.20 *Electrolytic capacitor*

4. Variable capacitor

This consists of a moveable set of metal vanes M which mesh with another fixed set *F* as shown in Fig. 10.21. The two sets are insulated from each other. The area of overlap can be varied by means of the knob thus changing the capacitance. A variable capacitor used for tuning in radio receiver has a maximum capacitance of about 500 pF and a minimum capacitance of about 50 pF.

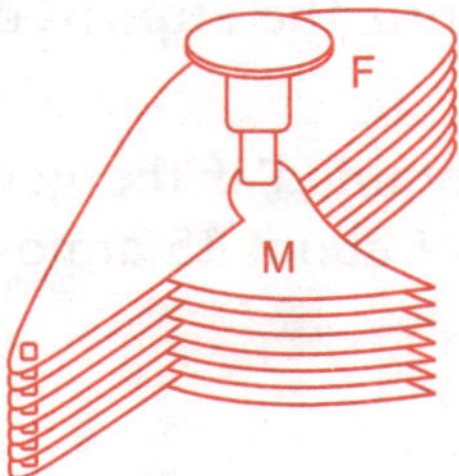

Fig. 10.21 *Variable capacitor*

10.4 VAN DE GRAAFF GENERATOR

Van de Graaff generator is an electrostatic machine, devised by Van de Graff, a French scientist, in 1931, which can generate voltages of several million volts. It was used in 1933, to produce several million volts for lighting a Chicago exhibition. It is used in atomic research as a high voltage generator. Van de Graaff generator is based on the phenomena of (*i*) the discharging action of points and (*ii*) the scattering of the charge given to a hollow conductor uniformly over its outer surface. The Van de Graaff generator is shown schematically in Fig. 10.22. It consists of a large hollow smooth metal sphere S mounted on insulating pillars. A belt B made of suitable insulator material moves between two pulleys P_1 and P_2 driven by a motor.

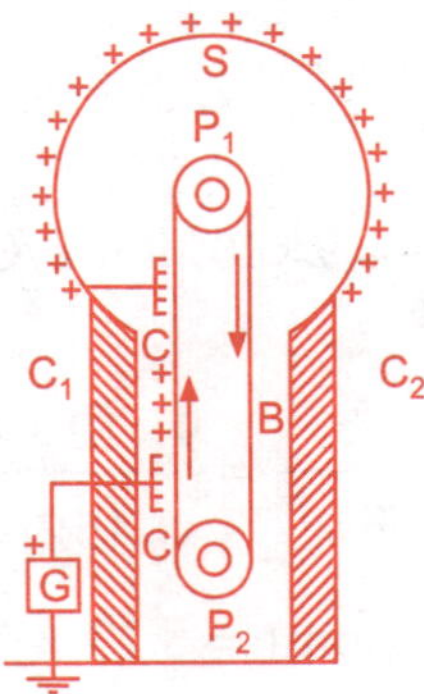

Fig. 10.22 *Van de Graaff generator*

C_1 and C_2 are metal combs which are a set of metal rods with a number of sharp points, facing the belt on one side. C is connected to the positive terminal of a DC generator of ten to twenty kilovolts, with the other terminal earthed. C_2 is connected to the sphere S. The comb C_1 connected to positive terminal of G, acquires a positive charge. Because of the points, positive charge is sprayed from C_1 to the belt B and the belt carries the positive charge upwards. Upon arriving opposite to the comb C_2, the charge on B induces a negative charge on C_2 and equal positive charge on S. Because of points, negative charge is sprayed from C_2 to the belt. The uncharged belt moves down to P_2 and the action is repeated. In this way, S continues to get charge and its potential grows continuously until the supply of charge and the loss of charge balance each other.

Very high voltages can be obtained if the generator is enclosed in steel tanks filled with nitrogen or freon gas at about 15 atmospheres pressure.

EXAMPLE 10.1

Obtain the field intensity at a point 0.1m from a point charge of $\dfrac{20 \times 10^{-9}}{3}C$.

SOLUTION

$$E = \frac{q}{4\pi\varepsilon_0 d^2} = \frac{9 \times 10^9 \times 20 \times 10^{-9}}{3 \times (0.1)^2} = 6 \times 10^3$$

Electric intensity at the point = 6×10^3 Newton/coulomb.

EXAMPLE 10.2

Find the intensity of the electric field at a distance of 5.3×10^{-9} cm from a proton of 1.6×10^{-19} C.

SOLUTION

Let the value of the constant, $\dfrac{1}{4\pi\varepsilon_0} = 9 \times 10^9$ Nm^2/C^2

$$E = \frac{q}{4\pi\varepsilon_0 d^2} = \frac{9 \times 10^9 \times 1.6 \times 10^{-19}}{\left(5.3 \times 10^{-9}\right)^2} = 5.13 \times 10^7$$

Intensity of electric field at the point = 5.13×10^{-7} N/C.

EXAMPLE 10.3

When two positive charges Q and 4Q are placed at a distance of 0.12 m from each other find the position of the neutral point.

SOLUTION

The neutral point has to be in between the charges. Let the position of the neutral point be x from the charge Q.

$$\frac{Q}{4\pi\varepsilon_0 x^2} = \frac{4Q}{4\pi\varepsilon_0 (0.12 - x)^2}$$

or
$$\frac{1}{x} = \frac{2}{0.12 - x}$$

i.e.
$$x = 0.04$$

The neutral point is at a distance 4 cm from Q and 8 cm from $4Q$.

EXAMPLE 10.4

A, B and C are the three corners of an equilateral triangle of side 5 cm. Two point charges +100 esu and −100 esu are placed at A and B respectively. Find the direction and magnitude of the electric field at C.

SOLUTION

The charges are $+\dfrac{100 \times 10^{-9}}{3}$ C and $-\dfrac{100 \times 10^{-9}}{3}$ C at A and B (Fig. 10.23).

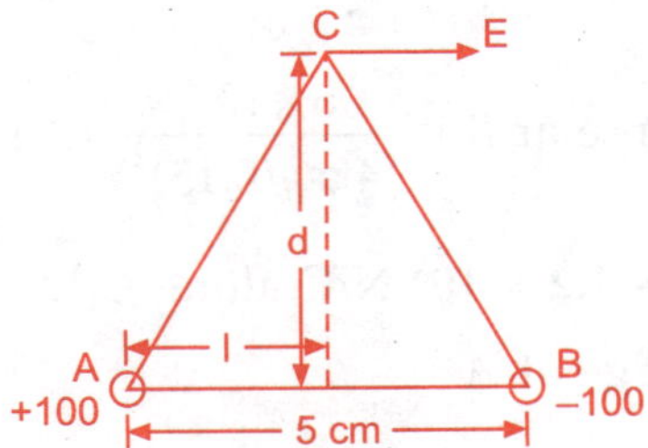

Fig. 10.23 *Three points of electric charges*

We have $1C = 3 \times 10^9$ esu of charge. This is similar to a magnet with north pole at A and south pole at B. The point C is on the right bisector of AB because ABC is an equilateral triangle. For a magnet,

$$F_B = \frac{M}{4\pi\mu_0 \left(d^2 + 1^2\right)^{\frac{3}{2}}}$$

where,
$$M = m \times 21$$

using q instead of m and ε_o in place of μ_o, intensity E at C is given by,

$$E = \frac{q \times 21}{4\pi\varepsilon_0 \left(d^2 + 1^2\right)^{\frac{3}{2}}} = \frac{9 \times 10^9 \times \left(100 \times 10^{-9}\right) \times 0.05}{3 \left(0.05^2\right)^{\frac{3}{2}}}$$

$$= 1.2 \times 10^5$$

Electric intensity at C = 1.2×10^5 Newton/coulomb in a direction parallel to AB.

EXAMPLE 10.5

Two small conductors 30 cm apart have positive charges 3×10^{-9} C and 6×10^{-9} C, respectively. Calculate (a) the electric intensity (b) the potential at a point midway between them.

In Fig. 10.24, A and B represent small conductors with charges 3μ C and 6μ C and C is the midpoint of AB.

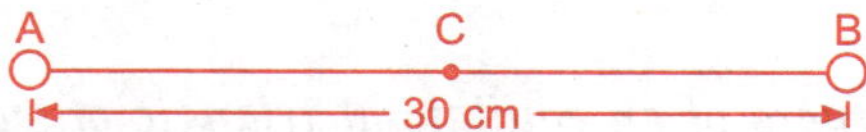

Fig. 10.24 *Two conductors at a distance*

SOLUTION

Intensity at C due to charge at A is

$$\frac{3\times10^{-9}}{4\pi\varepsilon_0 \times (0.15)^2} = \frac{9\times10^9 \times 3\times10^{-9}}{0.15^2} = 1.2\times10^3 \text{ N/C along } CB.$$

Intensity at C due to charge at B is $\dfrac{6\times10^{-6}}{4\pi\varepsilon_0 (0.15)^2} = 2.4\times10^3$ N/C along CA.

Resultant intensity at C is 1.2×10^3 N/C along CA.

Potential at C due to charge at A

$$= \frac{3\times10^{-9}}{4\pi\mu_0 \times 0.15} = \frac{9\times10^9 \times 3\times10^{-6}}{0.15} = 1.8\times10^2 V$$

Potential at C due to charge B

$$= \frac{6\times10^{-9}}{4\pi\varepsilon_0 \times 0.15} = 3.6\times10^2 V$$

Potential at $C = 1.8 \times 10^2 + 3.6 \times 10^2 = 540$ V

At the midpoint, the electric intensity$= 1.2 \times 10^3$ N/C and the electrical potential $= 540$ volts.

EXAMPLE 10.6

Positive charges of 2×10^{-9} C, 4×10^{-9} C and 8×10^{-9} C are placed at the corners A, B and C of a square ABCD of 20 cm side as shown in Fig. 10.25. Find the work required to transfer a unit positive charge from D to the centre of the square. O represents the centre of the square ABCD.

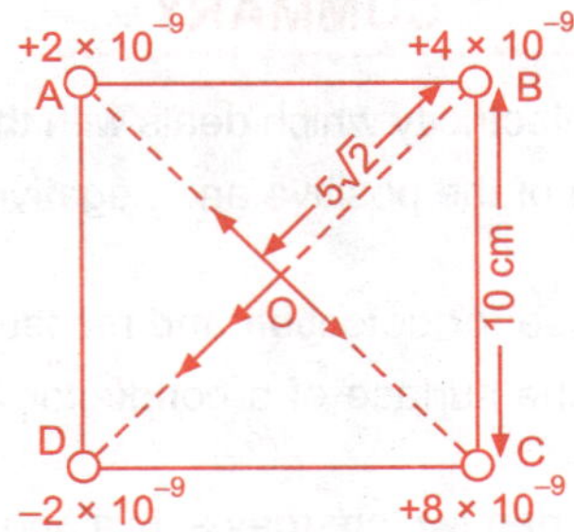

Fig. 10.25 *Electric charges at four corners of square*

SOLUTION

Potential at D = sum of the potentials due to the charges at A, B and C

$$V_D = \frac{1}{4\pi\varepsilon_0}\left[\frac{2\times10^{-9}}{20\times10^{-2}} + \frac{4\times10^{-9}}{20\sqrt{2}\times10^{-2}} + \frac{8\times10^{-9}}{20\times10^{-2}}\right]$$

$$= \frac{9\times10^9\times10^{-9}}{10^{-2}} = \left[\frac{2}{20} + \frac{\sqrt{2}}{20} + \frac{8}{20}\right] = 577.2 \text{ volts.}$$

Similarly, potential at O is given by

$$V_O = \frac{1}{4\pi\varepsilon_0}\left[\frac{2\times10^{-9}}{10\sqrt{2}\times10^{-2}} + \frac{4\times10^{-9}}{10\sqrt{2}\times10^{-2}} + \frac{8\times10^{-9}}{10\sqrt{2}\times10^{-2}}\right]$$

$$= 890.7 \text{ volts.}$$

Work required to transfer unit positive charge (one coulomb) from D to O is the difference between potential at these two points.

Work required = $890.7 - 577.2 = 313.5$ joules.

EXAMPLE 10.7

A parallel plate capacitor has two plates of sides 5.5 cm and 4 cm. Their distance apart is 0.7 mm. The dielectric constant of the medium in between is 4. Find the capacity of the capacitor.

SOLUTION

Capacity

$$C = \frac{\varepsilon A}{t} = \frac{\varepsilon_0 \varepsilon_r A}{t}$$

$$= \frac{8.85\times10^{-12}\times4\times5.5\times4\times10^{-4}}{0.7\times10^{-3}} = 27.7 \,\text{pF}$$

Capacity of the capacitor = $27.7 \,\text{pF}$.

SUMMARY

1. Electrostatics is the branch of electricity which deals with the study of stationary electric charge.
2. Electrification is the separation of the positive and negative charges which are already present on the body.
3. Leaf electroscope device is used for detection and measurement of charge.
4. The distribution of charge on the surface of a conductor is measured in terms of the surface density of charge.
5. **Coulombs law:** The force between two charges acts along the line between the charges, and is proportional to the product of the charges and to the inverse square of the distance between them. The force is repulsive for charges of the same sign and attractive for charges of opposite sign.
6. Electric intensity at a point is the force experienced by a unit positive charge placed at that point.
7. The electric intensity is reduced when air is replaced by a dielectric.
8. The potential at a point is reduced when air is replaced by a dielectric.
9. The capacity of a capacitor increases when air is replaced by a dielectric.
10. Electric polarization is the process by which the atoms or molecules in the dielectric are polarized.
11. Lines of magnetic force originate from a positive charge and terminate on a negative charge.
12. The total number of lines of induction passing normally through any surface is called the flux of electric induction or total normal electric induction (TNEI).
13. Gauss' theorem states that the total normal electric induction across any closed surface in an electric field is numerically equal to the total charge enclosed by the surface.
14. The potential at any point in an electrostatic field can be considered as that physical condition of the field, which determines the direction of flow of positive charge placed at that point.
15. Potential difference between two points is measured by the work done in moving unit positive charge from one point to the other.
16. Electrostatic potential at any point may be defined as a quantity, whose negative gradient in any direction gives the strength of the field in that direction.
17. Any line or surface passing through the points of constant potential is called as equipotential line or equipotential surface.
18. The dielectric strength refers to the breakdown voltage of a dielectric. It is the maximum voltage below which it behaves as a dielectric.
19. Capacitor or condenser is used to store charges.
20. If the capacitors are connected in parallel, the equivalent capacity is equal to the sum of the individual capacities of the capacitors. Parallel connection of a number of capacitors helps in storing larger quantity of electricity.
21. If the capacitors are connected in series, the reciprocal of the equivalent capacity is equal to the sum of the reciprocals of the individual capacities of the capacitors.
22. Van De Graff generator is used to generate high voltages of near about million volts. The generated voltages are used to speed up the particles like ions etc.

Current Electricity

11.1 INTRODUCTION TO CURRENT ELECTRICITY

11.1.1 Current

Electrostatics dealt with charges at rest in the last Chapter whereas current electricity deals with charges in motion. If a potential difference exists between two points of a conductor, there will be a flow of charge from the point of higher charge to the point of lower charge until the potential difference disappears. If the flow of charge between the two points can be maintained, such a flow of charge establishes an electric current. Rate of flow of charge is called the strength of the current. If the flow of charge is constant with respect to time, the current i is given by

$$i = \frac{q}{t}, \text{ or } q = it \tag{11.1}$$

However, if the rate of flow of charge with respect to time is not constant, then the current is given by

$$i = \frac{dq}{dt} \tag{11.2}$$

where dq is the amount of charge that flows through any cross section of the conductor in a time dt. The current i does not change if the cross section changes along the conductor, because the charge must be conserved. While electrons carry charge in metals, ions carry the charge in electrolytes or gaseous media.

When equal quantity of electricity flows across any cross section of a conductor in equal intervals of time, the current will be steady. Normally there are two ways of showing of current flow in the circuits, namely, conventional and conductional current. Conventional current flows from the positive terminal to the negative terminal although the movement of electron is in the opposite direction from the cathode to anode denoted as conductional current. The direction of flow of positive charge from a point at a higher potential to a point at a lower potential gives potential direction of current. Therefore, the direction of current is opposite to that of the flow of electrons. It must be remembered that the flow of positive charge in a given direction is equivalent to the flow of negative charge in the opposite direction.

In the Eq. (11.1), given by $i = \dfrac{q}{t}$, we have $i = 1$, when $q = 1$ and $t = 1$.

The strength of current through a conductor is one unit when the rate of flow of charge through any cross section of the conductor is unity. The practical unit of current is called ampere, named after Ampere, a French physicist. The strength of current is one ampere if the rate of flow of charge is one coulomb per sec.

Current is not measured directly by measuring charge and time but by the effects produced by the current such as the magnetic effects.

11.1.2 Electromotive Force (emf)

A continuous flow of current can be maintained with the help of an external agency that should supply energy to transfer the charge from one point to another point. In order to maintain a continuous current, a closed circuit must be established where, any other form of energy is converted into electrical energy. Such an external agency is a source of current or a source of electromotive force. An electromotive force causes differences of potential to exist between points in a circuit. The electromotive force of a source of current is the energy required to transfer or drive a unit charge through a closed circuit, in a reversible process.

Two Italian scientists named L. Galvani and A. Volta, discovered towards the end of 18th century that a device called cell could convert chemical energy into electrical energy. The phenomenon of electromagnetic induction, discovered independently by Michael Faraday, a British scientist and Joseph Henry, an American scientist, led to the development of electric generators for the conversion of mechanical energy into electrical energy. Other sources of emf were also discovered such as thermocouples in which heat is converted into electrical energy; photoelectric cells in which radiant energy is converted into electrical energy. A fuel cell utilises a continuous supply of hydrogen or hydrocarbon fuel to produce electrical energy. Piezo-electric effect is another phenomenon that can convert mechanical energy into electric energy or vice versa. A piezo-electric material produces electrical energy from mechanical energy when it is subjected to mechanical forces that deform the material.

11.1.3 Potential Difference

From Eq. (11.1), we know that current is related to charge by the equation,

$$i = \frac{q}{t}$$

Potential difference between the two ends of a conductor is the work done per second to maintain unit current between the ends.

The unit of potential difference is the volt, named after Volta, an Italian physicist. Potential difference between two points of a conductor is one volt if one joule of work has to be done per second to maintain a current of one ampere between the points.

11.1.4 Effects of Electric Current

When an electric current passes through a conductor, the following effects are produced:

1. **Magnetic effect:** When an electric current passes through a conductor, a magnetic field is produced around it. This will be discussed in more detail in the next section.

2. **Heating effect:** When electric current passes through a conductor, the conductor gets heated. The heating effect of current is utilized both for domestic and industrial purposes such as in electric lamps, electric heaters, arc welding etc.

3. **Chemical effect:** When an electric current passes through liquid conductors such as solutions of salts, bases and acids, dissociated radicals move under the influence of electric fields. Such chemical effect of current has extensive use in metallurgy, electrotyping, electroplating etc.

The above three effects may be demonstrated by an electrical circuit shown in Fig. 11.1 containing a thick copper wire MN, a glass vessel V with two platinum electrodes A and K dipped in acidulate water and a lamp L.

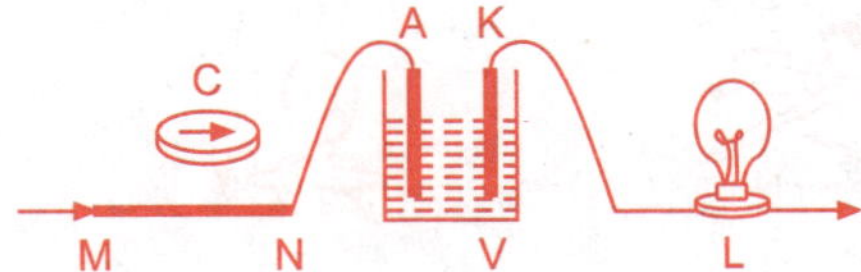

Fig. 11.1 *Effects of electric current*

A compass needle placed over the wire MN will deflect when a current is passed through the circuit, showing the magnetic effect of current. The filament of the lamp is heated to incandescence and the lamp glows indicating heating effect of current. Bubbles of gas are found to rise from the surfaces of the electrodes A and K. Hydrogen is given off at A and oxygen at K and water in the vessel V decreases showing the chemical effect of electric current.

11.2 MAGNETIC EFFECTS OF CURRENT

When current flows through a conductor, a magnetic field is induced around it. If a magnetic compass is kept near the conductor, its needle will be defected by the magnetic field. This magnetic effect was first discovered by the Danish scientist, Hans Christian Oersted, in 1820. Today there are many electrical appliances that operate based on the current induced magnetic field.

11.2.1 Direction of Magnetic Field

The direction of the magnetic field for a given direction of the current can be identified by any one of the three methods as follows. (*a*) According to Ampere's swimming rule, if a man swims along the wire carrying current with his face always towards the magnetic needle, and the current flowing from his feet towards his head then the North Pole of the magnetic needle is always deflected towards his left hand. The direction of motion of the North pole indicates the direction of the induced magnetic field. This is shown in Fig. 11.2 (*a*). (*b*) Imagine a right-handed corkscrew being rotated along the wire in the direction of the current. The direction of rotation of the thumb gives the direction of the magnetic lines of force. This is shown in Fig. 11.2 (*b*). (*c*) Imagine clasping a conductor by the right hand so that the extended thumb points in the direction of the current. The direction of fingers wrapped around the conductor as shown in Fig. 11.2 (*c*), gives the direction of the magnetic field.

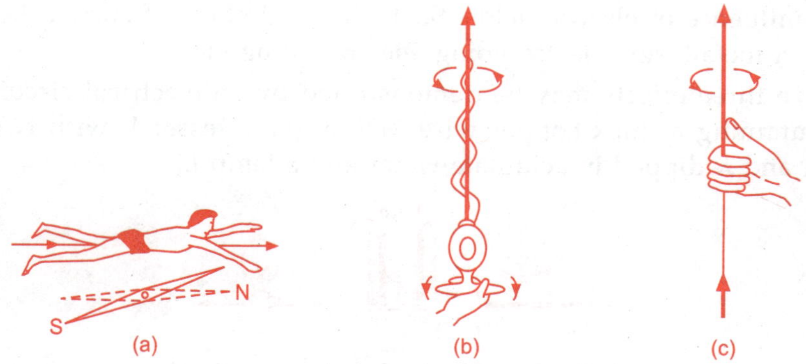

Fig. 11.2 *Direction of the current induced magnetic field*

When the conductor is in the form of a coil, the direction of the magnetic field can be obtained by following the 'clock rule' shown in Fig. 11.3. The end or face of the coil, in which current appears to flow clockwise acts like a magnetic south pole. If the current flows anticlockwise the face acts like a north pole.

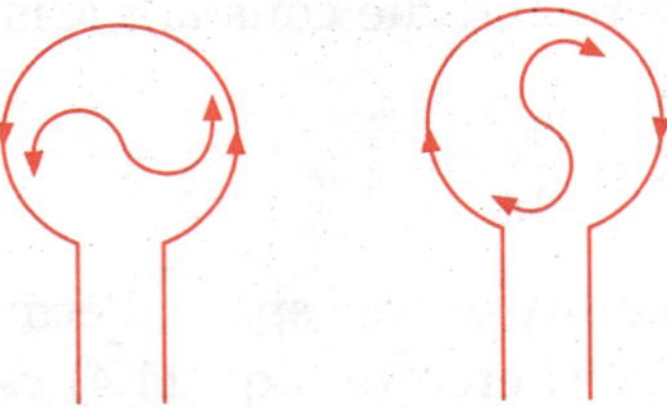

Fig. 11.3 *Magnetic field in a coil carrying current*

11.2.2 Laplace's Rule (or law)

The intensity of magnetic field around a current carrying conductor was first studied by Laplace, a French physicist and later by Ampere. The magnitude of the field is given by Laplace's law which was initially validated experimentally by Biot and Savart. Therefore, the law is sometimes called as Biot-Savart law.

The law describes the magnetic field generated by an electric current and relates the magnetic field to the magnitude, direction, length and proximity of the electric current. A conductor carrying a current I is shown in Fig. 11.4, where P is a point at a distance r from a small element (length δl) and θ is the angle between the direction of the element and the line joining P to it. The product $I\delta l$ is called a current element.

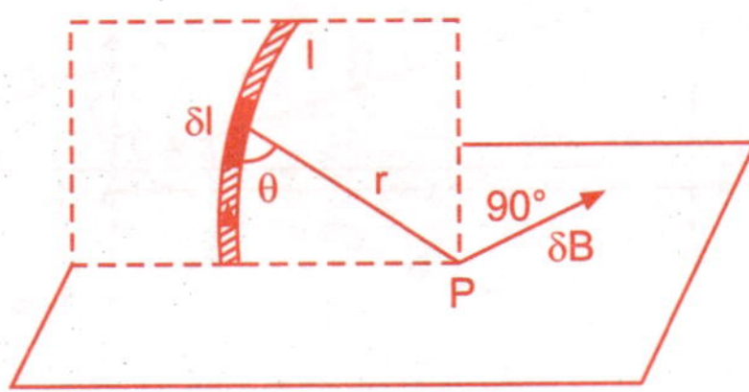

Fig. 11.4 *A current element*

Laplace's rule states that the magnetic field at a point due to a current element is (*i*) directly proportional to the strength of the current, (*ii*) directly proportional to the length of the element, (*iii*) directly proportional to the sine of the angle between the element and the line joining the point to it, and (*iv*) inversely proportional to the square of the distance of the point from the element.

If the magnetic field, at P due to the current element $I\delta l$ is δB then

$$\delta B \propto \frac{I\delta l \sin\theta}{r^2} \tag{11.3}$$

or,
$$\delta B = \frac{KI\delta l \sin\theta}{r^2} \tag{11.4}$$

where K is a constant whose value depends upon the system of units chosen. It must be noted that the direction of the field δB at P due to the current element $I\delta l$ is perpendicular to the plane containing the element l and the point P as shown in Fig. 11.4.

The unit of current is ampere and the constant K is given by

$$K = \frac{\mu_o}{4\pi}$$

for free space, μ_o being permeability of free space. Permittivity for air can be assumed as μ_o, for all practical purposes. Therefore, Eq. (11.4) can be written as

$$\delta B = \frac{\mu_o}{4\pi} \cdot \frac{I\delta l \sin\theta}{r^2} \tag{11.5}$$

Magnetic intensity, δF can be written as

$$\delta F = \frac{I\delta l \sin\theta}{4\pi r^2} \tag{11.6}$$

11.2.3 Field at any point along the Axis of a Circular Coil Carrying Current

In Fig. 11.5, P is a point at a distance x along the axis, from the center of a circular coil of n turns and radius r.

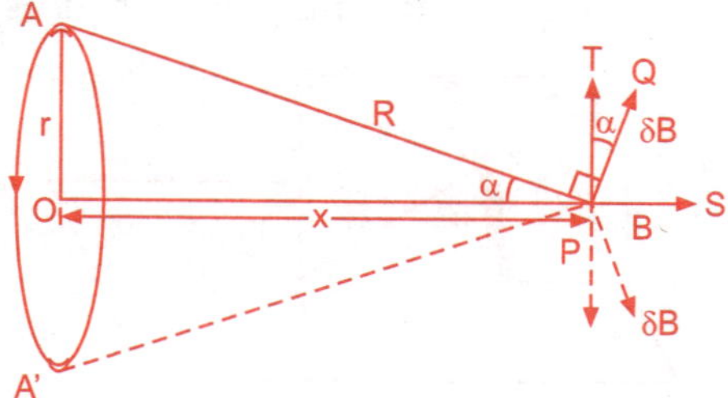

Fig. 11.5 *Field at any point along the axis of a circular coil carrying current*

If the coil carries a current I (ampere), the magnetic intensity δB at P due to a small element δl at A of one turn of the coil is given by Laplace equation,

$$\delta B = \frac{\mu_0}{4\pi} \cdot \frac{I\delta l \sin\theta}{R^2} \tag{11.7}$$

where $R = PA$ and θ, the angle made by PA with the elements is 90° since PA is perpendicular to δl. Therefore,

$$\delta B = \frac{\mu_0}{4\pi} \cdot \frac{I\delta l}{R^2} \tag{11.8}$$

The direction of δB is along PQ perpendicular to the plane containing the element dl and the point P, i.e., δB is in the plane PAO and acts perpendicular to the line PA. δB at P can be resolved into two components; (*i*) $\delta B \sin\alpha$ along OP or PS, and (*ii*) $\delta B \cos\alpha$ along PT parallel to OA where α = angle OPA.

The field due to another element δl at A' diametrically opposite to A, has a component $\delta B \sin \alpha$ along OP, but its component $\delta B \cos \alpha$ perpendicular to OP is in a downward direction. The two components $\delta B \cos \alpha$ and $\delta B \cos \alpha$ cancel each other as they are in opposite directions. Therefore, the field at P due to the diametrically opposite elements at A and A', acts along OP and is equal to $2\delta B \sin \alpha$. Hence, the resultant field at P due to the whole coil is directed along OP and its magnitude, B, is given by

$$B = \Sigma 2\delta B \sin \alpha \tag{11.9}$$

Here the summation is taken half way round the coil, because $2\delta B \sin \alpha$ includes contribution due to two diametrically opposite elements. Substituting the value δB from Eq. (11.8) into Eq. (11.9), we get

$$B = \Sigma 2\left(\frac{\mu_o}{4\pi} \cdot \frac{I\delta l}{R^2}\right) \sin \alpha = \frac{\mu_o}{2\pi} \cdot \frac{I \sin \alpha}{R^2} \Sigma \delta l \tag{11.10}$$

because I, R and α are constants. But $\Sigma \delta l$ = length of the conductor in half the coil and is equal to $\pi r n$ and hence

$$B = \frac{\mu_0}{2\pi} \frac{1}{\left(r^2 + x^2\right)} \frac{r}{\left(r^2 + x^2\right)^{1/2}} \pi r n \tag{11.11}$$

$$B = \frac{\mu_0 n r^2 I}{2\left(r^2 + x^2\right)^{\frac{3}{2}}} \tag{11.12}$$

Magnetic intensity, F, is given by

$$F = \frac{n r^2 I}{2\left(r^2 + x^2\right)^{\frac{3}{2}}} \tag{11.13}$$

The field is directed along the axis.

In the special case, when $x = 0$, the field is given by

$$B = \frac{\mu_o n I}{2r} \tag{11.14}$$

And the magnetic intensity F at the centre is given by

$$F = \frac{n I}{2r} \tag{11.15}$$

where, I = current in amperes, n = number of turns and r = radius of the coil. From Eq. (11.15), it can be noted that the intensity of a magnetic field can be expressed in amp-turns/meter or ampere/meter.

11.2.4 Tangent Galvanometer

A single coil tangent galvanometer is employed for the measurement of electric current. The operation of the instrument is based on the fact that current through a circular coil produces a magnetic field at the centre of the coil in a direction perpendicular to the plane of the coil, and the field is proportional to the current.

A single coil tangent galvanometer is shown in Fig. 11.6(*a*). It consists of a circular coil of insulated copper wire, having *n* number of turns. The coil is wound over a circular frame of wood or ebonite. The frame is fixed in a vertical plane on a horizontal base provided with levelling screws.

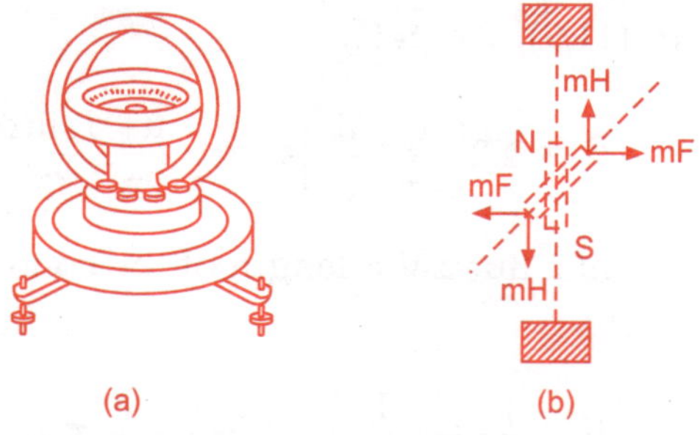

Fig. 11.6 *Tangent galvanometer*

The coil can be rotated about a vertical axis passing through the center of the coil. The ends of the coil are connected to binding screws provided at the base. A deflection magnetometer compass box is mounted horizontally at the center of the coil as shown in Fig. 11.6. The instrument must be levelled using the levelling screws so that the base is perfectly horizontal and hence the coil becomes vertical. Also, a spirit level can be used to ensure that the base is horizontal. The magnet swings freely and is perfectly parallel to its image.

The plane of the coil must be set in the magnetic meridian. This can be accomplished by first rotating the compass box until the 90 – 90 line is in the plane of the coil, and then rotating the coil, as a whole, so that the pointer reads 0° – 0°. The magnetic intensity at the centre of a circular coil carrying current is perpendicular to the plane of the coil and its magnitude is given by Eq. (11.15).

When there is no current through the coil, the magnetic needle will align in the direction of H which is horizontal intensity of earth's magnetic field. When a current I flows through the coil, with its plane in the magnetic meridian, a field of intensity F is set up at the center of the coil, in a direction perpendicular to the plane of the coil. The needle under the action of the two perpendicular fields F and H, is deflected through an angle θ (Fig. 11.6*b*).

The deflection θ is given by the relation

$$F = H \tan \theta \tag{11.16}$$

Using Eq. (11.15), we can write Eq. (11.16) in the form

$$\frac{nI}{2r} = H \tan \theta. \qquad (11.17)$$

Therefore, current I is obtained as

$$I = \frac{2rH}{n} \tan \theta \qquad (11.18)$$

or, the current is expressed as

$$I = K \tan \theta \qquad (11.19)$$

where the constant K is called the reduction factor of the tangent galvanometer, because it helps to reduce the deflection to current. The constant K depends on a given galvanometer at a particular place and varies from place to place as it depends on H. From Eq. (11.19), we have $I = K$, when $\theta = 45°$. This can be used to define the reduction factor as the current required to produce a deflection of 45°, when the coil is set in the magnetic meridian.

The current through the coil is proportional to the tangent of the deflection of the magnetic needle and hence the name for the galvanometer as tangent galvanometer. When the plane of the coil is perpendicular to the magnetic meridian, the needle will not be deflected, as the two fields will be along the same line.

The readings in a tangent galvanometer must be taken carefully with the following precautions in mind:

(*i*) readings must be taken at both the ends of the pointer in order to avoid error due to eccentricity, and (*ii*) both the ends of the pointer must be read again after reversing the current in order to avoid error due to improper mounting of the coil away from the magnetic meridian. The average of the readings must be taken as the deflection. For reasonable averaging of the errors, four readings may be considered for averaging.

11.2.5 Experiments with Tangent Galvanometer

In order to measure the current using a tangent galvanometer, it is necessary to first determine the reduction factor of the galvanometer. The reduction factor of a tangent galvanometer can be determined using an ammeter as follows.

In Fig. 11.7, G is the tangent galvanometer, A is the ammeter, B is the battery, K is the plug key, R is a rheostat and C is a commutator. The commutator is used in order to reverse the current as needed. The galvanometer is kept sufficiently away from the disturbing fields of the ammeter and the rheostat.

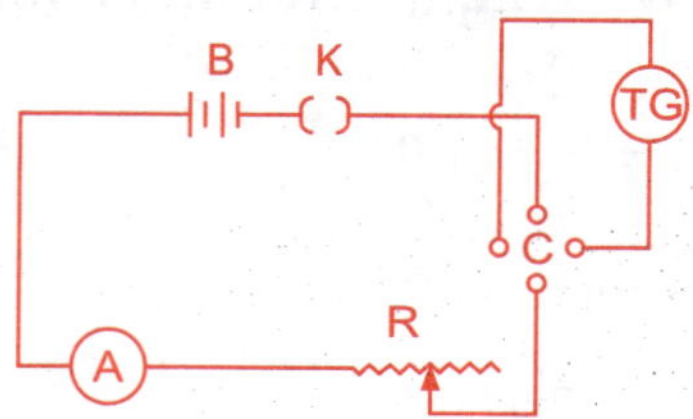

Fig. 11.7 *Determining reduction factor in tangent galvanometer*

The rheostat is adjusted so that the deflection is close to 45°. After noting the readings θ_1 and θ_2 of the pointer, the current is reversed using the commutator, and the readings θ_3 and θ_4 are noted. The average of the four readings constitutes the mean deflection θ. The current I in the ammeter is noted. The reduction factor K is obtained using the Eq. (11.19) as

$$K = \frac{I}{\tan\theta}.$$
(11.20)

Earth's horizontal intensity H at the plane can be calculated using Eq. (11.18) as

$$H = \frac{nK}{2r}.$$
(11.21)

After determining the reduction factor using Eq. (11.20), the galvanometer must be calibrated in order to determine whether any correction factor needs to be considered for the current measured using the galvanometer. The average deflection, θ is found as explained in the above experiment and the current I' as measured by the galvanometer is calculated using the relation

$$I' = K \tan\theta.$$
(11.22)

The difference $(I - I')$ gives the correction for the current measured by the galvanometer. The correction for different values of current are found, which should be used to correct the measurements taken using the tangent galvanometer.

The principle of operation of tangent galvanometer is based on the following factors:

1. The magnetic field F is at the centre of a circular coil carrying current along the axis.

2. When this field is at right angles to another field H in the same plane, a magnetic needle at the centre, in the plane of the fields, is deflected through an angle given by $F = H \tan\theta$.

Therefore, the following limitations must be kept in mind when using a tangent galvanometer.

(*i*) As F decreases with distance from the centre, the field is not uniform over the entire region of the needle.

(*ii*) As $I = K \tan \theta$, I depends on $\tan \theta$, $\tan \theta$ changes by a large amount, even for a small change in θ.

(*iii*) Measurement of I is accurate in the neighbourhood of $\theta = 45°$.

(*iv*) For accuracy, r must be small and n must be large to the extent practically possible.

11.3 FORCE ON A CONDUCTOR CARRYING CURRENT IN A MAGNETIC FIELD

Soon after the discovery of the magnetic effect, André-Marie Ampère, French mathematician and physicist, showed that a current carrying conductor placed in a magnetic field experiences a mechanical force. As a result, if the conductor is free to move, it will move in the direction of the force. It was observed that

(*i*) No force acts on the conductor when the magnetic field and the current are parallel

(*ii*) The force is maximum when the field and the current are perpendicular to each other, and

(*iii*) The force acts in a direction perpendicular to the plane containing the current and the field.

The force, f, acting on a conductor of length L, carrying a current I and placed in a uniform magnetic field of flux density B, is given by

$$f = BIL \sin \theta \qquad (11.23)$$

where, θ is the angle between I and B. The force has a maximum value at $\theta = 90°$. If the conductor is held perpendicular to the field, the direction of the mechanical force can be found using Fleming's left hand thumb rule (FLHT rule), stated as follows:

If the first three fingers of the left hand, identified as Main finger (thumb), Fore finger, and Central finger (middle finger), are held at right angles to each other such that the Central finger points in the direction of the Current and the Fore finger in the direction of the Field, then the Main finger points in the direction of Motion or Mechanical force, as shown in Fig. 11.8.

The Fleming's left hand thumb rule is also called motor rule.

It was already mentioned that a charge in motion constitutes a current. Therefore, a charged particle moving in a magnetic field experiences a mechanical force. If a positive charge q moves with a velocity v perpendicular to the direction of a uniform magnetic field B, it experiences a force f given by

$$f = qvB \qquad (11.24)$$

in a direction perpendicular to both velocity and the direction of B. In general f is given by

$$f = qvB \sin \theta \qquad (11.25)$$

where θ is the angle between the directions of v and B. We have $f = 0$ when $\theta = 0$. Therefore, direction of magnetic field at a point can be defined as the direction of the velocity of a positive charge in motion at that point such that it experiences no force.

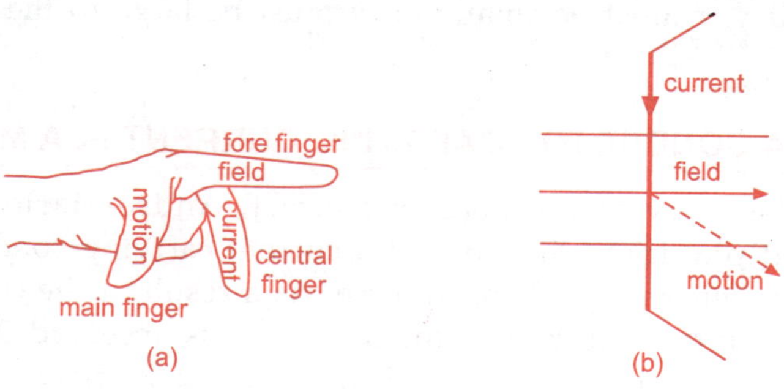

Fig. 11.8 *Fleming's left hand thumb rule*

11.3.1 Definition of Ampere

Ampere showed that, there was a force of attraction between two parallel wires carrying current in the same direction, and a force of repulsion between two wires carrying currents in the opposite direction. If I_1 and I_2 are the currents carried by two long parallel wires distance d apart, the force f acting on unit length of either conductor is given by

$$f = \frac{\mu_0 I_1 I_2}{2\pi d} \,.$$

(11.26)

When $I_1 = I_2 = 1$ ampere and $d = 1$ meter

$$f = \frac{\mu_0}{2\pi} = 2 \times 10^{-7} \text{ newton.}$$

The definition of ampere is based on the above relation. Ampere is the strength of the current that flows through two parallel infinitely long conductors of negligible cross section, placed in vacuum at a distance of one meter from each other, and produces between them a force of 2×10^{-7} newton per meter length.

11.3.2 Force on a Current Carrying Conductor Placed in a Magnetic Field

Current passing through a conductor produces a magnetic field around it. Referring to Fig. 11.9, the magnetic intensity at a point P due to a current element IdL is given by Eq. (11.6) as

$$\delta F = \frac{IdL \sin \theta}{4\pi r^2}$$

(11.27)

according to Laplace rule. If a magnetic pole of strength m is placed at P, the force on the pole is given by

$$f = \frac{mIdL\sin\theta}{4\pi r^2} \tag{11.28}$$

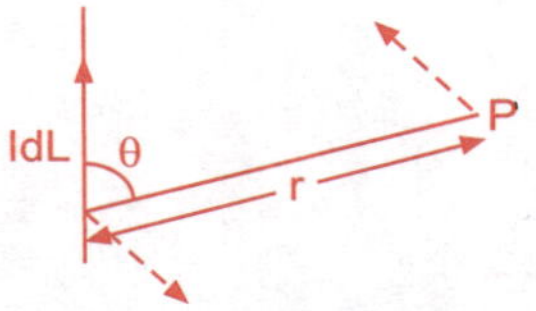

Fig. 11.9 *Force on a current carrying conductor placed in a magnetic field*

The pole m exerts an equal and opposite force on the conductor. If the conductor is free to move, it will move due to this force of reaction. At $\theta = 90°$, the force is given by Eq. (11.28) as

$$f = \frac{mIdL}{4\pi r^2} \tag{11.29}$$

The magnetic intensity, F, at the current element due to the pole is given by $\frac{m}{4\pi\mu r^2}$ where μ is the permeability of the medium. Consequently, force f can be written as

$$f = \frac{\mu mIdL}{4\pi\mu r^2} = \mu FIdL = BIdL \tag{11.30}$$

where $B = \mu F$ is the flux density of the magnetic field at the current element, due to the pole.

Therefore, a conductor of length L carrying current I, when placed at right angles to a uniform magnetic field of flux density B, experiences a force

$$f = BIL \tag{11.31}$$

in a direction given by Fleming's left hand thumb rule.

11.3.3 Moving Coil Galvanometer

Moving coil galvanometer is an electromechanical instrument which is used for the detection of electric currents through electric circuits. It is a sensitive instrument, and hence cannot be used for the measurement of heavy currents. We can measure very small currents by using galvanometer but the primary purpose of a moving coil galvanometer is the detection of electric current and not the measurement of current. The principle of operation of the moving coil galvanometer is based on the mechanical force experienced by a current carrying conductor when placed in a magnetic field. The coil of the galvanometer is suspended or mounted such that it can move freely in the magnetic field due to a fixed magnet. The instrument is quite sensitive and can detect very small currents of the order of 10^{-8} to 10^{-10} amps.

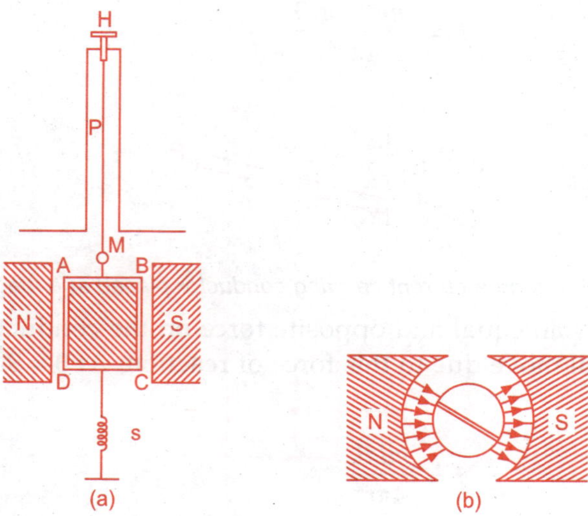

Fig. 11.10 *Moving coil galvanometer*

A schematic of a moving coil galvanometer is shown in Fig. 11.10. It is of the suspended type and consists of a rectangular coil *ABCD*, of several turns of insulated copper wire, suspended between the poles, *N* and *S*, of a powerful permanent magnet by means of a phosphor bronze strip *P*. The lower end of the coil has a small spring s and the upper end is attached to a torsion head *H* as shown in the Fig. 11.10. A circular strip of mirror *M*, attached to the suspension wire helps to measure the deflection of the coil by a lamp and scale arrangement.

A cylindrical piece of soft iron is mounted within the frame of the coil as shown in Fig. 11.10(*b*), in order to intensify the magnetic field. The pole faces of the magnet, *N* and *S*, are curved in such a way as to form segments of circles which are concentric with that of the cylinder. This makes the field to act in a radial direction in the gap between the pole faces and the cylinder. The plane of the coil lies parallel to the magnetic field for any position of the coil. This galvanometer is known as D'Arsonval galvanometer, named after the inventor.

The working of the galvanometer is described as follows. When there is no current flowing through the coil, the coil comes to rest in a position where the suspension wire is free from any twist. When current is passed through the coil, the vertical sides of the coil are subjected to forces which are equal and opposite and are parallel, acting in a direction perpendicular to the plane of the coil. The equal and opposite force pair constitute a couple, and rotate the coil producing a twist in the coil. When the moment of the couple due to the current and the couple due to the twist balance each other, the coil comes to rest. The moment of the couple due to the current is proportional to the current in the coil and the twisting couple is proportional to the deflection of the coil. Hence, the deflection in the coil gives a measure of the current through the galvanometer.

Referring to Fig. 11.11, B is the magnetic field produced by the magnet and I is the current flowing through the coil. As shown in Fig. 11.11, each vertical side of the coil experiences a force f which is perpendicular to the plane of the coil. The direction of the force is given by Fleming's left hand rule (FLHT rule).

Extending the Eq. (11.31), to the case of n coils, the magnitude of the force f is given by

$$f = nBIa \tag{11.32}$$

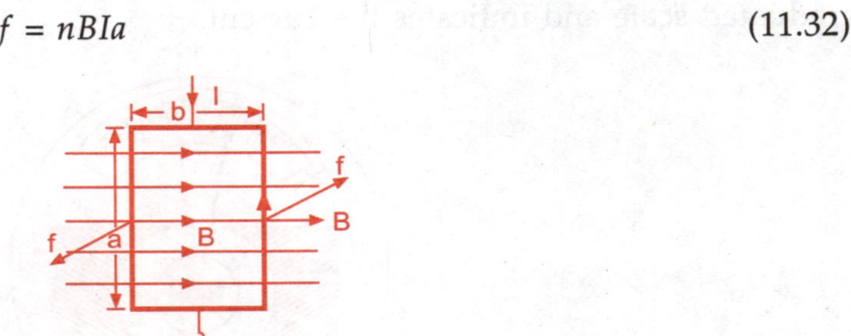

Fig. 11.11 *Directions of the field, current and the force in a coil*

The moment of the couple formed by the two equal and opposite parallel forces is given by the vector product of the force, f, and the moment arm, b, as

$$f \times b = nBIa \times b = nBIA \tag{11.33}$$

where $A = ab$, is the area of the coil. This couple is balanced by the couple due to twist in the suspension wire. If c is the couple per unit twist of the suspension wire, which is the torsional stiffness of the wire, the moment of twisting couple is given by $c\theta$. Consequently, for equilibrium, we have

$$nBIA = c\theta \tag{11.34}$$

And hence the current is given by

$$I = \frac{c\theta}{nBA} = K\theta \tag{11.35}$$

where $K = \dfrac{c}{nBA}$ is constant for the galvanometer. This shows that the current is proportional to the deflection. The constant K is called the current sensitivity of the galvanometer.

11.3.4 Table Galvanometer or Pointer Galvanometer

Suspended coil galvanometer is a sensitive instrument and can be used for the measurement of very small currents, however, it is not very convenient because it is not portable. Therefore, it cannot be used for field measurements. Pointer galvanometer is a modification of suspended coil galvanometer which is compact and portable.

The construction of the Pointer galvanometer is similar to that of a suspended coil galvanometer. Figure 11.12 shows a sectional view of the instrument. The coil, instead of being suspended by a wire, is mounted on jewelled bearings. Instead of phosphor bronze strip, two spiral springs are attached to the coil at its ends, one at the top and the other at the bottom. These provide the necessary restoring couple and also serve as leads for current. A pointer attached to the coil moves over a graduated scale and indicates the current.

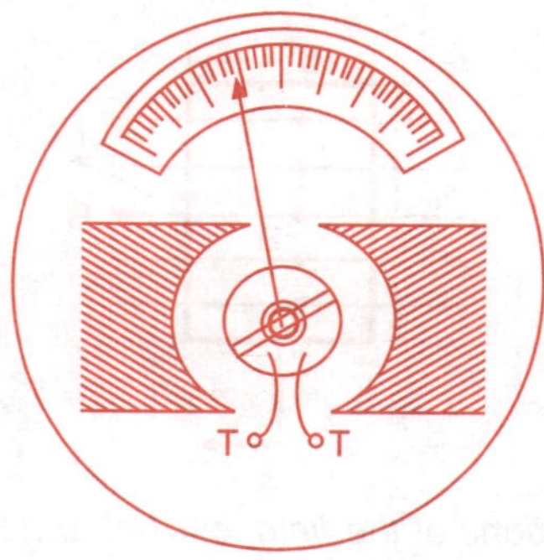

Fig. 11.12 *Table galvanometer or Pointer galvanometer*

11.3.5 Use of Galvanometer as Ammeter

Ammeters are used to measure current flowing through a circuit. An ammeter is essentially modified form of a galvanometer. In order to measure the strength of current in a circuit, it must be connected in series. An ideal ammeter must have zero internal resistance, so that the drop in voltage in the ammeter is as small as possible. This is to ensure that the ammeter connected in series does not change the current in the circuit. However, practical ammeters will have some resistance, which must be kept as small as possible. In principle, ammeter is a galvanometer with a suitable shunt which has low resistance connected in parallel to carry a large portion of the current as shown in Fig. 11.13.

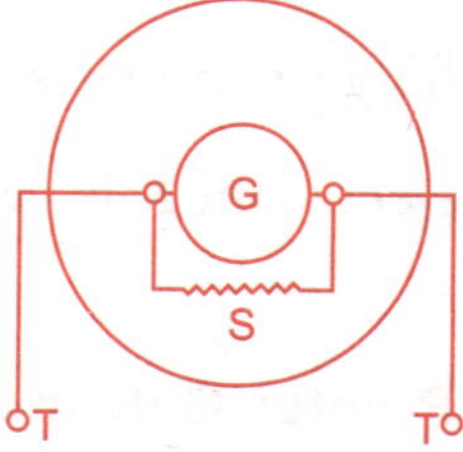

Fig. 11.13 *Galvanometer as ammeter to measure current*

When the current flowing through a circuit is I, let I_g be the portion of current flowing through the galvanometer and I_s be the portion of current flowing through the shunt resistance. Let the galvanometer produce a full scale deflection for the current I_g through it. Since, we want to measure the total current I, the reading on the galvanometer must be calibrated to read the total current I directly such that, an ammeter produces a full scale deflection for a current I (the scale must read in the range $0 - I$). In order to accomplish this, the value of the shunt S must be calculated as follows.

When a current I flows through the circuit (and hence through the instrument), only current I_g flows to the galvanometer coil giving full scale deflection, and the rest of the current should flow through the shunt. As the galvanometer coil and the shunt are in parallel, from Ohm's law we have equal voltage drops across the galvanometer and the shunt and consequently we have the relation

$$(I - I_g)S = I_g G \tag{11.36}$$

where G is the resistance of the galvanometer. Hence, the value of the shunt resistance is obtained as

$$S = \frac{I_g G}{I - I_g} \tag{11.37}$$

It follows from Eq. (11.37), that

$$I = \frac{I_g(G+S)}{S}. \tag{11.38}$$

Therefore, in order to get the circuit current, the galvanometer current should be multiplied by a factor K_{ga} given by

$$K_{ga} = \frac{G+S}{S} \tag{11.39}$$

Hence, the galvanometer scale can be graduated to read the main current I directly.

11.3.6 Use of Galvanometer as Voltmeter

A galvanometer can also be used as a voltmeter with suitable modifications in order to measure the potential difference between two points in an electrical circuit. In order to measure potential difference, it must be connected across the points between which potential difference is measured. The galvanometer is connected in parallel to the main circuit and any change in potential difference after connecting the instrument must be negligible. The current flowing in the main circuit should not be appreciably affected when the voltmeter is connected and the current drawn by a voltmeter must be negligibly small. Therefore, voltmeter must have a very high resistance. This is accomplished by adding a suitable high resistance in series to a galvanometer as shown in Fig. 11.14, where the main circuit passes through the two points denoted as T.

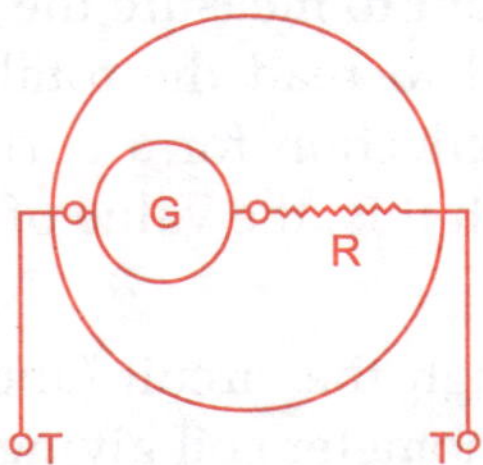

Fig. 11.14 *Use of galvanometer as voltmeter*

In order to convert a galvanometer, producing full scale deflection for a current I_g through it into a voltmeter, producing full scale deflection for a potential difference, V, (i.e. range $O - V$), the value of required high resistance, R, can be calculated as follows.

As the resistance in the galvanometer coil and the high resistance are in series, the total voltage drop across the instrument is given by

$$V = I_g (G + R) \qquad\qquad (11.40)$$

$$R = \frac{V}{I_g} - G \qquad\qquad (11.41)$$

where G is the resistance of the galvanometer and R is the external resistance connected in series with the galvanometer G. From Eq. (11.40), it is seen that in order to get the voltage, the galvanometer current should be multiplied by a factor given by

$$K_{gv} = (G + R) \qquad\qquad (11.42)$$

Hence, the scale may be graduated to give the potential difference in volts directly.

EXAMPLE 11.1

A current of 10 amperes passing through a coil of 2 turns produces a magnetic intensity of 25 amp-turn/meter at the centre. Find the radius of the coil.

SOLUTION

$I = 10$ amperes, $n = 2$ and $F = 25$ amp-turn/meter

$$F = \frac{nI}{2r}$$

Therefore, $\qquad\qquad r = \dfrac{nI}{2F} = \dfrac{2 \times 10}{2 \times 2.5} = 0.4$ m.

EXAMPLE 11.2

A circular coil of radius 0.10 m and having 10 turns of wire has a steady current in it. If a neutral point is obtained at its centre, when the plane of the coil is normal to the earth's magnetic meridian, calculate the current. (H = 25 amp-turn/meter).

SOLUTION

For a neutral point, $F = H$; $r = 0.1$ m, $n = 10$

$$\frac{nI}{2r} = H$$

$$I = \frac{2rH}{n} = \frac{2 \times 0.1 \times 25}{10} = 0.5$$

Hence, current through the coil = 0.5 amp.

EXAMPLE 11.3

A coil of 40 turns of mean radius 0.12 m carries a current of 10 amperes. Calculate the intensity of the magnetic field at a point 0.50 m along its axis from the centre of the coil.

SOLUTION

$$F = \frac{nr^2 I}{2\left(r^2 + x^2\right)^{\frac{3}{2}}} = \frac{40 \times 0.12^2 \times 10}{2 \times \left(0.12^2 + 0.5^2\right)^{\frac{3}{2}}} = 21.2$$

Therefore, intensity of magnetic field at the point = 21.2 amp-turn/meter.

EXAMPLE 11.4

A tangent galvanometer, has a coil of 22 turns and mean radius 0.10 m. Find the reduction factor at a place where H = 25 amp-turn/meter.

SOLUTION

$$K = \frac{2rH}{n} = \frac{2 \times 0.1 \times 25}{22} = \frac{5}{22} = 0.227$$

The reduction factor at the place is given by 0.227 A.

EXAMPLE 11.5

Two tangent galvanometers are identical in all respects except for the number of turns of their coils. Their deflections are 60° and 30°, respectively, when they are connected in series with a cell. Obtain the ratio of the number of their turns.

SOLUTION

$$I = K_1 \tan \theta_1 = K_2 \tan \theta_2$$

$$\frac{2rH}{n_1} \tan 60° = \frac{2rH}{n_2} \tan 30°$$

or,
$$\frac{n_1}{n_2} = \frac{\tan 60°}{\tan 30°} = 3$$

Therefore, the ratio of the number of turns = 3:1.

EXAMPLE 11.6

A tangent galvanometer of radius 0.10 m and 100 turns of wire shows a deflection of 45°. An ammeter in the same circuit shows a reading equal to 1/20th of an ampere. Find the value of horizontal component of earth t magnetic intensity.

SOLUTION

$$\theta = 45°, \; K = I = 1/20 \text{A}$$

$$\frac{2rH}{n} = K = I$$

$$H = \frac{1}{20} \times \frac{100}{2 \times 0.1} = 25$$

Hence, the horizontal component of earth's magnetic intensity = 25 amp-turn/meter.

EXAMPLE 11.7

A galvanometer gives full scale deflection with 0.2 mA. The resistance of its coil is 1000W. How can the galvanometer be converted into an ammeter to read currents upto 2 amperes?

SOLUTION

Since the ammeter is required to measure a high current, a shunt resistance must be used as shown in Fig. 11.15.

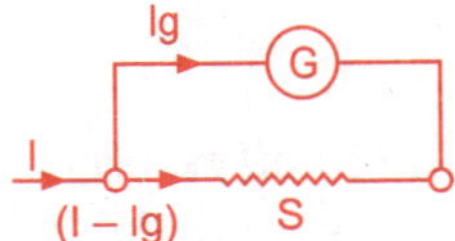

Fig. 11.15 *Galvanometer as ammeter*

The value of the shunt S is given by

$$S = \frac{I_g G}{I - I_g}$$

$$G = 1000\Omega, \; I_g = 0.2 \text{ mS}, \; I = 2A$$

$$S = \frac{0.2 \times 10^{-3} \times 1000}{2 - 0.0002} = \frac{0.2}{2} = 0.1\Omega$$

The conversion is effected by using a shunt of 0.1 Ω

EXAMPLE 11.8

A galvanometer has a resistance of 15 Ω, and gives full scale deflection with a current of one mA. It is required to convert it into a voltmeter with a range 15 volts. Find their resistance that should be connected in series Conversion to voltmeter.

SOLUTION

The high resistance must be connected in series as shown in Fig. 11.16.

Fig. 11.16 *Galvanometer as voltmeter*

$$V = 15 \text{ volts}, \; I_g = 1 \times 10^{-3} A, \; G = 15 \; \Omega$$

$$R = \frac{V}{I_g} - G = \frac{15}{10^{-3}} - 15 = 14985 \; \Omega$$

Therefore, a resistance of 14985 Ω has to be connected in series with the galvanometer.

EXAMPLE 11.9

A volt meter with a full scale deflection upto 10 volts has a resistance of 1000 W. Describe how to convert it into an ammeter reading upto 10 amperes.

SOLUTION

A shunt must be used as shown in Fig. 11.15. The value of the shunt is given by,

$$S = \frac{I_g G}{I - I_g}$$

$$I = 10 \text{ amps}, \ I_g G = V = 10 \text{ volts}, \ G = 1000\Omega$$

$$I_g = \frac{1}{100} = 0.01 \text{ amp.}$$

Hence,
$$S = \frac{10}{10 - 0.01} = 1$$

Therefore, a shunt of 1 ohm has to be used to convert the voltmeter into ammeter of desired range.

11.4 OHM'S LAW, RESISTANCE AND KIRCHCHOFF'S LAWS

Steady current through a conductor depends on the potential difference between its ends. This is similar to the flow of water in a pipe where, the flow velocity depends on the difference in water levels at the two ends of the pipe. Heat flow in a metal rod also depends on the temperature levels at the two ends of the rod.

G.S. Ohm, a German physicist, discovered in 1826, that for a given conductor, the strength of the current flow through the conductor is proportional to the potential difference between its ends, provided its temperature and other physical conditions remain unaltered.

The law may be symbolically written as
$$V = RI \tag{11.43}$$

where R is a constant of proportionality. The constant of proportionality, R, is called the resistance of the conductor.

Ohm's law holds good for metallic conductors, except for a few near absolute zero. Ohm's law is not applicable to conductors such as electron tubes, arcs, discharges through gases, semiconductors, ionic conductors. Ohm's law can be applied to a complete circuit or to part of a circuit.

11.4.1 Electrical Resistance

From the Eq. (11.43), the resistance of a conductor can be defined as the ratio of the potential difference across the two ends of the conductor to the current flowing through it, expressed as

$$\text{Resistance} = \frac{\text{Potential difference}}{\text{Current}}, \text{ or } R = \frac{V}{I} \tag{11.44}$$

Unit of resistance is called Ohm in honour of G.S. Ohm. It is normally denoted by the Greek letter Ω and is defined as the resistance of a conductor which carries a current of one ampere when a steady potential difference of one volt is applied across its ends.

11.4.2 Verification of Ohm's law

Ohm's law can be verified by ammeter–voltmeter method as shown in Fig. 11.17. The rheostat is adjusted for a small current so that the temperature of the standard resistance R does not change appreciably during the experiment. Current I in the circuit is read from ammeter A and the potential difference V, across the resistance R, is measured for different currents. The ratio $\dfrac{V}{I}$ is calculated in each case. The rheostat setting can be varied for different trials and the corresponding values of the voltage, V, and the current, I, are obtained. It will be found that the ratio, V/R is constant for all the trials.

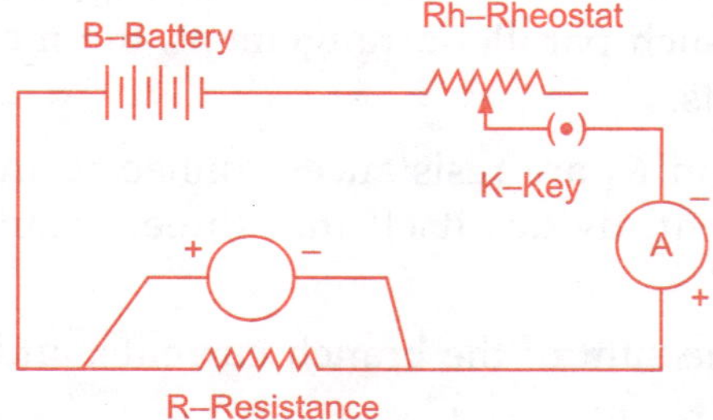

Fig. 11.17 *Verification of Ohm's law*

11.4.3 Resistances in Series

Series connection of resistances is shown in Fig. 11.18, where resistances are connected from end to end. The same current flows through all the resistances in series and the potential difference across the combination is equal to the sum of the potential differences across each resistance.

Fig. 11.18 *Resistances in series*

In Fig. 11.18, resistances R_1, R_2 and R_3 are connected in series. If the series combination is replaced by a single resistance R, producing the same effect as the combination, then R is called the effective or equivalent resistance of the series combination.

The current flowing through the series is I. If V_1, V_2 and V_3 are the potential differences across R_1, R_2 and R_3 respectively, then by Ohm's law, we have

$$V_1 = IR_1,\ V_2 = IR_2,\ V_3 = IR_3 \tag{11.45}$$

As the total potential difference across the combination is equal to the sum of the potential differences in the individual resistances, we have the relation

$$V = V_1 + V_2 + V_3 = IR_1 + IR_2 + IR_3 \tag{11.46}$$

Since the equivalent resistance is R, we have $V = IR$. Therefore, we can write

$$IR = IR_1 + IR_2 + IR_3 \qquad (11.47)$$

From Eq. (11.47), we get an expression for the equivalent resistance as

$$R = R_1 + R_2 + R_3 \qquad (11.48)$$

Therefore the effective resistance of a combination of resistances in series is equal to the sum of the individual resistances.

11.4.4 Resistances in Parallel

When resistances are connected across common terminals, such that the potential difference across each is the same as shown in Fig. 11.19, they are said to be connected in parallel. In such parallel arrangement, the main current is equal to the sum of the branch currents.

In Fig. 11.19, R_1, R_2 and R_3 are resistances connected in parallel, across common terminals. The main current divides itself into three streams of I_1, I_2 and I_3 flowing through R_1, R_2 and R_3.

The main current is the sum of the branch currents, and hence

$$I = I_1 + I_2 + I_3 \qquad (11.49)$$

The potential difference across each resistance is the same and can be expressed as

$$V = I_1R_1, \ V = I_2R_2, \ V = I_3R_3 \text{ and } V = IR \qquad (11.50)$$

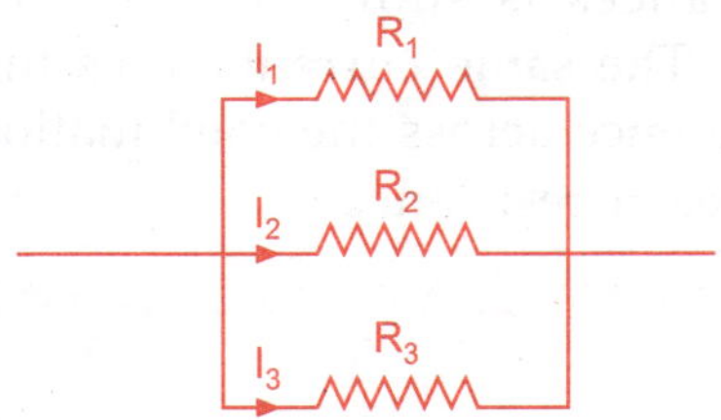

Fig. 11.19 *Resistances in parallel*

Substituting Eq. (11.50) into Eq. (11.49) we get

$$\frac{V}{R} = \frac{V}{R_1} + \frac{V}{R_2} + \frac{V}{R_3} \qquad (11.51)$$

From Eq. (11.51), we get the expression for the equivalent resistance for a parallel combination of resistances as

$$\frac{I}{R} = \frac{1}{R_1} + \frac{1}{R_2} + \frac{1}{R_3} \qquad (11.52)$$

Therefore for resistances that are connected in parallel, reciprocal of the effective resistance is equal to the sum of the reciprocals of the individual resistances.

For the particular case of two resistances, we have the relation

$$R = \frac{R_1 R_2}{R_1 + R_2} \tag{11.53}$$

Since the potential difference is the same across the resistances,

$$I_1 R_1 = IR = \frac{IR_1 R_2}{R_1 + R_2} \tag{11.54}$$

Therefore, for the case of two resistances in parallel, the current through the first resistance is given by

$$I_1 = \frac{IR_2}{R_1 + R_2} \tag{11.55}$$

In words, this relation can be written as

$$[\text{Current through one branch}] = \frac{(\text{main current})(\text{resistance of the other branch})}{(\text{sum of branch resistances})}$$

11.4.5 Internal Resistance

The resistance offered by an emf source to the flow of current through it is called the internal resistance of that device. Resistance in a circuit outside the source is called external resistance.

11.4.6 Ohm's Law Applied to a Circuit

A simple circuit is shown in Fig. 11.20, where the terminals A and B of a cell of emf, E, are connected in series to a resistance R. The cell supplies an energy E (in joules) for transferring one coulomb of charge around the complete circuit. A part of this energy, V is used in doing work to overcome the resistance R and a part, V' to overcome the cell internal resistance r. Therefore, using the principle of conservation of energy

$$E = V + V' \tag{11.56}$$

where V represents the potential difference across R and V' the potential difference across r. From Ohm's law, we have $V = IR$ and $V' = Ir$, where I is the current in the circuit. Therefore,

$$E = IR + Ir \tag{11.57}$$

Therefore, the total effective resistance in the circuit is given by

$$R + r = \frac{E}{I} \tag{11.58}$$

which can also be expressed as

$$[\text{Effective resistance in the circuit}] = \frac{\text{emf of the cell}}{\text{current in the circuit}}$$

Further, the current in the circuit can be expressed as

$$I = \frac{E}{R+r} \tag{11.59}$$

11.4.7 Terminal Potential Difference

Terminal potential difference is the potential difference across the terminals of a cell. It can be seen from Fig. 11.20, that the potential difference measured across the terminals A and B of the cell is the same as the potential difference across the ends of the resistance R. Therefore, the terminal potential difference is given by

$$V = IR = E - Ir \tag{11.60}$$

Thus, the terminal potential difference depends on the value of I and it can never exceed the value of E. Only when $r = 0$, $V = E$. It is impossible to have $r = 0$. However, it is possible to have a very low internal resistance as in an accumulator.

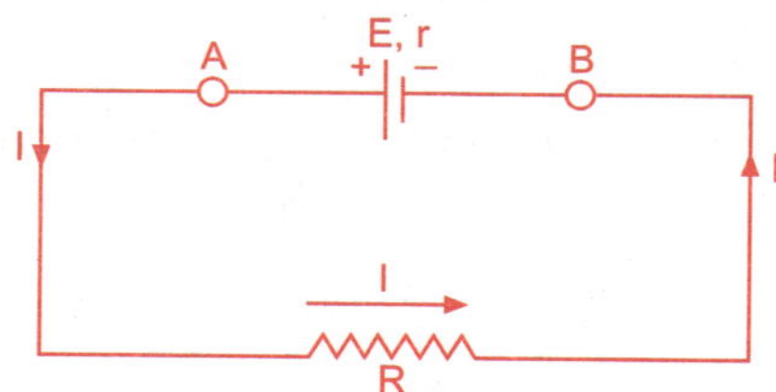

Fig. 11.20 *Effective resistance in the circuit*

Also when $I = 0$ it is possible to have $V = E$. It means that when there is no current drawn from the cell, the terminal potential difference of the cell is equal to the emf of the cell. A cell is said to be in open circuit when no current is drawn from the cell. Therefore, the emf of a cell is sometimes defined as the potential difference between the terminals of the cell when it is in open circuit.

11.4.8 Resistivity

Electrical resistance of a conducting wire depends on several factors such as (*i*) material of the wire (*ii*) the size of the wire and (*iii*) its temperature. Experimental studies have shown that the resistance of a wire is (*i*) proportional to its length, L, and (*ii*) inversely proportional to its area of cross section, A. Hence, R varies as $\dfrac{L}{A}$ and can be expressed as

$$R = \rho\frac{L}{A} \tag{11.61}$$

where ρ is a constant of proportionality. It is known as resistivity of the material of the wire, also called the specific resistance. For $L = 1$ and $A = 1$, $\rho = R$. Therefore, specific resistance of a material can be defined as the resistance of a wire of the material of unit length and unit area of cross section. It can also be defined as the resistance across opposite faces of a unit cube of the material.

From Eq. (11.61), it can be seen that the unit of specific resistance is ohm-meter. Copper, which is a good conductor, has a low resistivity of 1.7×10^{-8} ohm-meter. Carbon has a resistivity between 5×10^{-8} and 10^{-7} ohm-meter.

Reciprocal of resistance is called conductance. Its unit is mho ($\mho$). And the reciprocal of resistivity is called conductivity.

11.4.9 Temperature Coefficient of Resistance

Experimental studies have shown that the electrical resistance of metals increase with increase in temperature. The resistance varies approximately linearly with temperature. If R_o and R_t are the resistances of a metallic wire at temperatures 0°C and t°C, then

$$R_t = R_o (1 + \alpha t) \tag{11.62}$$

where α is a constant for a given substance and is known as temperature co-efficient of resistance which is given by

$$\alpha = \frac{R_t - R_o}{R_o t} \tag{11.63}$$

Temperature coefficient of resistance may be defined as the ratio of increase in resistance to its resistance at 0°C, per degree Celsius rise of temperature.

If R_1 and R_2 are the resistances at temperatures t_1 and t_2, respectively, then temperature coefficient of resistance α is given by the relation,

$$\alpha = \frac{R_2 - R_1}{R_1 t_2 - R_2 t_1} \tag{11.64}$$

The charge is carried in a metal by the free electrons. As the temperature is increased, the amplitude of vibration of atoms increases and there will be more collisions of drifting free electrons with the atoms. Consequently, the collisions increase the resistance of a pure metal at higher temperature.

The value of α is about 0.004/°C for pure metals. At very low temperatures, some metals, such as mercury, lead, tin, zinc, aluminium, offer practically no resistance. Resistance of mercury almost disappears at 4K (–269 °C). This phenomenon is known as superconductivity. Lead behaves similarly at 9 K (–263°C). Alloys of niobium and tin and of niobium and zirconium show superconductivity at about 18 K. The state in which a metal offers almost no resistance to the flow of current is called superconducting state.

Most alloys have very low temperature coefficients. Temperature coefficients of mangan in and constantan are extremely low (0.00001/°C). These alloys are used for standard resistances, as they are very slightly affected by change in their temperature when the current flows through them. German silver, nickelin, constantan, manganin etc., are extensively used in the construction of electrical appliances.

In the case of insulators like carbon, and pure semiconductors such as germanium and silicon, resistance decreases with rise in temperature and they have a negative temperature coefficient of resistance. Electrolytes also have a negative temperature coefficient.

In the case of a semiconductor, rise of temperature results in increase of thermal energy of valence electrons, and more and more valence electrons become free electrons. Hence, electrical resistance of a semiconductor decreases with rise in temperature. Semiconductors have a negative temperature coefficient of resistance. Temperature coefficient of resistance for carbon is about -5×10^{-4}/°C; α is about -4.8×10^{-2}/°C for germanium and about -7.5×10^{-2}/°C for silicon, at room temperature.

11.4.10 Thermistors

Resistors which are thermally sensitive are named as thermistors. Thermistors may have both positive temperature coefficient (PTC) and negative temperature coefficient (NTC). Thermistors are usually manufactured from semiconducting metal oxides, in the form of beads, discs or rods with a pair of platinum lead wires. The size may be as small as 0.1 mm for the bead type. This assembly is enclosed in a small glass bulb and sealed.

Thermistors are used to measure temperature changes as low as 0.01°C. They are used as surge current suppressors, as compensators for resistance variation of components in radio, television, and telecommunication and projection equipment. They are also used for temperature control, voltage stabilisation, remote sensing, in measurements of heat insulation, humidity, fluid velocity etc. PTC thermistors are intended primarily for detecting excessive temperature within industrial equipment.

11.4.11 Kirchhoff's Laws

Many electrical circuits can be modeled as combinations of sources and resistors connected in series and parallel arrangements. In such cases, Kirchhoff's laws can be applied in order to solve for the current flowing in the different parts of the circuit and the voltage across components. Kirchchoff's laws are applicable to any electrical circuit, both for steady currents as well as alternating currents, however, we will deal only with steady currents here. Kirchhoff's laws are stated as follows:

First law– The algebraic sum of the currents at a junction of an electrical circuit is zero.

Second law– In any closed mesh (or path) in an electrical network, the algebraic sum of the emfs is equal to the sum of all the voltage drops (products of current, I, and resistance, R) in that mesh.

The first law is a mathematical statement of the fact that charges do not accumulate at any point in an electrical circuit. By convention, the current flowing towards a junction is considered positive and the current flowing away from a junction is considered negative. A portion of an electrical circuit with five branches meeting at a junction at A is shown in Fig. 11.21(*a*).

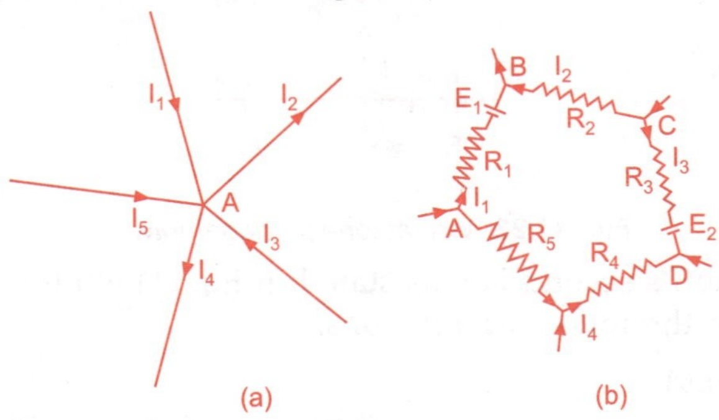

(a)   (b)

Fig. 11.21 *Kirchhoff's law (a) current flow to a junction (b) voltage drop in a mesh*

Applying the First law to the circuit, we have

$$I_1 - I_2 + I_3 - I_4 + I_5 = 0 \tag{11.65}$$

In general, if the junction has any number of branches, the corresponding mathematical relation is given by

$$\Sigma I = 0 \tag{11.66}$$

A portion of an electrical network with five junctions is shown in Fig. 11.21(*b*). Applying the Second law, we have

$$I_1 R_1 - I_2 R_2 + I_3 R_3 - I_4 R_4 - I_5 R_5 = E_1 - E_2 \tag{11.67}$$

In general, if there is more number of junctions, we have the relation

$$\Sigma IR = \Sigma E \tag{11.68}$$

In Eq. (11.67) or Eq. (11.68), the product of I and R are taken as positive for the current in one direction and negative in the other direction.

11.4.12 Wheatstone's Network

Wheatstone's network, named after C. Wheatstone, a professor of physics at King's College, London, is used for an accurate comparison of resistances. The network is as shown in Fig. 11.22 where P, Q, R and S are four resistances. It is connected to a source of emf, E, across one pair of opposite junctions A and C whose internal resistance is r. Across the other pair of opposite junctions a galvanometer of resistance G is connected. Applying Kirchhoff's first law given in Eq. (11.66) to different junction, currents in the different branches can be obtained as shown in the Fig. 11.22.

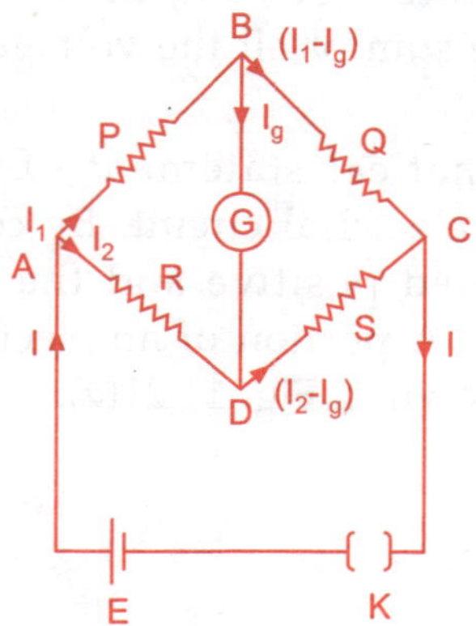

Fig. 11.22 *Wheatsone's bridge network*

Applying Kirchhoff's second law as stated in Eq. (11.68) for the different meshes in the circuit, we get the following relations:

For the mesh *ABDA*

$$I_1P + I_gG - I_2R = 0 \tag{11.69}$$

For the mesh *BCDB*

$$(I_1 - I_g)Q - (I_2 + I_g)S - I_gG = 0 \tag{11.70}$$

For the mesh *ADCEA*

$$I_2R + (I_2 + I_g)S + Ir = E \tag{11.71}$$

where $\qquad I_1 + I_2 = I$

If E and the resistances are known, the three equations can be solved for I_1, I_2 and I_g.

For the case of no current in the galvanometer, we have $I_g = 0$, and hence the first two of the above equations reduce to

$$I_1P - I_2R = 0 \tag{11.72}$$

or, $\qquad I_1P = I_2R$, and

$$I_1Q - I_2S = 0 \tag{11.73}$$

or, $I_1Q = I_2S$. Hence, we have the relation

$$\frac{P}{Q} = \frac{R}{S}. \tag{11.74}$$

When the resistances P, Q, R and S are adjusted such that no current flows through the galvanometer ($I_g = 0$), the circuit is balanced and the resistances are also balanced. If three resistances are known, the fourth resistance can be computed.

11.4.13 Meter Bridge

A meter bridge is used to determine the specific resistance of the material of a conducting metal wire. Wheatstone's network can be conveniently adapted for this purpose. The meter bridge consists of a uniform resistance wire AC, of length 1m, stretched on a wooden board as shown in Fig. 11.23.

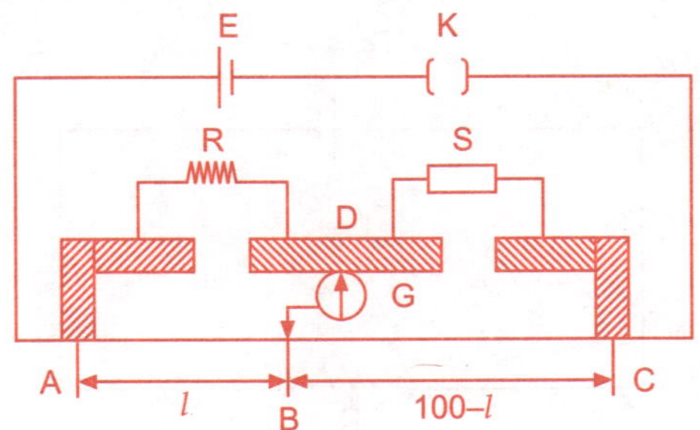

Fig. 11.23 *Set-up to measure specific resistance of a wire*

The ends of the wire are soldered to two thick L shaped copper strips. A copper strip, D, provided with terminals, is fixed between them in order to make electrical connections. Resistance R is connected in the left gap between the L shaped copper strip and D. A standard resistance box S is connected in the right gap. The circuit is closed and the position of the sliding contact, B, on the bridge wire is adjusted for zero deflection in the galvanometer, G. Let the balancing length $AB = l$.

The resistance R of the given wire is obtained using the Eq. (11.74) as

$$\frac{R}{S} = \frac{\text{Resistance of } AB}{\text{Resistance of } BC} = \frac{l}{100-l}. \tag{11.75}$$

Therefore, the unknown resistance is given by

$$R = \frac{Sl}{100-l} \tag{11.76}$$

The resistivity, ρ, of the material of the wire is calculated using Eq. (11.61) as

$$\rho = \frac{AR}{l} = \frac{\pi d^2 R}{4l} \tag{11.77}$$

where d is the mean diameter of the wire AC, l is the length of the wire, and R is the mean value of the measured resistance.

11.4.14 Grouping of Cells

When cells are used in electrical circuits for power supply, they have internal resistances in addition to providing the required emf. In applying the Ohm's law to such circuits where cells are used, the effective resistances and consequently, the current will be calculated; they depend on the way in which the cells are connected.

In the following, we will discuss the three ways in which cells may be connected in electrical circuits.

1. Cells in series: When cells are connected in series as shown in Fig. 11.24, their net emf is the arithmetic sum of the individual emfs. As the cells are in series, their effective internal resistance is also the arithmetic sum of the individual internal resistances.

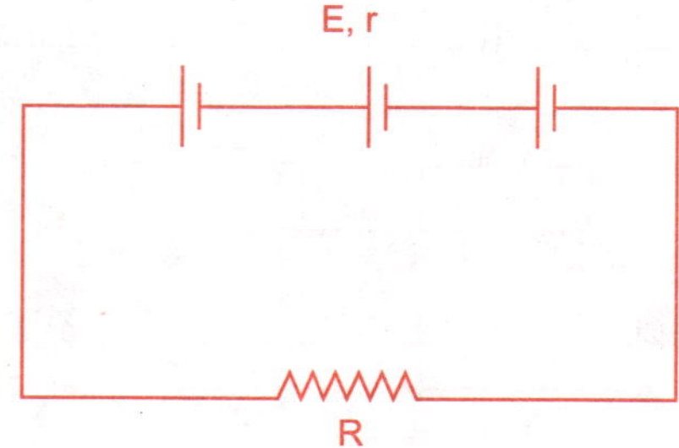

Fig. 11.24 *Cells connected in series*

If two cells are connected in series such that their emfs are in the same direction, they are said to be in conjunction with each other. If they are oppositely directed, the cells are said to be in opposition.

If n similar cells, each of emf, E, and internal resistance, r, are connected in series, in conjunction with each other, then net emf is equal to nE and their effective internal resistance is equal to nr. If these series combinations of cells are connected to an external resistance R, the current, I, set-up in the circuit, is given by

$$I = \frac{\text{Effective emf}}{\text{Effective resistance of the circuit}} = \frac{nE}{R + nr} \qquad (11.78)$$

When R is very high compared to r, we have $I \approx \dfrac{nE}{R}$.

2. Cells in parallel: A parallel arrangement of cells is shown in Fig. 11.25. If the emfs of all the individual cells are in the same direction, the combination behaves like one cell. If all the cells are similar, the same emf acts across each cell. Hence, the net emf, E, of the combination is the emf of one cell.

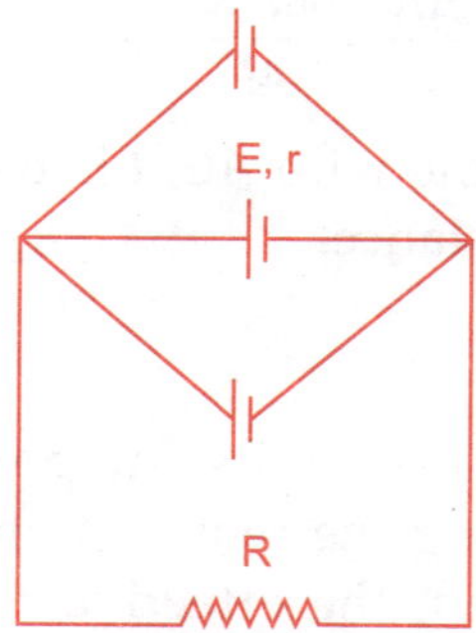

Fig. 11.25 *Cells connected in parallel*

As the cells are in parallel, effective internal resistance, r_e, is given by

$$\frac{1}{r_e} = \frac{1}{r}+\frac{1}{r}+\frac{1}{r}+\frac{1}{r}+\ldots \tag{11.79}$$

Therefore, if there are n cells, we have

$$\frac{1}{r_e} = \frac{n}{r} \tag{11.80}$$

The effective resistance is given by

$$r_e = \frac{r}{n} \tag{11.81}$$

If the combination is connected to an external resistance R, then the current I, flowing through R is given by

$$I = \frac{E}{R+r_e} = \frac{E}{R+\left(\dfrac{r}{n}\right)} = \frac{nE}{nR+r} \tag{11.82}$$

From Eq. (11.82), it follows that when nR is very small compared with r, we have

$$I \approx \frac{nE}{r}.$$

3. Cells in mixture of series and parallel: An arrangement of n branches of m cells each is shown in Fig. 11.26. An external resistance R is connected as shown in the Fig. 11.26.

The total internal resistance in each branch is mr, and therefore the effective resistance is given by

$$\frac{1}{r_e} = \frac{1}{mr}+\frac{1}{mr}+\frac{1}{mr}+\ldots n \text{ terms} = \frac{n}{mr} \tag{11.83}$$

Rewriting the relation, we have

$$r_e = \frac{mr}{n} \tag{11.84}$$

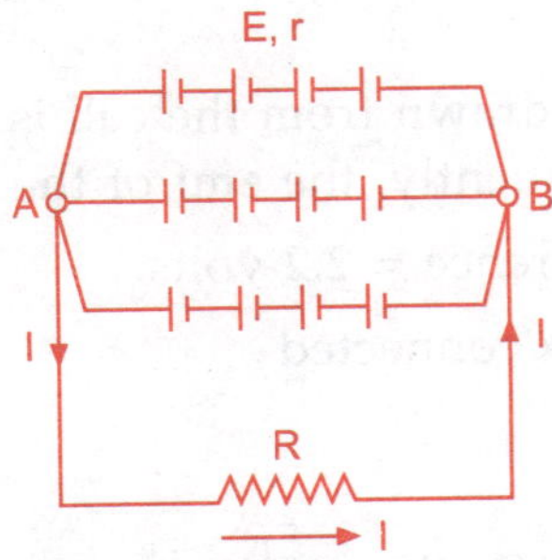

Fig. 11.26 *Cells in series – parallel combination connected*

The total emf in each branch is mE. Therefore, the current through the external resistance R is given by

$$I = \frac{mE}{R + \left(\dfrac{mr}{n}\right)} = \frac{mnE}{nR + mr} \tag{11.85}$$

For a given number of cells, mnE is constant. Therefore, the current I is maximum when $nR + mr$ is minimum. It can be shown that $nR + mr$ is minimum, when $nR = mr$ or when $R = (mr)/n$. Hence, the current drawn from a set of cells in a mixture of series and parallel arrangement will be a maximum when, the external resistance is equal to the effective internal resistance of the combination of cells.

EXAMPLE 11.10

The poles of a cell of emf 15 volts are connected to the ends of a 10 Ω coil. If the current in the circuit is 0.1 amp, find the internal resistance of the coil.

SOLUTION

Current in the circuit is given by the relation

$$I = \frac{E}{R + r}$$

Therefore, $\qquad 0.1 = \dfrac{1.5}{10 + r}$

or, $\qquad\qquad r = 5$

Internal resistance of the cell $= 5\ \Omega$.

EXAMPLE 11.11

A voltmeter connected to a cell reads 2.2 volts. When a resistance of 1 Ω is connected between the terminals of the cell, the reading of the voltmeter falls to 1.1 volt. Calculate the internal resistance of the cell.

SOLUTION

Let us assume that the current drawn from the cell is very small when the voltmeter is connected to the cell. Consequently, the emf of the cell is given by

E = terminal potential difference = 2.2 volts.

When one ohm resistance is connected

Potential difference = 1.1 V

Potential difference across $1\Omega = IR = \dfrac{ER}{R + r}$

$$1.1 = 2.2 \times \frac{1}{1+r}$$

or,
$$r = 1$$

Hence, the internal resistance of the cell = 1 Ω.

EXAMPLE 11.12

An electrical appliance has the specifications, 220 V, 550 W. What is the resistance when it is in use?

SOLUTION

Power P is given by the relation,

$$P = VI = \frac{V^2}{R}$$

$$R = \frac{V^2}{P} = \frac{220^2}{550} = 88$$

Therefore, resistance of appliance when in use is 88 Ω.

EXAMPLE 11.13

Two conductors having resistances of 4Ω and 8Ω are connected in parallel, with a cell of 2 volts with an internal resistance of 0.5Ω. Find the current in each conductor.

SOLUTION

The circuit is shown in Fig. 11.27.

Effective resistance of R_1 and R_2 is given by,

$$R = \frac{R_1 R_2}{R_1 + R_2} = \frac{4 \times 8}{12} = \frac{8}{3}\Omega$$

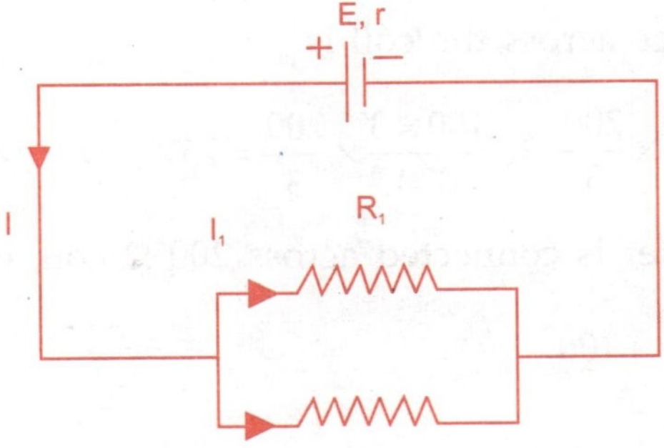

Fig. 11.27 *Resistances in parallel*

Main current

$$I = \frac{E}{R+r} = \frac{2}{(8/3)+0.5} = \frac{6}{9.5} \text{ amp}$$

$$I_1 = \frac{IR_2}{R_1+R_2} = \frac{6\times 8}{9.5\times 12} = 0.42 \text{ amp}$$

and

$$I_2 = \frac{6}{9.5} - \frac{4}{9.5} = \frac{2}{9.5} = 0.21 \text{ amp}$$

Current in the conductors are 0.42A and 0.21A.

EXAMPLE 11.14

Two resistance coils of 100 Ω and 200 Ω are connected in series across 100 volts. A moving coil voltmeter of 200 Ω is connected in turn across each coil. What will it read in each case?

SOLUTION

The connections are shown in Fig. 11.28.

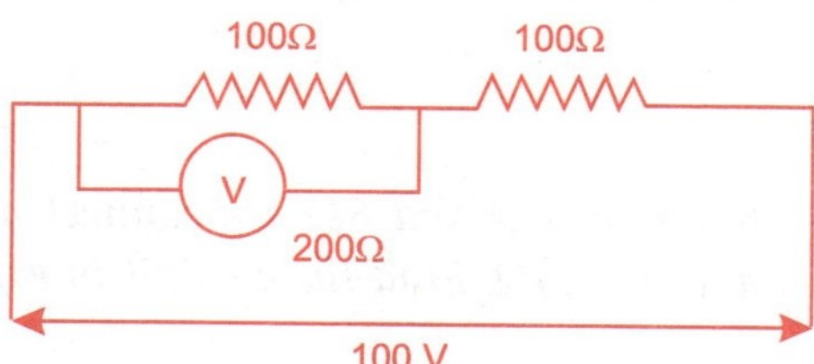

Fig. 11.28 *Two resistances in series*

(a) When the voltmeter is connected across 100 ohms coil as shown in Fig. 11.27, the effective resistance for the circuit is

$$\frac{100\times 200}{300} + 200 = \frac{800}{3}\Omega$$

Potential difference across the coil is

$$I \times \frac{200}{3} = \frac{100\times 3}{800} \times \frac{200}{3} = 25V.$$

(b) When the voltmeter is connected across 200 Ω coil, effective resistance of the circuit is $\dfrac{200\times 200}{400} + 100$

i.e. 200 Ω

Potential difference across the coil is

$$\frac{100}{200} \times 100 = 50 \text{ V}$$

The voltmeter reads 25 V and 50 V in the two cases.

EXAMPLE 11.15

What is the length of a manganin wire of diameter 0.44 mm required to construct a 10 ohm coil? The specific resistance of manganin = 44.3 × 10^{-8} ohm-meter.

SOLUTION

$$R = \frac{\rho L}{A}$$

$$R = 10 \ \Omega$$

$$\rho = 44.3 \times 10^{-8} \text{ Ohm-meter}$$

$$A = \text{Area of cross section} = \pi \ (0.22 \times 10^{-3})^2 \text{ m}^2$$

$$L = \frac{AR}{\rho} \pi \left(0.22 \times 10^{-3}\right)^2 \times \frac{10}{44.3} \times 10^{-8} = 3.432$$

Therefore, the length of the wire required = 3.542 m.

EXAMPLE 11.16

Two cells of emf 1.5 volts and 2 volts, respectively, and corresponding internal resistances 1Ω and 2Ω are connected in parallel to an external resistance of 5Ω. Find the current in each of the three branches of the network.

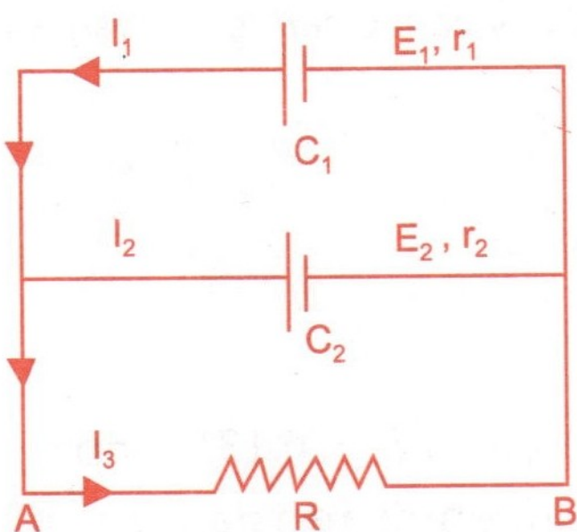

Fig. 11.29 *Two cells in parallel*

SOLUTION

In Fig. 11.29, emfs of the two cells C_1 and C_2 are assumed to be in the same direction. The external resistance is R. Currents are taken as indicated in the figure.

Applying Kirchhoff's laws, $I_3 = I_1 + I_2$ and for the mesh ABC_1A, $I_3R + I_1r_1 = E_1$

i.e. $\qquad (I_1 + I_2)5 + I_1 \times 1 = 1.5$

or $\qquad\qquad 6I_1 + 5I_2 = 1.5$ $\hfill (1)$

For the mesh ABC_2A, $I_3R + I_2r_2 = E_2$

i.e. $\qquad (I_1 + I_2)\, 5 + I_2 \times 2 = 2$

or $\qquad\qquad 5I_1 + 7I_2 = 2$ $\hfill (2)$

Solution of Eqs. (1) and (2) give, $I_1 = \dfrac{1}{34}$ amp, $I_2 = \dfrac{9}{34}$ amp and $I_3 = \dfrac{5}{17}$ amp.

EXAMPLE 11.17

Two cells of emfs. 1.6 volts and 2.4 volts have infernal resistances 2 ohms and 4 ohms. The two positive poles are connected to resistance of 6 ohms and negative poles connected to 8 ohms coil. If another wire of 10 ohms is connected between the midpoints of these wires, what is the potential difference across it?

SOLUTION

The network is shown in Fig. 11.30.

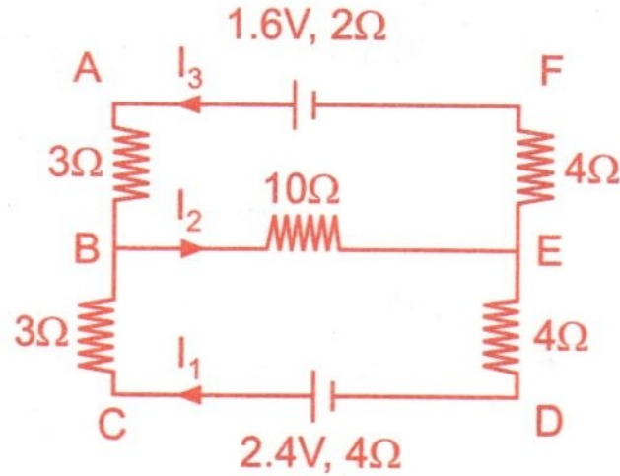

Fig. 11.30 *Cells connected to series-parallel resistors*

At B, $I_3 = I_2 - I_1$

For $BEDCB$, $10\,I_2 + 11I_1 = 2.4$

For $ABEFA$, $9I_3 + 10I_2 = 1.6$

or, $9(I_2 - I_1) + 10I_2 = 1.6$

$19I_2 - 9I_1 = 1.6$

Solution of Eqs. (1) and (2) gives $I_2 = 0.131$ amp.

Potential difference across 10 ohms resistance $= 10 \times 0.131 = 1.31$V.

EXAMPLE 11.18

Four resistances of 15Ω, 12Ω, 4Ω, and 10Ω, are connected in cyclic order to form a Wheatstone's network as shown in Fig. 11.31. Find out whether the network is balanced. If not, calculate the resistance to be connected in parallel with the resistance of 10Ω to balance the network.

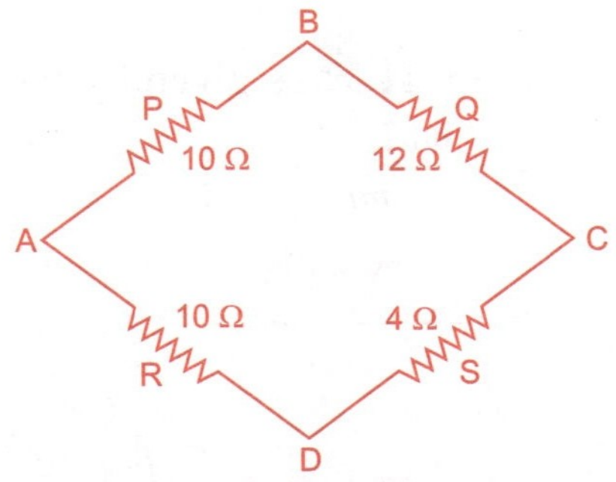

Fig. 11.31 *Four resistances forming Wheatstone's network*

SOLUTION

We have $\dfrac{P}{Q} = \dfrac{15}{12} = 1.25$

$$\frac{R}{S} = \frac{10}{4} = 2.5$$

$$\frac{P}{Q} \neq \frac{R}{S}$$

The network is not balanced.

If x is the resistance that has to be connected in parallel, then after connecting x parallel to 10 Ω

$$\frac{1}{R} = \frac{1}{x} + \frac{1}{10}$$

Where,

$$\frac{R}{S} = \frac{P}{Q}$$

$$R = \frac{P}{Q} S = (1.25)4 = 5\,\Omega$$

$$\frac{1}{5} = \frac{1}{x} + \frac{1}{10}$$

or,

$$\frac{1}{x} = \frac{1}{5} - \frac{1}{10} = \frac{1}{10}$$

Solving, $\quad x = 10\ \Omega$

Hence, 10 Ω resistance must be connected in parallel with the 10 Ω arm to balance the network.

EXAMPLE 11.19

Ten cells, each having a resistance of 1.5Ω and emf 2 volts, are connected in two parallel rows of five each. This battery is used to supply current to two lamps of 20Ω resistance connected in parallel. Find the strength of current in the circuit.

SOLUTION

Current in the circuit, shown in Fig. 11.32, is given by the relation

$$I = \frac{mnE}{nR + mr}$$

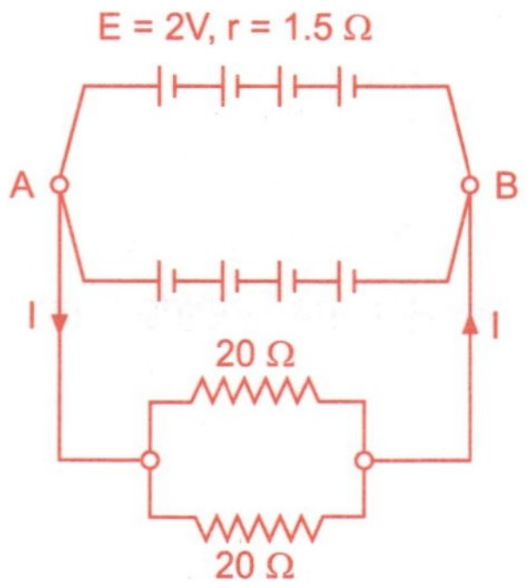

Fig. 11.32 *Series-parallel cells with parallel resistances*

where $m \rightarrow$ number of cells in series, $n \rightarrow$ number of rows, $E \rightarrow$ emf, $r \rightarrow$ internal resistance of each cell and $R \rightarrow$ external resistance. We have

$$\frac{1}{R} = \frac{1}{20} + \frac{1}{20}; \frac{1}{R} = \frac{2}{20}; R = 10\,\Omega$$

$$I = \frac{5 \times 2 \times 2}{2 \times 10 + 5 \times 1.5} = 0.73$$

Current in the circuit = 0.73 A.

EXAMPLE 11.20

A platinum resistance thermometer shows a resistance of 5.04 ohms at 0°C and 7.06 ohms at 100°C. When it is immersed in a boiling liquid, the resistance is found to be 6.63 ohms. Calculate the boiling point of the liquid.

SOLUTION

$R_o = 5.04\ \Omega$, $R_{100} = 7.06\ \Omega$ and $R_t = 6.63\ \Omega$

$$\alpha = \frac{R_t - R_o}{R_o t} = \frac{R_{100} - R_0}{R_0 \times 100}$$

$$t = \frac{(R_t - R_o)100}{R_{100} - R_0} = \frac{6.63 - 5.04}{7.06 - 5.04} \times 100 = 78.7°C$$

Boiling point of the liquid = 78.7°C.

EXAMPLE 11.21

When three wires of resistances 2 ohms, 6 ohms and 12 ohms respectively are connected in parallel and are included in a circuit with a cell of negligible internal resistance and a tangent galvanometer, the deflection is 60°. When the 2 ohms wire is removed the deflection become 45°. Calculate the resistance of the galvanometer.

SOLUTION

The arrangement is shown in Fig. 11.33. For the parallel connection of resistances we have

$$\frac{1}{R_1} = \frac{1}{12} + \frac{1}{6} + \frac{1}{2}; \; R_1 = \frac{4}{3}\Omega$$

$$I_1 = \frac{E}{\left(\frac{4}{3}\right)+G} = K\tan 60°$$

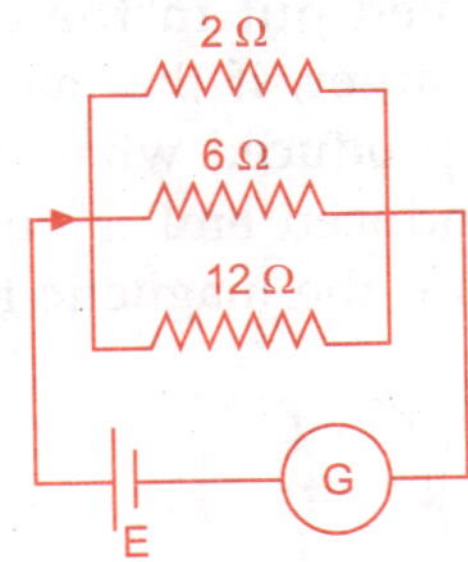

Fig. 11.33   Resistances in parallel

$$R_2 = \frac{12 \times 6}{18} = 4\,\Omega$$

$$I_2 = \frac{E}{4+G} = K\tan 45°$$

Solving for G between I_1 and I_2 gives. $G = 2.31\ \Omega$

Therefore, resistance of the galvanometer = 2.31 Ω.

11.5 ELECTROMAGNETIC INDUCTION

The discovery of magnetic effect of current by Oersted led Faraday to consider the possibility of the converse effect of getting an electric current in a circuit due to a magnetic field. Michael Faraday and Joseph Henry in the year 1831 independently proved that electromotive force is induced in a closed circuit due to change in the magnetic field. Electromotive force or current also gets induced, when a closed loop of wire or conductor is moved under the influence of magnetic field. The instantaneous emf that is induced is known as induced emf, and the phenomenon is known as electromagnetic induction. The discovery of electromagnetic induction started a revolution in the study of electrical science and engineering with far reaching effects leading to the development of electrical machines such as electrical motors, generators, transformers. The early studies on the phenomenon of electromagnetic induction were through simple experiments.

The following two experiments will give an appreciation of the phenomenon.

1. Magnet and coil experiment

If one end of a magnet is suddenly introduced into a coil connected to a sensitive galvanometer, as shown in Fig. 11.34, the galvanometer shows a momentary deflection.

This shows that introduction of a magnetic field induces a momentary current and hence an instantaneous emf is induced in the circuit. The direction of the current depends on whether the North Pole or the South Pole is introduced into the coil. If the magnet is withdrawn suddenly, a sudden deflection is again observed but in the opposite direction. Further, the deflection is found to be larger, if the magnet is introduced or withdrawn faster. The same effect is produced when the coil is moved in a stationary magnetic field. Further, induced emf is obtained only if there is a relative motion between the coil and the magnetic field.

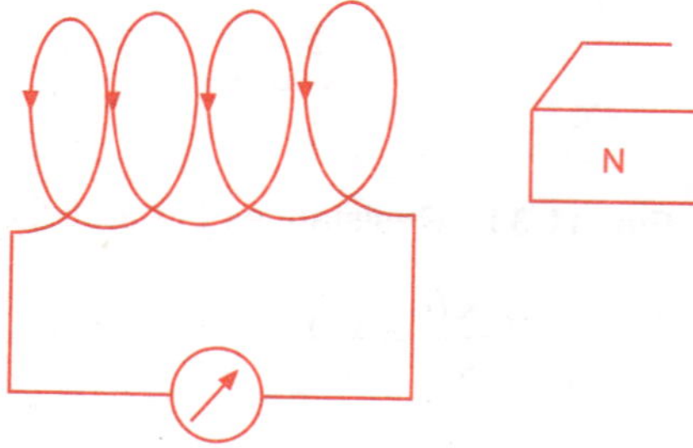

Fig.11.34 *Magnet moving into a coil*

2. Coil and coil experiment

The phenomenon of electromagnetic induction can also be demonstrated using two coils close together as shown in Fig. 11.35, or one coil inside the other. One of the coils is connected to a battery and the other to a sensitive galvanometer, as shown.

The former is called primary circuit and the latter is called the secondary circuit. When the battery circuit or the primary circuit is made and broken (switched on and off), the galvanometer shows deflection first in one direction and then in the opposite direction. A similar effect is obtained if the strength of the current in the primary is suddenly increased or decreased, or when the secondary coil is moved with respect to the current carrying primary.

Again, induced emf is obtained only when there is a relative motion between the coils with the steady current, or when the current is changing with the two coils at rest. No induced emf is generated when steady current flows through the primary, while the secondary coil is at rest.

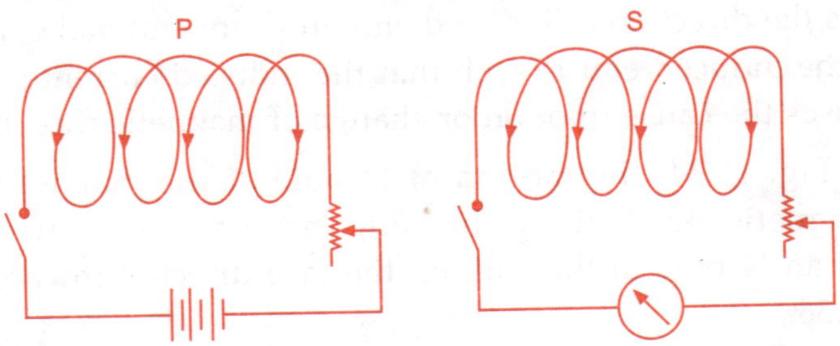

Fig. 11.35 *Electromagnetic induction due to current flow in a coil*

11.5.1 Laws of Electromagnetic Induction

Faraday concluded that an emf is induced in a closed circuit whenever there is a change of magnetic flux linking the coil. When one end of a magnet is introduced into a coil, the magnetic flux linking the coil increases. When the magnet is withdrawn the flux linkage decreases.

In the coil and coil experiment, magnetic flux is suddenly established during make, whereas during break, the flux suddenly disappears. When the primary current is increased or decreased, the magnetic flux associated with the secondary correspondingly increases or decreases.

Based on the results of these experiments, Faraday formulated what are known as Faraday's laws of electromagnetic induction.

1. Whenever magnetic flux linking a circuit changes, an instantaneous emf is induced in the circuit and this emf exists when the change is taking place.

2. The magnitude of the induced emf is directly proportional to the rate of change of magnetic flux linking the coil.

The second law may be expressed mathematically as,

$$\varepsilon \propto \frac{d\phi}{dt} \tag{11.86}$$

where ϕ represents the magnetic flux linking the coil and the ε represents the induced emf. This is a quantitative statement of Faraday's second law of electromagnetic induction and is sometimes known as the Faraday-Neumann law, in honour of both Faraday and the German physicist Franz Neumann.

11.5.2 Lenz's Law

Lenz's law is a common way of understanding how electromagnetic circuits obey Newton's third law and the conservation of energy. Lenz's law is named after Heinrich Lenz, a Russian physicist in 1833. Lenz's law states that:

"An induced electromotive force (emf) always gives rise to a current whose magnetic field opposes the original change in magnetic flux".

The law gives the direction of induced emf and conventional current. It says that, the direction of the induced emf is such that the induced current sets up a magnetic field which opposes the cause (motion or change of magnetic flux inducing the emf).

Considering Fig. 11.34, the motion of N pole of the magnet towards the coil increases the magnetic flux linking the coil, and induces a current. The induced current develops an N pole in the coil, on the face directed towards the magnet as shown in Fig. 11.36.

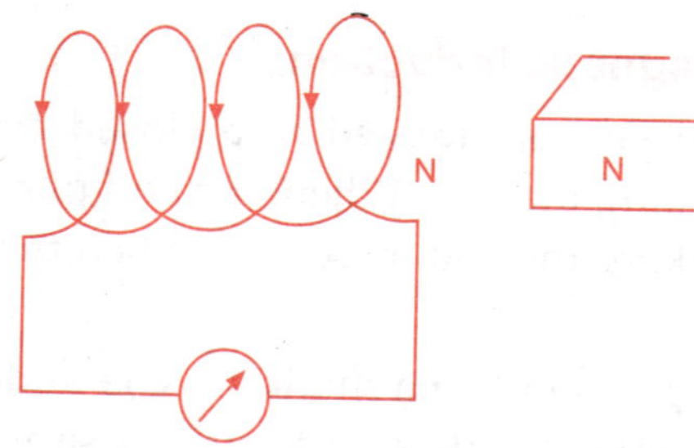

Fig. 11.36 *Directions of current and induced emf*

This tries to keep down the increase in the magnetic flux. Therefore, Eq. (11.86) can be written in the form

$$\varepsilon = -K\frac{d\phi}{dt} \tag{11.87}$$

where K is a constant of proportionality. If ε is in volts and the magnetic field is in tesla or Wb/m^2, K equals unity. Therefore, Eq. (11.87) can be written as

$$\varepsilon = -\frac{d\phi}{dt} \tag{11.88}$$

Faraday's law and Lenz's law combined together give both the magnitude and direction of the induced emf.

As mentioned earlier, Lenz's law follows from the principle of conservation of energy. Suppose the N pole of a magnet moving towards the coil as shown in

Fig. 11.36 is not opposed but helped, by developing a S pole in the coil at the face towards the magnet, then a magnet once set in motion towards a coil will continue to move in that direction. Such behaviour was not observed in experiments and is against the principle of conservation of energy. Therefore, the direction of the induced current is correctly described by Lenz's law.

11.5.3 Fleming's Right hand Thumb Rule

When a conductor is moved across a magnetic field, it cuts the lines of force and consequently an emf is induced across the conductor. The direction of induced emf and the current can be found by applying Fleming's right hand thumb rule, as shown in Fig. 11.37.

The rule states that, if the first three fingers of the right hand are held mutually perpendicular to each other, with the Fore finger in the direction of the Field and the Main finger (thumb) in the direction of Motion, then the Central finger (middle finger) points in the direction of the induced current. The rule is also known as generator rule.

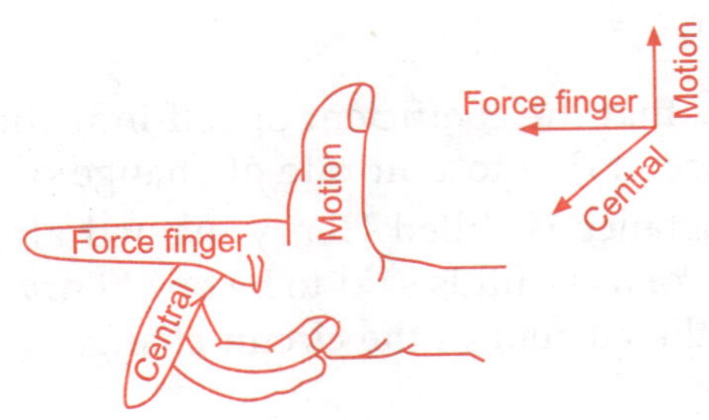

Fig. 11.37 *Fleming's right hand thump rule*

11.5.4 Self-induction

When a current is passed through a coil, it produces a magnetic field and the coil is linked with magnetic flux. Therefore, an instantaneous change in the magnetic flux induces an emf in the coil opposing the growth of current. This emf is called the back emf or inverse emf. Again, when the current is switched off, there is sudden fall of magnetic flux linking the coil. This sudden change in the magnetic flux induces an emf which opposes the decay of current. This emf is called forward emf or direct emf. This phenomenon of generation of emf in a coil due to a change of current in the coil is known as self-induction. This emf is called emf of self-induction and the coil is said to possess self-inductance. The effect will be more pronounced, if the coil is wound over a soft iron core instead of air in the core. It is observed in general that, the emf induced during break (switching the current off in the coil) is greater than the emf induced during a make (switching the current on in the coil).

11.5.5 Self-inductance

The magnetic flux ϕ linking with a coil, due to a current through it is proportional to the current I, expressed as $\phi \propto I$. This can be expressed in the form of an equation as

$$\phi = LI \tag{11.89}$$

where L is a constant of proportionality for the coil. It is called coefficient of self-induction or self-inductance. If $I = 1$, we have $L = \phi$.

Therefore, coefficient of self-induction of a circuit is equal to the magnetic flux linking the circuit due to unit current in the circuit.

From the laws of electromagnetic induction, we have the induced emf which is given by Eq. (11.88) as, $\varepsilon = -\dfrac{d\phi}{dt}$. Substituting for from Eq. (11.89), we have

$$\varepsilon = -\frac{d}{dt}(LI) \tag{11.90}$$

Since L is a constant, we have the induced emf given by

$$\varepsilon = L\frac{dI}{dt} \tag{11.91}$$

When $\dfrac{dI}{dt} = 1$, we have $\varepsilon = L$.

Therefore, we can say that the coefficient of self-induction of a circuit is equal to the emf induced in the circuit due to unit rate of change of current in the circuit.

Practical unit of inductance is called Henry, named after the American scientist Joseph Henry. Inductance of a circuit is said to be one henry, if the emf induced in the circuit is one volt, when the current in the circuit changes at the rate of one amp per sec.

Self-inductance of a coil depends on (*i*) the number of turns in the coil, (*ii*) the length of the coil, (*iii*) the area of cross section of the coil and (*iv*) the permeability of the medium inside the coil.

Any coil of wire will usually have both inductance and resistance. This is particularly undesirable in a coil used as a standard resistor. A non-inductively wound wire has one winding in one direction and one in the other direction. This design creates two resistors in parallel. Coils in standard resistance boxes are made non-inductive, by folding the particular length of insulated wire and then winding it doubled round a bobbin as shown in Fig. 11.38.

Fig. 11.38 *Resistances made non-inductive*

Self-induction can be demonstrated by the following simple experiment. A cylindrical coil (solenoid) of insulated thick copper wire, wound over a soft iron core, is connected in series with a battery B, an ammeter A, and a key K, in series. An electric bulb L is connected in parallel with the coil as shown in Fig. 11.39.

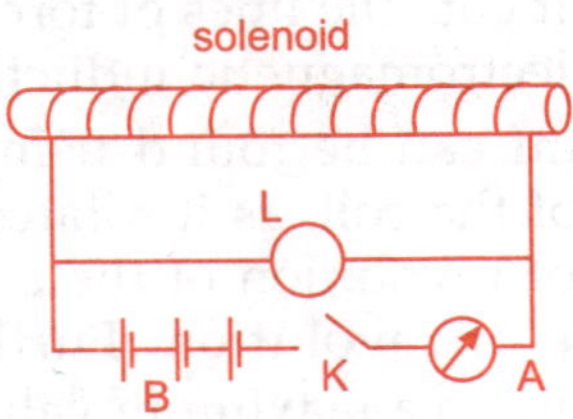

Fig. 11.39 *Self induction*

When the key is closed, the ammeter shows a large current instantaneously and becomes steady after a short interval. The lamp will also glow dimly for a moment. This is due to the induced back emf which opposes the growth of current. When the key is opened and the current is switched off, however, the current does not become zero instantaneously, but takes a little time. This is because the induced direct emf opposes the decay of current. The lamp will glow more luminously, for a moment. This shows that the emf induced at the break is larger than that at the make of the circuit.

11.5.6 Mutual Induction

When there are two coils in close vicinity, variation of current in one coil (called primary coil) will induce an emf in the second coil (called secondary coil) due to mutual induction.

The induced emf is given by

$$\varepsilon = -\frac{M dI}{dt} \tag{11.92}$$

where M denotes coefficient of mutual induction or mutual inductance. It is equal to the emf induced in one circuit due to unit rate of change of current in the other. Mutual inductance depends on (*i*) the number of turns in the primary coil, (*ii*) the number of turns in the secondary coil, (*iii*) area of cross section of the secondary coil and (*iv*) the permeability of the medium in the primary. In addition to these, the coefficient of mutual induction also depends on the degree of coupling, which is the extent to which the magnetic flux due to the primary current links the secondary. Coefficient of mutual induction for a pair of coils is said to be one henry, if the emf induced in one coil is one volt, when the current in the other coil varies at the rate of one ampere per second.

The construction of induction coil and transformer, which has primary and secondary coils, is based on mutual induction.

11.5.7 Alternating Current

A current with magnitude varying periodically with time, and reversing in direction every half a cycle is called an alternating current, normally referred to as *AC*.

When a coil rotates in a magnetic field, with its axis of rotation perpendicular to the field as shown in Fig. 11.40, it cuts the lines of force and induces an emf in the coil according to the principles of electromagnetic induction.

The direction of induced emf can be found using Fleming's right hand thumb rule. Four successive positions of the coil, as it rotates, are shown in Figs. 11.40(*b*) to (*e*). During one complete cycle of revolution of the coil, the induced emf reverses or alternates in every half a cycle of revolution. Further, during each of these half cycles, the emf increases from zero to a maximum value and then decreases gradually

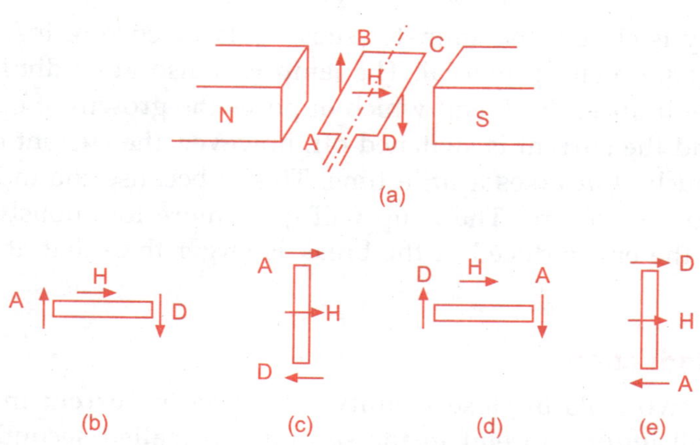

Fig. 11.40 *Alternating current*

from the maximum value to zero again. When the magnetic field and the speed of revolution of the coil are uniform, it can be shown that the emf varies sinusoidally or harmonically. In other words, a plot of the variation of the induced emf against time will be a sine curve. The current also varies in a similar manner. This type of harmonically varying or sinusoidal current is the simplest type of alternating current. Frequency of *AC* is the number of such cycles per sec, represented by *f*.

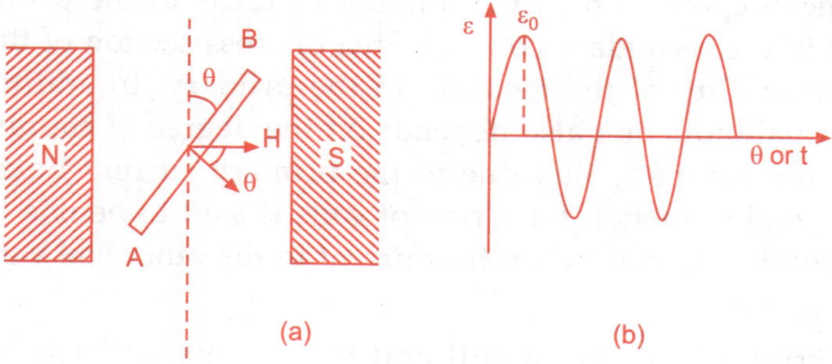

Fig. 11.41 *Coil rotating in magnetic field with uniform speed*

The magnitude of the emf induced at any instant is obtained as follows:

If the coil rotates with a uniform angular speed ω, ($\omega = 2\pi f$) then θ, the angle of rotation in a time t is given by $\theta = \omega t$. The angle θ measured from the instant when the coil is vertical as shown in Fig. 11.41.

The component of H perpendicular to the coil is $H \cos \theta$ and the flux linking the coil in this position is given by

$$\phi = nA\mu_0 H \cos \theta = nA\mu_0 H \cos \omega t \qquad (11.93)$$

where n is the number of turns and A is the area of the coil. The induced emf at the instant is given by

$$\varepsilon = -\frac{d\phi}{dt} = \omega nA\mu_0 H \sin \omega t \qquad (11.94)$$

If the coil is wound over a core of relative permeability μ_r, the expression for magnetic flux is obtained as

$$\phi = nA\mu H \cos \omega t = nAB \cos \omega t \qquad (11.95)$$

where $B = \mu H$, and $\mu = \mu_0 \mu_r$. Substituting we get the induced emf as

$$\varepsilon = \omega n\, AB \sin \omega t \qquad (11.96)$$

This can be expressed as

$$\varepsilon = \varepsilon_0 \sin \omega t \qquad (11.97)$$

where $\qquad \varepsilon_0 = \omega n\, AB.$

Since the maximum value of $\sin \omega t$ or $\sin \theta$ is 1, ε_o represents the maximum or peak value of the emf. The alternating emf ε is sinusoidal. The variation of emf ε with angle θ or time t is shown in Fig. 11.41(b).

If the circuit has only resistance R, then the current flowing through is given by

$$I = \frac{\varepsilon}{R} = \frac{\varepsilon_0}{R}\sin \omega t = I_0 \sin \omega t \qquad (11.98)$$

where $I_0 = \dfrac{\varepsilon_0}{R}$, is the maximum or peak value of the current. Hence current varies in exactly the same manner as the voltage. Current and the emf reach their maximum or minimum values always together or they are said to be in phase.

Since the average value of $\sin \theta$ or $\cos \theta$ over a complete cycle (from 0 to 2π) is zero, the average values of alternating voltage and alternating current over a complete cycle are zero. For this reason DC instruments measuring AC voltage or current will indicate zero deflection. In AC, the average value of current is defined as average taken over half a cycle. It can be shown that

$$I_{av} = \frac{2}{\pi}I_o \qquad (11.99)$$

$$V_{av} = \frac{2}{\pi}V_o. \qquad (11.100)$$

Therefore, *AC* meters are based upon the measurement of root mean square value (rms value). Mean square value is defined as the average of the squared values over one complete cycle.

The root mean square (rms or effective) value of an alternating current is defined as that value of steady current which would dissipate heat at the same rate in a given resistance as the alternating current. Hot wire ammeter used for measuring current depends on the square of the current because the heat developed by a current, is proportional to the square of the current. It can be shown that

$$I_{rms} = \frac{I_o}{\sqrt{2}} \qquad\qquad (11.101)$$

and

$$V_{rms} = \frac{V_o}{\sqrt{2}} \qquad\qquad (11.102)$$

The rms value is also called the virtual value or effective value which is shown by *AC* instruments. In all the electrical appliances, the voltage marked on is nothing but the rms voltage to represent the voltage capacity of the instrument.

11.5.8 Circuit Elements

In an alternating current circuit, if the circuit contains only resistance, then the current and the voltage are always in phase. This means that when the voltage reaches a maximum, the current also reaches its maximum value, and when the voltage reaches a minimum value, the current also reaches its minimum value. Such a situation is expressed as $I = V/R$. If the resistance is larger, the current in the circuit gets smaller, and hence the resistance tries to impede the flow of current. There are two other kinds of circuit elements which impede the flow of current in specific ways. In one of them, the current always lags behind the voltage by 90°. Such a circuit element is called an inductor. In the second, the current always leads the voltage by 90°. Such an element is called a capacitor. In general, an alternating current circuit contains resistance, inductance and capacitance. These are called circuit elements.

(*a*) When a circuit contains a pure resistance R, and the applied voltage is $V = V_0$ sin ωt, then the current I is given by $I = I_0$ sin ωt where $I = V/R$ and $I_0 = V_0/R$. Therefore, the current and the voltage are in phase. This is shown in Fig. 11.42.

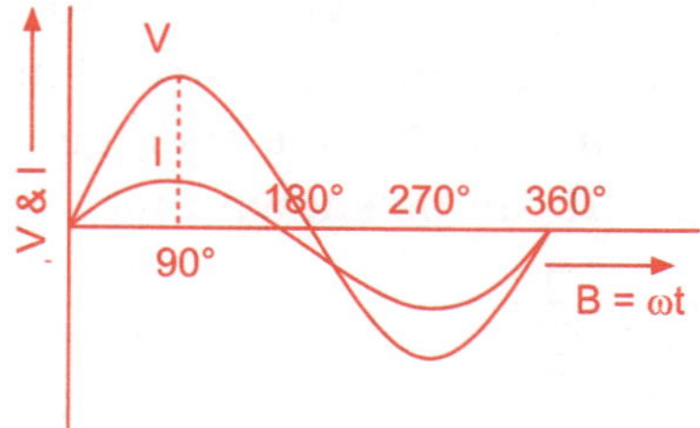

Fig. 11.42 *Current and voltage in phase in a resistance*

(b) If the circuit contains a pure inductance L and the applied voltage is $V = V_0$ sin ωt then the induced emf,

$$\varepsilon = -V = -V_0 \sin \omega t \tag{11.103}$$

However, we have

$$\varepsilon = -L\frac{dI}{dt} \tag{11.104}$$

Hence,

$$L\frac{dI}{dt} = V_0 \sin \omega t \tag{11.105}$$

This can be written as

$$\frac{dI}{dt} = \frac{V_0}{L}\sin \omega t \tag{11.106}$$

Integrating this expression we get

$$I = -\frac{V_0}{\omega L}\cos \omega t \tag{11.107}$$

which can be written as

$$I = I_o \sin\left(\omega t - \frac{\pi}{2}\right)$$

where $I_o = \dfrac{V_o}{\omega L}$ is the maximum value of current. The quantity ωL is called effective *AC* resistance or inductive reactance or inductance. It is normally represented by X_L. When L is in henrys and f is in cycles per sec., inductive reactance $X_L = \omega L = 2\pi f L$ is expressed in ohms. It can be seen that the current in an inductance lags the voltage by 90°. Figure 11.43 shows the variation of current and voltage with time.

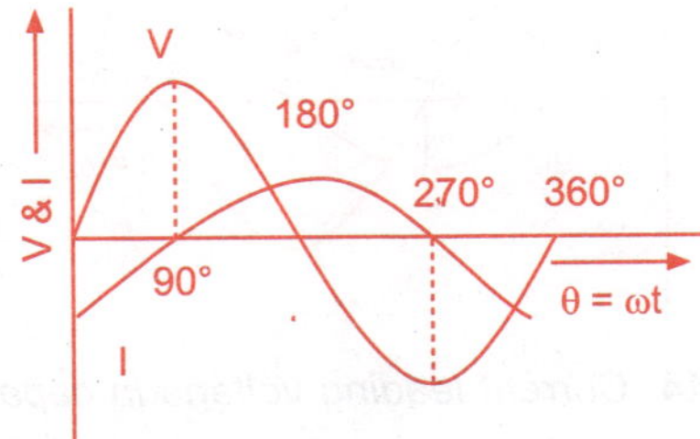

Fig. 11.43 *Current lagging voltage in inductance*

It can be seen from the figure that the maximum current occurs after the maximum voltage. The time interval between the two corresponds to 90°, and we say that the current lags the voltage.

Resistance is a measure of the opposition to the flow of a steady current, whereas inductance is a measure of the opposition to variation of current.

(c) If an *AC* voltage $V = V_o \sin \omega t$ is applied to a capacitor of capacitance *C*, the instantaneous charge on the capacitor is given by

$$q = CV = CV_o \sin \omega t \qquad (11.109)$$

The instantaneous current *I* is given by

$$I = \frac{dq}{dt} = CV_o \omega \cos \omega t = \frac{V_o}{1/\omega C} \cos \omega t \qquad (11.110)$$

which can be written as

$$I = I_o \cos \omega t = I_o \sin\left(\omega t + \frac{\pi}{2}\right) \qquad (11.111)$$

where $I_o = V_o\left(\dfrac{1}{\omega C}\right)$ gives the maximum value of current. The quantity $\dfrac{1}{\omega C}$ is called effective *AC* resistance or capacitive reactance of the capacitor. It is normally represented by X_C. The capacitive reactance $X_C = \dfrac{1}{\omega C} = \dfrac{1}{2\pi f C}$ is given in ohms, when *C* is in farads and *f* is in cycles/sec. From Eq. (11.111) it is seen that, the current in a capacitor leads its voltage by 90° or the potential difference across a capacitor lags the current through it by 90°. The variation of current and voltage with time is shown in Fig. 11.44, where it is seen that the maximum for the current occurs earlier than the maximum voltage. The time interval between the two corresponds to 90°, or we say that the current leads the voltage by 90°.

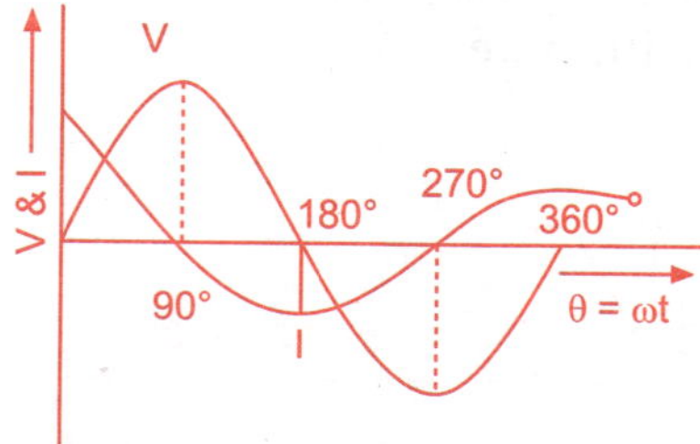

Fig. 11.44 *Current leading voltage in capacitance*

It must be noted that both inductive and capacitive reactances depend on the frequency of *AC*, because $X_L = \omega L$ and $X_C = \dfrac{1}{\omega C}$. Inductive reactance increases with frequency, whereas capacitive reactance decreases with increase of frequency.

Both inductive reactance and capacitive reactance are forms of impedance, which impede the flow of current in an *AC* circuit. Impedance is measured in ohms, when *V* is in volts and *I* is in amperes. If an *AC* circuit contains resistance, an inductance and a capacitance, as shown in Fig. 11.45, the instantaneous voltage consists of (*i*) the voltage V_R across the resistance, (*ii*) the voltage V_L across the inductance and (*iii*) the voltage V_C across the capacitor.

The phase relations among the voltages V_R, V_L and V_C in the three forms of impedances are shown in Fig. 11.45(*b*) in the form of a phase diagram, or vector diagram. It can be seen that V_R is in phase with the current and is represented by *OA* along the positive X-axis. The voltage V_L across the inductance leads the current by 90° and is shown along the positive Y-axis. The voltage V_C across the capacitor lags

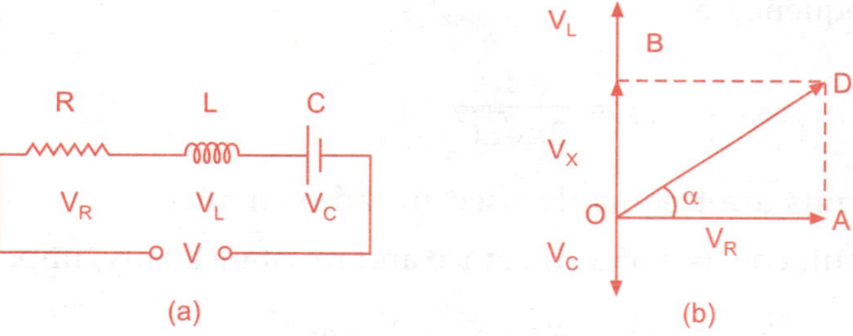

Fig. 11.45 *An R − L − C circuit*

the current by 90° and is shown along the negative Y-axis. If V_L is greater than V_C then the sum of V_L and V_C is $V_X = V_L - V_C$ and is represented by *OB*. The diagonal *OD* of the rectangle *OADB* represents the vector sum of V_X and which is the same as the vector sum of V_L, V_C and V_R. From the right angle triangle *OAD*, we get the relation

$$V^2 = V_x^2 + V_R^2 = (V_L - V_C)^2 + V_R^2 \qquad (11.112)$$

But $V_L = IX_L$, $V_C = IX_C$ and $V_R = IR$. Therefore

$$V^2 = (IX_L - IX_C)^2 + I^2 R^2 = I^2[(X_L - X_C)^2 + R^2] \qquad (11.113)$$

From this we can get an expression for the current as

$$I = \frac{V}{\sqrt{R^2 + (X_L - X_C)^2}} \qquad (11.114)$$

The above equation can be written as $I = \dfrac{V}{Z}$. Therefore, the impedance Z is given by

$$Z = \sqrt{R^2 + (X_L - X_C)^2} \qquad (11.115)$$

From Fig. 11.45(b), the angle α between V and V_R is given by

$$\tan \alpha = \frac{AD}{OA} = \frac{V_L - V_C}{V_R} = \frac{IX_L - IX_C}{IR} = \frac{X_L - X_C}{R} \qquad (11.116)$$

Therefore, the voltage leads the current by an angle α given by Eq. (11.116). The impedance Z is minimum and hence I is maximum when $X_L = X_C$. When the current in the circuit is maximum, the circuit is said to be at resonance. At resonance conditions, current becomes very high if the resistance is very small.

At resonance, $X_L = X_C$ implies that $\omega L = \dfrac{1}{\omega C}$ or $\omega^2 = \dfrac{1}{LC}$. Therefore, we get an

expression for the frequency as $\omega = \dfrac{1}{\sqrt{LC}}$ and hence $2\pi f = \dfrac{1}{\sqrt{LC}}$. From this we can get

the resonance frequency as

$$f = \frac{1}{2\pi\sqrt{LC}} \qquad (11.117)$$

Resonant circuits are frequently used in radio circuits.

When the circuit consists of inductance and resistance only, impedance is given by

$$Z^2 = R^2 + X_L^2 = R^2 + \omega^2 L^2 \qquad (11.118)$$

This can be written in the form

$$L^2 = \frac{Z^2 - R^2}{\omega^2} \quad \text{or} \quad L = \frac{\sqrt{Z^2 - R^2}}{\omega} = \frac{\sqrt{Z^2 - R^2}}{2\pi f}. \qquad (11.119)$$

11.5.9　Determination of Self-inductance of a Coil

The self-inductance of a coil can be determined employing a set-up as shown in Fig. 11.46. AC/DC instruments are used to measure the current and the potential difference.

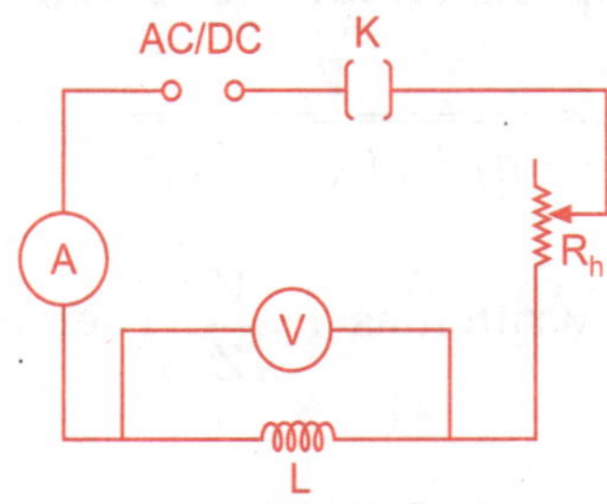

Fig. 11.46 *Self inductance of a coil*

A battery of cells is used for *DC* supply. A suitable current is passed through the circuit by adjusting the rheostat, R_h. The current, I, and the voltage, V, are used to get the resistance, R, of the coil from the relation, $R = \dfrac{V}{I}$. The procedure is repeated using *AC* voltage supplied through a step down transformer. In this case, the ratio of the voltage to current, $Z = \dfrac{V}{I}$ gives the impedance Z of the coil.

11.5.10 Choke

The mean power consumed in *AC* circuits is given by

$$P = V_e I_e \cos \phi$$

where V_e and I_e are rms values of the voltage and the current, and ϕ is the phase angle between them or the angle by which the current lags or leads the voltage.

For a pure inductance or for a pure capacitance, we have $\phi = 90°$ and hence $P = 0$, or no power is drawn. A coil made of a large number of turns of thick copper wire has very little resistance in *DC* circuit. However, because of its high inductance, it offers large impedance in *AC* circuit and allows only a small current through it. With a variable iron core, the impedance may be varied over a wide range. Such inductance coils having large inductance but negligible resistance are called choke coils. Choke coil is used in *AC* circuits to regulate the current, instead of a rheostat. If a rheostat is used to limit the current, power is absorbed from the source and wasted in the form of heat. Instead, a choke coil has the important advantage that it cuts down current without absorbing appreciable power. Since, the resistance of a choke is quite small, heat loss is negligible. The only loss is the hysteresis loss in the core of the inductance. But this is very much less than the heat loss in resistance used in place of choke.

The choke coils used at low and audio-frequency have an iron (magnetic) core whereas, those used on high or radio-frequency have an air core. Magnetic core coils find extensive use in resonance circuits at radio frequency. Choke coils are also used in mercury lamps and sodium vapour lamps in *AC* circuits.

Capacitors can also be used as current limiters. A parallel combination of inductance and capacitance with negligible resistance act as a perfect choke at a particular frequency.

11.5.11 Transformer

Transformer is used for increasing (stepping up) or decreasing (stepping down) the alternating voltages. It is used in a number of electrical appliances where the supply voltage must be changed to suit the appliance. It is also used in electrical transmission over long distances in order to minimize power losses. A transformer is based on the principle of mutual induction. It consists of two coils (primary and secondary) wound over a closed, laminated iron core as shown in Fig. 11.47.

If an alternating current flows through the primary coil, the magnetic flux through the primary varies continuously. The varying flux through the primary links the secondary. If all the flux produced by the primary links the secondary, the flux linking the coils is proportional to the respective number of turns in the coils.

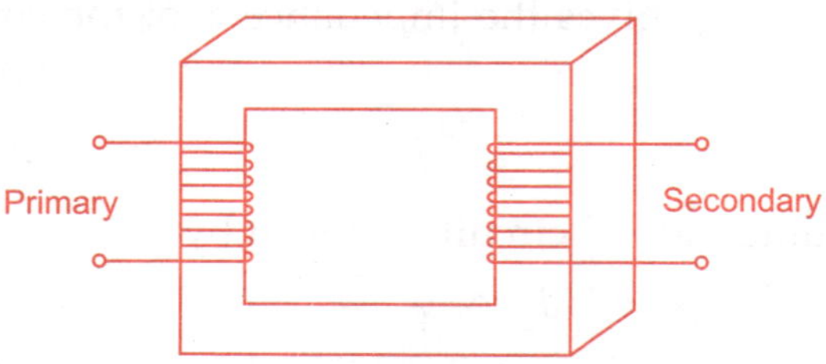

Fig. 11.47 *Transformer*

But, induced emf is proportional to the rate of change of flux. Hence, the induced emf is proportional to the respective number of turns. If E_p and E_s are the induced emfs, N_p and N_s the number of turns in the primary and secondary coils respectively, then we have

$$\frac{E_s}{E_p} = \frac{N_s}{N_p} = T \qquad (11.120)$$

where T is the turns ratio. In practice, however, induced emf, E_p will be less than the voltage applied to the primary and the secondary voltage at the terminals is slightly less than the emf, E_s, induced in the secondary.

Under ideal conditions with no power losses, power input to the primary is equal to the power output from the secondary. Therefore,

$$E_p I_p = E_s I_s \text{ or } \frac{I_p}{I_s} = \frac{E_s}{E_p} \qquad (11.121)$$

which implies that the ratio between the currents is inversely proportional to the ratio between the respective voltages.

When $N_s > N_p$, we have $E_s > E_p$, and the transformer is known as step up transformer. When $N_s < N_p$, then $E_s < E_p$ and the transformer is known as step down transformer.

A step up transformer steps up the primary voltage to give a higher secondary voltage, but steps down the current in the same ratio in the secondary. A step down transformer has a smaller emf in the secondary than in the primary, but has a larger current in the secondary than in the primary.

Electricity is transmitted over long distances at high voltage and low currents in order to minimise power losses, since, power loss will be small for low currents. This is accomplished using a step up transformer at the starting point, and stepping down again using a step down transformer at the consumer locations.

There are number of sources of losses in a transformer:

(*i*) Flux leakage loss—All the flux produced by the primary may not link the secondary. This is minimized by special types of winding.

(*ii*) Eddy current loss—This is minimized by using laminated cores of special alloys such as permalloy.

(*iii*) Hysteresis loss—This is minimized by using core material having a small hysteresis loop.

(*iv*) Power loss due to heating—This is minimized by using thick copper wires of low resistance. Heat produced by eddy currents and hysteresis loss raises the temperature of the transformer. Transformer is kept cool by circulating special types of oils in tanks enclosing the transformer.

In *AF* (Audio Frequency) transformers which operate in the frequency range 20-16000 cycles per sec., iron core is used. *RF* (Radio Frequency) transformers which operate at high frequencies use air core. As eddy current losses are very high at very high frequencies, iron core transformers are not preferred at high frequencies.

Alternating current is advantageous over direct current in many practical applications:

(*i*) *AC* voltages can be changed from one value to another value for different applications using appropriate transformers.

(*ii*) In long distance transmission of electrical energy, energy loss can be minimized in *AC* transmission.

11.5.12 AC Meters

Since the average value of *AC* over a complete cycle is zero, ordinary ammeters and voltmeters show zero reading when used in *AC* circuits.

In order to conveniently read the *AC* voltages and currents, the *AC* meters must have steady pointer deflections while the current or voltage reverse their direction. The fact that the square of the *AC* current or voltage is always positive is used in *AC* meters. They are so designed that the deflection depends on the square of current or voltage. Moving iron and hot wire meters are based on this property. *AC* voltmeters are similar in construction to *AC* ammeters but are equipped with series resistances.

Although the deflection in an *AC* meter is proportional to the average of the square of the current, the scale is not graduated to read the square values. For convenience, the meters are graduated to indicate the rms value. Consequently the scale is not linear and the graduations are closely spaced at the beginning and farther apart, away from the zero end.

Hot wire meters are based on heating effect of current. A hot wire ammeter basically consists of a fine resistance wire between fixed terminals A, B as shown in Fig. 11.48.

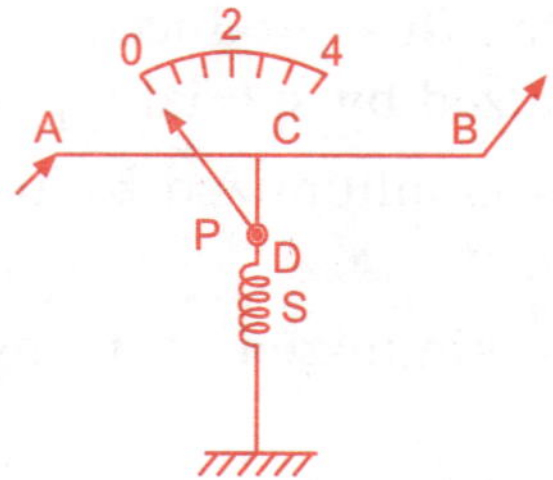

Fig. 11.48 *Hot wire ammeter*

A phosphor bronze wire *CD*, attached to *AB*, passes round a pulley *P* and is held to it by a spring *S*. The pulley is attached to the pointer of the instrument. When a current flows through *AB*, the wire sags owing to expansions and the sag is taken up by the wire *CD* and the pointer is deflected. The wire attains a temperature at which, the rate of loss of heat is equal to the rate of heat generation. The deflection is proportional to the average rate of generation of heat in the wire *AB* and this is proportional to the value of current.

Moving Iron meter is a common type of *AC* meter. It is based on the force of repulsion between two iron rods inside a coil, through which *AC* current is passed. In principle, a moving iron meter, consists of two iron rods R_1 and R_2, inside a coil shown in Fig. 11.49.

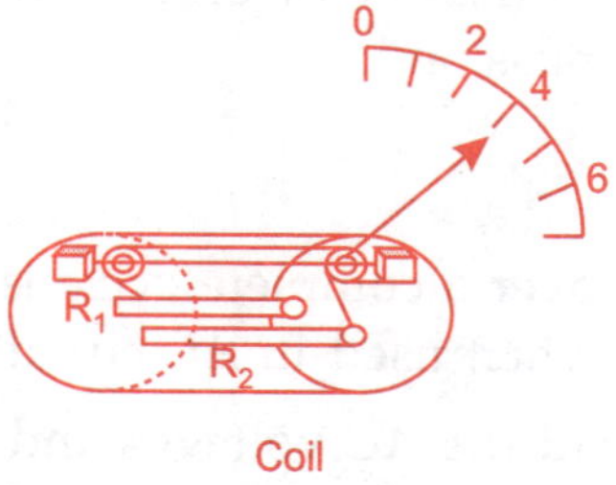

Fig. 11.49 *Moving iron meter*

The coil and one rod R_1 are fixed. The other rod R_2, is attached to an axle, carrying a pointer. The motion of the pointer is controlled by springs. When a current passes through the coil, the rods are magnetized with their neighbouring ends having the same polarity irrespective of the direction of current. Consequently, the average force is in the same direction and the pointer deflects through an angle proportional to this force. The magnetization of the rods being approximately proportional to the current, the force of repulsion between the rods is proportional to the square of the current. Hence, the deflection is proportional to mean square value of the current.

EXAMPLE 11.22

A coil of area 0.1 sq. m with 50 turns is acted upon by an uniform field of 2×10^{-2} weber/m^2 at right angles. If the coil is removed from the field in one hundredth of a second obtain the average emf induced in it.

SOLUTION

Magnetic flux linking the coil in the field is given by

$$\phi = NAB$$

where N = number of turns, A = area of the coil and B = magnetic field strength

$$\phi = 50 \times 100 \times 10^{-4} \times 2 \times 10^{-2} = 10^{-2} \text{ Weber}$$

When the coil is removed, the flux is reduced to zero.

$$\text{Induced emf} = \frac{\text{change of flux}}{\text{time}} = \frac{10^{-2}}{10^{-2}} = 1 \text{ volt}$$

Therefore, the induced emf = 1 volt.

EXAMPLE 11.23

A 2 m long conductor moves at the rate of 50 m/sec perpendicular to the earth's horizontal field which is 5×10^{-5} weber/m^2. Find the emf induced along the conductor.

SOLUTION

Distance travelled by the conductor in one sec = 50 m. Length of the conductor is 2 m.

Area swept out by the conductor = $50 \times 2 = 10^2 \text{m}^2$.

Number of magnetic lines of induction cut by the conductor per sec = $(5 \times 10^{-5}) \times 10^2 = 5 \times 10^{-3}$ volt.

EXAMPLE 11.24

Two adjacent coils have a mutual inductance of 0.25 H. If the current in the primary changes from zero to 2 amp in 0.05 sec. Find the average induced emf in the secondary.

SOLUTION

$$E = M\frac{dI}{dt} = 0.25 \times \frac{2}{0.05} = 10$$

Induced emf in the secondary = 10 volts.

EXAMPLE 11.25

A rectangular coil of 300 turns has an average area 0.25 m × 0.15 m. The coil rotates at a speed of 50 cycles/sec in a uniform magnetic field of strength 0.04 weber/m², about an axis perpendicular to the field. Find the maximum value of induced emf.

SOLUTION

$N = 300$, $A = 25 \times 15 \times 10^{-4}$ m², $B = 0.04$

$$\omega = 2\pi f = 2\pi \times 50 = 100\pi$$

$$E_{max} = NAB\omega = 300 \times 25 \times 15 \times 10^{-4} \times 0.04 \times 100\pi$$

$$E_{max} = 141.3$$

Hence the induced emf = 141.3 volts.

EXAMPLE 11.26

If the AC mains supply voltage is 220 volts at 50 cycles/sec, find the current passing through a bulb of 44 watts. Find also the peak value of voltage and current.

SOLUTION

The *AC* voltage of 220 volts represents rms value or effective value.

$$V_{rms} = \frac{V_o}{\sqrt{2}} = 220 \text{ volt}$$

$$V_o = 220\sqrt{2} = 311.2 \text{ volt}$$

$$\text{Peak value} = V_{rms} I_{rms}; \quad 44 = 220 I_{rms}$$

$$I_{rms} = \frac{44}{220} = 0.2 \text{ amp}$$

Peak value of current $\quad = I_{rms}\sqrt{2} = 0.2\sqrt{2} = 0.283$ amp

Current through the bulb $\quad = 0.2$ amp

Peak value of voltage $\quad = 311.2$ volt

Peak value of current $\quad = 0.283$ amp.

EXAMPLE 11.27

Obtain the impedance of a coil of self inductance 0.1 henry and resistance 20 ohms when it is connected to an AC of 60 Hz.

SOLUTION

$L = 0.1H$, $R = 20$ ohms, $f = 60$ Hz

$$Z^2 = R^2 + X_L^2 = R^2 + \omega^2 L^2$$

where,

$$\omega = 2\pi f$$

$$Z = \left(20^2 + 4\pi^2 \times 60^2 \times 0.1^2\right)^{\frac{1}{2}} = \left(400 + 1422\right)^{\frac{1}{2}}$$

$$Z = 1.$$

Therefore, impedance of the circuit = 42.7 ohms.

EXAMPLE 11.28

A primary coil of transformer is connected to 220 volts AC source. The primary has 660 turns of wire. The secondary coil supplies 6.3 volts to filaments of tubes. Find the number of turns in the secondary.

SOLUTION

$$\frac{N_s}{N_p} = \frac{V_s}{V_p}$$

$$N_s = \frac{V_s}{V_p} N_p = \frac{6.3}{220} \times 660 \approx 19$$

Hence the number of turns in the secondary = 19.

11.6　THERMOELECTRICITY

Thermoelectric effect is the direct conversion of temperature differences to electric voltage and vice versa. A thermoelectric device creates a voltage when there is a difference in temperature at two points in a closed circuit. Conversely, when a voltage is applied across the two points, it creates a temperature difference between the two points. An applied temperature gradient causes charge carriers in the material to diffuse from the hot side to the cold side.

This effect can be used to generate electricity, measure temperature or change the temperature of objects. Because, the direction of heating and cooling is determined by the polarity of the applied voltage, thermoelectric devices are efficient temperature controllers.

The thermoelectric effect encompasses three separately identified effects: the (*i*) Seebeck effect, (*ii*) Peltier effect and (*iii*) Thomson effect. Sometimes it is referred to as the Peltier–Seebeck effect because the effect was independently discovered by the French physicist Jean Charles Athanase Peltier and Balt-German physicist

Thomas Johann Seebeck in 1826. The other method called as, Joule heating, the heat that is generated whenever a voltage is applied across a resistive material, is related though it is not generally termed a thermoelectric effect. The Peltier–Seebeck and Thomson effects are thermodynamically reversible, whereas Joule heating is not.

11.6.1 Seebeck Effect

When two wires of dissimilar metals are joined together at their ends to form a loop and the junctions are maintained at different temperatures, a current is found to flow in the circuit. An emf is set-up in the circuit and electrical energy is produced in the circuit. There is a gain of heat energy at the hot junction. The production of current and hence emf by maintaining the junctions of two dissimilar metals at different temperatures is known as Seebeck effect. The emf is called thermoelectric emf or thermo emf and the current produced is called thermoelectric current. The pair of dissimilar metals with a pair of junctions is called a thermocouple.

In Fig. 11.50, two wires of dissimilar metals such as copper and iron, are joined at their ends, with a sensitive galvanometer in the circuit.

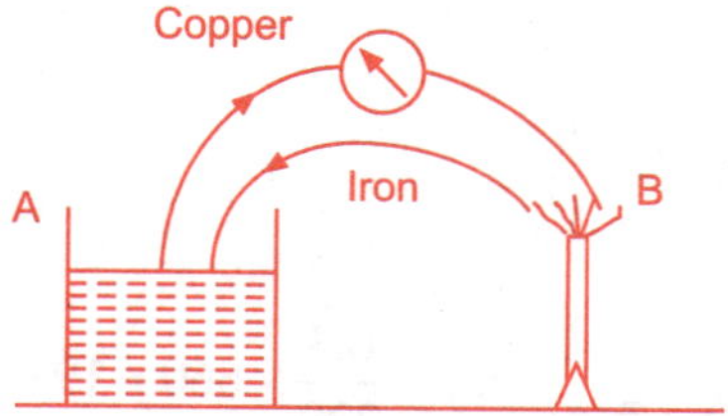

Fig. 11.50 *Seebeck effect*

If the junction *B* is heated by a flame and the other junction is kept cold, the galvanometer needle will show a deflection, indicating a flow of current in the circuit. If the junction *A* is heated and the junction *B* is kept cold, the current is found to flow in the opposite direction.

In the case of copper-iron pair, current flows from copper to iron through the hot junction. In the case of antimony-bismuth pair, current flows from antimony to bismuth through the cold junction. Seebeck studied the phenomenon with various metals and arranged them in a series known as thermoelectric series. If any pair of metals in this series is used, current flows through the cold junction from the metal higher in the series to the metal lower in the series at normal temperatures.

A few of the metals, in order, in the series are antimony, iron, zinc, lead, copper, nickel and bismuth. The separation between the metals in the series represents the magnitude of the emf produced for a given temperature difference. For example, bismuth-antimony pair gives a higher emf than the copper-iron pair. Thermoelectric emfs are of the order of hundred microvolts per degree C. The emf for antimony-bismuth couple is about 100 microvolts per degree C and for copper-iron it is about 15 microvolts per degree C.

An explanation to the Seebeck effect can be provided as follows. Free electrons move about in a metal in a random manner. Their motion resembles the motion of the molecules of a gas in a container. The density of free electrons in a metal is a function of temperature. It is different for different metals at a given temperature. When two metals are brought in contact with each other, some of the free electrons diffuse across the junction. The densities of free electrons in two metals being different, electrons flow from the metal of higher electron density to the metal with lower electron density, until the densities are equalized. Consequently, an emf is developed at the junction. When both the junctions of a thermocouple are at the same temperature, the emf at one junction is equal and opposite to the emf at the other junction, and therefore the net emf is zero. When the junctions are at different temperatures, the emfs at the junctions are different and there is a net emf in the circuit. This accounts for the Seebeck effect. Seebeck effect is a reversible phenomenon.

In 1834, Peltier, a French physicist, observed that when an electric current flows through a thermocouple of two dissimilar metallic conductors, the junctions being at the same temperature, one junction is cooled and the other junction is heated. Heat is absorbed at one junction and heat is generated at the other. The cooling or heating of a junction depends on the direction of current and the type of metal pairs. This phenomenon is the inverse of the Seebeck effect and is known as Peltier effect.

The Peltier effect is zero if an alternating current is used. However, the alternating current produces Joule heating. The Joule heating effect is used in the construction of thermometers. Peltier effect must be distinguished from Joule effect. The major difference between the two effects is that Peltier effect is reversible while Joule effect is irreversible. In Joule effect, heat is produced irrespective of the direction of the current flow.

With the discovery of Peltier effect, it was easy to find an explanation of Seebeck effect. The Seebeck effect sets up a net emf when there is a difference in temperatures of the junctions. The resulting current flow due to the Peltier effect will try to cool the hot junction and heat the cold junction. As a result, heat is absorbed at the hot junction, a part is converted into electrical energy and the rest is rejected at the cold junction. The thermocouple, in effect, acts like a heat engine.

Twenty years later, William Thomson (became Lord Kelvin) advanced a third thermoelectric effect, known as the Thomson effect where, the heat is absorbed or produced when current flows in a material with a temperature gradient. The heat is proportional to both the electric current and the temperature gradient. The proportionality constant, known as the Thomson coefficient is related by thermodynamics to the Seebeck coefficient.

11.6.2 Laws of Thermoelectric Effect

1. Law of intermediate metals

If the junction of thermocouple of elements A and B at temperature θ is opened and a third element X is introduced, the emf remains the same as before, provided both the junctions of X are at the same temperature θ. This is shown in Fig. 11.51.

Essentially the law states that the introduction of an additional metal into any thermoelectric circuit does not alter the emf of the circuit provided both the ends of the metal introduced are at the same temperature. The law may be generalized for any number of intermediate metals whose ends are at the same temperature. This is a very important property of thermo couple circuits because the introduction of copper connecting wires and measuring instruments, such as a galvanometer, will not affect the emf provided the introduced parts are at the same temperature as the adjacent parts.

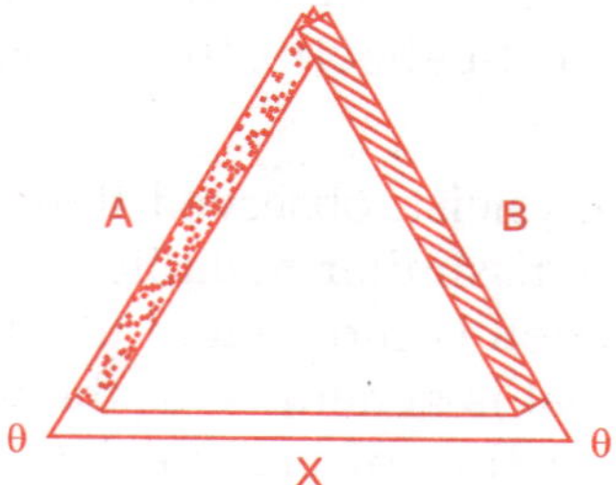

Fig. 11.51 *Insertion of intermediate metal in a junction*

2. Law of intermediate temperatures

For a given thermocouple, the emf between two temperatures is equal to the sum of the emfs corresponding to any number of successive intervals into which the given range of temperatures may be divided.

Therefore, if e_1 and e_2 are the thermoelectric emf of a couple for temperatures (θ_1, θ_2) and (θ_2, θ_3), the emf e_3 for the range (θ_1, θ_3) is given by

$$e_3 = e_1 + e_2 \tag{11.122}$$

This law may be generalized to any number of intermediate temperatures.

11.6.3 Variation of Thermoelectric emf with Temperature

The thermoelectric emf developed in a thermocouple depends on the difference in temperatures of the two junctions. If the temperature of the cold junction is maintained at a constant temperature, and the temperature of the hot junction is gradually increased, the emf also is found to increase.

The variation of thermoelectric emf with temperature is shown in Fig. 11.52.

The curve of thermoelectric emf against temperature shown in Fig. 11.52 is known as thermoelectric curve. The emf increases until it reaches a maximum value at a particular temperature. This maximum temperature is known as neutral temperature. If the temperature is increased further, the emf decreases until it is zero after which it reverses in direction. The thermoelectric curve resembles a parabola, with its axis parallel to the axis of emf.

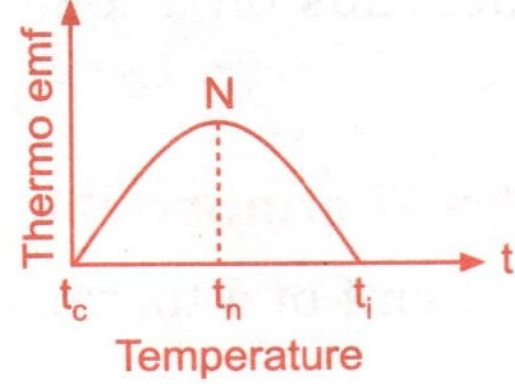

Fig. 11.52 *Variation of thermocouple with temperature*

The neutral temperature of the hot junction of a thermocouple where the emf becomes maximum, is a constant for a given pair of metals. Neutral temperature does not depend upon the temperature of the cold junction. The temperature at which the emf reverses direction is called temperature of inversion. The temperature of inversion depends on the temperature of the cold junction. The difference between the temperature of inversion and the neutral temperature is always same as the difference between the neutral temperature and the cold junction temperature. As shown in Fig. 11.52, if t_n, t_i and t_c represent neutral temperature, temperature of inversion and temperature of cold junction, respectively, then, we have the relation

$$t_i - t_n = t_n - t_c \qquad (11.123)$$

For copper-iron thermocouples, the neutral temperature is 270°C and the temperature of inversion is 540°C, when cold junction is maintained at 0°C.

11.6.4 Seebeck Coefficients

When the cold junction is kept at constant temperature and the temperature of the hot junction is varied, the thermoelectric emf is almost a parabolic function of the temperature. Therefore, thermoelectric emf, e, is given by the relation

$$e = a\theta + b\theta^2 \qquad (11.124)$$

where a and b are constants for a given pair of metals and θ is the temperature difference between the hot and cold junctions. The constants a and b are called Seebeck coefficients. They can be determined by finding the values of emf for two different values of θ. The constant a is expressed in volts per degree C and b in volts per degree C^2. It is usual practice to keep the cold junction at 0°C.

If e_1 and e_2 are the emf at θ_1 and θ_2, respectively, then from Eq. (11.124), we have

$$e_1 = a\theta_1 + b\theta_1^2 \qquad (11.125)$$

$$e_2 = a\theta_2 + b\theta_2^2 \qquad (11.126)$$

By solving Eqs. (11.125) and (11.126), the values of "a" and "b" can be obtained. These values represent values of "a" and "b" in the range θ_1 and θ_2.

The Eq. (11.124), is an approximate relation between "e" and "θ". Seebeck coefficients vary not only from metal to metal, but also vary with temperature, for a given metal. In fact, constant "a" depends on Peltier coefficient which varies with temperature, and constant *"b"* depends on Thomson coefficient which also varies with temperature.

11.6.5 Measurement of emf of a Thermocouple using a Potentiometer

The set-up for the measurement of emf of a thermocouple is shown in Fig. 11.53(*a*).

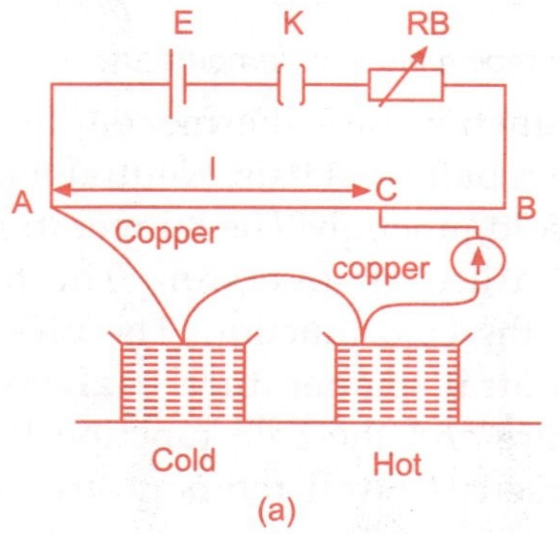

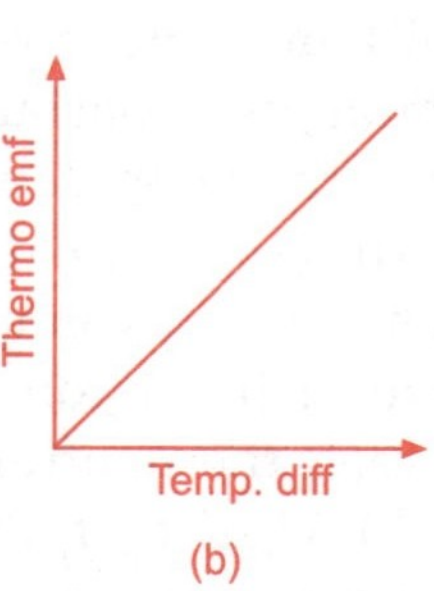

Fig. 11.53 *Measurement of emf using potentiometer*

The potentiometer wire AB is connected in series with a storage cell of emf E, a resistance box and a plug key. The resistance of the potentiometer wire is measured to be R. A suitable resistance R_1 is unplugged from resistance box so that the potential difference between A and B is about 2×10^{-4} volt

$$\left(\text{Potential difference between } A \text{ and } B = \frac{ER}{R+R_1} \right).$$

One of the junctions of the thermocouple is immersed in ice cold water contained in a beaker and the other junction in hot oil or boiling water contained in another beaker. The lead of positive junction of the thermocouple is connected to the positive end A of the potentiometer and the other lead is connected to a sliding contact through a sensitive galvanometer.

The position of the sliding contact on the potentiometer wire is adjusted such that the spot of light, obtained on the scale by reflection in the mirror of the galvanometer is not deflected. The balancing length $AC = l$ is noted and the corresponding temperature θ_2 of the hot junction is noted. The experiment is repeated for different temperatures of the hot junction, as it cools, in steps of about 10°C. If θ_1 is the temperature of the cold junction, temperature difference θ between the junctions is given by

$$\theta = \theta_2 - \theta_1$$

Thermo emf, e, for different values of θ is calculated using the relation

$$e = \frac{ER_1}{100}(R+R_1) \tag{11.127}$$

A graph of e against θ can be drawn. For moderate temperature differences the graph is found to be a straight line as shown in Fig. 11.53(b).

11.6.6 Thermoelectric Power

In Eq. (11.124), the thermoelectric emf is given as $e = a\theta + b\theta^2$. Therefore, the rate of change of emf with the temperature is given as

$$P = \frac{de}{d\theta} = a + b\theta \tag{11.128}$$

In Eq. (11.128), $P = \frac{de}{d\theta}$, is the slope of the thermoelectric curve at any temperature. It is called the thermoelectric power of the thermocouple at the particular temperature. The Eq. (11.128) is the equation of a straight line and hence we can conclude that the, thermoelectric power varies linearly with temperature. Thermoelectric power of a thermocouple at any temperature is the change in thermo emf per degree change in the temperature of the hot junction.

11.6.7 Thermoelectric Diagram

Thermoelectric diagram is a plot of the thermoelectric power against the temperature of the hot junction, when the cold junction is maintained at a constant temperature. Thermoelectric diagrams are also called thermoelectric power lines or Tait diagrams in honor of Peter Guthrie Tait, a British scientist and a good friend of Maxwell. It follows from Eq. (11.128), that thermoelectric diagram is a straight line as shown in Fig. 11.54. The slope of the straight line is $2b$ and the intercept with the thermoelectric power axis gives the constant a.

At the neutral temperature, (θ_n), we have, $\dfrac{de}{d\theta} = 0 = a + 2b\theta_n$, and hence

$$\theta_n = -\frac{a}{2b} \tag{11.129}$$

If the cold junction is at 0°C, then θ_n gives the neutral temperature. The temperature of inversion can be obtained as follows. Substituting $e = 0$ in Eq. (11.124), we get

$$0 = a\theta + b\theta^2 = \theta(a + b\theta) \tag{11.130}$$

Solving this equation we get either $\theta = 0$, or $\theta = -\dfrac{a}{b}$.

The first root is the temperature of the cold junction whereas the second value gives the temperature of inversion θ_I, where

$$\theta_i = -\frac{a}{b} \tag{11.131}$$

Thermoelectric diagrams are normally drawn with lead as a standard of reference, by taking the particular metal and lead as elements of the thermocouple.

The thermo emf e_{AB} for a pair of metals A and B is given by the relation

$$e_{AB} + e_{APb} + e_{PbB} = 0 \tag{11.132}$$

where e_{APb} and e_{PbB} are the thermo emfs for the pairs of metal A-Lead and Lead-metal B combinations, for the same range of temperatures. The Eq. (11.132) follows from the law of intermediate metals.

The two lines in Fig. 11.54 are for copper-lead and iron-lead thermocouples. The point of intersection of the two lines gives the neutral temperature for copper-iron couple.

In Fig. 11.54, the areas $ABCFA$ and $ABDEA$ represent, respectively, the thermo emfs of copper-lead and iron-lead thermocouples for the temperature range $\theta_1 - \theta_2$.

The difference in area, $FCDEF$ represents the corresponding thermo emf for copper-iron thermocouple.

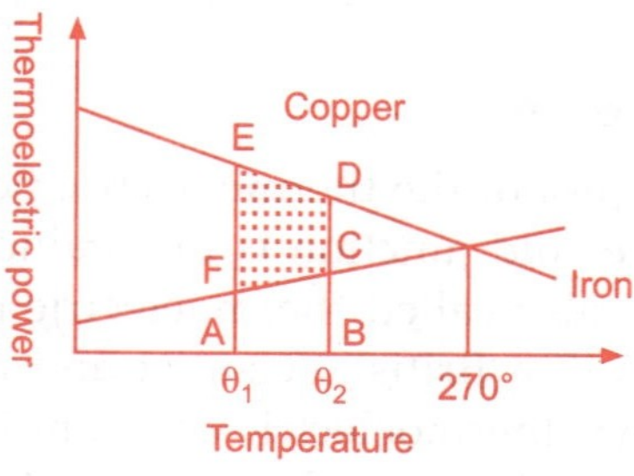

Fig. 11.54 *Tait diagram*

11.6.8 Applications of Thermoelectric Effect

1. Temperature measurement

Since the emf of a thermocouple depends upon the difference in temperature at the junctions, the thermo emf can be used as a measure of temperature. Different pairs of junctions must be used for measuring temperature in different ranges. A copper-constantan couple may be used for a temperature range of –200°C to 400°C. Thermocouples of platinum and platinum-rhodium wires enclosed in a quartz tube or in porcelain tube are used upto 1400°C. Molybdenum-tungsten couples can be used upto 1700°C. Such type of thermometers is called thermoelectric pyrometers. Thermocouple thermometers can be used for measurement of temperatures, both high and low and for detection of radiations from a source. Thermoelectric thermometers are calibrated using platinum resistance thermometers. Thermopile consisting of a series arrangement of similar thermocouples, insulated from each other, is used for measuring very small temperature differences. Thermopile can be used to measure and detect radiation from distant sources.

2. Boys' radio micrometer

Boys' radio micrometer is one of the most sensitive instruments for the detection of radiation. It is essentially a sensitive galvanometer and a thermocouple combined together.

3. Thermo ammeters

Thermo ammeter is essentially a sensitive pointer galvanometer and a thermocouple combined together. Thermo ammeters are used for the measurement of small alternating or direct currents. Specially designed thermo ammeters are used for the measurement of weak radio frequency currents. The instrument can be converted into a thermocouple voltmeter.

4. Thermoelectric generators

Thermoelectric effect can be used for the direct conversion of heat into electrical energy without the need for any moving parts. Although the efficiency is low, thermopile can be a very convenient electric generator when a small amount of power is required. Semiconductor thermocouples have been developed, in which the cooling of the junctions of two dissimilar semiconductors can be achieved by the passage of direct current, down to a temperature of about −80°C. This is reversible, and hence the same arrangement can be used both for cooling and heating.

EXAMPLE 11.29

Find the neutral temperature, the temperature of inversion and the total emf for copper-iron thermocouple between the temperatures 0°C and 100°C given that a = −13.89 μv per °C and b = 0.026 μv per °C².

SOLUTION

Neutral temperature

$$\theta_n = \frac{a}{2b} = \frac{13.89 \times 10^{-6}}{2 \times 0.026 \times 10^{-6}} = 267°C$$

Temperature of inversion = 2 × 167 = 534°C

Total emf, $e = a\theta + b\theta^2$

$$= -13.89 \times 10^{-6} \times 100 + 0.026 \times 10^{-6} \times 100^2$$

Neutral temperature = 267°C.

Temperature of inversion = 534°C

Total emf = − 1129 micro volts.

EXAMPLE 11.30

The thermoelectric power of iron-copper thermocouple at 100°C is 6.68 micro volts per °C and the neutral temperature for the couple is 282°C. Find the thermoelectric power at 0°C.

SOLUTION

Thermoelectric power, $\dfrac{de}{d\theta} = a + 2b\theta$

$$6.68 \times 10^{-6} = a + 200\,b$$

At neutral temperature, $\dfrac{de}{d\theta} = 0 = a + 2b(282)$

$$0 = a + 564b \tag{2}$$

From Eq. (1) and (2)

$$b = \frac{6.68 \times 10^{-6}}{354}$$

$$a = \frac{6.68 \times 10^{-6} \times 564}{764} = -10.35$$

Thermoelectric power at 0°C = a = −10.35 micro volts per degree C.

EXAMPLE 11.31

The thermoelectric powers for iron-lead couple and copper-lead couples are given by $P_{iron} = 1734 - 0.45\theta$ and $P_{copper} = 1376 + 0.85\theta$, the cold junction being at 0°C. Calculate the neutral temperature for copper-iron thermocouple.

SOLUTION

At the neutral temperature, thermoelectric power for the couple is zero.

$$1376 + 0.85\theta = 1734 - 0.45\theta$$

or

$$1.30\theta = 358$$

$$\theta = \frac{358}{1.30} = 275.4°C.$$

EXAMPLE 11.32

The thermoelectric power of cadmium is 3 μV per °C at 0°C and i5 μV per °C at 300°C. Calculate the values of the constants a and b.

SOLUTION

$$\frac{de}{d\theta} = a + 2b\theta$$

$$3 \times 10^{-6} = a + 2(b \times 0)$$

$$a = 3 \times 10^{-6}$$

$$15 \times 10^{-6} = a + 2b \times 300$$

$$b = \frac{12 \times 10^{-6}}{2 \times 300} = 2 \times 10^{-4} = 0.02 \times 10^{-6}$$

$$a = 3 \ \mu V \text{ per } °C$$

$$b = 0.02 \ \mu V \text{ per } °C^2.$$

SUMMARY

1. The direction of flow of positive charge from a point at a higher potential to a point at a lower potential gives potential direction of current.

2. The electromotive force of a source of current is the energy required to transfer or drive a unit charge through a closed circuit, in a reversible process.

3. Laplace rule states that the magnetic field at a point due to a current element is (*i*) directly proportional to the strength of the current, (*ii*) directly proportional to the length of the element, (*iii*) directly proportional to the sine of the angle between the element and the line joining the point to it, and (*iv*) inversely proportional to the square of the distance of the point from the element.

4. **Fleming's left hand thumb rule:** If the first three fingers of the left hand, identified as Main finger (thumb), Fore finger, and Central finger (middle finger), are held at right angles to each other such that the Central finger points in the direction of the Current and the Fore finger in the direction of the Field, then the Main finger points in the direction of Motion or Mechanical force.

5. A galvanometer can also be used as an ammeter with a suitable shunt which has low resistance connected in parallel to carry a large portion of the current.

6. A galvanometer can also be used as a voltmeter with a suitable shunt which has high resistance connected in series to carry a small portion of the current.

7. Ohm's law states that, the strength of the current flow through the conductor is proportional to the potential difference between its ends, provided its temperature and other physical conditions remain unaltered.

8. The resistance of a conductor can be defined as the ratio of the potential difference across the two ends of the conductor to the current flowing through it.

9. When the resistances are connected in series, the same current flows through the series of resistances and the potential difference across the combination is equal to the sum of the potential differences across each resistance.

10. When the resistances are connected in parallel, the main current is equal to the sum of the branch currents.

11. Electrical resistance of a conducting wire depends on (*i*) material of the wire (*ii*) the size of the wire and (*iii*) its temperature.

12. Experimental studies have shown that the electrical resistance of metals increases with increase in temperature.

13. Thermally sensitive resistors are named as thermistors. They can measure temperature changes as low as 0.01°C.

14. Kirchhoff's First law: The algebraic sum of the currents at a junction of an electrical circuit is zero.

15. **Kirchhoff's Second law:** In any closed mesh (or path) in an electrical network, the algebraic sum of the emfs is equal to the sum of all the voltage drops (products of current, I, and resistance, R) in that mesh.

16. Wheatstone's network is used for an accurate comparison of resistances.

17. Electromotive force or current also gets induced, when a closed loop of wire or conductor is moved under the influence of magnetic field. The instantaneous emf that is induced is known as induced emf, and the phenomenon is known as electromagnetic induction.

18. **Faraday's laws of electromagnetic induction:** Whenever magnetic flux linking a circuit changes, an instantaneous emf is induced in the circuit and this emf exists when the change is taking place.

19. The magnitude of the induced emf is directly proportional to the rate of change of magnetic flux linking the coil.

20. **Lenz's law:** An induced electromotive force (emf) always gives rise to a current whose magnetic field opposes the original change in magnetic flux

21. **Fleming's right hand thumb rule:** The first three fingers of the right hand are held mutually perpendicular to each other, with the Fore finger in the direction of the Field and the Main finger (thumb) in the direction of Motion, then the Central finger (middle finger) points in the direction of the induced current. The rule is also known as generator rule.

22. **Back emf:** When a current is passed through a coil, it produces a magnetic field and the coil is linked with magnetic flux. Therefore, an instantaneous change in the magnetic flux induces an emf in the coil opposing the growth of current.

23. **Mutual induction:** When there are two coils in close vicinity, variation of current in one coil (called primary coil) will induce an emf in the second coil (called secondary coil) due to mutual induction.

24. The current always lags behind the voltage by 90°. Such a circuit element is called an inductor.

25. The current always leads the voltage by 90°. Such an element is called a capacitor.

26. Transformer is used for increasing (stepping up) or decreasing (stepping down) the alternating voltages.

27. The thermoelectric effect is the direct conversion of temperature differences to electric voltage and vice versa.

28. **Seebeck effect:** When two wires of dissimilar metals are joined together at their ends to form a loop and the junctions are maintained at different temperatures, a current is found to flow in the circuit.

29. **Laws of thermoelectric effect:**

   (*a*) **Law of intermediate metals:** The introduction of an additional metal into any thermoelectric circuit does not alter the emf of the circuit provided both the ends of the metal introduced are at the same temperature.

   (*b*) **Law of intermediate temperatures:** For a given thermocouple, the emf between two temperatures is equal to the sum of the emfs corresponding to any number of successive intervals into which the given range of temperatures may be divided.

30. Thermoelectric diagram is a plot of the thermoelectric power against the temperature of the hot junction, when the cold junction is maintained at a fixed temperature.

❑❑❑

Modern Physics   12

12.1 INTRODUCTION TO ATOMIC PHYSICS AND NUCLEAR PHYSICS

12.1.1 Structure of Matter

The structure of matter has been a subject of great curiosity since very early times. Early Greeks and Indians had considered matter to be disconti-nuous. In the 4th and 5th centuries people regarded the matter to consist of small indivisible and invisible particles (spermata). Kanada, an Indian philosopher, assumed matter to consist of small particles. However, real progress regarding the structure of matter started only in the nineteenth century. In 1808 an English Chemist, John Dalton, advanced his atomic theory that could explain certain experimental facts about chemical combinations. The atomic theory was supported by the successful development in the kinetic theory of gases.

12.1.2 Atomic Structure and Electrons

During the decade 1895 to 1905, the idea that the atom is indivisible was discarded. It was conclusively shown that atom has a definite structure and that it can be broken into its constituent particles. Faraday's laws of electrolysis (1838) suggested the existence of an elementary charge. Faraday found that a charge of 96,500 coulombs is required to liberate one gm equivalent of a substance. If N denotes the Avagadro number, then 96,500/N, gives the amount of charge per ion as 1.6×10^{-19} coulomb. A charge which is lower than 1.6×10^{-19} coulomb is not observed means that electricity is atomic in nature. In 1891, Dr. Johnstone Stoney, a British physicist, established the existence of these atoms of electricity. In 1897, J. J. Thomson, an English physicist, established by his experiments using discharge of electricity through gases that electrons are common constituents of all matter. Thomson and Millikan established through their experiments that electron is a negatively charged particle having a charge $e = 1.6 \times 10^{-19}$ coulomb and a mass $m = 9.1 \times 10^{-31}$ kg. The discovery of radioactivity and the electron, rekindled the interest in the atomic structure of matter. Since the atom is electrically neutral, it was concluded that an atom must contain positively charged particles to balance the negative charge of electrons. Further, since the mass of an electron is extremely small, the mass of the atom depends on its positively charged particles. In 1911, Lord Rutherford, a British physicist, advanced the basic structure of nuclear model of the atom.

An Atom consists of three fundamental particles, which are electron, proton and neutrons. The atomic number, Z, of an atom represents the number of protons in the atom. The mass number, A, of an atom represents the total number of nucleons (protons and neutrons) in its nucleus. An element, X, is represented by $_Z X^A$, which provides information about its atomic number and the atomic mass number. It is possible to remove one or more electrons from an atom through ionisation. For example, if one electron is removed from a Helium atom, it develops a unit positive charge and it is said to be singly ionized. If the atoms carry excess of positive or negative charge, they are called positive ions or negative ions, respectively.

12.2 SPECTRA

The visible light consists of waves with wavelengths ranging from about 4×10^{-7} m for the extreme violet to about 7.5×10^{-7} m for the deep red. By passing light through a glass prism it is possible to see the different colors in the white light. Spectrum is a band of colours resulting from the dispersion of composite light. Spectrum consists of a regular arrangement of the components of light, according to their wave lengths. The spectra can be grouped together based on how they are produced such as flame spectra, arc spectra, spark spectra and discharge tube spectra. Also, spectra can be grouped as emission spectra and absorption spectra.

12.2.1 Emission Spectra

Light from a luminous source consists of emission spectrum. Emission spectra are classified as continuous, line and band spectra depending on their appearance.

Continuous spectrum: Continuous spectrum has a continuous luminous band of all wavelengths over a wide range. The intensity is maximum at a certain point in the spectrum and gradually decreases on both sides of this point. This position of maximum intensity shifts towards the shorter wavelength region with increase of temperature.

Solids which are heated to the white hot state, and liquids and vapours under high pressure emit continuous spectra. Electric lamps, gas flames emitting white light, a platinum wire heated to incandescence (white hot state), and an electric arc producing white light are examples that emit continuous spectra.

Line spectrum: Line spectrum has many bright lines separated by dark spaces. Line spectra are emitted by the atoms of elements. Hydrogen, iron, sodium etc., always give the same characteristic lines in the same positions in their spectra. Lights from a gas or vapour excited by electric discharge, or a flame into which a volatile salt is introduced emit line spectrum. The characteristic line spectra from atoms of elements help in the identification of elements and detection of unknown elements.

Band spectrum: Band Spectrum consists of a series of broad luminous bands. They may be sharp at one edge, shading off gradually towards the other end. Band spectrum is characteristic of molecules of an element or compound in the diatomic state. For example, cyanogen (CN) emits band spectrum. Mercury vapour may be made to emit a band spectrum. A rarefied gas such as nitrogen in a vacuum tube, emits a band spectrum when subjected to a weak discharge.

12.2.2 Absorption Spectra

Light from a source emitting a continuous range of wavelengths gives absorption spectrum when passed through an absorbing material. Absorption spectra may be broadly classified into: continuous absorption spectra, line absorption spectra and band absorption spectra.

Continuous absorption spectrum: Continuous absorption spectrum is produced by passing light of continuous spectrum through matter in the solid or liquid state. When white light is passed through coloured glass, a wide range of wavelengths is found to be completely removed from the continuous emission spectrum. For example, a piece of ruby glass absorbs all colours except some portion of the red. In the resultant absorption spectrum only red region will be visible and the rest will be dark. A magenta coloured piece of glass can absorb the whole central part of the visible spectrum.

Line absorption spectrum: When characteristic lines are absorbed from a continuous spectrum it results in a line absorption spectrum. Line absorption spectrum is obtained by passing white light from an incandescent substance through some relatively cooler vapour. Certain wavelengths are absorbed and the continuous spectrum is crossed by dark lines. According to Kirchhoff's law of radiation, any substance which emits certain radiations at a certain temperature can absorb the very same radiations at the same temperature. The production of line absorption spectra follows Kirchhoff's law of radiation.

When white light emitting continuous spectrum is passed through a sodium flame, the continuous spectrum will lose two D lines (Fig. 12.1) which are characteristic of sodium light. Solar spectrum is also an example of line absorption spectrum.

Band absorption spectrum: When light from an incandescent source is passed through a coloured solid or liquid, some parts of the continuous spectrum are lost. The resulting spectrum is called band absorption spectrum.

Absorption spectrum for blue cobalt glass has three dark bands in the region from red to green. A dilute aqueous solution of potassium permanganate gives five absorption bands in the green region. Some gases such as vapour of iodine gives a large number of narrow dark bands.

Absorption is different for different substances. In some cases, absorption depends on the temperature of the absorbing medium. For example, the absorption bands in the absorption spectrum of cobalt chloride slightly widen themselves when the temperature is raised. In the case of a solution, the absorption bands change with change of concentration. A dilute aqueous solution of potassium permanganate shows five absorption bands in the green region. With increase of concentration, the bands widen and there is a reduction in the number of bands.

12.2.3 Fraunhofer Lines

A careful examination of the solar spectrum will show that it is crossed by a large number of dark lines. They were first observed in 1802 by an English scientist, Wollaston, but they were studied in detail by Fraunhofer, a German physicist, in 1814. The lines are called Fraunhofer lines in his honour. These lines are denoted by letters of the alphabet, and some of these lines are shown qualitatively in Fig. 12.1.

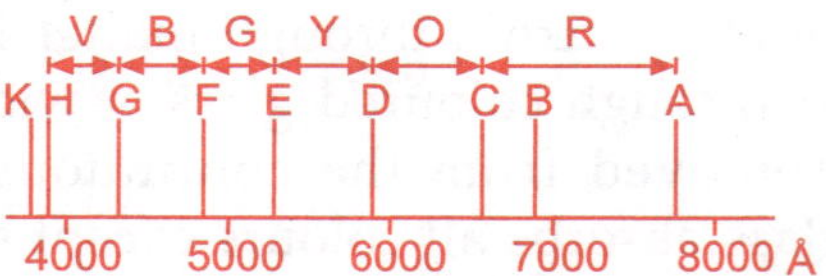

Fig. 12.1 *Fraunhofer lines*

In Fig. 12.1, line *A* lies in the extreme red on the right, *B* and *C* in the red region, *D* in the yellow-orange, *E* in the green, *F* and *G* in the blue, and *H* and *K* in the violet region. In 1859, Kirchhoff, a German physicist, provided an explanation of Fraunhofer lines as follows. The intensely hot central portion of the sun, called photosphere, is surrounded by an envelope of relatively cooler gases, called chromosphere. When intensely hot white light from the centre of the sun passes through the chromosphere, lines emitted by the gaseous elements in the chromospheres are absorbed. The Fraunhofer lines provides information on the elements present in the sun's atmosphere. Through Fraunhofer lines more than sixty elements have been found in the sun's atmosphere. Helium was discovered for the first time by the study of Fraunhofer lines.

During a total solar eclipse, the Fraunhofer lines are seen if a part of the photosphere is exposed. Just when the eclipse becomes total, the line absorption spectrum suddenly changes into a bright line emission spectrum.

12.2.4 Fluorescence

When an atomic or molecular system is excited into a higher energy state by absorption of energy, it returns to its original lower energy state in a time less than 10^{-5} second. When this happens, the system is found to glow brightly by emitting radiation of longer wavelength. When ultraviolet light is incident on certain substances, they emit visible light. Fluorescence lasts as long as the fluorescing substance remains exposed to incident ultraviolet light and re-emission of light stops as soon as incident light is cut-off.

12.2.5 Phosphorescence

Phosphorescence is delayed fluorescence. When the molecules of some substances are excited by the absorption of incident ultraviolet light, they do not return to their original state immediately. The emission of light continues even after the exciting radiation is removed. The continued fluorescent reaction is called phosphorescence.

12.2.6 Electromagnetic Spectrum

The wave theory of light considers that light consists of electromagnetic waves of extremely small wavelengths. In fact visible light is only one kind of electromagnetic radiation. There are other kinds of electromagnetic radiations such as gamma rays, X-rays, ultraviolet rays, infrared rays and radio waves. The entire group of such radiations are grouped as electromagnetic spectrum and is shown in Fig. 12.2. All of these radiations travel through space with the velocity of light.

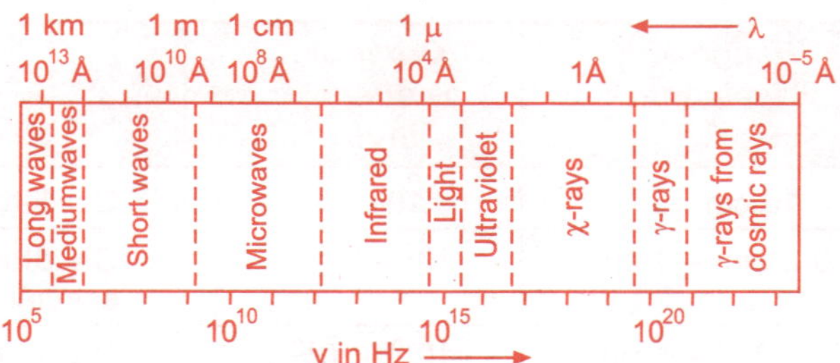

Fig. 12.2 *Electromagnetic spectrum*

The change in the wavelengths, and the corresponding frequencies, of radiations at different parts of the electromagnetic spectrum differ from one another which gives them their characteristic properties.

The wavelengths are expressed in length units as follows:

1 micron (1μ) = 10^{-6} m

1 milli micron ($1m\mu$, or nanometer) = 10^{-9} m

1 Angstrom (1Å) = 10^{-10} m

1 picometer = 10^{-12} m

1 femtometer = 10^{-15} m

The electromagnetic spectrum is very wide in comparison to the visible range of the spectrum which has a very narrow range of wavelengths from about 4000 Å to about 7500 Å. Visible light is emitted by incandescent bodies. Infrared rays are emitted by hot bodies and they range from about 7500 Å to 0.4 mm. Longer infrared rays are emitted by quartz mercury lamps. Waves longer than the far infrared region can be generated by electrical oscillations. When the wavelengths are in the range from 0.2 mm to about 1 cm, they are called microwaves. They are used in communications. The waves having wavelengths longer than 10 cm are called radio waves. Radio waves are classified as short, medium and long waves. They are used in communications and wireless broadcasting.

There are electromagnetic waves with wavelengths shorter than those of visible light and are called ultraviolet rays which extend upto 40 Å. These can be produced by vacuum sparks. X-rays occur beyond the ultraviolet rays and their wavelengths extend upto 1 Å. They are produced when fast moving electrons are suddenly stopped by matter. The waves with wavelengths that are in the region of 0.1 Å to 0.01 Å, are called γ-rays. They are emitted by radioactive materials. The γ-rays from cosmic rays extend upto 10^{-5} Å. As the very word indicates their origin lies in cosmos.

There is no clearcut demarcation between the above ranges and each region of the spectrum overlaps the adjoining region at both ends. The following table provides the different regions in the electromagnetic spectrum.

TABLE 12.1 Wavelengths and Frequencies of Some Electromagnetic Waves

Name	Range of wavelength	Order of frequency in Hz	Source
Radio waves	> 10 cm	< 10^9	Oscillating electric circuits
Microwaves	0.2 mm – 1 cm	$10^9 - 10^{12}$	Oscillating currents in special vacuum tubes
Infrared	7500 Å – 4 mm	$10^{12} - 10^{15}$	Outer electrons in atoms and molecules
Visible	4000 Å – 7500 Å	≈ 10^{15}	Outer electrons in atoms
Ultraviolet	40 Å – 4000 Å	$10^{15} - 10^{17}$	Outer electrons in atoms
X-rays	0.1 Å – 100 Å	$10^{17} - 10^{19}$	Inner electrons in atoms and sudden stopping of high energy free electrons
γ-rays	0.01 Å – 0.1 Å	$10^{19} - 10^{21}$	Nuclei of atoms, and sudden deceleration of high energy particles from accelerators
γ-rays from cosmic rays	< 0.01 Å	> 10^{21}	Cosmos

12.2.7 Infrared Radiation

The part of the spectrum with frequencies below the red region is known as infrared (IR) and the part above violet is known as ultraviolet (UV) as shown in Fig. 12.2.

Infrared radiation was first detected in 1800 by W. Herschel, by the heating effect of the radiations. Herschel discovered that a sensitive thermometer recorded maximum temperature at a point below the red end of the visible spectrum.

Infrared radiation may be divided into three parts. (*i*) Photoelectric infrared $\rightarrow$ 0.75μ to 3μ (1μ = 1 micron = 10^{-6} m) (*ii*) near infrared $\rightarrow$ 3μ to 25μ and (*iii*) far infrared 25μ to 1 mm.

Sources of infrared radiations: Infrared radiations are emitted by all hot bodies irrespective of whether they emit visible light or not hot enough to produce visible light. Sun is obviously a source of visible light, but sun is also a powerful source of infrared radiations.

For continuous infrared radiations in selected regions of the spectrum, some incandescent substances that have high selective emissivity in the desired region can be used. In the range upto about 2μ a Point olite Lamp can be used as a source, which is an arc struck between tungsten electrodes in an evacuated glass bulb. In the region above 2μ, a Nernst filament or a Globar is used. A Nernst filament is a high resistance brittle element composed of the oxides of zirconium, thorium and cesium, held together by a binding material. A Globar is a rod of silicon carbide that emits infrared radiations strongly beyond 20μ, when heated to incandescence.

Properties of infrared radiations: Infrared rays are similar to visible light, except that their wavelengths are different from those of visible light. Infrared rays also travel with the velocity of light. Reflection and refraction of infrared rays follow the same behaviour of the visible light. They also exhibit interference, diffraction and polarization. They can easily penetrate deep into fog and smoke. Infrared radiations can be detected by such instruments as a thermopile or a bolometer. Infrared radiations in the range beyond 2μ in wavelength are strongly absorbed by glass. Therefore, in the study of infrared spectra, lenses are replaced by stainless steel mirrors and glass prism is replaced successively by prisms of quartz (below 3.5μ), Fluorite (upto 8μ), Rock salt (upto 16μ), sylvine (15 to 20μ), Potassium bromide (20 to 30μ) and Cesium bromide (10 to 50μ). For radiations in the far infrared region, gratings are used instead of prisms.

Applications of infrared rays: The infrared radiations from a compound are characteristic of that compound and may be used for identification of the compound. Infrared spectra are also useful in the determination of molecular structure of compounds.

Infrared rays are well suited for long distance photography since they can easily pass through atmosphere even in the presence of fog or smoke. Their high penetrating power makes them well suited for examining old paintings and in the detection of erasure and forgery in documents, etc. Infrared search lights and telescopes were developed during Second World war to reach far into enemy camps.

Infrared rays find many applications in the medical field, where they are useful in diagnosis of superficial tumours, dislocations of bones, in curing sprains etc. Exposure to infrared rays produces heating effect on deeper tissues and stimulates blood circulation.

12.2.8 Ultraviolet Radiation

The ultraviolet radiations are electromagnetic radiations with wavelengths shorter than those of the violet spectra in the visible light. Ultraviolet spectra were discovered in 1801 by the German physicist Johann Wilhelm Ritter, when he found that the photographic action extends beyond the violet end of the visible spectrum. When a piece of uranium glass coated with Vaseline is moved across the spectrum, it glows even when it is beyond the violet end of the visible spectrum indicating the presence of ultraviolet rays.

Thomas Young, an English physicist, photographed Newton's rings in ultraviolet radiations and confirmed the nature of these radiations with light in 1811. The wavelength of ultraviolet rays are shorter than that of visible light, and extends from 4000 Å to 40 Å. Ultraviolet region may be broadly divided into three parts: (*i*) near ultraviolet (4000 Å to 3000 Å) (*ii*) Far ultraviolet 3000 Å to 2000 Å and (*iii*) extreme ultraviolet 2000 Å to 40 Å.

Sources of ultraviolet radiations: Sunlight is rich in ultraviolet radiations. Electric arc of carbon, iron or other materials, mercury vapour lamps, discharge of electricity through hydrogen contained in a quartz tube are some of the sources emitting ultraviolet radiations. A quartz mercury lamp emits ultraviolet radiations down to 2537 Å in wavelength. For even shorter wavelengths, cadmium arc is commonly used.

Properties of ultraviolet radiations: Ultraviolet radiations exhibit similar characteristics as those by visible light in terms of reflection, refraction, interference, diffraction, polarization etc. Detection of ultraviolet rays can be done in four different ways using their properties: (*i*) Photographic effect (*ii*) Fluorescence (*iii*) Phosphorescence and (*iv*) Photoelectric effect. Ultraviolet rays are particularly active in decomposing salts of silver. They may have beneficial or harmful effects on living tissues depending on their wavelengths. In general, ultraviolet rays with shorter wavelengths are harmful and those with longer wavelengths 1 > 2900 Å are beneficial.

Ultraviolet radiations are absorbed by ordinary glass and this property is conveniently used in the study of ultraviolet spectra. Quartz prisms are used in the region from 4000 Å to 2000 Å; and fluorite prisms may be used upto 1200 Å. For ultraviolet radiations with wavelengths below 1200 Å, gratings are used instead of prisms. Since oxygen and water vapour show strong absorption below 1850 Å, evacuated chamber must be used for work in this region. Ultraviolet region with wavelengths below 1850 Å is called vacuum ultraviolet.

Applications of ultraviolet radiations: Ultraviolet rays have amny applications in several areas. They activate certain chemical reactions such as decomposing hydrogen iodide in aqueous solution into hydrogen and iodine. They excite fluorescence in many substances which has led to the development of fluorescent tubes. They are used to test the purity of gems and to distinguish between real gems and artificial gems. They are used as efficient sterilizers in places where blood plasma, drugs, vaccines, etc., are prepared. They are also used for sterilisation of water. Ultraviolet radiations are used in microscopes with high resolving power, known as ultra microscopes. They are used in photo electric alarms. They are useful in the production of Vitamin D and those suffering from Vitamin D deficiency benefit by exposure to ultraviolet rays.

12.2.9 Microwaves

In 1887, Hertz, a German physicist, successfully produced electromagnetic waves of wavelengths in the range of 10 metres. Microwaves are much longer in wavelength and have frequencies well below the infrared radiations. The longest wavelength of infrared is about a millimetre long. Microwaves travel with the velocity of light.

Microwaves also obey the laws of reflection and refraction. They show diffraction at a metal straight edge. Microwaves are long enough to show interference effects very well. They are used in communication.

12.2.10 Radio Waves

Radio waves are electromagnetic radiations with wavelengths longer than that of infrared waves in the electromagnetic spectrum. Radio waves have frequencies from 300 GHz to as low as 3 kHz, and corresponding wavelengths range from 1 mm to 100 km. They travel at the speed of light. Radio waves are generated when there is lightning, or are also produced by astronomical objects. Artificially generated radio waves are used for fixed and mobile radio communication, broadcasting, radar and other navigation systems, communications satellites, computer networks. Different frequencies of radio waves have different propagation characteristics in the Earth's atmosphere; long waves may cover a part of the earth very consistently but do not reflect well while shorter waves can reflect off the ionosphere and travel around the world. Radio waves with shorter wavelengths bend or reflect very little and travel straight.

12.2.11 X-rays

Wavelengths of X-rays have shorter wavelength than ultraviolet rays and extend up-to about 1 Å in wavelength. They are produced when fast electrons are suddenly stopped by a suitable solid target of heavy metal. When they were first discovered by Rontgen in 1895, X-rays were believed to be charged particles. But they exhibit all the properties of light and are in fact very short electromagnetic waves.

X-rays are produced using Rontgen's gas filled tube and Coolidge tube. X-rays affect photographic plate, ionise gases, produce fluorescence in many substances and produce photoelectric effect. X-rays can penetrate and can pass through many solids such as flesh, thin sheets of metal, wood etc.

X-rays are employed in medicine, material testing in industry and scientific research. Fracture of bones can easily be located by X-ray photograph. X-rays are used to locate foreign bodies such as bullets, pins, coins etc., in human body. X-rays are used on living organisms for the treatment of cancer and some skin diseases. They are used for non-destructive testing of materials and detecting defects such as cavities in castings and cracks in welds. They are used to locate minute flaws in the parts of aircraft. They are used to study the structure of crystals.

12.3 QUANTUM THEORY OF RADIATION

Quantum theory: A body that is heated to incandescence will emit radiations with a continuous range of wavelengths. A portion of these radiations will be in the visible range while the remaining will be distributed throughout the regions of longer and shorter wavelengths as shown in Fig. 12.3. The intensity of radiation increases with increase in temperature. Also with increasing temperature, the wavelength of the most intense radiation shifts towards the shorter wavelength (from the red end towards the blue end). Electromagnetic theory could not explain these experimental observations.

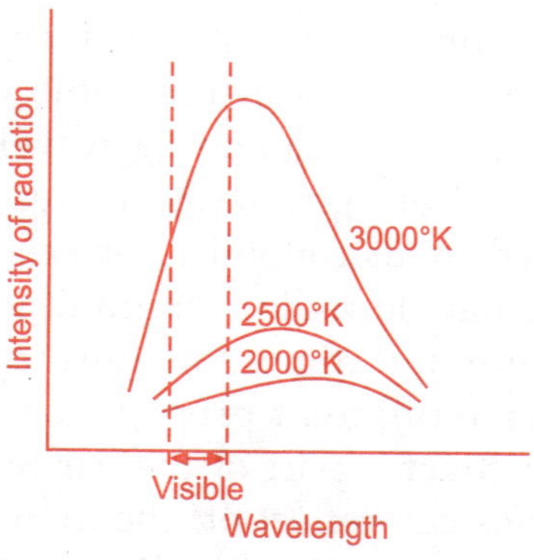

Fig. 12.3 *Wavelength vs Intensity of radiation*

Towards the end of nineteenth century, attempts were made by Wien, a German physicist, and by the British physicists Rayleigh and Jeans, to explain the distribution of energy in the spectrum of black body radiation. Black body radiation is the type of electromagnetic radiation within or surrounding a body in thermodynamic equilibrium with its environment, or emitted by a black body, which is an opaque and non-reflective body held at constant, uniform temperature. The radiation has a specific spectrum and intensity that depends only on the temperature of the body. Wien derived an expression for the distribution of energy which agrees with experimental results for short wavelengths while Rayleigh-Jeans formula holds good for longer wavelengths.

Planck's hypothesis: In 1900, Max Planck advanced a hypothesis based on his experimental results to explain the distribution of energy in the spectrum of black body radiation. Planck's hypothesis was that the electromagnetic radiation of a particular frequency, ν, is emitted and absorbed in packets of energy, E, given by the relation,

$$E = h\nu \tag{12.1}$$

where ν is frequency and h is a universal constant called Planck's constant and its value is given by 6.625×10^{-34} Joule-second. Planck called these packets of energy as quanta. He hypothesized that the exchange of energy between radiation and matter does not take place continuously, but in discrete packets of energy.

Einstein's hypothesis: Planck's hypothesis could explain exchanges of energy in certain situations. Einstein extended Planck's hypothesis in order to explain the phenomenon of photoelectric effect. Einstein stipulated that light exists as tiny oscillating packets of electromagnetic energy, which he called as quanta or photons. Einstein assumed that light comes from a source in the form of streams of photons. Each photon has a definite amount of energy, depending on frequency. The quantum theory of light, as described by Einstein's hypothesis, is able to explain the photoelectric effect, Raman effect, Compton effect etc. This clearly shows that light has a particle nature. However, the phenomena of interference, diffraction and polarisation can be explained by the wave nature of light and therefore the wave theory cannot be discarded. The same light beam can diffract around an obstacle and then impinge on a metal surface to eject photoelectrons. These two processes occur separately. Therefore, the behavior of light can be explained sometimes considering them as particles and sometimes as waves. In order to fully explain the nature of light both theories must be adopted depending on the context.

Einstein used quantum theory to explain the photoelectric effect and to explain the variation in specific heat of elements with temperature. Quantum theory of light has been successfully applied to explain Compton effect discovered in 1923. The Compton effect (also called Compton scattering) is the result of a high-energy photon colliding with a target, which releases loosely bound electrons from the outer shell of the atom or molecule. The scattered radiation experiences a wavelength shift, and this shift cannot be explained on the basis of classical wave theory. However, it can be explained on the basis of Einstein's photon theory. The effect was first demonstrated in 1923 by Arthur Holly Compton. The quantum theory is also able to explain atomic spectra by Bohr in 1913.

12.3.1 Photoelectric Effect

Photo emission: In 1887, Hertz discovered that the intensity of the spark between two high voltage electrodes increased when they were exposed to ultraviolet light. Further experiments showed that when light is incident on certain materials such as selenium, electrons are ejected. The emission of electrons under the influence of light is called photoelectric effect. The electrons emitted are called photoelectrons. Zinc and magnesium respond photoelectrically to ultraviolet. Lithium, sodium, potassium, rubidium and caesium are sensitive even to visible light. Photoelectrons are emitted by radiations of short wavelengths such as ultraviolet rays, X-rays and γ-rays.

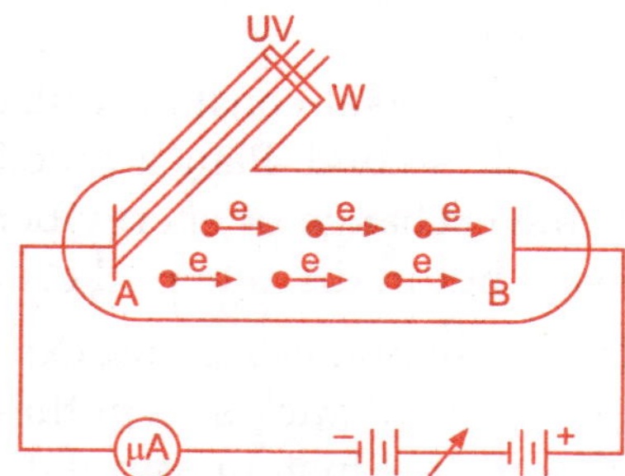

Fig. 12.4 *Photoelectric effect*

Experimental study of photoelectric effect: The apparatus shown in Fig. 12.4, can be used for the experimental study of photoelectric effect. A and B are two plates inside an evacuated tube T. The two plates are connected to source of variable DC voltage and a micro ammeter. Plate A is the photo emissive surface and is connected to the negative terminal while plate B is the collector plate, and is connected to the positive terminal. When light is incident on the plate A, electrons are emitted from this plate and a current flows as seen in the micro ammeter. A quartz window, W, is sealed to the tube which allows ultraviolet light to pass through and fall on the plate A. The apparatus is used to study the dependence of current on (*i*) the intensity of light (*ii*) the frequency of light (*iii*) the voltage between the plates and also (*iv*) the effect of using different materials as photo emissive surfaces.

If plate B is made negative, the electrons are retarded or slowed down. If the negative potential, also known as retarding potential, is sufficiently high, no electrons reach the plate B and the current becomes zero. The retarding potential just sufficient to stop the electrons from reaching the plate B is called the stopping potential. The dependence of the stopping potential on the frequency of the incident radiation was studied by Millikan in 1916.

The photoelectric effect is characterized by the following:

1. Photoelectric effect is initiated instantaneously and the photoelectric current begins to flow as soon as the radiation falls on the emitter.

2. There is a certain limiting frequency below which there is no photoelectric effect. This minimum frequency is called threshold frequency, which is different for different materials.

3. The number of photoelectrons released or the photoelectric current is directly proportional to the intensity of the radiation, for the given frequency of incident radiation when the frequency is greater than the threshold frequency.

4. If the frequency of incident radiation is greater than the threshold frequency, the maximum kinetic energy of the photoelectrons depends only on the frequency of incident radiation and is independent of the intensity of radiation. The graph of kinetic energy of photoelectrons against the frequency of radiation is a straight line as shown in Fig. 12.5. The straight lines are different for different materials with the same slope irrespective of the emitter. Electromagnetic theory of light and wave theory of light were insufficient to explain these experimental observations whereas the quantum theory is able to explain.

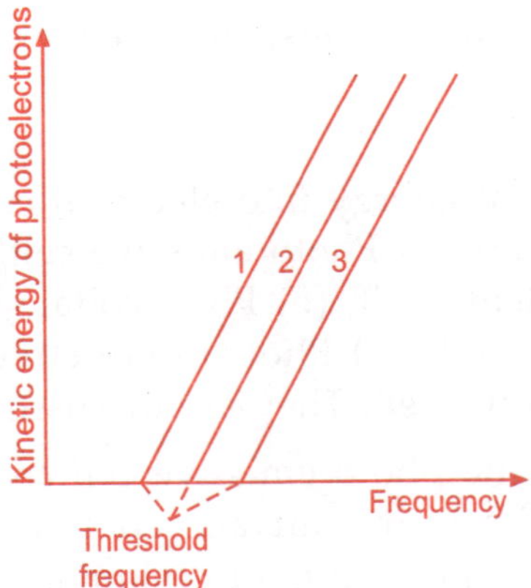

Fig. 12.5 *Kinetic energy of photoelectrons against the frequency of radiation*

Einstein's explanation of photoelectric effect: Einstein assumed that when a photon interacts with matter it behaves as a particle and transfers its energy completely to an individual electron. A part of the energy is used by the electron to free itself from the material. The minimum energy necessary to free the electron from a material is called photoelectric work function of the material. Photoelectric work function is usually expressed in electron volts.

Einstein concluded that the kinetic energy of a photoelectron must be equal to the difference between the energy of the incident photon and the photoelectric work function of the material. This condition is expressed as

$$\left[\begin{array}{c}\text{Kinetic energy of}\\ \text{photoelectron}\end{array}\right] = \text{Photon energy-work function}$$

$$\frac{1}{2}mv^2 = h\nu - w$$

$$h\nu = \frac{1}{2}mv^2 + w \tag{12.2}$$

where v is the velocity of the photoelectron and h is the Planck's constant. The Eq. (12.2) is called Einstein's photoelectric equation which explains all the observed facts about photoelectric effect. An examination of Eq. (12.2) shows that:

For velocity, $v = 0$, $h\nu_0 = w$, represents the minimum energy required for emission, where ν_0 is called threshold frequency.

If $h\nu < w$, there is no emission. This corresponding frequency is the threshold frequency.

The kinetic energy of photoelectrons depends only on the frequency of incident photon and not on the intensity of incident light. Number of photoelectrons and hence photoelectric current depends on intensity and not on frequency of incident light.

Millikan's experiments validated Einstein's theory of photoelectric effect.

12.3.2 Photoelectric Cells

Photoelectric cell converts light energy into electrical energy. There are three types of photoelectric cells, which are, (*i*) Photo emissive cell in which the cathode emits electrons when light is incident on it, (*ii*) Photovoltaic cell where a current results when light falls on the cell, and (*iii*) Photo conductive cell, where the resistance decreases when light falls on the cell. They are discussed in detail below.

1. Photo Emissive Cell: The photo emissive cell consists of an evacuated glass bulb or quartz tube with its inner surface coated with a thin layer, C, of a photosensitive material such as potassium or cesium as shown in Fig. 12.6.

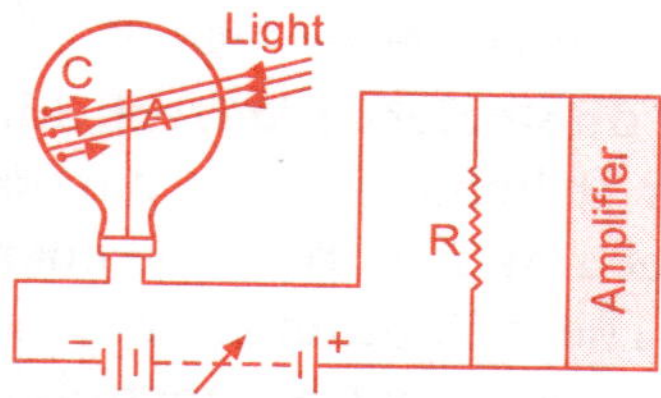

Fig. 12.6 *Photo emissive cell*

A small portion of the glass bulb is transparent in order to allow light to enter the tube. A thin rod A of platinum or nickel is mounted inside the bulb. The coating C and the rod A are insulated from each other and the leads from these are brought out through the base of the bulb and are connected to two terminals at the base. The lead from C is connected to the negative and that from A to the positive of a *DC* voltage source of 80 V to 100 V through a high resistance R. When light of frequency, greater than the threshold frequency, falls on the cell, electrons are emitted from the photo metal and are collected by the collector A. The current is quite feeble, and in the order of just a few micro amps. An amplifier is used to amplify the current signal.

12.3.3 Photovoltaic Cell

Photovoltaic cell, schematically shown in Fig. 12.7, consists of a thin semiconducting layer of cuprous oxide (Cu_2O) formed on a metal base. The semiconducting layer is covered with a thin film of sputtered silver or gold which is practically transparent. When light is incident on the cell, it ejects electrons at the boundary between the metal film and the semiconducting layer. The electrons travel against light and the voltage across the resistance R is proportional to the intensity of light. These cells do not require an external source of emf and they are sturdy, compact, inexpensive and they have a high sensitivity. They are used in light meters, exposure meters, etc.

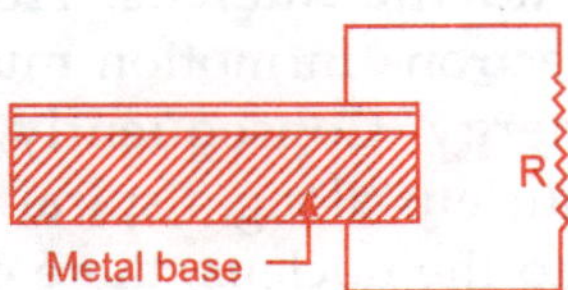

Fig. 12.7 *Photovoltaic cell*

12.3.4 Photoconductive Cell

Photoconductive cell consists of a slab of a semiconductor such as selenium whose electrical resistance decreases when light is incident on it. Resistance decreases with increase in the intensity of incident light. As the change in resistance is not directly proportional to the intensity of light, cells can be used only for detection of light and in automatic relays.

12.3.5 Applications of Photoelectric Effect

Some applications of the photoelectric cells are given below:

(*i*) In light meters to measure the intensity of illumination at a place for lighting installations.

(*ii*) In order to compare the luminous intensities of two sources of light.

(*iii*) For automatic on and off switching of street lights.

(*iv*) In identification of colours.

(*v*) In counting machines.

(*vi*) For photoelectric control of relays in switching on an alarm such as a fire alarm.

(*vii*) In Temperature indicator and temperature control in order to make them more sensitive.

(*viii*) In smoke detectors.

(*ix*) For detection of minor flaws in metal sheets.

(*x*) For transmission of pictures as in Television.

(*xi*) Transmission of voice and musical sound to a distance of several kilometres over a light beam can be made using photoelectric cells.

12.3.6 Bohr's Theory of Atomic Structure

In order to explain the large angle scattering of particles, Rutherford concluded that most of the mass of an atom is concentrated in a central core called nucleus having positive charge and surrounded by electrons having negative charge. The size of the nucleus is about 10^{-5} of the atomic size. Since atom as a whole must be neutral, the positive charge of the nucleus must be equal to the negative charge of the surrounding electrons.

In order for atoms to be mechanically stable, Rutherford assumed that the electrons must revolve around the nucleus. However, according to classical electromagnetic theory, the electrons in motion must continuously radiate energy, thereby losing some of its energy. Consequently, the electron must slow down eventually and approach the nucleus along a spiral path and fall ultimately into the nucleus. If the electrons fall into the nucleus, there can be no mechanical stability of the atom.

The deficiency in Rutherford's nuclear atom model was removed by Niels Bohr, a Danish physicist in 1913 with his model of the atom. Bohr proposed some modifications to Rutherford's model on the basis of Plank's quantum theory of radiation.

Bohr's theory of atomic structure is based on the following:

1. Atom consists of a positively charged nucleus at the centre. Electrons, the negatively charged particles, move around the nucleus in certain selected orbits called stationary orbits of specific energy.

2. The radii of the orbits are such that the angular momentum of the electron is an integral multiple of $\dfrac{h}{2\pi}$, where h is Planck 's constant.

3. When an electron jumps from a higher energy level orbit to a lower energy level orbit, it gives out electromagnetic radiation of a particular frequency.

   The frequency of the radiation is given by quantum condition

   $$h\nu = E_1 - E_2$$

4. The Coulombian and Newtonian forces are applicable in the region of the atom.

According to the first postulate, when electrons revolve in an orbit for any length of time they do not emit any radiation. Each orbit is associated with a definite amount of energy and the orbits are called energy levels.

According to the second postulate, angular momentum in any orbit is given by

$$mvr = \frac{nh}{2\pi} \tag{12.3}$$

where m is mass of electron, v is orbital velocity, r is radius of orbit, n is a positive integer (1, 2, 3...) called Principal quantum number and h is Plank's constant. The angular momentum is said to be quantized.

According to the third postulate,

$$E_{n2} - E_{n1} = h\nu \tag{12.4}$$

where E_{n1} and E_{n2} are the energies in the orbits n_1 and n_2 and ν is the frequency of emitted radiation. Equation (12.4) is referred to as Bohr's frequency condition.

Bohr's theory of atomic structure was successful in explaining the origin of spectra.

12.3.7 Hydrogen Atom Spectrum

There were several early attempts to form a mathematical relation between the wavelengths or frequencies of various spectral lines of an atom. Johann Jakob Balmer, a Swiss mathematician and mathematical physicist, was successful in 1885 in establishing such a relation for the spectral lines of hydrogen atom in the visible region. He arrived at a relation between the wavelength λ of these lines in the visible region of hydrogen spectrum in the form,

$$\lambda = \frac{3647n^2}{n^2 - 4}$$

where $n = 3, 4...$ etc.

Balmer's equation is expressed in terms of wave numbers, $\bar{\nu}$, and is in the form

$$\bar{\nu} = R\left(\frac{1}{2^2} - \frac{1}{n^2}\right) \tag{12.5}$$

where R is a constant called Rydberg's constant with a value of 1.09678×10^7 m^{-1}. The lines in Balmer series are called H_α, H_β, H_γ etc.

In 1906, Theodore Lyman, a Harvard physicist, discovered a series of lines in the ultraviolet region represented by the formula

$$\bar{\nu} = R\left(\frac{1}{1^2} - \frac{1}{n^2}\right) \tag{12.6}$$

where $n = 2, 3$ and so on.

Two series of lines in the infrared were discovered by Paschen and Brackett and are respectively represented by the relation

$$\bar{V} = R\left(\frac{1}{3^2} - \frac{1}{n^2}\right) \tag{12.7}$$

with $n = 4, 5, 6\ldots$, and

$$\bar{V} = R\left(\frac{1}{4^2} - \frac{1}{n^2}\right) \tag{12.8}$$

with $n = 5, 6, \ldots$ and so on.

A series of lines in the far infrared is called Pfund series, named after the American born physicist August Herman Pfund, and is represented by

$$\bar{V} = R\left(\frac{1}{5^2} - \frac{1}{n^2}\right) \tag{12.9}$$

where $n = 6, 7$ and so on.

A general formula that can represent all the series is given by

$$\bar{V} = R\left(\frac{1}{m^2} - \frac{1}{n^2}\right) \tag{12.10}$$

where m and n are integers. Further, $m = 1, 2, 3, 4$ and 5 for the Lyman, Balmer, Paschen, Brackett and Pfund series, respectively, with n taking integral values starting from $(m + 1)$.

12.3.8 Bohr's Theory of Hydrogen Atom Spectrum

Considering the nucleus having a positive charge, e, to be at rest, the force of attraction on the electron of charge e is equal to $\dfrac{e^2}{4\pi\varepsilon_0 r_n^2}$ where r_n is the radius of the nth circular orbit in which the electron moves around the nucleus.

If m is the mass and v is the orbital speed of the electron, the centripetal force $\dfrac{mv^2}{r_n}$ is provided by the Coulombian force. Equating the Coulombian force to the centripetal force results in

$$\frac{mv^2}{r_n} = \frac{e^2}{4\pi\varepsilon_0 r_n^2}$$

This may be rewritten as

$$mv^2 r_n = \frac{e^2}{4\pi\varepsilon_0} \tag{12.11}$$

Following Bohr's second postulate, the angular momentum is given by

$$mvr_n = \frac{nh}{2\pi} \qquad (12.12)$$

Squaring on both sides of Eq. (12.12), we get

$$m^2v^2r_n^2 = \frac{n^2h^2}{4\pi^2} \qquad (12.13)$$

Comparing Eqs.(12.11) and (12.13) gives

$$r_n = \frac{\varepsilon_0 n^2 h^2}{\pi m e^2} \qquad (12.14)$$

Therefore $r_n \, \alpha \, n^2$.

The radii of the different orbits are in the ratio 1:4:9:16 etc. The radius of the first orbit for hydrogen atom is called Bohr radius and is equal to 0.53 Å or 5.3×10^{-11} m. From Eq. (12.11) and (12.12), we have

$$V = \frac{e^2}{2nh\varepsilon_0} \qquad (12.15)$$

Potential energy of electron in the nth orbit is equal to $-\dfrac{e^2}{4\pi\varepsilon_0 r_n}$ and kinetic

energy, $\dfrac{1}{2}mv^2$, is equal to $\dfrac{e^2}{8\pi\varepsilon_0 r_n}$ from Eq. (12.11). Hence, the total energy, E, is given by

$$E = -\frac{e^2}{4\pi\varepsilon_0 r_n} + \frac{e^2}{8\pi\varepsilon_0 r_n}$$

Consequently,

$$E_n = -\frac{e^2}{8\pi\varepsilon_0 r_n}$$

Substitution of the value of r_n from Eq. (12.14) gives

$$E_n = \frac{-me^4}{8\pi\varepsilon_0^2 n^2 h^2}. \qquad (12.16)$$

Negative sign implies that the electron is bound to the nucleus and energy must be supplied to remove the electron. As the quantum number n increases or the radius of the orbit increases, the numerical value of energy decreases; but the actual value increases in view of the negative sign. Thus an electron in an outer orbit has more energy than an electron in an inner orbit.

When an electron jumps from a higher orbit n_2 of energy E_{n2} to a lower orbit n_1 of energy E_{n1}, it emits energy equal to $(E_{n2} - E_{n1})$ and this appears as radiation of frequency ν given by

$$h\nu = E_{n2} - E_{n1} = \frac{me^2}{8\varepsilon_0^2 h^2}\left(\frac{1}{n_1^2} - \frac{1}{n_2^2}\right)$$

or, the frequency is given by

$$\nu = \frac{me^4}{8\varepsilon_0^2 h^3}\left(\frac{1}{n_1^2} - \frac{1}{n_2^2}\right). \tag{12.17}$$

The wave number is given by

$$\bar{\nu} = \frac{\nu}{c} = R\left(\frac{1}{n_1^2} - \frac{1}{n_2^2}\right) \tag{12.18}$$

where $R = \dfrac{me^4}{8\varepsilon_0^2 ch^3}$ is the Rydberg's constant. The Eq. (12.18) is identical with Eq. (12.10).

There is a good agreement between the value of R calculated by the substitution of numerical values for m, e, ε_o, c and h and that obtained from spectroscopic studies, and is equal to $.0917 \times 10^7$ m^{-1}. When the Bohr's theory was proposed, only the Balmer and Paschen series for hydrogen atom were known. The other series are proposed theoretically in Eq. (12.18) and were verified experimentally. The Lymann series in 1916, the Brackett series in 1922 and Pfund series in 1929 studied experimentally and arrived at the same wave numbers theoretically; thus validating the Bohr's theory.

When an electrical discharge is passed in a discharge tube the different electrons in the atoms of hydrogen absorb different amounts of energy. Consequently, the electrons in different atoms are raised to different levels and they come down to different lower levels, emitting radiations of different energy. Consequently all the lines are seen.

Since the nucleus is not stationary, the reduced mass, $\dfrac{mM}{m+M}$, where M, the mass of nucleus, has to be replaced with m, the mass of the electron. Further, for an atom like hydrogen having one electron, the Rydberg constant $R = \dfrac{mZ^2 e^4}{8\varepsilon_0^2 ch^3}$, with correction for m. Therefore, Rydberg constant is different for different elements.

12.3.9 Energy Level Diagrams

The notion of several orbits is able to explain the existence of different energy levels for the electrons in an atom. Energy level diagram for the electrons is shown in Fig. 12.8. The energy values corresponding to the various orbits (quantum states) are calculated from Eq. (12.16).

The lowest energy level ($n = 1$) corresponds to the normal or the ground state. The higher energy levels are called the excited states. In the energy level diagram, the energy states are represented by horizontal lines and the electron jumps between these states along vertical lines.

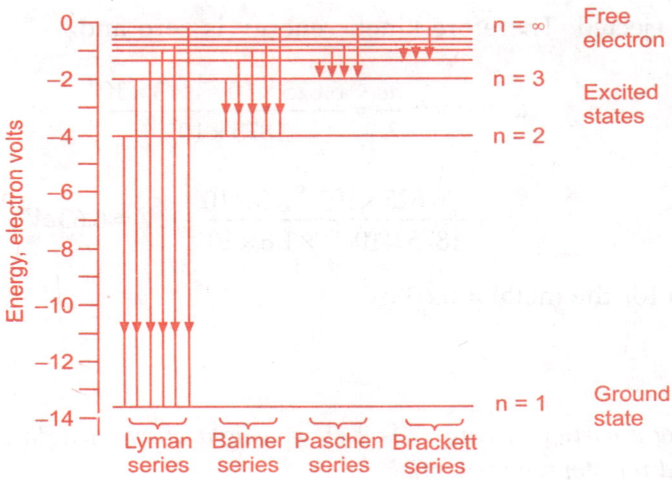

Fig. 12.8 *Energy level diagram*

Energy of electron in the ground state is -21.76×10^{-19} J or -13.6 eV; this is the amount of energy required to remove the electron completely from hydrogen atom. When an electron is completely removed from an atom, the atom is said to be ionized. Ionization potential of hydrogen is 13.6 electron volts (eV). Energy in the first excited state is -3.4 eV. Energy required to raise the electron from the first orbit to the second orbit (from the ground state to the first excited state) is 10.2 eV; first excitation potential of hydrogen is said to be 10.2 electron volts.

EXAMPLE 12.1

What will be the maximum kinetic energy of the photoelectrons ejected from magnesium (work function = 3.7 eV) when irradiated by UV light of frequency 1.5×10^{15}/sec ($h = 6.625 \times 10^{-34}$ J sec).

SOLUTION

Energy of photon = $h\nu$

$$= 6.625 \times 10^{-34} \times 1.5 \times 10^{15} \text{ J}$$

$$\approx 6.21 \text{ eV}$$

Maximum KE $= (h\nu - w) = 6.21 - 3.7 = 2.51$ eV.

EXAMPLE 12.2

Light of wavelength 1875 Å can just eject electrons from a metal. Calculate the work function for the metal.

SOLUTION

Electrons are just ejected. Therefore kinetic energy is zero and

$$w = h\nu = \frac{hc}{\lambda} = \frac{6.625 \times 10^{-34} \times 3 \times 10^8}{1875 \times 10^{-10}}$$

$$= \frac{6.625 \times 10^{-34} \times 3 \times 10^8}{1875 \times 10^{-10} \times 1.6 \times 10^{-19}} eV = 6.63 \, eV$$

Work function for the metal = 6.63 eV.

EXAMPLE 12.3

The work function of a particular emitter is 2 eV and light of wavelength 3000 Å is used to cause emission. Find the stopping potential.

SOLUTION

$$\text{Energy of Photon} = h\nu = \frac{hc}{\lambda} = \frac{6.625 \times 10^{-34} \times 3 \times 10^8}{3000 \times 10^{-10}}$$

$$= \frac{6.625 \times 10^{-34} \times 3 \times 10^8}{3000 \times 10^{-10} \times 1.6 \times 10^{-19}} eV$$

$$= 4.14 \text{ eV}$$

Kinetic energy $= eV = h\nu - w = 4.14 - 2 = 2.14$ eV

Stopping potential = 2.14 volts.

EXAMPLE 12.4

Calculate the radius, velocity and energy of the electron in the nth orbit in hydrogen atom, from the following data.

SOLUTION

$$e = 1.6 \times 10^{-19} C, \ m = 9.1 \times 10^{-31} \text{ Kg,}$$
$$h = 6.625 \times 10^{-34} \text{ Js}, \ \varepsilon_0 = 8.85 \times 10^{-12} \text{ F/m}$$
$$C = 3 \times 10^8 \text{ m/s}$$

Radius
$$r_n = \frac{n^2 h^2 \varepsilon_0}{\pi m e^2}$$

$$= \frac{n^2 \times \left(6.625 \times 10^{-34}\right)^2 \left(8.85 \times 10^{-12}\right)}{\pi \times \left(9.1 \times 10^{-34}\right)\left(1.6 \times 10^{-19}\right)^2}$$

$$= 5.3 \times 10^{-11} \ n^2 \text{ meter}$$

$$v_n = \frac{e^2}{2nh\varepsilon_0}$$

$$= \frac{\left(1.6 \times 10^{-19}\right)^2}{2n \times \left(6.625 \times 10^{-34}\right)\left(8.85 \times 10^{-12}\right)}$$

$$= \frac{2.18 \times 10^6}{n} \text{ m/s}$$

$$E_n = \frac{me^4}{8\varepsilon_0^2 n^2 h^2}$$

$$= \frac{\left(9.1 \times 10^{-31}\right) \times \left(1.6 \times 10^{-19}\right)^4}{n^2 \times 8 \times \left(8.85 \times 10^{-12}\right)^2 \times \left(6.625 \times 10^{-34}\right)^2}$$

$$= \frac{21.76 \times 10^{-19}}{n^2} - J$$

$$= \frac{13.6}{n^2} eV$$

Radius $= 5.3 \times 10^{-11} \ n^2$ meter, velocity $\approx \dfrac{2.18 \times 10^6}{n}$ m/s and energy $= -\dfrac{13.6}{n^2}$ eV.

EXAMPLE 12.5

Calculate the wave number, wavelength and frequency of H_a line of hydrogen (R = 1.097 \times 10^7 \ m^{-1}$).

SOLUTION

H_α line is Balmer series

Wave number,
$$\bar{v} = R\left(\frac{1}{2^2} - \frac{1}{3^2}\right)$$

$$= 1.097 \times 10^7 \times \frac{5}{36}$$

$$= 1.524 \times 10^6 \text{ m}^{-1}$$

$$= 1.097 \times 10^7 \times \frac{5}{36}$$

$$= 1.524 \times 10^6 \text{ m}^{-1}$$

Wavelength,
$$\lambda = \frac{1}{\bar{v}} = \frac{1}{1.524 \times 10^6} = 6.561 \times 10^{-7} \text{ m}$$

Frequency,
$$v = \frac{c}{\lambda} = \frac{3 \times 10^8}{6.561 \times 10^{-7}} = 4.572 \times 10^{14} \text{ Hz}.$$

EXAMPLE 12.6

Calculate the value of Rydberg constant given m = 9.1 × 10⁻³¹ Kg, e = 1.6 × 10⁻¹⁹ C, ε₀ = 8.85 × 10⁻¹² F/m, c = 3 × 10⁸ m/s, h = 6.625 × 10⁻³⁴ Js.

SOLUTION

$$R = \frac{me^4}{8\varepsilon_0^2 c^2 h^3}$$

$$= \frac{\left(9.1 \times 10^{-31}\right) \times \left(16 \times 10^{-19}\right)^4}{8 \times \left(8.85 \times 10^{-12}\right)^2 \times \left(3 \times 10^8\right)^2 \times \left(6.625 \times 10^{-34}\right)^3}$$

$$= 1.097 \times 10^7 \text{ m}^{-1}.$$

EXAMPLE 12.7

Calculate the value of the Rydberg constant for hydrogen, given the wavelength of H_β line is 6553 Å.

SOLUTION

The wave number $\bar{v}$ of H_β line is given by

$$\bar{v} = R\left(\frac{1}{2^2} - \frac{1}{3^2}\right) = R\left(\frac{1}{4} - \frac{1}{9}\right) = R\frac{5}{36}$$

$$R = \frac{36 \times \bar{v}}{5} = \frac{36 \times 1}{5 \times 6553 \times 10^{-10}} = 1.098 \times 10^7 \, m^{-1}.$$

EXAMPLE 12.8

The radius of the first Bohr orbit of electron in hydrogen atom is 0.529Å. Calculate the radius of the second Bohr orbit in singly ionized helium atom.

SOLUTION

$$r \propto \frac{n^2}{Z}$$

$$\text{for hydrogen} \rightarrow \quad 0.529 \propto \frac{1^2}{1} \qquad\qquad Z = 1, n = 1$$

$$\text{for helium} \rightarrow \quad r \propto \frac{2^2}{2} \qquad\qquad Z = 2, n = 2$$

$$\frac{r}{0.529} = \frac{2}{1}, \text{ or } r = 1.058\text{Å}.$$

EXAMPLE 12.9

Calculate the maximum energy of the electrons emitted from a nickel surface by ultraviolet light of frequency 1.5×10^{15}/sec. If the work function of nickel is 5eV, determine the K.E. of the electron and also the maximum wavelength at which the effect takes place.

SOLUTION

$$v = 1.5 \times 10^{15}/\text{sec}; \omega = 5 \text{ eV};$$

$$hv = 6.625 \times 10^{-34} \times 1.5 \times 10^{15} \text{ J}$$

$$= \frac{6.625 \times 10^{-34} \times 1.5 \times 10^{15}}{1.6 \times 10^{-19}} = 6.21 \text{ eV}$$

Kinetic energy $= hv - \omega = 6.21 - 5 = 1.21$ eV

Kinetic energy of electron $= 1.21$ eV

$$\omega = hv_0 = \frac{hc}{\lambda_0}$$

$$\lambda_0 = \frac{hc}{w} = \frac{6.625 \times 10^{-34} \times 3 \times 10^8}{5 \times 1.6 \times 10^{-19}} = 2.484 \times 10^{-7} \text{ m}$$

Maximum wavelength = 2.484×10^{-7} m.

12.4 RADIOACTIVITY

Radioactivity was discovered by Henri Becquerel, a French physicist, in 1896. While investigating the relationship between X-rays and fluorescence, he found that photographic plate was blackened when a uranium compound was placed near the plate. Further investigations showed that this property had no relation to fluorescence of X-rays, but was a property of the element uranium. The photographic plate was affected due to the radiations emitted by uranium. The emission of these radiations is spontaneous, and not caused by external agencies such as temperature, pressure, electric field and magnetic field. These radiations were named after him as Becquerel rays.

Marie Curie and Pierre Curie discovered in 1898 a new element polonium and later on radium which were many times more radioactive than uranium. Substances which emit Becquerel rays are said to be radioactive. Experiments showed that atoms of radioactive substances disintegrate into other atoms. All the elements from atomic number 81 to 92 are found to be radioactive. The phenomenon of spontaneous disintegration of heavy nuclei with the emission of certain radiations is called radioactivity. Radioactivity was also found in naturally occurring elements and is known as natural radioactivity.

12.4.1 Nature of Radioactive Rays

Lord Rutherford, a British physicist, found three types of radiations while studying radioactive substances. A simple arrangement for the study of radioactive radiations is shown in Fig. 12.9. A sample of radioactive material is placed in a small hole in a lead block. The radiations are subjected to an electric field produced by two parallel metal plates P_1 and P_2. One type of radiation is bent slightly towards the negative plate, another is bent strongly towards positive plate and a third kind is not bent at all. These are designated as α (alpha), β (beta) and γ (gamma) rays, respectively. The effect of a strong magnetic field, which is perpendicular to the plane of the paper and directed inwards is shown in Fig. 12.9(*b*). This also establishes the existence of the three types of radiations.

The experiments demonstrate that α-rays consist of positively charged particles, β-rays consist of negatively charged particles and γ-rays are electrically neutral. All radioactive atoms emit either α-rays or β-rays, but never both, and this may be accompanied by γ-rays.

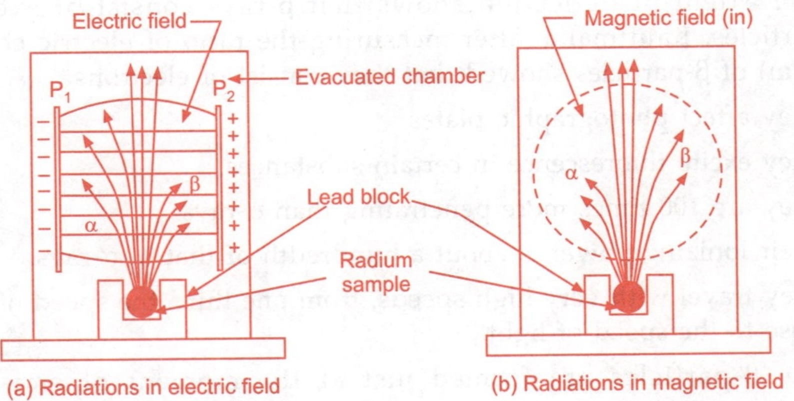

Fig. 12.9 *Radioactive radiations*

Properties of radioactive rays:

The following properties are attributed to the α, β and γ-rays.

The α-rays:

1. The α-rays are deflected by electric and magnetic fields. The sense of deflection shows that they are positively charged particles.

2. Rutherford and Geiger carried out a detailed study of the extent of deflection, and concluded that the charge on the α-particles is +2 and its mass is 4 atomic mass units. Hence, they are identified with helium nuclei.

3. They affect photographic plates.

4. They excite fluorescence in certain substances such as zinc sulphide and when the fluorescence is examined under a microscope, it is found to consist of individual scintillations. This shows that α-rays consist of discrete particles.

5. They are easily absorbed by thin metal foils for example, 0.01 cm thick aluminium foil. They are stopped by 2 cm to 8 cm thickness of air. These show that the range of α-particles is very small and hence their power of penetration is very small.

6. They are very good ionizers producing thousands of ions per cm along their track. The ionizing power of α-rays is about one hundred times greater than that of β-rays.

7. Certain materials such as glass, crystals, upon irradiation with intense α-rays change their colour and may become brittle.

8. They travel with velocities as high as 10^7 m/sec.

The β-rays:

1. The β-rays are deflected by electric and magnetic fields. The sense of deflection shows that they are negatively charged particles.

2. The extent of deflection shows that β-rays consist of extremely light particles. Kauffmann, after measuring the ratio of electric charge to mass (e/m) of β-particles showed that they consist of electrons.

3. They affect photographic plates.

4. They excite fluorescence in certain substances.

5. They are 100 times more penetrating than α-rays.

6. Their ionizing power is about a hundredth of that of α-rays.

7. They travel with very high speeds, from one third the speed of light to very close to the speed of light.

8. The β-particles are formed just at the moment of emission due to transformation of neutrons into protons and electrons. They do not exist as regular parts of the nucleus.

The γ-rays:

1. The γ-rays are not deflected by electric field or magnetic field and hence they are not charged. They are electromagnetic radiations of short wavelength, varying from 0.1Å to 0.00 1Å.

2. They affect photographic plates.

3. They excite fluorescence in certain substances.

4. They are the most penetrating of the three radiations. Their power of penetration is about 100 times that of β-particles. They can pass through 30 cm of iron.

5. Their ionization power is the least among the three radiations and is about one hundredth of that of β-rays.

6. They destroy living tissues. This property is employed in the treatment of cancer where γ-rays emitted from radium are used.

7. The γ-rays originate from the nucleus and are emitted only after the nucleus ejects either an α or β particle.

The most common methods of measuring radioactive radiations are using electroscopes, electrometers, cloud chambers, bubble chambers, ionization chambers, proportional counters, Geiger-Muller counter, scintillation counter, photo-emulsion technique and spark chambers.

12.4.2 Radioactive Decay

When a radioactive element emits α or β particles, the original atom, called the parent atom, changes into another called the daughter atom. The parent atom does not exist anymore and is said to disintegrate. If the daughter atom is also radioactive, the disintegration continues till a stable element (usually lead) is reached finally. The process gives rise to a series of radioactive elements. There are three main series namely uranium, thorium and actinium series of radioactive elements. Results of α and β ray emissions can be described using Soddy's group displacement law which states that:

"When an α-particle is ejected out from a nucleus, the atomic weight of the new element is reduced by 4 and its atomic number reduces by 2 and it falls in a group of the periodic table two columns to the left of the original element. When a β-particle is emitted from the nucleus, the resulting element has the same atomic weight as the parent element and the atomic number is increased by 1, but lies one column to the right in the periodic table.

Examples for the above are:

$$_{88}Ra^{226} \xrightarrow{\alpha} {}_{86}Rn^{222} \qquad\qquad _{82}Ra(B)^{214} \xrightarrow{\beta} {}_{83}Ra(C)^{214}$$

$$_{92}U^{238} \xrightarrow{\alpha} {}_{90}Th^{234} \qquad\qquad _{90}Th^{234} \xrightarrow{\beta} {}_{91}Pa^{234}.$$

When a radioactive element disintegrates by emitting α or β particles, the number of atoms of the parent element goes on decreasing with passage of time. This is called radioactive decay. All the atoms of a radioactive element do not disintegrate at the same time and the disintegration is a continuous process. In 1902, Rutherford and Soddy established experimentally the law of radioactive disintegration or the law of radioactive decay. According to this law, the rate of disintegration, or the number of atoms that undergo disintegration, depends on the number of atoms present at that time. Since the disintegration is a continuous process, the number of atoms present decreases. Hence the rate of disintegration decreases with time.

If N is the number of atoms of a radioactive element present at any given instant t and dN is the number of atoms that disintegrate during the time interval dt (from t to $t + dt$), then, the rate of disintegration $\left(\dfrac{dN}{dt}\right)$ is proportional to N, the number of atoms present. This can be mathematically expressed as $\dfrac{dN}{dt} \propto N$.

or, it can be expressed in the form

$$\frac{dN}{dt} = -\lambda N \qquad\qquad\qquad (12.19)$$

where λ is a constant which depends on the element. It is called the decay constant or disintegration constant or radioactive constant. The negative sign indicates that the number of atoms decreases with time. Rewriting Eq. (12.19) in the form,

$$\lambda = -\frac{dN/dt}{N} = \frac{-(dN/N)}{dt}$$

It can be seen that λ represents the chance (or probability) of disintegration of radioactive atom in one second. It must be noted that the disintegration is random in nature and hence radioactivity is a statistical phenomenon. A radioactive element having a higher rate of disintegration is considered to be more active. Rate of disintegration is called activity. Unit of activity is Curie, which is defined as the quantity of any radioactive substance in which number of disintegrations per sec. is 3.7×10^{10}.

From Eq. (12.19), we have $\dfrac{dN}{N} = -\lambda dt$. Integrating this expression we get, $\log_e N = -\lambda t + K$.

where K is constant and e is the Napierian base.

At $t = 0$, $N = N_o$, the initial number of atoms.

$$\log_e N_0 = -0 + K$$

$$K = \log_e N_0$$

$$\log_e N = -\lambda t + \log_e N_0$$

$$\log_e N - \log_e N_0 = -\lambda t$$

$$\log_e \left(\frac{N}{N_0}\right) = -\lambda t$$

$$\frac{N}{N_0} = e^{-\lambda t}$$

$$N = N_0 e^{-\lambda t}. \tag{12.20}$$

From Eq. (12.20), it is seen that the number of atoms of a given radioactive element decreases exponentially with time.

When
$$\lambda = \frac{1}{t}, N = N_0 e^{-(1/t)t} = N_0 e^{-1}$$

and
$$N = \frac{N_0}{e}.$$

Therefore, the decay constant can be defined as the reciprocal of the time during which the number of atoms of the original element decreases to $\dfrac{1}{e}$ of its initial value.

Half-life: In Eq. 12.20, we see that $N = 0$ when $t \rightarrow \infty$. Therefore, theoretically, an infinite time is required for all atoms of a radioactive element to disintegrate completely. How fast the disintegration of radioactive elements takes place can be specified by using a quantity called half-life period or half-life (T). The half-life period of a radioactive element is defined as the time during which 50% of the original atoms disintegrate. T is characteristic of radioactive element and can be used to distinguish radioactive elements. The half-life period for uranium is 4.5 billion years, for radium it is about 1,600 years, for radon it is 3.8 days and for Ra (C') it is only 1.64×10^{-4} sec. For a given radioactive element 50% of atoms remain without disintegration at the end of time T, 25% at the end of $2T$, 12.5% at the end of $3T$ etc., and only about 0.1 % after a time 10T. Therefore absolute life of a radioactive element is finite for practical purposes.

Disintegration constant and half-life of a radioactive element are related to each other. Putting $N = \dfrac{N_0}{2}$, when $t = T$, in Eq. (12.20), we have

$$\frac{N_0}{2} = N_0 e^{-\lambda t}$$

i.e.

$$e^{-\lambda t} = \frac{1}{2}$$

Taking logarithms on both sides of the equation, we have

$$-\lambda T \log_e e = \log_e 1 - \log_e 2$$
$$-\lambda T(1) = 0 - \log_e 2$$
$$\lambda T = 0.693$$

$$T = \frac{0.693}{\lambda} . \tag{12.21}$$

It is seen from Eq. (12.21), that the half-life of a radioactive element is inversely proportional to its disintegration constant.

Average life: Since radioactive disintegration is a statistical phenomenon, the average life time or life expectancy of a radioactive elements is one of the parameters. It is independent of the age of radioactive atom. For example, Ra^{226} atom formed by the α-decay of Th^{230} has the same chance of decaying into Rn^{222}, whether Ra^{226} was formed just at the moment or a million years ago. Average life or mean life T_{av} of a radioactive element is defined as the average time for which the atoms of a radioactive element exist. It is equal to the average of the lives of all atoms of the radioactive sample.

Starting from Eq. (12.20), we have

$$N = N_0 e^{-\lambda t}$$

The mean life or the average life is given by

$$T_{av} = \frac{1}{A} \int_0^\infty tN(t)dt = \frac{N_0}{A} \int_0^\infty t e^{-\lambda t} dt$$

where A is the area of the curve $N(t)$ given by

$$A = \int_0^\infty N(t)\,dt = N_0 \int_0^\infty e^{-\lambda t}\,dt = \frac{N_0}{\lambda}$$

Solving for the average life we get

$$T_{av} = \frac{1}{\lambda} \tag{12.22}$$

which means that the average life of a radioactive atom is the reciprocal of its disintegration constant. The half-life T, the decay constant λ, and the average life T_{av}, are related as follows

$$T = \frac{0.693}{\lambda} = 0.693\,T_{av} \tag{12.23}$$

Artificial radioactivity or induced radioactivity:

Frederick Joliot and Irene Curie Joliot discovered in 1934 in the course of their experiments that when Aluminium was bombarded with α-particles, the target continued to emit radiations even after the source of α-particles was removed. Further studies showed that the radiations consist of positrons. The behaviour was similar to that of natural radioactivity. They explained this phenomenon as follows: when α-rays bombard Aluminium, an unstable nucleus is formed and this nucleus disintegrates spontaneously. This is known as artificial radioactivity or induced radioactivity. The nuclear reaction in the case of Aluminium is given by

$$_2He^4 +_{13} Al^{27} \rightarrow_{15} P^{30} +_0 n^1 .$$

The resulting phosphorus is radioactive and is called radio phosphorus. It emits a positron and is converted into silicon.

$$_{15}P^{30} =_{14} Si^{30} + e^+$$

The half-life period T_{av}, of radio phosphorus is 2.5 minutes. Protons, deuterons and neutrons also produce artificially radioactive substances, when they are used as bombarding agents. For example, when neutrons bombard magnesium, radio sodium is produced.

$$_0n^1 +_{12} Mg^{24} \rightarrow_{11} Na^{24} +_1 H^1$$

$$_{11}Na^{24} =_{12} Mg^{24} + e^- \quad (T = 14.8 \text{ hours})$$

Based on the above, artificial radioactivity can be defined as a process of producing radioactive elements by bombarding non-radioactive elements with accelerated particles. Isotope of any element which is radioactive is called radioisotope.

The main difference between natural radioactivity and artificial radioactivity is that besides α, β and γ rays in both the cases, positrons are emitted in the case of artificial radioactivity.

12.4.3 Detection of Radiations

Radiations can be detected by the effects they produce such as (i) ionisation of a gas through which they are passed, (2) fluorescence due to radiations in certain materials, (3) effect of radiations on photographic plate.

Ionization effect of radiations on a gas is mostly used for detection of radiations in devices such as gas counters, semiconductors, counters, Wilson cloud chamber, bubble chamber and spark chamber. Scintillation counters using the fluorescence effect and photographic emulsions using the effect on a photographic plate are also used in some devices.

A gas counter, consists of a metal cylinder with a suitable gas. A metal rod or wire, insulated from the surroundings, is mounted along the axis of the cylinder. A potential difference is applied between the electrodes. The radiation to be detected is passed through the chamber resulting in the ionization of the gas. The free charges move towards the respective electrodes which results in an ionization current. The ionization chamber, proportional counter and Geiger-Muller counter belong to this category and their operation depends on potential difference.

In an ionization chamber, the potential difference is low compared to that of proportional counter and G-M counter and has a suitable value so that the ionization current is saturated. Proportional counters operate with a range of potential differences for which the ionization current is proportional to the operating voltage. The region of potential difference is comparatively higher for Geiger-Muller counter and the potential difference is such that the final ionization is independent of the magnitude of the primary ionization.

Alpha particles and protons can be detected using ionization chambers. Ionization chambers are not suitable for β-rays and γ-rays. Proportional counters are quite suitable for soft radiations such as low energy β-rays and X-rays. Ionization chambers and proportional counters are not suitable for high energy particles. G-M counters are very useful in detecting nuclear particles.

12.4.4 Geiger-Muller Counter

A Geiger-Muller counter is a device that can detect the presence of ionizing particles emitted by a substance and is used to determine the presence of radioactivity. It consists of a hollow cylindrical tube of copper, C, about 0.5 m long. It is enclosed in a cylindrical tube of glass, G. The tube is filled with a suitable gas at low pressure depending on the particle to be detected. A tungsten wire, W, is mounted along the axis of the tube as shown in Fig. 12.10. A potential difference of about 1000 to 3000 volts is maintained between the tube and the wire through a high resistance R. One end of R is earthed and the voltage is adjusted such that there is no spontaneous discharge.

When a charged particle enters the tube, the enclosed gas is ionized by the extraction of electrons from the molecules of the gas. During the motion of these electrons towards the anode, more and more ionization takes place. This forms a small current pulse through the resistance R. The potential difference across R is amplified and made to operate an automatic counting device. Each particle passing through the tube is registered by the counter. If a loud speaker is connected in the place of the counting device, a click can be heard as each particle enters the tube. Geiger Muller counter, or the G-M counter, can detect X-rays and γ-rays also.

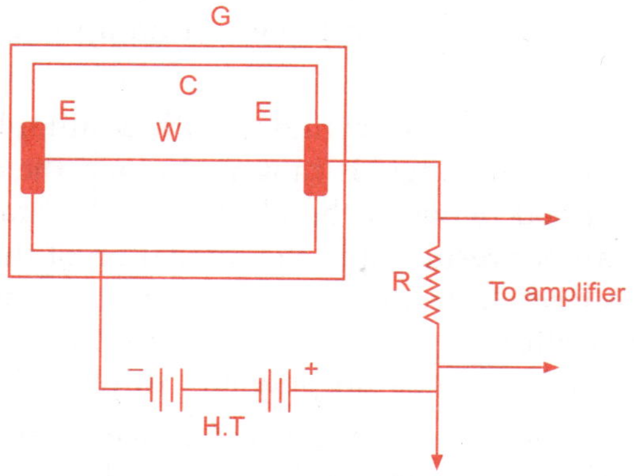

Fig. 12.10 *Geiger-Muller Counter*

Solid State detector: Solid-state detector, also called Semiconductor Radiation Detector, is a radiation detector in which a semiconductor material, such as a silicon or germanium crystal, constitutes the detecting medium. It consists of a silicon p-n junction diode, with the flat surface of n-type coated with a thin gold film. A diode is used with a reverse bias which produces a depletion layer. When an energetic particle enters, it produces an electron hole pair. These pairs are carried away by the electric field. The corresponding voltage pulse in a load resistance is amplified to operate a counting device. Solid state detectors are particularly suitable for the study of high energy particles. Germanium detectors are better suited than silicon detectors for gamma (γ) ray detection.

Scintillation Counter: Scintillation counter is used for the detection and measurement of high energy radiations. It is one of the oldest types of radiation detectors. Its operation is based on the principle that radiations striking a fluorescent material (such as zinc sulphide) produce a tiny flash of light. These flashes, called scintillations, can be seen or detected with the help of a photomultiplier tube in conjunction with a counting device. Counting rate is very high in the region of 10^9/second.

Photographic emulsion: The principle of operation is similar to that of ordinary photography. Special types of emulsions, called nuclear emulsions, containing about ten times the usual proportion of silver halide grains are used. They are sensitive to all types of charged particles. The tracks produced in nuclear emulsions are very short and they are studied with high power microscopes. Cosmic rays have been studied with the help of photographic emulsions.

12.4.5 Applications of Radioisotopes

Radio isotopes are produced when elements are placed in nuclear reactors. Some isotopes can be produced by neutron bombardment and some by neutron capture. Some of the important radioisotopes so produced are tritium (H^3), radio carbon (C^{14}), radio-sodium (Na^{24}), radio phosphorus (P^{32}), radio sulphur (S^{35}), radio cobalt (Co^{60}), and radio iodine (I^{131}). Radioisotopes are widely used as tracers in medicine, where a radioisotope is added to the system under study and the course of the isotope is followed to understand the system behaviour. Radioactive tracers are also used in agriculture, industry and research.

(1) **Medicine:** Radio isotopes are very useful in medical diagnosis. Iodine is absorbed more strongly by thyroid glands, than by other tissues. By administering radio iodine to a patient and using sensitive detectors of β-rays, information about thyroid glands can be obtained.

Radio iodine and some other isotopes are useful in the diagnosis of brain tumours. If the blood circulation in a person is restricted, the problem locations can be detected using radio sodium.

(2) **Agriculture:** Radio phosphorus is used in crops to give better yields. Radio phosophorus is used along with fertilizers in order to select a fertilizer that is the most suitable to a particular soil.

Perishable cereals are kept fresh for a longer time by exposing them to radiations. Onions, potatoes and gram are kept unspoiled by small doses of radiation.

(3) **Industry:** Radiations from radio isotopes are used (*i*) to check the thickness of paper, rubber or metal sheets (*ii*) to detect leakage in burned pipes carrying petrol and (*iii*) to find the level of liquid in a closed reservoir.

(4) **Radioactive dating:** Estimation of age using radioactive dating is employed by anthropologists and archaeologists. One of the isotopes used is radio carbon-14, which has a radioactive equilibrium in the atmosphere. After the death of an organism, amount of C^{14} goes on decreasing. For example old bones contain less C^{14} than new bones and old wood contains less C^{14} than new wood. Comparison of the concentration ratio of C^{14} to C^{12} in the remnants of dead organisms, with the normal ratio in corresponding living organisms helps to estimate the age. The method is used in estimating the age of the earth from measurements on the relative amounts of U^{238} and Pb^{206} in geological specimens.

(5) Research: In scientific research, radiations from radio isotopes are used to test materials. The material to be tested is placed between the source of radiations and radiation detector. The change in counting rate helps to study the physical properties and the structure of the material.

EXAMPLE 12.10

The half life of a radioactive element is 4×10^8 years. Calculate its decay constant and mean life.

SOLUTION

$$\text{Decay constant, } \lambda = \frac{0.693}{T} = \frac{0.693}{4 \times 10^8}$$

$$= 1.73 \times 10^{-9} \text{ per year}$$

$$\text{Mean life} = \frac{1}{\lambda} = \frac{1}{1.73 \times 10^{-9}} = 5.77 \times 10^8 \text{ years}$$

$$\text{Decay constant} = 1.73 \times 10^{-9} \text{ per year}$$
$$\text{Mean life} = 5.77 \times 10^8 \text{ years.}$$

EXAMPLE 12.11

A radioactive element has a half-life of 20 days. How much of the element would be left after 80 days, if the original amount was 4 gm?

SOLUTION

In every half-life, the amount is reduced to half.

Half-life = 20 days. Give time 80 days = 4 half-lives

$$\text{Amount left after 80 days} = \left(\frac{1}{2}\right)^4 \times 4 = \frac{4}{16} = 0.25 \text{ gm.}$$

In general amount left after n half-lives would be $\left(\frac{1}{2}\right)^n$ of initial amount.

EXAMPLE 12.12

A radioactive sample has a strength of 8 micro curies. If the half-life of the sample is 5 days what will be the strength after 15 days?

SOLUTION

$$\text{Strength} = \text{activity,}$$

$$\text{Activity } a = \frac{dN}{dt} = -\lambda N_1$$

In the beginning $a_1 = 8 = -\lambda N_1$

After 15 days, $a_2 = -\lambda N_2$

$$\frac{a_2}{a_1} = \frac{N_2}{N_1}.$$

After 15 days, i.e., 3 half-lives, (half-life = 5 days) the amount of the sample that remains is $\left(\frac{1}{2}\right)^3$ of N_1

$$N_2 = \left(\frac{1}{2}\right)^3 N_1$$

Hence

$$\frac{a_2}{8} = \left(\frac{1}{2}\right)^3$$

or,

$$a_2 = 1$$

Strength after 15 days is 1 micro curie.

EXAMPLE 12.13

A radioactive sample of half-live T contains 1000 nuclei. How many will be left after a time $\frac{T}{2}$?

SOLUTION

$$N_0 = 1000, \lambda = \frac{0.693}{T}, t = \frac{T}{2}$$

$$N = N_0 e^{-\lambda t}$$

$$2.303 \log N = 2.303 \log N_0 - \lambda t = 2.303 \times 3 - \frac{0.693}{2}$$

$$\log N = \frac{6.5625}{2.303}$$

or

$$N = 707$$

Number of atoms left after a time $\frac{T}{2}$ is 707.

EXAMPLE 12.14

What is the mass in g of 10 micro curies of cobalt 60? (half-life = 5.3 years)

SOLUTION

$$1 \text{ year} = 365 \times 86400 \text{ sec.}$$

$$1 \text{ curie} = 3.7 \times 10^{10}$$

$$\lambda = \frac{0.693}{T} = \frac{0.693}{5.3 \times 365 \times 86400}$$

$$\text{Activity} = \frac{dN}{dt} = \lambda N \qquad \text{(leaving negative sign)}$$

$$10 \times 10^{-6} \times 3.7 \times 10^{10} = \frac{0.693}{5.3 \times 365 \times 8640} N$$

$$N = 8.925 \times 10^{13}$$

For 6.02×10^{23} atoms $\rightarrow$ mass is 60g

For 8.925×10^{13} atoms $\rightarrow \dfrac{60 \times 8.925 \times 10^{13}}{6.02 \times 10^{23}}$

$$= 8.896 \times 10^{-9} \, g$$

Mass of 10 μ curies of Co^{60} is $8.896 \times 10^{-9} \, g$.

12.5 ATOMIC NUCLEUS

12.5.1 Whole Number Rule

In 1815 and 1816, William Prout, an English chemist, discovered that the atomic weights of the elements appeared to be whole multiples of the atomic weight of hydrogen. He hypothesized that the hydrogen atom was the fundamental object, and that the atoms of other elements were actually groups of various numbers of hydrogen atoms. He proposed that hydrogen was the fundamental building block of matter. Prout's hypothesis is known as whole number rule. According to this rule, the atomic weights of all elements must be integers when the atomic weight of hydrogen is taken as unity. But experiments showed that the atomic weights of many elements differed sharply from whole numbers. A few examples are chlorine of atomic weight 35.46, magnesium (24.32), nickel (58.69) and zinc (65.38). The study of natural radioactivity and experiments with positive rays indicated that all atoms of the same element need not be identical. Frederick Soddy, an English radiochemist, gave the name Isotopes to the different atoms of the same element, but with different atomic weights. Prout's hypothesis was discarded because it could not explain fractional atomic weights of some elements and also the existence of isotopes. However, the whole number rule was revived later in a new form, when the investigations of Thomson, Aston and other workers showed that the atomic weights of the isotopes of all elements are whole numbers. Determination of atomic masses of isotopes of an element using mass spectrograph shows that the departure of mean atomic weights from the integral values was due to the existence of isotopes. For example, neon of atomic weight 20.2 has two isotopes of atomic weights 20 and 22. The value 20.2 for the atomic weight is due to the fact that the gas is normally a mixture of the isotopes in the ratio 9:1. Mass spectrographic measurement shows that chlorine has two main isotopes of masses 35 and 37 with ratio of 3.07:1. This leads to a mean atomic weight of 35.46. The whole number rule assumed importance again because it showed the way to a rational theory of the structure of atomic nucleus.

Subsequent discoveries of electron, proton and neutron made the whole number rule to appear so natural that the atomic nuclei were considered to consist of protons and neutrons.

12.5.2 Atomic Nucleus

The centre of the atom is called the nucleus. In order to explain the experimental observations of large angle scattering of α-particles by thin metallic foils, Rutherford assumed that most of the mass of the atom is concentrated at the centre of the atom. The nucleus has a diameter of about one fermi (1 fermi = 10^{-15} m). The nucleus carries positive charge and is surrounded by electrons, making the atom as a whole neutral. The distribution of electrons is within a radius of about 10^{-10} m (atomic radius). This means that electrons occupy the space in between 10^{-15} m and 10^{-10} m. Only a very small portion of this space is occupied by electrons and hence it may be considered to be almost empty.

Based on his nuclear model of atom, Rutherford developed a theory for the scattering of α-particles. The theory was verified experimentally in 1913 by Geiger and Marsden in England, which confirmed Rutherford's nuclear atom model. In order to explain the mechanical stability of the atom, Rutherford assumed that the electrons revolve around the nucleus in certain orbits just like planets around the sun. Rutherford's nuclear atom model was modified in 1913 by Niels Bohr, a Danish scientist, in order to explain the observed hydrogen spectrum. Later developments show that the nucleus consists of neutrons, mesons, pions in addition to protons.

Nuclear charge: The charge on the nucleus, called nuclear charge, is positive and is equal to the total negative charge of orbital electrons. Nuclear charge is represented by Ze where e is the magnitude of electronic charge and Z is the number of electronic charges around the nucleus. Z is called the atomic number of the particular element. Because hydrogen is considered as a building block of atoms of other elements, the atomic nucleus of hydrogen was given a special name, proton, which means first. Proton is an essential constituent of all matter and its mass is equal to 1.67×10^{-24} g or 1.67×10^{-27} kg.

In 1919, Rutherford found that a reaction between α-particles and nitrogen resulted in the creation of oxygen and hydrogen. The reaction can be written as follows

$$2He^4 + 7N^{14} \rightarrow 8O^{17} + {}_1H^1 + \text{energy.}$$

Nuclear size: The size of the nucleus depends on the mass number. The first measurements of nuclear radii were made by α-ray scattering experiments. Nuclear radii are found to be in the range 1 to 10 fermi. Nuclear radius R is approximately given by the relation

$$R = R_0 A^{1/3} \tag{12.24}$$

where A is mass number and R_o, a constant with a value 1.2 to 1.4 fermi. Nuclear density is about 2.5×10^{14} gm/cc or 2.5×10^{17} kg/m. Matter in some stars has extremely high density.

Atomic mass unit: The atomic masses are extremely small, and hence a small unit called atomic mass unit (amu) is used to express atomic masses and masses of subatomic particles. The atomic mass unit is defined as one twelfth of the mass of one atom of carbon-12 isotope. The number of atoms in one gm of carbon-12 isotope (or 12 gm of carbon-12) is 6.02×10^{23}, the Avagadro number.

Therefore, mass of one atom of carbon $\dfrac{12}{6.02 \times 10^{23}}$

$$1 \text{ amu} = \frac{12}{6.02 \times 10^{23} \times 12} = 1.66 \times 10^{-24} \text{ gm}$$

mass of neutron = 1.0087 amu

mass of proton = 1.0073 amu.

Nuclear mass: The nuclear mass is the mass of the nucleus. Since the mass of electrons is negligibly small in comparison with those of the proton and the neutron, atomic mass can be taken as nuclear mass. Atomic masses are obtained from mass spectrography. Atomic masses are measured in comparison with the atomic mass of carbon-12 isotope, which is taken as 12 amu.

The nucleus spins about an axis and because of its charge and spin, the nucleus possesses a magnetic moment.

Constituents of nucleus: Before neutrons were discovered, it was assumed that nuclear matter consisted of protons and electrons. Such a proton-electron hypothesis was put forward in order to account for the mass number and atomic number, and agreed with the fact that some radioactive elements emit electrons (α-rays). Experimental facts about nuclear spin, nuclear magnetic moment and energy inside a nucleus did not support the existence of electrons inside the nucleus.

Around 1920, Rutherford predicted the existence of a neutral particle with mass close to that of proton. Bothe and Becker, in 1930, discovered that a highly penetrating radiation was produced when beryllium was bombarded with α-particles. They thought that these radiations were hard γ-rays. Later Curie and Joliot found out that these radiations could expel high speed protons from hydrogenous matter, for example, paraffin. In 1932, Chadwick, an English physicist, showed that these radiations consisted of neutral particles with mass close to that of protons which are called neutrons.

When the neutrons were discovered, the proton-electron hypothesis was discarded. Following this Werner Heisenberg, a German physicist suggested the proton-neutron hypothesis. According to the proton-neutron hypothesis, the nucleus is composed of protons and neutrons. Protons and neutrons are called 'nucleons'. If the number of nucleons is equal to A, (the mass number) and if the number of protons is Z, (the atomic number) then the number of neutrons is equal to $(A - Z)$. But for normal hydrogen, neutron enters as an essential constituent of matter. The mass of proton is very close to that of a neutron. Mass of proton = 1.0073 amu and mass of neutron = 1.0087 amu.

With the proton-neutron hypothesis it is possible to explain the existence of isotopes. The nuclei of the isotopes of a given element differ only in the number of neutrons in them. The hypothesis is consistent with the known facts of radioactivity. The emission of a β-particle by a nucleus is explained by assuming that a neutron changes into a proton and a β-particle. This is in agreement with the experimental fact that during β-emission, atomic number increases by one unit while its mass number remains unchanged.

In view of the fact that all the known nuclear properties are explained in terms of proton-neutron hypothesis, this structure of nucleus came to be accepted.

Mass defect: The atomic mass of any stable nucleus is found to be less than the sum of the masses of the constituent particles such as protons and neutrons. The atomic masses are not whole integer numbers. The difference between the sum of the masses of the constituent particles of an atom and its atomic mass is known as mass defect. The mass defect ΔM of an atom of mass M, atomic number Z and mass number A, is given by

$$\Delta M = \left[ZM_p + (A - Z)M_n + ZM_e \right] - M$$

$$\approx \left[ZM_p + (A - Z)M_n \right] - M \tag{12.25}$$

where M_n, M_p and M_e represent the masses of neutron, proton and electron, respectively.

Packing Fraction: Mass number of an atom is the number of nucleons (protons and neutrons) in its nucleus. The atomic mass is the total mass of all the constituent particles in an atom after subtracting any mass defect. In an atom if M is the atomic mass and A the mass number of an atom, then the deviation of atomic mass from the mass number, or mass correction is equal to $(M - A)$, and the ratio $\left(\dfrac{M-A}{A}\right)$ represents the packing fraction p. Packing fraction may be positive or negative. The packing fraction is negative for elements having mass numbers between 20 and 200 and it is positive for other elements.

Packing fraction, and hence mass correction, is different from mass defect. For C-12, packing fraction is zero, however, the mass defect is not zero. A nucleus having positive packing fraction has its mass M greater than its mass number A and a nucleus having negative packing fraction has its mass less than its mass number. Negative p indicates stable nuclei. On the other hand, positive p suggests that the nucleus is less stable. He^4, Be^8, C^1 and O^{16} are very stable as compared with other elements in the same mass number region.

Einstein's mass-energy relation: Einstein established from his theory of relativity that mass and energy are interchangeable. Thus, energy E, produced by conversion of a mass m, is given by the relation

$$E = mc^2$$

where c is the velocity of light. E is in joules when m is in kg and c is in m/sec. The relation $E = mc^2$ is called Einstein's mass-energy relation, or Einstein's principle of equivalence of mass and energy.

Accordingly, 1 kg of mass is equivalent to 9×10^{16} joules of energy.

The relation $E = mc^2$ holds good for all forms of energy. Therefore, mass and energy may be considered as the two forms of the same physical entity and the laws of conservation of mass and energy are unified into one law. Nuclear fission and nuclear fusion and production of electron positron pairs in cosmic rays provide experimental evidence for Einstein's mass-energy relation.

12.5.3 Binding Energy

Einstein's mass-energy relation could explain the mass defect. The nucleons in a nucleus are held together by strong nuclear forces. Work must be done in separating them from each other. Therefore, energy must be supplied to the nucleus in order to break it down completely into its constituent particles. This energy that binds the nucleons together is called the binding energy of the nucleus. Einstein's mass-energy relation, $E = mc^2$, confirms that the binding energy of a nucleus is equivalent to its mass defect. Binding energy of a nucleus is the energy required to break the nucleus into its constituent particles. The mass defect accounts for the amount of mass which is converted into energy in order to hold the nucleons together.

Binding energy for any nucleus can be obtained from the relation, $BE = \Delta E = (\Delta M)c^2$. Binding energy may be expressed in amu or MeV, where the eV is defined below.

Electron volt: It is usual to measure the energy of charged or uncharged particles in terms of electron volt units. Electron volt is a very small unit of energy. Electron Volt is defined as the energy acquired by an electron in falling through a potential difference of 1 volt. It is represented by eV.

$$1 \text{ joule} = 1 \text{ coulomb} \times \text{volt}$$

$$1 \text{ eV} = (\text{electronic charge in coulomb}) \times \text{volt joule}$$

$$= 1.602 \times 10^{-19} \text{ joule}$$

$$1 \text{ MeV} = 1.602 \times 10^{-13} \text{ joule}$$

Energy equivalent of 1 amu = 931 MeV

BE in MeV = 931 × (mass defect in amu).

The ratio of the BE to the mass number is the BE per nucleon. BE per nucleon is used to compare the binding energy for different nuclei. A plot of BE per nucleon against mass number A is shown in Fig. 12.11.

The binding energy per nucleon starts with a low value for deuteron (nucleus of deuterium, heavy hydrogen) and then increases sharply. The BE per nucleon is 1.1 MeV for hydrogen, 7 MeV for helium, 8 MeV for oxygen and nearly 8.8 MeV for copper. Most of the elements, except very heavy and very light, have BE per nucleon equal to about 8 MeV. The nuclei having maximum value of BE per nucleon are more difficult to break apart. Higher the BE per nucleon, greater the stability of the nucleus. Some nuclei (for example, He^4, Be^8, C^{12}, O^{16}...) have extremely high values of BE per nucleon and, they are exceptionally stable compared with their neighbours. In any nuclear change, the resulting nucleus will lie in the middle portion of the curve, corresponding to maximum stability.

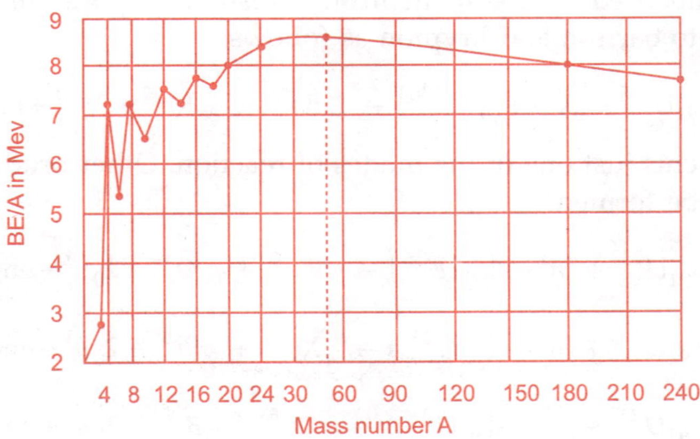

Fig. 12.11 *Binding energy graph*

The fission of light elements produces a heavier nucleus while the fission of heavy nucleus gives two or more lighter nuclei. In these cases the binding energy is increased with consequent release of energy.

12.5.4 Nuclear Fission

When a heavy nucleus splits up into two nuclei of comparable masses the process is called nuclear fission. In 1934, Enrico Fermi, an Itaian physicist, suggested that new elements (with atomic number greater than 2) could be expected when U^{235} was bombarded with slow neutrons in order to break down the nucleus. Two German scientists, Otto Hahn and Fritz Strassmann, found in 1939 that one of the elements formed by the bombardment of U^{235} with neutrons was a radioactive isotope of the element Ba^{141}. This meant that there must be another element to account for the original atom, uranium. German physicists Lise Meitner and Otto Robert Frisch used the concept of fission as an explanation. The other product was shown to be krypton. The new elements formed are called fission fragments as shown in Fig 12.12.

Lise Meitner(1878–1968) was an Austrian physicist, later Swedish, who worked on radioactivity and nuclear physics. Meitner was part of the team that discovered nuclear fission, an achievement for which her colleague Otto Hahn was awarded the Nobel Prize. Meitner is often mentioned as one of the most glaring examples of women's scientific achievement overlooked by the Nobel committee. Lise Meitner was the aunt of Frisch.

During fission, pairs of elements in the middle of the periodic table were obtained as fission fragments. The products fall into two groups: one of atomic number in the range 35 to 43 and mass number in the range 80 to 110, and the other of atomic number in the range 51 to 57 and mass number from 125 to 160. The fission fragments are accompanied by two or three neutrons and an enormous amount of energy is released in the process which is due to the conversion of a portion of atomic mass into energy.

If U^{235} is bombarded by a slow neutron, an isotope of uranium is formed which may break up into barium and krypton as follows:

$$_{92}U^{235} + {}_0n^1 \rightarrow \left({}_{92}U^{236} \right) \rightarrow {}_{56}Ba^{141} + {}_{36}Kr^{92} + 3\,{}_0n^1 + \text{energy}$$

which represents just one of the modes of reaction. Other products as mentioned below may also be formed:

$$_{92}U^{235} + {}_0n^1 \rightarrow \left({}_{92}U^{236} \right) \rightarrow {}_{56}Ba^{140} + {}_{36}Kr^{94} + 2\,{}_0n^1 + \text{energy}$$

$$_{92}U^{235} + {}_0n^1 \rightarrow \left({}_{92}U^{236} \right) \rightarrow {}_{54}Xe^{139} + {}_{38}Sr^{95} + 2\,{}_0n^1 + \text{energy}$$

$$_{92}U^{235} + {}_0n^1 \rightarrow \left({}_{92}U^{236} \right) \rightarrow {}_{57}La^{153} + {}_{35}Br^{81} + 2\,{}_0n^1 + \text{energy}$$

There exists a mass defect of 0.216 amu in this reaction, which is equivalent to about 200 MeV. This magnitude of energy released by fission of 1 gm atom of U^{235} is phenomenal because this is equal to that released by the combustion of 50 million tons of coal.

During each fission, about 2 to 3 neutrons are released. The neutrons ejected during fission may cause further fission. Consequently the reaction is self-sustained and is known as a chain reaction. In one type of chain reaction, neutrons are built up to a certain level and thereafter the number of fission producing neutrons is kept constant. Because this reaction is controlled, atomic power is generated at a definite rate. This property is used in atomic reactors. The other type of chain reaction is used in atom bomb, where the number of neutrons is allowed to multiply indefinitely and energy is released in a very short interval of time, resulting in a violent explosion.

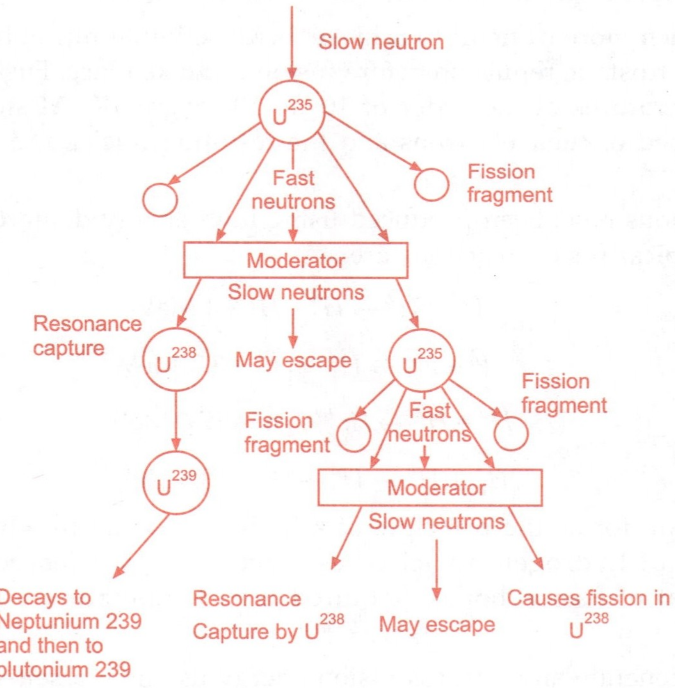

Fig. 12.12 *Controlled chain reaction*

In order to sustain a chain reaction, the size of fissionable material must have a particular value called critical size.

A heavy nucleus such as that of uranium, is unstable on account of low binding energy per nucleon. In such a nucleus which is already unstable, the addition of a neutron makes it even more unstable and results in considerable agitation in the form of a series of rapid oscillations in the compound nucleus. If the oscillations become violent, fission takes place. Average binding energy per nucleon, or specific binding energy of the parent nucleus being less than the specific binding energy for the fission fragments, enormous energy is released during fission.

12.5.5 Nuclear Fusion

Nuclear fusion is a nuclear reaction in which two or more atomic nuclei join together, or "fuse", to form a single heavier nucleus. During this process, some of the mass of the fusing nuclei is converted into energy which is released. Fusion is the process that powers active stars.

The fusion of two nuclei with lower masses than iron (which, along with nickel, has the largest binding energy per nucleon) generally releases energy, while the fusion of nuclei heavier than iron absorbs energy. The opposite happens for the nuclear fission process. This means that fusion generally occurs for lighter elements only, and fission normally occurs only for heavier elements.

Fusion is much more difficult to achieve, because initial nuclei have to overcome their mutual electrostatic repulsion before fusion can take place. Fusion takes place at very high temperatures of the order of 10^7 to 10^8 degree K. At such temperatures atoms are stripped of their electrons and the resulting nuclei and electrons form a plasma.

Fusion reactions have been produced using high energy deuterons from particle accelerators. Typical fusion reactions are:

$$H^2 + H^2 \rightarrow H^3 + H^1 + 4 \text{ MeV}$$

$$H^2 + H^2 \rightarrow H^3 + {}_0n^1 + 3.3 \text{ MeV}$$

$$H^3 + H^2 \rightarrow He^4 + {}_0n^1 + 17.6 \text{ MeV}$$

$$H_e^3 + H^2 \rightarrow H_e^4 + H^1 + 18.3 \text{ MeV}$$

Nuclear fusion forms the principle of a hydrogen bomb in which uncontrolled fusion reaction of hydrogen nuclei takes place. Energy released is many times greater than that of fission bomb. Required high temperature is produced using fission process.

In order to generate and harness fusion energy usefully, a self-sustaining fusion chain reaction must be created. There are accompanying problems in this regard such as containment of plasma, instability of plasma etc. Although problems in the extraction of fusion energy have not been overcome at present, its potential for peaceful use of energy is enormous.

Bethe and Weizsacker provided a satisfactory explanation of the origin of energy in stars in 1939 on the basis of fusion of hydrogen nuclei. There are two different cycles of processes – one is the proton-proton cycle and the other one is carbon cycle. The net result is the same, namely, four hydrogen atoms getting converted into one helium atom with the liberation of energy of about 25 MeV. Although the energy released per fusion is less than that released per fission, in view of the lightness of the nuclei involved, the energy released per unit mass in fusion is very much greater than that released per unit mass in fission.

12.5.6 Nuclear Reactor

A nuclear reactor is a device in which a sustained nuclear chain reaction may be maintained thereby generating a controlled supply of nuclear energy. All the reactors today are fission type reactors.

A fundamental distinction is in the fuel used and the associated equipment and devices. The essential features of a reactor including the fuel and the associated control equipment are:

(1) Type of fuel – This is the material containing the fissile isotope and is called reactor fuel;

(2) Moderator – This is material used to control or slow down neutrons. Graphite, heavy water, beryllium or its oxide serve as good moderators;

(3) Reflector – This is used to minimize the escape of neutrons (reflector of high mass number for reactors using fast neutrons);

(4) Coolant – This is the material employed to remove the heat generated as a result of fission (ordinary water, heavy water, liquid metals, organic liquids and gases);

(5) Control material – This is the material used for controlling the rate of fission. In thermal reactors, neutron absorbing materials such as silver alloy with cadmium and indium or boron with steel, Aluminium or carbon can be used;

(6) Reactor shield – This is used for the protection from intense neutrons and gamma radiations emitted during reaction. It usually consists of a layer of concrete about 2 to 3 meters in thickness. In high power reactors, there will be another shield close to the core. This consists of a few cm thick iron or steel.

Reactors may be classified according to their purpose such as:

(a) Research reactors – They are used primarily to supply neutrons for research and radio isotope manufacture;

(b) Breeder reactors – The purpose of these reactors is to convert a fertile material (U^{238}) into fissile material (Pu^{239}) for use as a fuel in other reactors;

(c) Power reactors – The primary purpose of a power reactor is the utilization of the fission energy into useful power.

In a reactor, enriched uranium or plutonium in the form of rods are inserted in a calculated manner in a huge pile of graphite blocks, so that fast neutrons are slowed down and their speed becomes critical. Control rods of silver alloy with cadmium or boron incorporated in stainless steel are inserted between uranium rods so that they can be raised or lowered. The reactor is enclosed in a thick concrete wall. Fission is initiated by raising the control rods. A power reactor is schematically shown in Fig. 12.13. The reaction can be controlled by adjusting the position of the rods. A coolant is pumped rapidly through the reactor in order to carry away the heat produced during fission. The coolant passes through a heat exchanger which transfers the heat to the water flowing through the heat exchanger. The steam produced this way may be used to generate power.

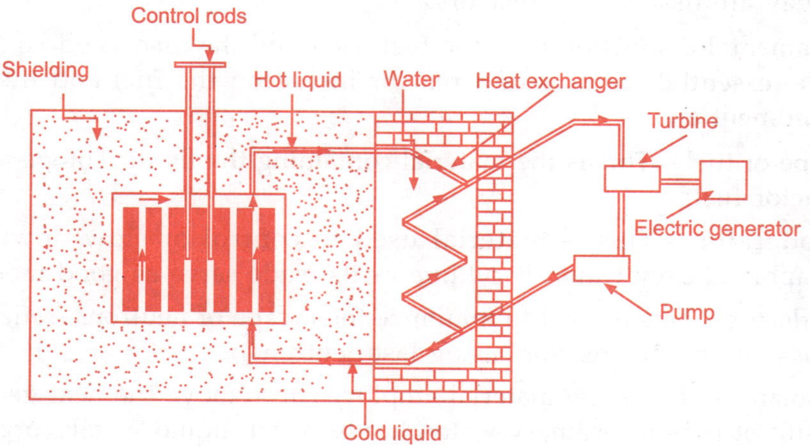

Fig. 12.13 *Nuclear reactors*

In 1942, Enrico Fermi, scientist at the University of Chicago, was successful in building and operating a nuclear reactor for the first time.

At present, there are many power reactors in the world used for power generation. The following countries have successfully harnessed nuclear power to meet their energy needs. USA, Russia, UK, France, China and India etc., possess many power reactors.

TABLE 12.2   Mass, (in amu) for particles and atoms

electron	$\rightarrow$	0.00055
neutron	$\rightarrow$	1.00867
proton	$\rightarrow$	1.00728
$_1H^1$	$\rightarrow$	1.00783
$_1H^2$	$\rightarrow$	2.01410
$_2He^4$	$\rightarrow$	4.00260
$_3Li^7$	$\rightarrow$	7.01601
$_6C^{12}$	$\rightarrow$	12.00000
$_7N^{14}$	$\rightarrow$	14.00307
$_8O^{16}$	$\rightarrow$	15.99492

EXAMPLE 12.15

Calculate the binding energy for deuterium

SOLUTION

Mass of proton and electron	=	1.00783 amu
Mass of neutron	=	1.00867 amu
Total mass	=	2.01650 amu
Mass of deuterium	=	2.01410 amu
Mass defect	=	0.0024 amu
1 amu	=	931 MeV
Binding energy = 0.0024 × 931	=	2.23 MeV.

EXAMPLE 12.16

Calculate the binding energy per nucleon for C^{12}

SOLUTION

Mass of 6 proton + 6 electrons	=	1.00783 × 6 amu
Mass of 6 neutrons	=	1.00867 × 6 amu
Total	=	12.09900 amu
Mass of carbon 12 isotope	=	12.000 amu
Mass defect = 12.099 – 12.000	=	0.099 amu
Binding energy	=	0.099 × 931

$$\text{Binding energy per nucleon} = \frac{0.099 \times 931}{12}$$

$$= 7.68 \text{ MeV}$$

EXAMPLE 12.17

Calculate the binding energy of an α-particle given that mass of a proton = 1.0078 amu, mass of a neutron = 1.00867 amu and mass of an α-particle = 4.003 amu.

SOLUTION

$$\text{Mass defect} = 2(1.0078 + 1.00867) - 4.003$$

$$= 4.03294 - 4.003 \approx 0.03 \text{ amu}$$

$$\text{Binding energy} = 0.03 \times 931 = 27.93 \text{ MeV}$$

$$\text{Binding energy of α-particle} = 27.9 \text{ MeV}.$$

EXAMPLE 12.18

In the fission of $_{92}U^{235}$ by thermal neutrons, the two fragments were identified as $_{56}Ba^{141}$ and $_{36}Kr^{92}$. Calculate the energy released when one nucleus of uranium undergoes fission.

SOLUTION

The fission reaction may be represented by the equation

$$U^{235} + n^1 \rightarrow Ba^{141} + Kr^{92} + 3n^1 + energy$$

Rest of mass of U^{235} = 235.0439 amu

Rest mass of n = 1.0087 amu

Total mass of reactants = 236.0526 amu

Rest mass of Ba^{141} = 140.9178 amu

Rest mass of Kr^{92} = 91.8854 amu

Rest of mass of 3 neutrons = 3.0261 amu

Total mass of products = 235.8293 amu

Mass difference = 0.2233 amu

Energy released = 0.2233 × 931 = 207.9 MeV.

EXAMPLE 12.19

U^{235} nucleus undergoes fission as follows:

$$_{92}U^{235} + _0n^1 \rightarrow _{56}Ba^{141} + _{36}Kr^{92} + 3_0n^1 + energy$$

How many kJ of energy is released when 1 gm of U^{235} undergoes fission completely?

SOLUTION

From Example 12.18, the mass difference = 0.2233 amu

Energy released per atom = (0.2233 × 1.66 × 10^{-27} kg) × (3 × 10^8 m/s)^2 = 3.34 × 10^{-11} J

Number of atoms in 1 gm atom of U^{235} is 6.023 × 10^{23}

Number of atoms in 1 gm is $\dfrac{6.023 \times 10^{23}}{235}$

Energy released by 1 gm is $\dfrac{6.023 \times 10^{23}}{235} \times 3.34 \times 10^{-11}$

$$= 8.6 \times 10^{10} \text{ J.}$$

12.6 SOLID STATE ELECTRONICS

Thomas A. Edison, the American inventor, was studying of methods to improve the electric bulb in 1883. He sealed a metal plate inside the bulb and applied a potential difference between the plate and the filament. He connected a galvanometer in the circuit and found that when the plate was made positive with respect to the filament there was a current in the circuit whereas there was no deflection in the galvanometer when the plate was made negative. This effect is known as Edison effect. The Edison effect was applied by J. A. Fleming, Lee de Forest and others, in a variety of electron tubes. It opened up a new branch of electrical science and engineering called electronics. The development of semiconductor devices begun in 1940's, and transistors have completely revolutionised life today. A tremendous advance has been made in miniaturization of electronic devices through the introduction of integrated circuits (called IC). This has led to the branch called solid state electronics.

Thermionic emission

A filament that is heated to incandescence gives off large quantities of electrons. This phenomenon of emission of electron by a hot body is called thermionic emission. Electrons emitted from hot metal surface are called thermo electrons or thermions.

In 1901, Sir Owen Willans Richardson, a British physicist, made a systematic study of thermionic emission of electrons. He showed that the number of electrons emitted depends on the nature of the filament and its temperature. In order to explain the thermionic emission, Richardson assumed that a metallic conductor consists of a large number of free electrons. At ordinary temperatures, these free electrons do not possess sufficient energy to escape out of the surface of the conductor. The minimum energy that an electron needs in order to come out of the surface of a metal is called thermionic work function. When the metal is heated, some of the free electrons acquire sufficient energy and escape out of the surface. Thermionic emission is used to obtain a copious supply of electrons in electron tubes, Coolidge X-ray bulbs, cathode ray oscillographs, etc.

12.6.1 Cathodes

Cathodes are the source of electrons, or emitters, in devices based on thermionic emission. The cathodes can be classified as directly heated (filament type) cathode (Fig. 12.14*a*) and indirectly heated cathode (Fig. 12.14*b*). Directly heated type of cathode consists of a tungsten or thoriated tungsten filament. An indirectly heated cathode consists of a sleeve of nickel, coated with oxides of alkaline earth elements. This is heated by an insulated filament of tungsten or thoriated tungsten kept inside the sleeve. Indirectly heated cathodes are many times more efficient than directly heated cathodes and are widely used.

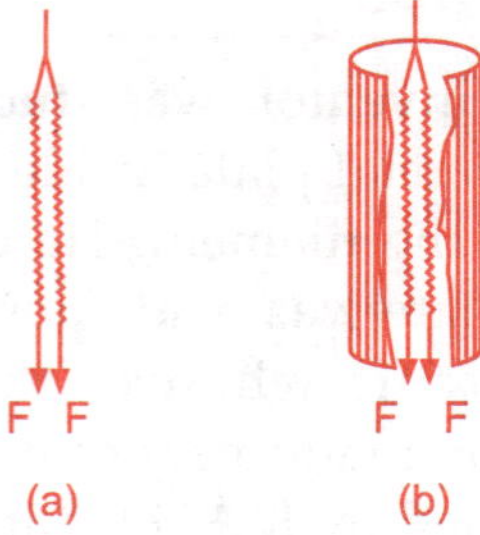

Fig. 12.14 *Cathodes*

12.6.2 Diode

In 1904, John Ambrose Fleming, an English electrical engineer and physicist, constructed a thermionic valve called Fleming valve or Diode valve. The valve essentially consists of two electrodes, a cathode K and a plate P as shown in Fig. 12.15(*a*). The cathode serves as a source of electrons. The plate is normally made of nickel, molybdenum or iron in the form of a hollow metallic cylinder, coaxial with the cylindrical cathode. The electrodes are enclosed in an evacuated glass or metal envelope. The plate, the cathode and the two ends of the filament are connected to pins P, K and FF, respectively, through the tube base. Fig. 12.15(*b*) shows symbolic representation of a diode valve.

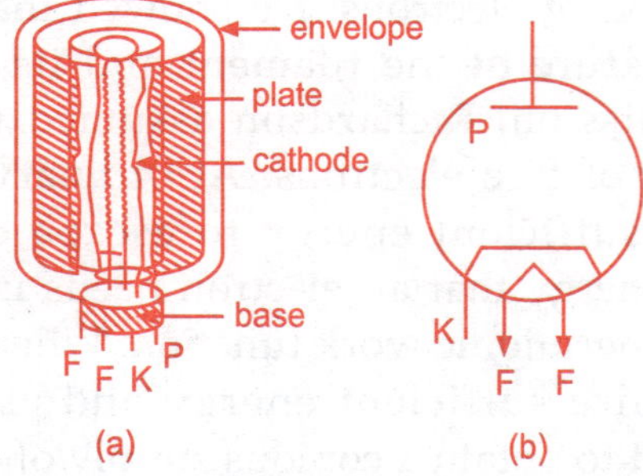

Fig. 12.15 *Diode*

When the filament is heated by a low tension (LT) voltage, DC or AC, the cathode gets heated and emits electrons. When the potential of the plate is very low, the electrons emitted by the cathode form a cloud around it, and this is known as the space charge. This stops further emission of electrons. When the plate is maintained at a higher positive potential, by means of a high tension (HT) DC voltage, it attracts electrons from the space charge towards itself. This constitutes a small current called plate current from plate to cathode. When the plate is maintained at a negative potential with respect to the cathode, it repels the electrons and no current flows. Thermionic tubes are called valves, in view of the fact that electrons can flow only from cathode to plate and not in the reverse direction. This phenomenon is used in rectification of AC.

12.6.3 Triode

In 1907, Lee de Forest an American inventor, introduced a third electrode called grid G between the plate P and the cathode K as shown in Fig. 12.16(a). This valve is known as triode valve. The grid is in the form of a wire mesh or spiral and surrounds the cathode. Electrons can pass through it towards the plate. This grid, usually called control grid, is placed closer to the cathode than to the plate. Five terminals come out through the ebonite base, two for filament, one each for the plate, the grid and the cathode. Fig. 12.16(b) shows a symbolic representation of a triode.

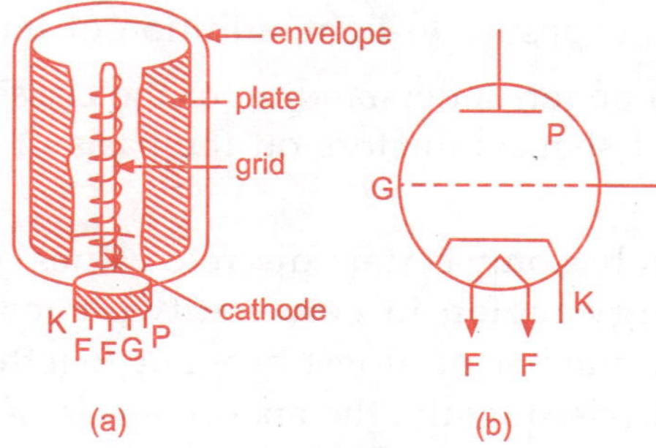

Fig. 12.16 *Triode*

The grid is given a positive or negative potential. When triode is used in a circuit, the grid is usually given a negative potential, called grid bias. The grid enables a delicate and effective control of the plate current. Even a small change in the grid voltage can cause a large variation in the plate current. This property is used in amplification. The triode is basically an amplifier although it is used as an oscillator, a rectifier and a detector.

12.6.4 Conductors, Insulators and Semiconductors

In the study of electronics, materials are classified into three categories, namely, conductors, insulators and semiconductors, depending on the resistance offered by the materials to the flow of current.

In a conductor, electrons in the outermost shells of atoms are loosely bound to the nuclei. They are called valence electrons or conduction electrons. In a metallic solid, the valence electrons do not remain attached to any particular atom, but become free electrons. These free electrons move from atom to atom throughout the crystal. This flow of electrons is responsible for the flow of current under the action of an electric field. The resistivity of a good conductor is very low and is of the order of 10^{-8} ohm-meter. In an insulator, the valence electrons are firmly attached to their nuclei and there are hardly any free electrons and no electrons available for electrical conduction. The resistivity of an insulator is very high and is of the order of 10^{12} to 10^{20} ohm-meter. There is a class of materials whose conductivity lies in between conductors and insulators. These are called semiconductors. Their electrical resistivity is generally in the range 10^{-4} to 10^{7} ohm-meter. Silicon and germanium are typical examples of basic semiconductors. Important compound semiconductors are cadmium sulphide (CdS), lead sulphide (PbS), lead telluride (PbTe), indium antimonide (InSb), gallium arsenide (GaAs) etc. Of these, CdS is used in light meters; PbS and PbTe have been used in infrared detectors. GaAs has been used in the manufacture of transistors, solid state lasers and some special high frequency devices.

At absolute zero, a pure and perfect crystal of most semiconductors would behave as an insulator. The characteristic semiconducting properties are usually brought about by heating, exposure to light, addition of impurities etc.

It is possible to have a better understanding of the classification of materials into conductors, insulators and semiconductors on the basis of energy band concept on solids.

An electron in an atom has only certain discrete values of energy; this means that there is a forbidden energy region in between two consecutive allowed energy values. When there are a number of atoms brought together (as in a crystal), they interact with each other. Consequently, the energy levels of each atom are disturbed slightly, and each of the energy level splits into a number of levels corresponding to the number of atoms, the split levels being around the original level. In other words, each energy level of a single atom is broadened forming a group of very close split energy levels which can be regarded as a continuous group. Such an energy group is called an energy band.

The group of energy levels for valence electrons is called valence band; this band is common to all the atoms of the solid. The forbidden energy region directly above the valence bond is called the forbidden band (or simply energy gap). The band of allowed split energy levels above the forbidden band is called conduction band. This band corresponds to the first excited states and is normally almost empty. Electrons in the valence band jump into the conduction band when they get sufficient energy. The energy band diagrams of a metal, an insulator and a semiconductor are shown in Figs. 12.17 (a), (b) and (c) respectively.

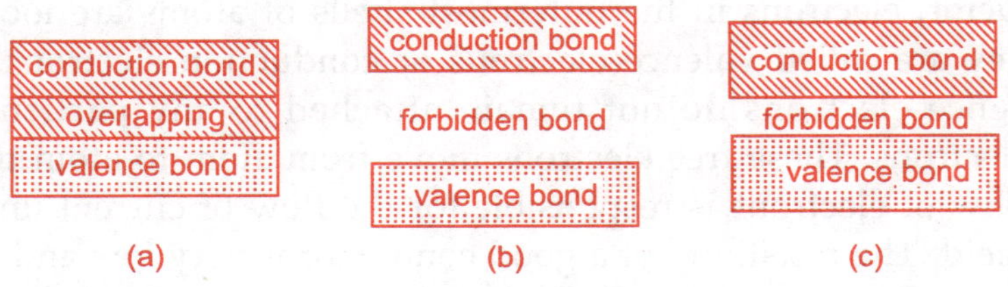

Fig. 12.17 *Energy band diagrams*

The electrical conductivity of a material is determined by the number of electrons present in the conduction band. The width of the forbidden band is called the energy gap, E_g. In the case of conductors there is an overlapping of conduction band and valence band. In the case of insulators the energy gap, E_g, between the conduction band and valence band is so wide that electrons cannot be given sufficient energy to get into conduction band. In the case of semiconductors the gap E_g is narrow, so that the electrons, by external stimulation can acquire sufficient energy to get to conduction band.

The energy gap of semiconductors varies from 0.2 eV to 2.5 eV compared to 6 eV in the case of diamond (typical insulator). E_g is about 0.7 eV for Ge, 1.1 eV for Si and 1.3 eV for Gallium arsenide.

The conductivity of semiconductors increases markedly with temperature. In silicon the number of electrons increases by a factor of 106 between 250 K and 450 K. The conduction properties of semiconductors can be enhanced by the addition of minute traces of certain impurities (other elements) called doping agents or dopants. The process of adding dopants is called doping. The semiconductor in which impurity is added is known as doped semiconductor. It is interesting to note that the degree of conductivity of a semiconductor can be controlled whereas it is not so in the case of conductors.

Pure semiconductors are called intrinsic semiconductors whereas doped semiconductors are called extrinsic semiconductors. There are two types of extrinsic semiconductors depending upon the types of impurities. One type of impurity called the donor which produces a surplus of electrons which constitutes a current. Such semiconductors are called n-type semiconductors. The other type of impurity called acceptor causes a deficiency of electrons creating vacant sites of electrons called holes (positive charge). Holes can conduct current by their movement. Such semiconductors are called p-type semiconductors.

The atomic structure of tetravalent germanium is such that in the pure state, there are no free electrons to act as conductors. If a very small amount of pentavalent impurity, for e.g., antimony or arsenic be introduced into the crystal of germanium, four out of five electrons of each impurity atom enter into bonds with the nearest germanium atoms to form covalent bonds. The remaining valence electrons move through the crystal and act as current carriers as shown in Fig. 12.18(a). In other words, the added pentavalent impurity donates electrons and the material becomes an n-type semiconductor. The name n-type indicates that the majority charge carriers are electrons which carry negative charge. If a trivalent impurity such as indium or boron is introduced, each added atom takes away one electron from neighbouring germanium atoms to form covalent bonds as shown in Fig. 12.18(b). This results in the creation of holes. Electrons from neighbouring atoms can fill these holes, resulting in an apparent movement of holes. Thus, holes act as current carriers. In other words, added trivalent impurity accepts electrons and the material becomes a *p*-type semiconductor. The name *p*-type indicates that the majority charge carriers are holes which carry positive charge. Both *n*-type and *p*-type semiconductors are used for the manufacture of semiconducting diodes and transistors.

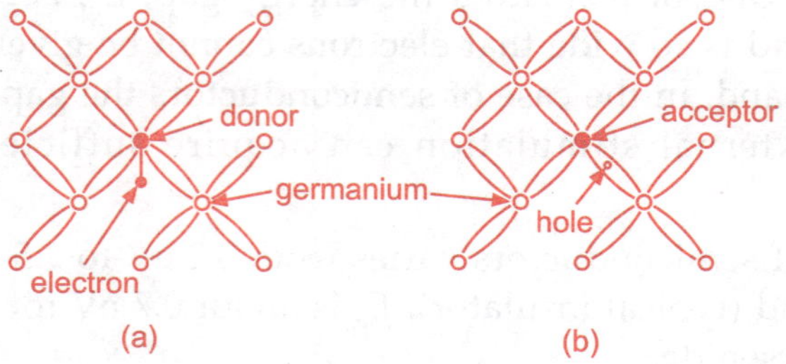

Fig. 12.18 *n-type and p-type semiconductors*

12.6.5 Semiconductor Diode

The directional nature of current flow across a boundary between *n*-type and *p*-type semiconductors is used semiconductor devices. By adding a donor impurity to one side and an acceptor impurity to the other side of a pure semiconductor, the first side becomes *n*-type and the other side becomes *p*-type. As a result a *p-n* junction is formed as shown in Fig. 12.19. There will be a diffusion of electrons and holes across the boundary and under equilibrium conditions, a potential difference is established across the *p-n* junction.

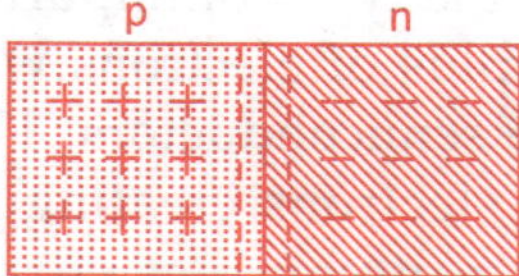

Fig. 12.19 *p-n junction*

By introducing a battery power supply across the *p-n* junction with the *p*-type connected to the positive terminal as shown in Fig. 12.20(*a*), an internal potential difference at the *p-n* junction adds to the external potential difference and assists the current.

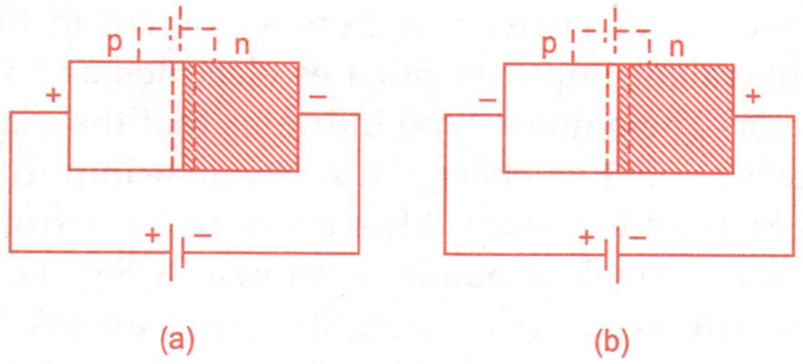

Fig. 12.20 *Forward biased and Reverse biased*

This is because the effective resistance is reduced providing an easy path to the current. This condition of junction is said to be forward biased. If the battery connections are reversed as shown in Fig. 12.19(*b*) the potential difference across the boundary opposes the external potential difference and blocks the flow of current. The junction is now said to be reverse biased. Thus, the *p-n* junction passes current in one direction only and the *p-n* junction acts as a rectifier or a diode. Fig. 12.21 shows the schematic of a diode.

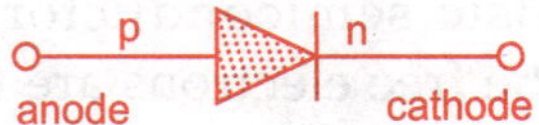

Fig. 12.21 *Diode*

The *p*-side is called the anode and the *n*-side is called the cathode. The symbol for the diode looks like an arrow that is directed from the *p*-side to the *n*-side. This signifies that conventional current flows easily from the *p*-side to the *n*-side.

Semiconductor diodes are a huge improvement over diode valves in view of the following:

(*i*) they do not require heater filament,

(*ii*) they have long life,

(*iii*) they are small in size, rugged and easy to handle,

(*iv*) they are economical and highly efficient, and

(*v*) they consume less power.

Formation of depletion layer: There is a high concentration of free holes on the *p*-side of the *p-n* junction, while there is a high concentration of free electrons on *n*-side. Free electrons from the *n*-side tend to diffuse towards the *p*-side and the free holes diffuse towards the *n*-side. These current carriers recombine with each other in the vicinity of the junction. Consequently, the region at the junction is devoid of free current carriers and the region is called depletion layer or barrier region. This region contains immobile positive and negative ions as shown in Fig.12.22. However, there will be holes on the *n*-side of the boundary and electrons on the *p*-side. Further crossing is prevented by repulsion of the same type. The depletion layer behaves like an insulator.

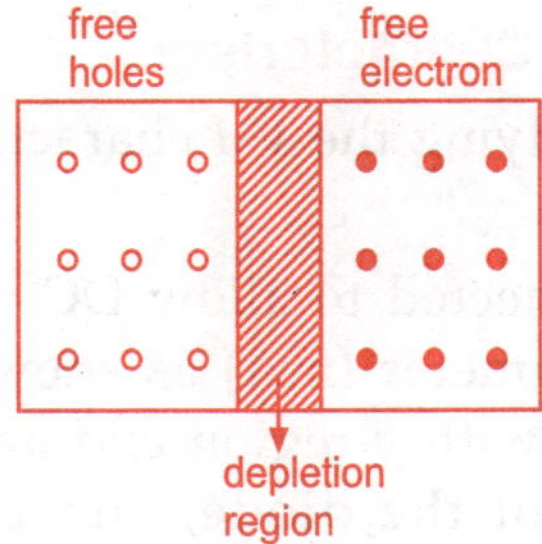

Fig. 12.22 *Depletion layer in a p-n junction*

Majority and Minority charge carriers: Silicon and germanium have all their valence electrons in covalent bonds. At absolute zero, there are no excess electrons that are free to drift through as current. Ideally, this represents a perfect insulator. However, at room temperature, some of the valence electrons possess sufficient thermal energy to break their bonds and jump from valence band to conduction bands, and become free electrons. These free electrons create equal number of electron vacant sites called holes in the broken covalent bonds. Such a pure semiconductor is called intrinsic semiconductor. In other words, an intrinsic semiconductor is one in which the free electrons are those which are excited from the valence band and the number of free electrons is equal to the number of free holes. Electric current in the intrinsic semiconductor is thus constituted by both, the electrons which are elevated into the conduction band as well as the holes that are left behind in the valence band.

At room temperature, the intrinsic concentrations for silicon and germanium are 1.4×10^{16} [1/m] to 2.5×10^{16} [1/m]. In terms of temperature T and energy E_g, the free electron concentration is given by

$$n_i = Ae^{-Eg/2kT}$$

where k = Boltzmann's constant and A is almost a constant. In extrinsic semiconductors, at room temperature, all the electrons from donor atoms are excited on to the conduction band. Hence at room temperature, density of free electrons, n, is given by

$$n = n_i + n_D$$

where n_i is the intrinsic electron concentration and n_D is the concentration of donor impurity. Usually $n \approx n_D$. Due to donor type doping, free electron concentration increases and those free electrons can occupy the holes. Therefore, the number of holes decreases.

In a n-type semiconductor at room temperature, number of free electrons are more than that of holes, and electrons carry most of the current. Therefore, electrons are majority carriers and holes are minority carriers. In p-type semiconductors the situation is reversed. Hence in p-type, holes are majority carriers and electrons are minority carriers.

12.6.6 Semiconductor Diode Characteristics

The circuit arrangement for studying the V-I characteristic of a semiconductor diode is shown in Fig. 12.23.

A p-n junction diode is connected to allow DC voltage source using a potential divider in series with a milliammeter (mA) as shown in Fig. 12.23(a). The positive end of the battery is connected with p-region end and the negative of the battery is connected with n-region end of the diode, and therefore the diode is forward biased. The voltage V across the diode is read by the voltmeter, the corresponding current being recorded by the milliammeter.

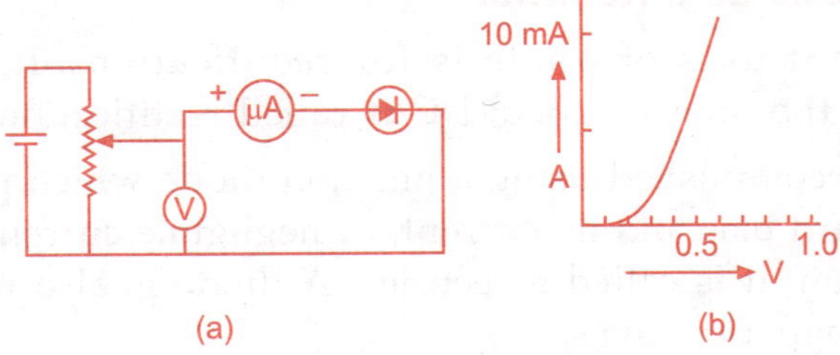

Fig. 12.23 *V-I characteristic of a semiconductor diode*

A voltage of about 0.1 volt is sufficient to overcome the potential barrier at the junction and allow a current flow. Thereafter the current increases rapidly with increasing voltage and a voltage of one to two volts permit a current of 20 to 100 mA as shown in Fig. 12.22(*b*). The variation is almost exponential.

When the battery voltage is reversed in polarity i.e., the *p-n* junction is reverse biased as shown in Fig. 12.24(*a*), the flow of current stops almost completely. However, a small current of the order of microamperes flows across the junction. When the voltage is sufficiently high, the current increases abruptly to a large value as shown in Fig. 12.24(*b*). This always occurs at a fixed reverse potential for a given diode and is known as zener potential.

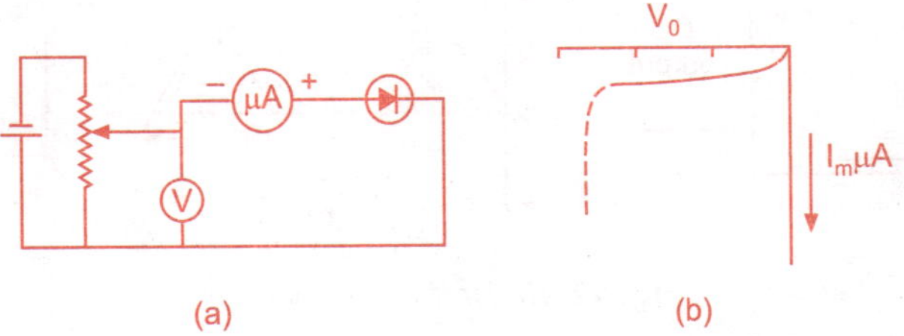

Fig. 12.24 *p-n junction in reverse biased*

When a diode is operated at a constant forward bias voltage, V_f, the corresponding current is I_f, and the ratio $\dfrac{V_f}{I_f}$ is called forward DC resistance, R_s, of the diode. If ΔI_f is the change in current for a change ΔV_f in the forward bias voltage, the ratio $\dfrac{\Delta V_f}{\Delta I_f}$ is called forward AC resistance, r_s, of the diode.

In the case of reverse bias, the ratio $\dfrac{V_r}{I_r}$ is called reverse resistance, R_o. Forward resistance is of the order of tens of ohms, whereas the reverse resistance is very high of the order of hundreds of kilo-ohms.

12.6.7 Junction Diode as a Rectifier

One of the important uses of diode is for rectification of AC. The process of conversion of AC to the unidirectional DC is called rectification.

Rectification is accomplished using a junction diode which permits large current to flow during forward bias and no current, or negligible current under reverse bias. In such an application, it is called a rectifier. A diode is also used as a detector of modulated electromagnetic waves.

Half wave rectifier: The half wave rectification is schematically shown in Fig. 12.25. The AC voltage to be rectified is applied to the primary of a transformer, and the secondary is connected to the diode with a resistance R (called load) in series. During the half of the cycle when the diode is forward biased, current flows through the load R. During the other half cycle, when the diode is reverse biased, no current or negligible current flows through R. Thus the output voltage is DC. This is known as half wave rectification, because half the wave is suppressed. The output voltage is not steady and varies with time. The input voltage and the output voltage are shown in Fig. 12.25(b) and (c), respectively. V represents input voltage and V_R represents output voltage across R.

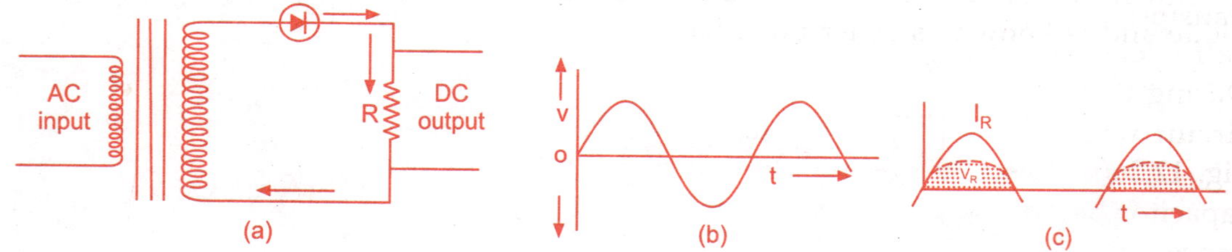

Fig. 12.25 *Half wave rectifier*

Full wave rectifier: The scheme for full wave rectification is shown in Fig. 12.26. Two diodes and a centre tap transformer are used. The diodes are connected across the secondary with a resistance R (called load) as shown in the figure. The AC voltage to be rectified is applied to the primary.

The centre tap T of the secondary coil can be considered as a reference and is taken as zero for the AC voltage in the secondary. At any instant, when one end (A) of the secondary has a positive potential with respect to T, the other end (B) has an equal negative potential with respect to T.

During the half cycle of AC when A is positive, diode D_1 is forward biased and diode D_2 is under reverse bias. Therefore, the current flows through R along AD_1RTA as D_1 conducts and D_2 does not.

During the next half cycle of AC, B becomes positive and A is negative. Therefore, diode D_2 is forward biased and diode D_1 is under reverse bias. Consequently, current flows through R along BD_2RTB, which is in the same direction as before. Hence the output voltage across R is unidirectional. Figures 12.26(a) and (b)

show the input voltage and output voltage, respectively. V represents input voltage and VR represents output voltage. In Fig 12.26(b), IR represents current through R. In Fig. 12.26(b) it is seen that the output voltage is in the form of periodic half sine pulses. In order to reduce the variation in output voltage, and make it steady output voltage, a filter circuit arrangement consisting of inductance and capacitance of suitable values, is used.

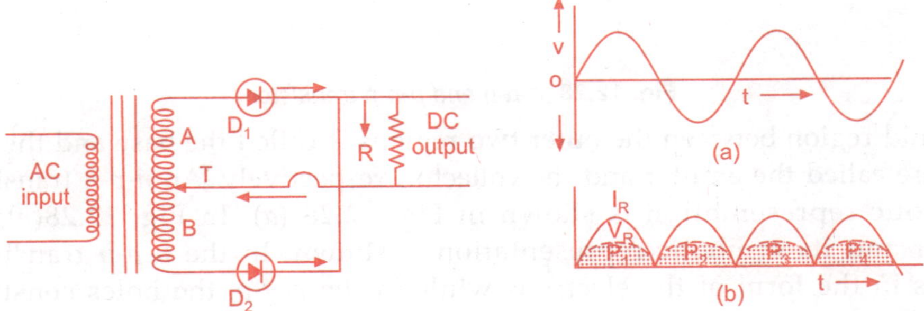

Fig. 12.26 *Full wave*

A bridge circuit which gives full wave rectification, without using a centre tap transformer is shown in Fig. 12.27. Four diodes are used in this device. During the half cycles, when A is positive with respect to B, diodes D_1 allow current flow. During the next half cycles, diodes D_2 allow current flow. The output of the bridge circuit without the smoothing capacitor, C, will be similar to the one shown in Fig. 12.26(b). With the addition of a smoothing capacitor, as shown in Fig. 12.27, the capacitor becomes charged to practically the peak value of varying DC voltage and hence a much better DC out voltage is obtained.

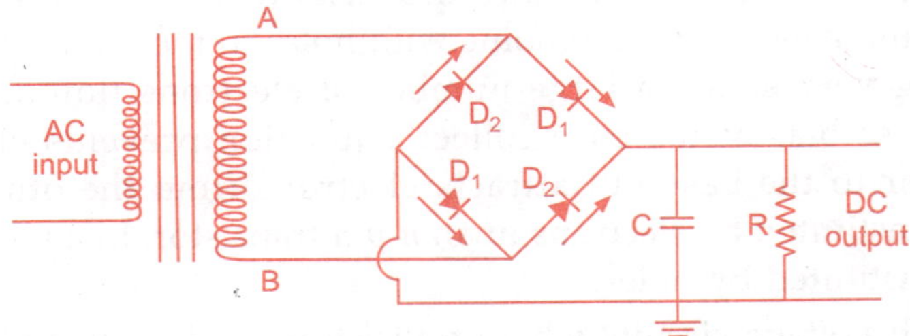

Fig. 12.27 *Bridge circuit*

12.6.8 Transistor

The transistor is a semiconductor device which replaces the triode valve. Transistor was first constructed by Bardeen, Shockley and Brattain in 1948. It is a device which consists of either one p-type region sandwiched between two n-type region (n-p-n transistor) as shown in Fig. 12.28(a) or one n-type region sandwiched between two p-type regions (p-n-p transistor) as shown b in Fig. 12.28(b).

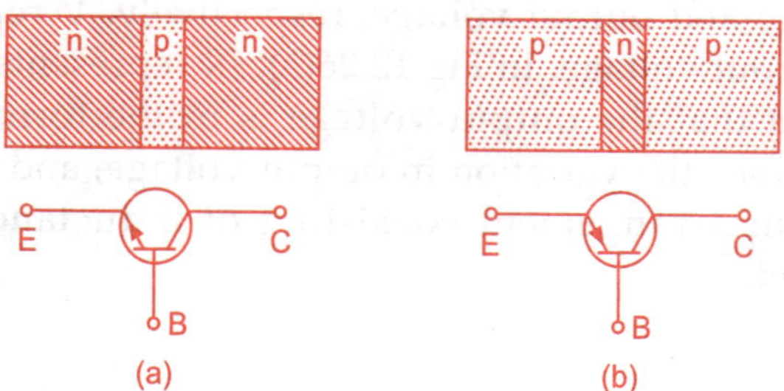

Fig. 12.28 *n-p-n and p-n-p transistor*

The mid region between the outer two regions is called the base and the two end regions are called the emitter and the collector, respectively. An *n-p-n* transistor and its symbolic representation is shown in Fig. 12.28 (*a*). In Fig. 12.28(*b*), a *p-n-p* transistor and its symbolic representation is shown. In the *n-p-n* transistor, the current is in the form of the electrons while in the *p-n-p*, the holes constitute the current. The emitter, the collector and the base in a transistor perform the same function as the cathode, the anode and the grid, respectively, as in a triode valve. The emitter region is more doped than the collector region. The base is lightly doped compared to the emitter and the collector, with the base layer being usually very thin (nearly 0.002 cm). The area of contact of collector with the base is more than that of the emitter with the base.

In a transistor, the emitter-base junction is usually forward biased while the collector base junction is reverse biased. In an *n-p-n* transistor the majority carriers of the *n*-type region are electrons. With emitter base junction forward biased, the potential barrier height at this junction is greatly reduced and the electrons from the heavily doped emitter region surge into the base, which is thin as well as lightly doped. A few of these electrons recombine with holes in the *p*-type region. Thickness of the base being very small, a large number of electrons flow into the collector region. The reverse bias at the base collector junction prevents electrons flowing from the collector to the base but attracts electrons from the other side. The net current is thus constituted by electrons in an *n-p-n* transistor. In a *p-n-p* transistor, the net current is constituted by holes.

Except in high voltage circuits where transistors tend to breakdown, transistors have completely replaced triode valves. The advantage of using transistors are:

    (*i*) they are extremely small in size, light in weight, compact and rugged

    (*ii*) they do not need a heating filament

   (*iii*) they have longer life

   (*iv*) they operate at low voltages, and

    (*v*) they consume less power.

One more distinction between the valves and transistors is that the transistors work on the basis of current while the valves work on the basis of voltage.

Transistor operation: Normally, the emitter base junction is forward biased while the collector base junction is reverse biased in a transistor. Thus the current enters the transistor through the emitter terminal in the case of a *p-n-p* transistor, whereas it leaves the transistor through the emitter terminal for an *n-p-n* transistor. The emitter, base and collector currents are denoted by IE, IB and IC. They are taken to be positive when the currents flow into the transistor. The voltage drops from emitter to base, collector to base and collector to emitter are represented by the symbols V_{EB}, V_{CB} and V_{CE}, respectively. These voltages are taken to be positive when the terminal indicated by the first subscript is positive with respect to the terminal indicated by the second subscript. As the emitter base junction is normally forward biased, E is negative for *n-p-n* transistor and is positive for a *p-n-p* transistor. Further, collector base being normally reverse biased, V_{CB} is positive for *n-p-n* transistor and is negative for *p-n-p* transistor.

Depending upon the common terminal between the input and the output circuits, a transistor can be operated in any one of the three modes, namely, common base (CB) mode, common emitter (CE) mode and common collector (CC) mode. The common emitter provides satisfactory current amplification, and is widely used.

12.6.9 Transistor Characteristics

Transistor characteristics are curves showing the relations between the various current and voltage variables of a transistor. The curve obtained by plotting input current against the input voltage at constant output voltage, is called input characteristic and the plot of output current against the output voltage at constant input current, is called output characteristic. The curve obtained by plotting output current against input current at constant output voltage is called transfer characteristic.

A circuit for obtaining characteristics of a *n-p-n* transistor in the CE mode is shown in Fig. 12.29.

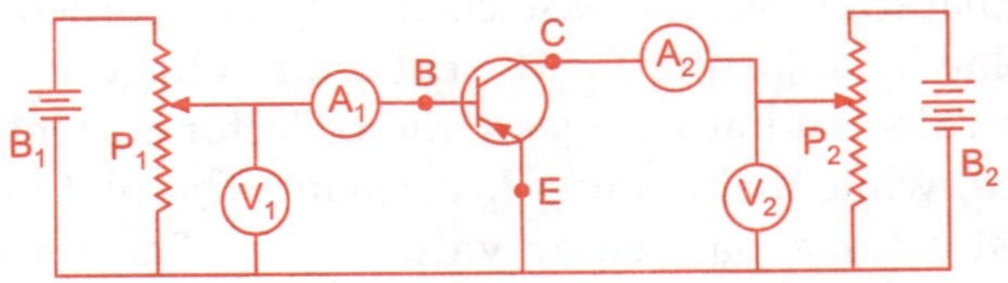

Fig. 12.29 *n-p-n transistor in the CE mode*

In Fig. 12.29, B_1 and B_2 are battery power supplies of 1.5 V and 4.5 V, respectively, connected to potentiometers P_1 and P_2 (typically 1 $k\Omega$ and 0.5 $k\Omega$, respectively). The potentiometers help to vary the base emitter voltage V_{BE} or V_B and the collector emitter voltage V_{CE} or V_C. The voltages are measured by DC voltmeters, V_1 and V_2, capable of measuring in steps of 10 mV.

A_1 is a DC micro ammeter to measure base current I_B and A_2 is a DC milliammeter to measure collector current I_C. In order to get the input characteristics, potentiometer P_1 is set for minimum base voltage V_B (V_{BE}) and P_2 is adjusted for a desired value of collector voltage V_C (V_{CE}). The base voltage is then increased in steps and at each step, the base current IB and the base voltage V_B are noted, V_C being held constant.

The plot of I_B against V_B gives input characteristics for a particular value of V_C.

Typical characteristic curves are shown in Fig. 12.30(a). The ratio $\left(\dfrac{\Delta V_B}{\Delta I_B}\right)$ gives input resistance, r, which is of the order of kilo-ohms.

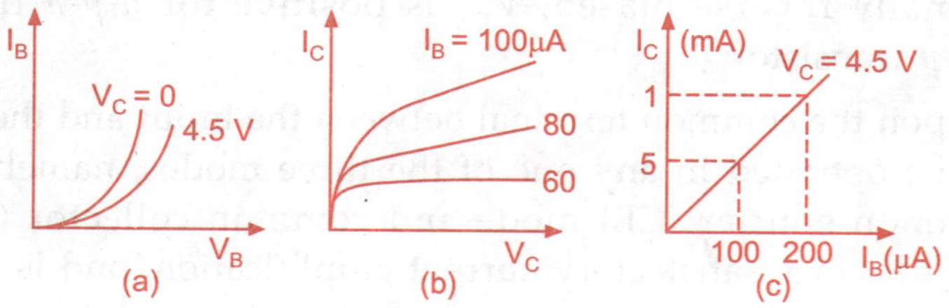

Fig. 12.30 *Transistor characteristics*

The collector current I_C is directly related to the emitter current I_E. The ratio of I_C to I_E is called DC current gain of a transistor and is denoted by α. In general, current gain or current amplification factor, α, is defined as the ratio of small change in the collector current to the change in the emitter current at constant collector voltage, and is given by

$$\alpha = \left(\frac{\Delta I_C}{\Delta I_E}\right)V_C$$

The values of α are in the range of 0.95 to 0.99.

For the output characteristic, the base current I_B as recorded by A_1 is maintained at a desired low value by adjusting P_1. The collector voltage V_C is increased in steps starting from a low value and at each step, the collector current I_C and the collector voltage V_C are noted, while maintaining I_B constant. The plot of I_C against V_C gives output characteristic for a particular value of I_B. The knee of the curves in Fig. 12.30(b) correspond to low voltage of V_C of the order of 0.2V. For higher

voltages, I_C varies linearly with V_C. The ratio $\left(\dfrac{\Delta V_C}{\Delta I_C}\right)$ along the straight part of the curve is called output resistance, r_o, which is of the order of hundreds of kilo-ohms.

The plot of collector current I_C against base current I_B at constant collector voltage, V_C, gives transfer characteristic which is shown in Fig. 12.30(c). This is usually considered under AC signal conditions, with transistor as an amplifier. The ratio $\left(\dfrac{\Delta I_C}{\Delta I_B}\right) V_C$ is called current transfer ratio and is represented by β. The ratios (or quantities) α and β are called transistor parameters. It must be noted that is a more sensitive factor to measure transistor quality. Further, α and β are related by the equation

$$\alpha = \frac{\beta}{1+\beta}$$

or,

$$\beta = \frac{\alpha}{1-\alpha}$$

For the value, $\alpha = 0.98$, the value of $\beta = 49$. As the value of tends to unity, β becomes very large. Typical values of β lie in the range 20 to 1000 in practice.

12.6.10 Transistor as an Amplifier

A transistor can also be used as voltage amplifier. The collector circuit is normally reverse biased, and has very high internal resistance. This permits a high load resistance in the collector circuit without affecting the output current. A small AC potential in the input circuit alters the current entering the device and results in a large current or slightly smaller change of current in the load resistance. This results in both voltage amplification and power amplification.

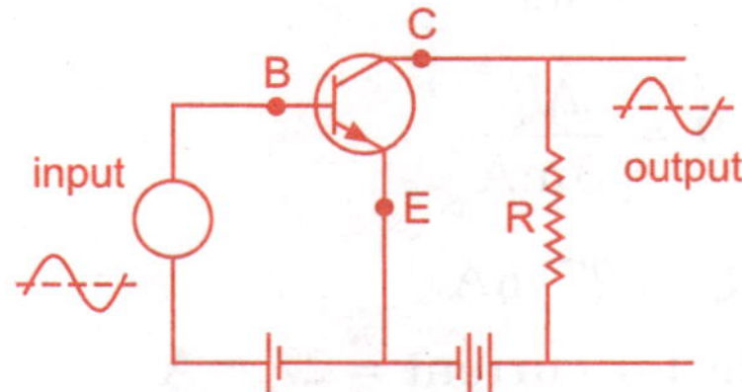

Fig. 12.31 *Common emitter amplifier circuit*

A simple common emitter amplifier circuit of an *n-p-n* transistor is shown in Fig. 12.31. The emitter base circuit is forward biased while the collector base circuit is reverse biased. During the positive half cycle of the AC signal that is fed into the input circuit, the forward bias is opposed which results in a decrease of collector current and hence a decrease in potential difference across the load R.

Therefore, as the input cycle varies through a positive half cycle, the output varies through a negative half cycle and vice versa, with the input and output voltages being out of phase by 180°.

In the common emitter circuit, the collector current I_C is controlled by the base current. If the input signal voltage ΔV_i, is fed to the emitter-base circuit having low input resistance r_i, then the change in emitter current is given by

$$\Delta I_e = \frac{\Delta V_i}{r_i}$$

However, we have

$$\Delta I_C = \alpha \Delta I_e = \alpha \frac{\Delta V_i}{r_i}$$

Change in output voltage across R_L is given by

$$\Delta V_o = \Delta I_C R_L = \alpha \frac{\Delta V_i R_L}{r_i}$$

Therefore, the Voltage gain $= \dfrac{\Delta V_o}{\Delta V_i} = \alpha \dfrac{R_L}{r_i}$.

EXAMPLE 12.15

A transistor has $\alpha = 0.9$. If the transistor is connected with its emitter grounded, what change in collector current will flow if the change in base current is 3 mA.

SOLUTION

$$\beta = \frac{\alpha}{1-\alpha} = \frac{0.9}{1-0.9} = 9$$

$$\beta = \frac{\Delta I_C}{\Delta I_B}$$

i.e.

$$9 = \frac{\Delta I_C}{3\,\text{mA}}$$

$$\Delta I_C = 27 \text{ mA.}$$

Therefore, change in collector current = 27 mA.

EXAMPLE 12.16

A given transistor has $\beta = 50$. What is the change in collector current corresponding to a change of 2 mA in emitter current at constant collector voltage?

SOLUTION

$$\alpha = \left(\frac{\Delta I_C}{\Delta I_E}\right)V_c \text{ and } \alpha = \frac{\beta}{1+\beta} = \frac{50}{51} = 0.98$$

$$0.98 = \frac{\Delta I_C}{2 \text{ mA}} \text{ or } \Delta I_C = 0.196 \text{ mA.}$$

Change in collector current = 0.196 mA

EXAMPLE 12.17

A transistor with α = 0.98 is operated as an amplifier with load resistance of 5 kΩ. If the input resistance is 70 kΩ, calculate the voltage and power gains by the circuit.

SOLUTION

$$\text{Voltage gain} = \frac{\alpha R_L}{r_i} = 0.98 \times \frac{5 \times 10^3}{70} = 70.$$

The current gain is α.

The power gain   = (current gain) × (voltage gain)

$$= 0.98 \times 70 = 68.6.$$

EXAMPLE 12.18

The dc forward current transfer ratio β for a transistor is 20. What current must flow at the base, if the current in the collector circuit is 30 mA?

SOLUTION

$$\beta = \frac{I_C}{I_B}$$

$$20 = \frac{30 \text{ mA}}{I_B} \text{ or } I_B = 1.5 \text{ mA}$$

Base current = 1.5 mA.

EXAMPLE 12.19

A transistor has α = 0.9. Its input resistance is 0.5 kW and output resistance is 5 kΩ. If a current of 2 mA flows in the emitter circuit, what is the power gain?

SOLUTION

$$\text{Power gain} = \frac{P_0}{P_i} = \frac{\text{output power}}{\text{input power}} = \frac{I_E^2 R_L}{I_C^2 r_1}$$

$$= \frac{\left(\alpha^2 I_C^2\right) R_L}{I_C^2 r_i}$$

for $I_E = \alpha I_C$

$$= \frac{\alpha^2 R_L}{r_i} = \frac{0.9^2 \times 5 \times 10^3}{0.5 \times 10^3} = 8.1$$

Power gain = 8.1

$$\text{Voltage gain} = \text{current gain} \times \frac{\text{load resistance}}{\text{input resistance}}.$$

Power gain in a transistor is expressed as the ratio of output power to the input power. Thus,

$$\text{power gain} = \frac{(\text{output current}) \times (\text{output voltage})}{(\text{input current}) \times (\text{input voltage})}$$

$$\text{power gain} = \text{current gain} \times \text{voltage gain}.$$

Current gain being less than 1, the power gain is slightly less than voltage gain.

SUMMARY

1. Atoms of each element are stable and emit characteristic spectrum. The spectrum consists of a set of isolated parallel lines termed as line spectrum. It provides useful information about the atomic structure.

2. Absorption spectra are of three kinds: continuous absorption spectra, line absorption spectra and band absorption spectra.

3. Light from a luminous source gives emission spectrum. Emission spectra are further classified according to their appearance as continuous, line and band spectra.

4. The spectrum of electromagnetic waves stretches, in principle, over an infinite range of wavelengths. Different regions are known by different names: γ-rays, X-rays, ultraviolet rays, visible rays, infrared rays, microwaves and radio waves in the order of increasing wavelength from 10^{-2} Å or 10^{-12} m to 10^6 m. They interact with matter via their electric and magnetic fields which set in oscillation charges present in all matter. The detailed interaction and hence the mechanism of absorption, scattering, etc., depend on the wavelength of the electromagnetic waves, and the nature of the atoms and molecules in the medium.

5. Infrared waves, with frequencies lower than those of visible light, vibrate not only the electrons, but entire atoms or molecules of a substance. This vibration increases the internal energy and consequently, the temperature of the substance. This is why infrared waves are often called *heat waves*.

6. The centre of sensitivity of our eyes coincides with the centre of the wavelength distribution of the sun. It is because humans have evolved with visions most sensitive to the strongest wavelengths from the sun.

7. Radioactivity is an indication of the instability of nuclei. Stability requires the ratio of neutron to proton to be around 1:1 for light nuclei. This ratio increases to about 3:2 for heavy nuclei. (More neutrons are required to overcome the effect of repulsion among the protons.) Nuclei which are away from the stability ratio, i.e., nuclei which have an excess of neutrons or protons are unstable. In fact, only about 10% of known isotopes (of all elements), are stable.

8. After Einstein showed the equivalence of mass and energy, $E = mc^2$, we cannot any longer specify separate laws of conservation of mass and conservation of energy, but we have to state

an unified law of conservation of mass and energy. The most convincing evidence that this principle operates in nature comes from nuclear physics. It is central to our understanding of nuclear energy and harnessing it as a source of power. Using the principle, Q of a nuclear process (decay or reaction) can be expressed also in terms of initial and final masses.

9. The nature of the binding energy (per nucleon) curve shows that exothermic nuclear reactions are possible when two light nuclei fuse or when a heavy nucleus undergoes fission into nuclei with intermediate mass.

10. For fusion, the light nuclei must have sufficient initial energy to overcome the coulomb potential barrier. That is why fusion requires very high temperatures.

11. Although the binding energy (per nucleon) curve is smooth and slowly varying, it shows peaks at nuclides like 4He, 16O etc. This is considered as evidence of atom-like shell structure in nuclei.

Analysis of Motion 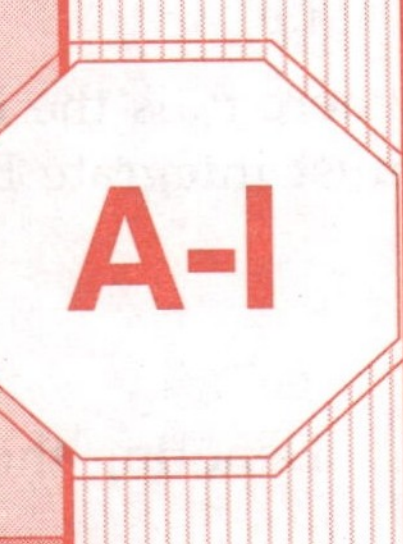

A-I.1 UNIFORM RECTILINEAR MOTION

In uniform motion the acceleration is zero for every value of time t. Hence the velocity, v, is constant, i.e.

$$\frac{dx}{dt} = v = \text{Constant.} \tag{AI.1}$$

The position coordinate x is obtained by integrating this equation. Denoting the initial value of x at $t = 0$ as x_0, we can write

$$\int_{x_0}^{x} dx = \int_{0}^{t} dt$$

i.e.
$$x - x_0 = vt \tag{AI.2}$$

i.e.
$$x = x_0 + vt$$

This equation may be used only if the velocity of the particle is constant.

A-I.2 UNIFORMLY ACCELERATED RECTILINEAR MOTION

In uniformly accelerated rectilinear motion, the acceleration, a, of the particle is constant, i.e.

$$\frac{dv}{dt} = a = \text{Constant.} \tag{AI.3}$$

The velocity, v, is obtained by integrating this equation

$$\int_{v_0}^{v} dv = a\int_{0}^{t} dt$$

i.e.
$$v - v_0 = at \tag{AI.4}$$

i.e.
$$v = v_0 + at$$

where v_0 is the initial velocity. In order to obtain the corresponding acceleration, we must integrate Eq. (AI.4) again with respect to time. From Eq. (AI.4) we can write,

$$v = \frac{dx}{dt} = v_0 + at \tag{AI.5}$$

Denoting the initial value of x by x_0, and integrating, we get

$$\int_{x_0}^{x} dx = \int_{0}^{t} (v_0 + at)dt$$

i.e.
$$x - x_0 = v_0 t + \frac{1}{2}at^2 \tag{AI.6}$$

i.e.
$$x = x_0 + v_0 t + \frac{1}{2}at^2 .$$

We can get another relation connecting the acceleration, velocity and displacement, by eliminating the time variable as

$$a = \frac{dv}{dt} = \frac{dv}{dx}\frac{dx}{dt} = \frac{dv}{dx}v \tag{AI.7}$$

where the chain rule of differentiation has been used. We can write

$$v\frac{dv}{dx} = a = \text{Constant} \tag{AI.8}$$

i.e.
$$v\, dv = a\, dx.$$

Integrating both sides with appropriate limits, we get

$$\int_{v_0}^{v} v\, dv = a \int_{x_0}^{x} dx$$

i.e.
$$\frac{1}{2}\left(v^2 - v_0^2\right) = a\,(x - x_0) \tag{AI.9}$$

i.e.
$$v^2 = v_0^2 + 2a\left(x - x_0\right).$$

It is very important to remember that Eqs. (AI.4), (AI.6) and (AI.9) are applicable only when the acceleration of the particle is constant.

□□□

Principal SI Units used in Mechanics

Quantity	Symbol(Unit)	Dimension
Acceleration	....	m/s^2
Angle	....	rad
Angular acceleration	....	rad/s^2
Angular velocity	....	rad/s
Area	....	m^2
Density	....	kg/m^3
Energy	J (Joule)	N.m
Force	N (Newton)	N or $kg.m/s^2$
Frequency	Hz (Hertz)	s^{-1}
Impulse	....	N.s
Length	....	m
Mass	....	kg
Moment of a force	....	N.m
Power	W (Watt)	J/s
Pressure	Pa (Pascal)	N/m^2
Stress	Pa (Pascal)	N/m^2
Time	....	s
Velocity	....	m/s
Volume, solids	....	m3
Volume, liquids	L (litre)	$10^{-3}.m^3$
Work	J (Joule)	N.m

FPS Units and their SI Equivalents

Quantity	FPS Unit	SI Equivalent
Acceleration	ft/s^2	$0.3048\ m/s^2$
Area	ft^2	$0.0929\ m^2$
Energy	ft.lb	1.356 J
Force	lb	4.448 N
Impulse	lb.s	4.448 N.s
Length	ft	0.3048 m
Mass	lb mass	0.4536 kg
	slug	14.59 kg
Moment of a force	lb.ft	1.356 N.m
Moment of inertia	$lb.ft.s^2$	$1.356\ kg.m^2$
Momentum	lb.s	4.448 kg.m/s(N.s)
Power	ft.lb/s	1.356 W
Pressure or stress	lb/ft^2	47.88 Pa
Velocity	ft/s	0.3048 m/s
Volume, solids	ft^3	$0.02832\ m^3$
Volume, liquids	gal	3.785 L
Work	ft.lb	1.356 J

1 ft	=	12 in (inch)
1 mile	=	5280 ft
1 lb	=	16 oz (ounce)
1 kip	=	1000 lb
1 ton	=	2000 lb
1 hp(horse power)	=	745.7 W
1 lb/in2(psi)	=	6.895 kPa (kilo Pascals)
1 gal(gallon)	=	qt (quart)

ft → foot	lb → pound	gal → gallon	m → metre	kg → kilogram
N → Newton	J → Joule	Pa → Pascal	L → Litre	W → Watt

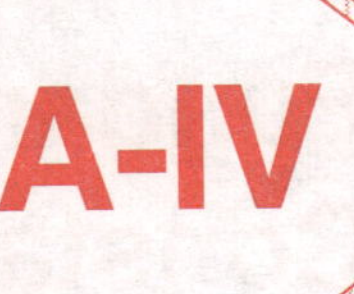

Multiplication Factor	Prefix	Symbol
10^{12}	tera	T
10^{9}	giga	G
10^{6}	mega	M
10^{3}	kilo	k
10^{2}	hecto	h
10	deka	da
10^{-1}	deci	d
10^{-2}	centi	c
10^{-3}	milli	m
10^{-6}	micro	μ
10^{-9}	nano	n
10^{-12}	pico	p
10^{-15}	femto	f
10^{-18}	atto	a

Fundamental Physical Constants

Name	Symbol	Value
Speed of light	c	2.99792458×10^8 m/s
Planck constant	h	$6.6260755 \times 10^{-34}$ J.s
Planck constant	h	$4.1356692 \times 10^{-15}$ eV.s
Planck hbar	$\hbar$	$1.0545727 \times 10^{-34}$ J.s
Planck hbar	$\hbar$	6.582121×10^{-16} eV.s
Gravitation constant	G	6.67259×10^{-11} m^3.kg^{-1}.s^{-2}
Boltzmann constant	k	1.380658×10^{-23} J/K
Boltzmann constant	k	8.617385×10^{-5} eV/K
Molar gas constant	R	8.314510 J/mol.K
Avagadro's number	N_A	6.0221×10^{23} mol^{-1}
Charge of electron	e	$1.60217733 \times 10^{-19}$ C
Permeability of vacuum	μ_0	$4\pi \times 10^{-7}$ N/A^2
Permittivity of vacuum	ε_0	$8.854187817 \times 10^{-12}$ F/m
Coulomb constant	$1/4\pi\varepsilon_0 = K$	8.987552×10^9 N.m^2/c^2
Faraday constant	F	9645.309 C/mol
Mass of electron	m_e	$9.1093897 \times 10^{-31}$ kg
Mass of electron	m_e	0.51099906 MeV/c2
Mass of proton	m_p	$1.6726231 \times 10^{-27}$ kg
Mass of proton	m_p	938.27231 MeV/c^2
Mass of neutron	m_n	$1.6749286 \times 10^{-27}$ kg

Name	Symbol	Value
Mass of neutron	m_n	939.65563 MeV/c^2
Atomic mass unit	u	1.6605402 × 10^{-27} kg
Atomic mass unit	u	931.49432 MeV/c^2
Avagadro's number	N_A	6.0221367 × 10^{23}/mol
Stefan-Boltzmann constant	σ	5.67051 × 10^{-8} W/m^2.K^4
Rydberg constant	R_∞	10973731.534 m^{-1}
Bohr magneton	μ_B	9.2740154 × 10^{-24} J/T
Bohr magneton	μ_B	5.788382 × 10^{-5} eV/T
Flux quantum	ϕ_0	2.067834 × 10^{-15} T/m^2
Bohr radius	a_0	0.529177249 × 10^{-10}m
Standard atmosphere	atm	101325 Pa
Wien displacement constant	b	2.897756 × 10^{-3} m.K

Astronomical Units/DATA

Name	Symbol	Number	Exp.	CGS Units
Astronomical unit	AU	1.496	13	cm
Parsec	pc	3.086	18	cm
Light year	ly	9.463	17	cm
Solar mass	M_o	1.990	33	g
Solar radius	R_o	6.960	10	cm
Solar luminosity	L_o	3.900	33	erg s^{-1}
Solar Temperature	T_o	5.780	3	K

ASTRONOMICAL CONSTANTS

Quantity	Value
Astronomical unit (A.U.)	149,597,870.691 kilometers
Light year (ly)	$9.460536207 \times 10^{12}$ km = 63,240 A.U.
Parsec (pc)	$3.08567802 \times 10^{13}$ km = 206,265 A.U.
Sidereal year	365.2564 days
Tropical year	365.2422 days
Gregorian year	365.2425 days
Earth mass	5.9736×10^{24} kilograms
Sun mass	1.9891×10^{30} kg = 332,980 $\times$ Earth
Mean Earth radius	6371 kilometers
Sun radius	6.96265×10^{5} km = 109 $\times$ Earth
Sun luminosity	3.827×10^{26} watts

References

1. Resnick, R. and Halliday, D. : "Physics", Wiley Eastern Limited, Madras, 1990.

2. Gamow, G. and Cleveland, J. M. : "Physics: foundations & frontiers", 3rd ed. Englewood Cliffs, N.J. : Prentice-Hall, 1976.

3. Serway, R. A. and Jewett, J. W. : "Physics for scientists and engineers", 6th ed.,Thomson, Brooks,Cole, Belmont, CA, 2004.

4. Jha, D. K. : "Textbook of Heat", Discovery Publishing House, New Delhi, 2004.

5. Muncaster, R. : "A-Level Physics" 4th ed., Nelson Thornes Ltd., Cheltenham, UK, 1993.

6. Ramachandra Rao, M.A. : "Pre-University Physics", Fifteenth Revised Edition, Geetha Book House, Mysore, 1985.

7. Ramachandra, R., Rama Rao, A.H. and Aswathanarayana Rao, T. : "A Text Book of Physics", Excellent Educational Enterprises, Bangalore, 1987.

8. Gomber, K.L. and Gogia, K.L. : "Fundamentals of Physics", Pradeep Publications, 2012.

9. Beiser, Arthur : "The Concepts of Modern Physics", 6th Edition, McGraw Hill Publications, New York, 2008.

10. Halliday, David and Resnick, Robert and Walker, Jearl : "Fundamentals of Physics, 7th Edition, John wiley sons publications, 2005.

11. Rolnick, William B. : "The Fundamental Particles and Their Interactions", Addison-Wesley, 1994.

12. David Griffiths, "Introduction to Elementary Particles", John Wiley-VCH, 2nd Edition, 2008.

13. Carlsmith, Duncan : "Particle Physics", Pearson Education, 2013.

14. Kane, Gordon : "Modern Elementary Particle Physics", Estview Press, 1993.

15. Jerry D. Wilson and Anthony J. Buffa : "College Physics", 7th Edition, Prentice Hall, 2009.

16. Frederick Bueche and Hecht, Eugene : "Schaum's Outline of College Physics", 11th Edition, McGraw Hill Publications, 2011.

17. Raymond, A. : Serway and Jewett, John W. : "Principles of Physics", Brooks Cole publication, 2003.

18. Giancoli Douglas C. : "Physics: Principles with Applications", 5th Edition, Prentice Hall, 2004.

19. Randall D. Knight : "Physics for Scientists and Engineers with Modern Physics: A Strategic Approach", Addison Wesley, 2012.

20. Edwin R. Jones, Richard L. Childers : "Contemporary College Physics", 3rd Edition, McGraw-Hill, 2001.

Index